Geometric Gradient

Geometric Series Present Worth:

To Find P $(P/A,g,i,n)$
Given A_1, g When $i = g$
$$P = A_1[n(1+i)^{-1}]$$

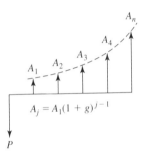

$$A_j = A_1(1+g)^{j-1}$$

To Find P $(P/A,g,i,n)$
Given A_1, g When $i \neq g$
$$P = A_1\left[\frac{1-(1+g)^n(1+i)^{-n}}{i-g}\right]$$

Capital Cost Allowance Formulas

Capital Salvage Factor: $\text{CSF} = \left[1 - \frac{td}{(i+d)}\right]$

Capital Tax Factor: $\text{CTF} = \left[1 - \left(\frac{td}{i+d}\right)\left(\frac{1+i/2}{1+i}\right)\right]$

Capital Cost Allowance (year 1): $\text{CCA}_1 = P\left(\frac{d}{2}\right)$ for $n = 1$

Capital Cost Allowance (year n): $\text{CCA}_n = Pd\left(1 - \frac{d}{2}\right)(1-d)^{n-2}$ for $n \geq 2$

Undepreciated Capital Cost: $\text{UCC}_n = P\left(1 - \frac{d}{2}\right)(1-d)^{n-1}$ for $n \geq 2$

Disposal Tax Effect (year n):
$$DTE = t \times IF[(S > P), (UCC_n - P), (UCC_n - S)] - \frac{1}{2} \times t \times MAX\{(S - P), 0\}$$

Symbols

i = Interest rate per interest period*.

n = Number of interest periods.

P = A present sum of money.

F = A future sum of money. The future sum F is an amount, n interest periods from the present, that is equivalent to P with interest rate i.

A = An end-of-period cash receipt or disbursement in a uniform series continuing for n periods, the entire series equivalent to P or F at interest rate i.

G = Uniform period-by-period increase or decrease in cash receipts or disbursements; the arithmetic gradient.

g = Uniform *rate* of cash flow increase or decrease from period to period; the geometric gradient.

t = Marginal income tax rate.

d = Capital Cost Allowance rate.

D1126488

ENGINEERING ECONOMIC ANALYSIS

CANADIAN EDITION

Donald G. Newnan
San Jose State University

John Whittaker
University of Alberta

Ted G. Eschenbach
University of Alaska Anchorage

Jerome P. Lavelle
North Carolina State University

OXFORD
UNIVERSITY PRESS

OXFORD
UNIVERSITY PRESS

70 Wynford Drive, Don Mills, Ontario M3C 1J9
www.oup.com/ca

Oxford University Press is a department of the University of Oxford.
It furthers the University's objective of excellence in research, scholarship,
and education by publishing worldwide in

Oxford New York
Auckland   Cape Town   Dar es Salaam   Hong Kong   Karachi
Kuala Lumpur   Madrid   Melbourne   Mexico City   Nairobi
New Delhi   Shanghai   Taipei   Toronto

With offices in
Argentina   Austria   Brazil   Chile   Czech Republic   France   Greece
Guatemala   Hungary   Italy   Japan   Poland   Portugal   Singapore
South Korea   Switzerland   Thailand   Turkey   Ukraine   Vietnam

Oxford is a trade mark of Oxford University Press
in the UK and in certain other countries

Published in Canada
by Oxford University Press

Library and Archives Canada Cataloguing in Publication

Engineering economic analysis / Donald G. Newnan . . . [et al.]. —
Canadian ed.
Includes bibliographical references and index.
ISBN–13: 978–0–19–541925–2.–   ISBN–10: 0–19–541925–1
1. Engineering economy. I. Newnan, Donald G.

TA177.4.E527 2005   658.15   C2005–903723–7

Cover Design: Brett Miller

2 3 4 - 09 08 07
This book is printed on permanent (acid-free) paper ⊗
Printed in Canada

Contents

Preface and Acknowledgements

In 1976, Don Newnan prefaced the first edition of this book with the statement:

> This book is designed to teach the fundamental concepts of engineering economy to engineers. By limiting the intended audience to engineers it is possible to provide an expanded presentation of engineering economic analysis and do it more concisely than if the book were written for a wider audience.

Now, three decades and nine editions later, the book continues to be a leading textbook in the United States in the field of engineering economy.

It is my privilege to adapt this text for Canadian engineering audience. In this endeavour I have tried to be true to the earlier authors' goal of presenting an easy-to-understand, up-to-date presentation of economic analysis. This was the goal for the completely Canadianized Chapters 11 and 12, which discuss and explain depreciation and taxation in Canada. At the same time, I have retained some of the American material to help explain the differences and peculiarities that have characterized the engineering relationship between the United States and Canada. Canadian engineers have always had to understand cross-border communication because we calculate in US gallons, Imperial gallons, and cubic metres while driving cars that have speedometers calibrated in miles and kilometres. Our dollars have different purchasing power, but the rules of economic decision are equally valid on both sides of the border. Thus some of the chapter-end questions have been left in American units.

The power of personal computers and the versatility of spreadsheet programs have heightened both the understanding and the application of engineering economic techniques. No longer are present values, annual cost calculations, and replacement studies obscure practices of engineering economists—they are now available to be studied and dissected on the manager's laptop. Sensitivity analysis in the form of asking 'What if?' questions has enhanced the utility and acceptance of the techniques.

This edition maintains the approach to spreadsheets that was established in the previous edition. Rather than relying on spreadsheet templates, the emphasis is on helping students learn to use the enormous capabilities of software that is available on every computer. This approach reinforces the traditional engineering economy factor method since the equivalent spreadsheet functions (PMT, PV, RATE, etc.) are used frequently. For those students who would benefit from a refresher on or an introduction to writing good spreadsheets, there is an appendix to introduce spreadsheets. In Chapter 2, spreadsheets are used to draw cash flow diagrams. Then, from Chapter 4 to Chapter 15, every chapter has a concluding section on spreadsheets. Each of these sections is designed to reinforce the other material in the chapter and to add to the student's knowledge of spreadsheets. If spreadsheets are used, the student will be very well prepared to apply this tool to real-world problems after graduation. This approach is designed to support a range of approaches to spreadsheets. If they wish, professors and students can rely on the traditional tools of engineering economy and, without loss of continuity, completely ignore the material on spreadsheets. Or at the other extreme, professors can introduce the concepts and require all computations to be done with spreadsheets. Or a variety of approaches may be used, depending on the professor, students, and particular chapter.

This book contains a number of special features:

- Chapter 2 ('Engineering Costs') introduces engineering cost estimating and sets the stage for the economic analysis to follow.
- Chapter 8 ('Incremental Analysis') combines spreadsheet graphing capabilities and internal rate of return calculations so in order to simplify and clarify the usually confusing concept of using the incremental IRR to chose from a group of alternatives.
- Chapters 9 and 10 emphasize how the uncertainty that is part of every engineering application can be considered by using sensitivity analysis and probability concepts.
- Chapter 12 ('Income Taxes') illustrates the comparative advantages of spreadsheet and manual methods for developing after-tax cash flow.
- Chapter 18 ('Accounting and Engineering Economy') explains the relationships between the two practices.

This textbook is the result of the hard work of a lot of people for over thirty years. To recognize these people and this outstanding effort, the Acknowledgements from the ninth edition are reproduced below. For this, the first Canadian edition, I would like to thank all the people at Oxford University Press (Canada), especially David Stover, for bringing the book to Canada; my editor Phyllis Wilson for listening to me; Nick Durie, for producing the book in record time; and Freya Godard, who taught me things I didn't know about Canadian English. Finally, I want to thank my wife, Nancy, who is my greatest supporter, and my daughter and copyeditor, Annthea.

John Whittaker
University of Alberta, 2005

Acknowledgments (from the ninth edition) by Ted G. Eschenbach and Jerome P. Lavalle

Many people have directly or indirectly contributed to the content of the book in its ninth edition. We have been influenced by our Stanford and North Carolina State University educations, our university colleagues, and students who have provided invaluable feedback on content and form. We are particularly grateful to the following professors for their work on previous editions:

Dick Bernhard, North Carolina State University
Charles Burford, Texas Tech University
Jeff Douthwaite, University of Washington

Utpal Dutta, University of Detroit, Mercy
Lou Freund, San Jose State University
Vernon Hillsman, Purdue University
Oscar Lopez, Florida International University
Nic Nigro, Cogswell College North
Ben Nwokolo, Grambling State University
Cecil Peterson, GMI Engineering & Management Institute
Malgorzata Rys, Kansas State University
Robert Seaman, New England College
R. Meenakshi Sundaram, Tennessee Technological University
Roscoe Ward, Miami University
Jan Wolski, New Mexico Institute of Mining and Technology
and particularly Bruce Johnson, U.S. Naval Academy

We would also like to thank the following professors for their contributions to this edition:

Mohamed Aboul-Seoud, Rensselaer Polytechnic Institute
V. Dean Adams, University of Nevada Reno
Ronald Terry Cutwright, Florida State University
Sandra Duni Eksioglu, University of Florida
John Erjavec, University of North Dakota
Ashok Kumar Ghosh, University of New Mexico
Scott E. Grasman, University of Missouri-Rolla
Ted Huddleston, University of South Alabama
B.J. Kim, Louisiana Tech University
C. Patrick Koelling, Virginia Polytechnic Institute and State University
Hampton Liggett, University of Tennessee
Heather Nachtmann, University of Arkansas
T. Papagiannakis, Washington State University
John A. Roth, Vanderbilt University
William N. Smyer, Mississippi State University
R. Meenakshi Sundaram, Tennessee Technological University
Arnold L. Sweet, Purdue University
Kevin Taaffe, University of Florida
Robert E. Taylor, Virginia Polytechnic Institute and State University
John Whittaker, University of Alberta

Our largest thanks must go to the professors (and their students) who have developed the supplements for this text. These include:

Thomas Lacksonen, University of Wisconsin-Stout
David Mandeville, Oklahoma State University
William Peterson, Old Dominion University
William Smyer, Mississippi State University
R. Meenakshi Sundaram, Tennessee Technological University
Ed Wheeler, University of Tennessee, Martin

Engineering Economic Analysis

CHAPTER 1

Making Economic Decisions

Could the World Trade Center Have Withstood the 9/11 Attacks?

In the immediate aftermath of the terrorist attacks of September 11, 2001, most commentators assumed that no structure, however well built, could have withstood the damage inflicted by fully fuelled passenger jets travelling at top speed.

But questions soon began to be raised. Investigators scrutinizing the towers' collapse noted that they had withstood the initial impact with amazing resilience. What brought them down was the fires that followed. Knowledgeable investigators noted that the rapid progress of the Twin Towers fires showed similarities with earlier high-rise blazes that had resulted from more mundane causes, suggesting that better fire prevention measures could have saved the buildings from crumbling.

In spring 2002, a report drafted by the Federal Emergency Management Agency and the American Society of Civil Engineers suggested that the light, fluffy spray-applied fire-proofing used throughout the towers might have been particularly vulnerable to damage from an impact or bomb blast. Sturdier material had been available, but it would have added significant weight to the building.

After Completing This Chapter...

The student should be able to:

- Distinguish between simple and complex problems.
- Discuss the role and purpose of engineering economic analysis.
- Describe and give examples of the nine steps in the *economic decision-making process*.
- Select appropriate economic criteria for use with different type of problems.
- Solve simple problems associated with engineering decision making.

QUESTIONS TO CONSIDER

1. How did the cost and weight of fireproofing material affect the engineers' decision making when the Twin Towers were being constructed?
2. How much should a builder be expected to spend on improved fireproofing, given the unlikelihood of an attack on the scale of 9/11?
3. How have perceptions of risk changed since the 9/11 attacks, and how might this affect future building design decisions?

This book is about making decisions. **Decision making** is a broad topic, for it is a major aspect of everyday human existence. This book develops the tools for analyzing and solving the economic problems that are commonly faced by engineers. Even very complex situations can be broken down into components from which sensible solutions are produced. If one understands the decision-making process and has tools for obtaining realistic comparisons between alternatives, one can expect to make better decisions.

Although we will focus on solving problems that confront firms in the marketplace, we will also use examples of how these techniques may be applied to the problems faced in daily life. Let us start by looking at some problems.

A SEA OF PROBLEMS

A careful look at the world around us clearly demonstrates that we are surrounded by a sea of problems. There does not seem to be any exact way of classifying them, simply because they are so diverse in complexity and 'personality'. One approach would be to arrange problems by their *difficulty*.

Simple Problems

On the lower end of our classification of problems are simple situations.

- Should I pay cash or use my credit card?
- Do I buy a semester parking pass or use the parking meters?
- Shall we replace a burned-out motor?
- If we use three crates of an item a week, how many crates should we buy at a time?

These are pretty simple problems, and good solutions do not require much time or effort.

Intermediate Problems

At the middle level of complexity we find problems that are primarily economic.

- Shall I buy or lease my next car?
- Which equipment should be selected for a new assembly line?
- Which materials should be used as roofing, siding, and structural support for a new building?
- Shall I buy a one- or two-semester parking pass?
- Which printing press should be purchased? A low-cost press requiring three operators, or a more expensive one needing only two operators?

Complex Problems

At the upper end of our classification system we discover problems that are indeed complex. They represent a mixture of *economic, political,* and *humanistic* elements.

- The decision of Mercedes-Benz to build an automobile assembly plant in Tuscaloosa, Alabama, illustrates a complex problem. Beside the economic aspects, Mercedes-Benz had to consider possible reactions in the American auto industry. Would the

German government pass legislation to prevent the overseas plant from being built? What about German labour unions?

- The selection of a girlfriend or a boyfriend (who may later become a spouse) is obviously complex. Economic analysis can be of little or no help.
- The annual budget of a corporation is an allocation of resources, but the budget process is heavily influenced by non-economic forces such as power struggles, geographical balancing, and impact on individuals, programs, and profits. For multinational corporations there are even national interests to be considered.

THE ROLE OF ENGINEERING ECONOMIC ANALYSIS

Engineering economic analysis is most suitable for intermediate problems and the economic aspects of complex problems. Such problems have the following characteristics:

1. The problem is *important enough* to justify our giving it serious thought and effort.
2. The problem can't be worked out in one's head—that is, a careful analysis *requires that we organize* the problem and all the various consequences, and this is just too much to be done all at once.
3. The problem has *economic aspects* important in reaching a decision.

When problems have those three characteristics, engineering economic analysis is a useful technique for seeking a solution. Since one will encounter vast numbers of such problems in the business world (and in one's personal life), engineering economic analysis is often required.

Examples of Engineering Economic Analysis

Engineering economic analysis focuses on costs, revenues, and benefits that occur at different times. For example, when a civil engineer designs a road, a dam, or a building, the construction costs occur in the near future; the benefits to the users begin only when construction is finished, but then the benefits continue for a long time.

In fact nearly everything that engineers design calls for spending money in the design and building stages, and after completion revenues or benefits occur—usually for years. Thus the economic analysis of costs, benefits, and revenues occurring over time is called *engineering* economic analysis.

Engineering economic analysis is used to answer many different questions.

- *Which engineering projects are worthwhile?* Has the mining or petroleum engineer shown that the mineral or oil deposit is worth developing?
- *Which engineering projects should have a higher priority?* Has the industrial engineer shown which factory improvement projects should be funded with the available dollars?
- *How should the engineering project be designed?* Has the mechanical or electrical engineer chosen the most economical motor size? Has the civil or mechanical engineer chosen the best thickness for insulation? Has the aeronautical engineer made the best trade-offs between 1) lighter materials that are expensive to buy but cheaper to fly and 2) heavier materials that are cheap to buy and more expensive to fly?

Engineering economic analysis can also be used to answer questions that are personally important.

- *How to achieve long-term financial goals:* How much should you save each month to buy a house, retire, or fund a trip around the world? Is going to graduate school a good investment—will your additional earnings in later years balance your lost income while you are in graduate school?
- *How to compare different ways to finance purchases:* Is it better to finance your car purchase by using the dealer's low interest rate loan or by taking the rebate and borrowing money from your bank or credit union?
- *How to make short- and long-term investment decisions:* Is a higher salary better than stock options? Should you buy a one- or two-semester parking pass?

THE DECISION-MAKING PROCESS

Decision making may take place by default; that is, a person may not consciously recognize that an opportunity for decision making exists. This fact leads us to a first element in a definition of decision making. To have a decision-making situation, there must be at least two alternatives available. If only one course of action is available, there can be no decision making, for there is nothing to decide. There is no alternative but to proceed with the single available course of action. (It is rather unusual to find that there are no alternative courses of action. More frequently, alternatives simply are not recognized.)

At this point we might conclude that the decision-making process consists of choosing from among alternative courses of action. But this is an inadequate definition. Consider the following situation:

At a race track, a bettor was uncertain about which of the five horses to bet on in the next race. He closed his eyes and pointed his finger at the list of horses printed in the racing program. Upon opening his eyes, he saw that he was pointing to horse number 4. He hurried off to place his bet on that horse.

Does the racehorse selection represent the process of decision making? Yes, it clearly was a process of choosing among alternatives (assuming the bettor had already ruled out the "do-nothing" alternative of placing no bet). But the particular method of deciding seems inadequate and irrational. We want to deal with rational decision making.

Rational Decision Making

Rational decision making is a complex process that contains nine essential elements, which are shown in Figure 1-1. Although these nine steps are shown sequentially, it is common for a decision maker to repeat steps, take them out of order, and do steps simultaneously. For example, when a new alternative is identified, then more data will be required. Or when the outcomes are summarized, it may become clear that the problem needs to be redefined or new goals established.

The value of this sequential diagram is to show all the steps that are usually required, and to show them in a logical order. Occasionally we will skip a step entirely. For example, a new alternative may be so clearly superior that it is immediately adopted at Step 4 without further analysis. The following sections describe the elements listed in Figure 1-1.

FIGURE 1-1 One possible flow chart of the decision process.

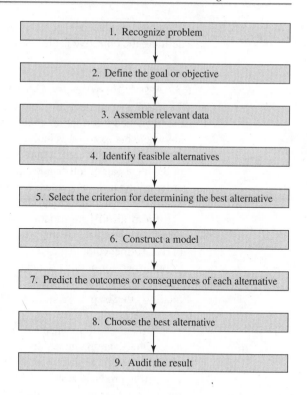

1. Recognize problem

2. Define the goal or objective

3. Assemble relevant data

4. Identify feasible alternatives

5. Select the criterion for determining the best alternative

6. Construct a model

7. Predict the outcomes or consequences of each alternative

8. Choose the best alternative

9. Audit the result

1. Recognize the Problem

The starting point in rational decision making is recognizing that a problem exists.

Some years ago, for example, it was discovered that several species of ocean fish contained substantial concentrations of mercury. The decision-making process began with this recognition of a problem, and the rush was on to determine what should be done. Research revealed that fish taken from the ocean decades before and preserved in labouratories also contained similar concentrations of mercury. Thus, the problem had existed for a long time but had not been recognized.

In typical situations, recognition is obvious and immediate. A car accident, a cheque that bounces, a burned-out motor, an exhausted supply of parts all produce the recognition of a problem. Once we are aware of the problem, we can solve it as best we can. Many firms establish programs for total quality management (TQM) or continuous improvement (CI) that are designed to identify problems so that they can be solved.

2. Define the Goal or Objective

The goal or objective can be a grand, overall goal of a person or a firm. For example, a personal goal could be to lead a pleasant and meaningful life, and a firm's goal is usually to operate profitably. The presence of multiple, conflicting goals is often the foundation of complex problems.

But an objective need not be a grand, overall goal of a business or an individual. It may be quite narrow and specific: 'I want to pay off the loan on my car by May', or 'The plant must produce 300 golf carts in the next two weeks', are more limited objectives. Thus, defining the objective is the act of describing exactly the task or goal.

3. Assemble Relevant Data

To make a good decision, one must first assemble good information. In addition to all the published information, there is a vast quantity of information that is not written down anywhere but is stored as individuals' knowledge and experience. There is also information that remains ungathered. A question like 'How many people in your town would be interested in buying a pair of left-handed scissors?' cannot be answered by examining published data or by asking any one person. Market research or other data gathering would be required to obtain the desired information.

From all this information, what is relevant in a specific decision-making process? Deciding which data are important and which are not may be a complex task. The availability of data further complicates this task. Some data are available immediately at little or no cost in published form; other data are available from specific knowledgeable people; still other data require surveys or research to assemble the information. Some data will be of high quality—that is, precise and accurate, while other data may rely on individual judgment for an estimate.

If there is a published price or a contract, the data may be known exactly. In most cases, the data are uncertain. What will it cost to build the dam? How many vehicles will use the bridge next year and in year 20? How fast will a competing firm introduce a competing product? How will demand depend on growth in the economy? Future costs and revenues are uncertain, and the range of likely values should be part of assembling relevant data.

The problem's time horizon is part of the data that must be assembled. How long will the building or equipment last? How long will it be needed? Will it be scrapped, sold, or shifted to another use? In some cases, such as for a road or a tunnel, the life may be centuries with regular maintenance and occasional rebuilding. A shorter time period, such as 50 years, may be chosen as the problem's time horizon, so that decisions can be based on more reliable data.

In engineering decision making, an important source of data is a firm's own accounting system. These data must be examined quite carefully. Accounting data focus on past information, and engineering judgment must often be applied to estimate current and future values. For example, accounting records can show the past cost of buying computers, but engineering judgment is required to estimate the future cost of buying computers.

Financial and cost accounting are designed to show accounting values and the flow of money—specifically **costs** and **benefits**—in a company's operations. When costs are directly related to specific operations, there is no difficulty; but there are other costs that are not related to specific operations. These indirect costs, or **overhead,** are usually allocated to a company's operations and products by some arbitrary method. The results are generally satisfactory for cost-accounting purposes but may be unreliable for use in economic analysis.

To create a meaningful economic analysis, we must determine the *true* differences between alternatives, and that might require some adjustment of cost-accounting data. The following example illustrates this situation.

EXAMPLE 1-1

The cost-accounting records of a large company show the average monthly costs for the three-person printing department. The wages of the three department members and benefits, such as vacation and sick leave, make up the first category of **direct** labour. The company's indirect or overhead costs—such as heat, electricity, and employee insurance—must be distributed to its various departments in *some* manner and, like many other firms, this one uses *floor space* as the basis for its allocations.

Direct labour (including employee benefits)	$ 6,000
Materials and supplies consumed	7,000
Allocated overhead costs:	
200 m^2 of floor area at $25/m^2	5,000
	$18,000

The printing department charges the other departments for its services to recover its $18,000 monthly cost. For example, the charge to run 1000 copies of an announcement is:

Direct labour	$ 7.60
Materials and supplies	9.80
Overhead costs	9.05
Cost to other departments	$26.45

The shipping department checks with a commercial printer that would print the same 1000 copies for $22.95. Although the shipping department needs only about 30,000 copies printed a month, its foreman decides to stop using the printing department and have the work done by the outside printer. The in-house printing department objects to this. As a result, the general manager has asked you to study the situation and recommend what should be done.

SOLUTION

Much of the printing department's output reveals the company's costs, prices, and other financial information. The company president considers the printing department necessary to prevent disclosing such information to people outside the company.

A review of the cost-accounting charges reveals nothing unusual. The charges made by the printing department cover direct labour, materials and supplies, and overhead. The allocation of indirect costs is a customary procedure in cost-accounting systems, but it is potentially misleading for decision making, as the following discussion indicates.

	Printing Department		Outside Printer	
	1000 Copies	30,000 Copies	1000 Copies	30,000 Copies
Direct labour	$ 7.60	$228.00		
Materials and supplies	9.80	294.00	$22.95	$688.50
Overhead costs	9.05	271.50		
	$26.45	$793.50	$22.95	$688.50

The shipping department would reduce its cost from $793.50 to $688.50 by using the outside printer. In that case, how much would the printing department's costs decline? We will examine each of the cost components:

1. *Direct Labour.* If the printing department had been working overtime, then the overtime could be reduced or eliminated. But, assuming no overtime, how much would the saving be? It seems unlikely that a printer could be fired or even put on less than a 40-hour work week. Thus, although there might be a $228 saving, it is much more likely that there will be no reduction in direct labour.

2. *Materials and Supplies.* There would be a $294 saving in materials and supplies.

3. *Allocated Overhead Costs.* There will be no reduction in the printing department's monthly $5000 overhead, for there will be no reduction in department floor space. (Actually, of course, there may be a slight reduction in the firm's power costs if the printing department does less work.)

The firm will save $294 in materials and supplies and may or may not save $228 in direct labour if the printing department no longer does the shipping department work. The maximum saving would be $294 + 228 = $522. But if the shipping department is permitted to obtain its printing from the outside printer, the firm must pay $688.50 a month. The saving from not doing the shipping department work in the printing department would not exceed $522, and it probably would be only $294. The result would be a net increase in cost to the firm. For this reason, the shipping department should be discouraged from sending its printing to the outside printer.

Gathering cost data presents other difficulties. One way to look at the financial consequences—costs and benefits—of various alternatives is as follows.

- *Market Consequences.* These consequences have an established price in the marketplace. We can quickly determine raw material prices, machinery costs, labour costs, and so forth.

- *Extra-Market Consequences.* There are other items that are not directly priced in the marketplace. But by indirect means, a price may be assigned to these items. (Economists call these prices **shadow prices.**) Examples might be the cost of an employee injury or the value to employees of going from a five-day to a four-day 40-hour week.

- *Intangible Consequences.* Numerical economic analysis probably never fully describes the real differences between alternatives. The tendency to leave out consequences that do not have a significant impact on the analysis itself, or on the conversion of the final decision into actual money, is difficult to resolve or eliminate. How does one evaluate the potential loss of workers' jobs due to automation? What is the value of landscaping around a factory? These and a variety of other consequences may be left out of the numerical calculations, but they should be considered in conjunction with the numerical results in reaching a decision.

4. Identify Feasible Alternatives

One must keep in mind that unless the best alternative is considered, the result will always be suboptimal.[1] Two types of alternatives are sometimes ignored. First, in many situations a do-nothing alternative is feasible. This may be the 'Let's keep doing what we are now doing' or the 'Let's not spend any money on that problem' alternative. Second, there are often feasible (but unglamorous) alternatives, such as 'Patch it up and keep it running for another year before replacing it.'

There is no way to ensure that the best alternative *is* among the alternatives being considered. One should try to be certain that all conventional alternatives have been listed and then make a serious effort to suggest innovative solutions. Sometimes a group of people considering alternatives in an innovative atmosphere—**brainstorming**—can be helpful. Even impractical alternatives may lead to a better possibility. The payoff from a new, innovative alternative can far exceed the value of carefully selecting between the existing alternatives.

Any good listing of alternatives will produce both practical and impractical alternatives. It would be of little use, however, to seriously consider an alternative that cannot be adopted. An alternative may not be feasible for a variety of reasons. For example, it might violate fundamental laws of science, require resources or materials that cannot be obtained, or not be available in time. Only the feasible alternatives are retained for further analysis.

5. Select the Criterion for Determining the Best Alternative

The central task of decision making is choosing from among alternatives. How is the choice made? Logically, to choose the best alternative, we must define what we mean by *best*. There must be a **standard,** or set of **standards,** to enable us to judge which alternative is best. Now, we recognize that *best* is a relative term on one end of the following relative subjective judgment:

Worst	Bad	Fair	Good	Better	Best

relative subjective judgment spectrum

Since we are dealing in *relative terms,* rather than *absolute values,* the choice will be the alternative that is relatively the most desirable. Consider a driver found guilty of speeding and given the alternatives of a $175 fine or three days in jail. In absolute terms, neither alternative is good. But on a relative basis, one simply makes the best of a bad situation.

There may be an unlimited number of ways that one might judge the various alternatives. Several possible standards are the following:

- Must create minimal disturbance to the environment.
- Must improve the distribution of wealth among people.

---

[1] A group of techniques called value analysis is sometimes used to examine past decisions. With the goal of identifying a better solution and, hence, improving decision making, value analysis re-examines the entire process that led to a decision viewed as somehow inadequate.

- Must minimize the expenditure of money.
- Must ensure that the benefits to those who gain from the decision are greater than the losses of those who are harmed by the decision.[2]
- Must minimize the time needed to accomplish the goal or objective.
- Must minimize unemployment.
- Must maximize profit.

Selecting the standard by which to choose the best alternative will not be easy if different groups support different standards and desire different alternatives. The criteria may conflict. For example, minimizing unemployment may increase the expenditure of money. Or minimizing environmental disturbance may conflict with minimizing time to complete the project. The disagreement between management and labour in collective bargaining (concerning wages and conditions of employment) reflects a disagreement over the objective and the standards for choosing the best alternative.

The last standard—must maximize profit—is the one normally selected in engineering decision making. All problems then fall into one of three categories: fixed input, fixed output, or neither input nor output fixed.

Fixed Input.   The amount of money or other input resources (like labour, materials, or equipment) is fixed. The objective is to use them effectively.

Examples:
- A project engineer has a budget of $350,000 to overhaul a portion of a petroleum refinery.
- You have $300 to buy clothes for school.

For economic efficiency, the benefits or other outputs must be maximized.

Fixed Output.   There is a fixed task (or other output objectives or results) to be accomplished.

Examples:
- A civil engineering firm has been given the job of surveying a tract of land and preparing a 'record of survey' map.
- You wish to purchase a new car with no optional equipment.

If economic efficiency is the criterion for a situation of fixed output, then the costs or other inputs must be minimized.

Neither Input nor Output Fixed.   The third category is the general situation, in which the amount of money or other inputs is not fixed, nor is the amount of benefits or other outputs.

Examples:
- A consulting engineering firm has more work available than it can handle. It is considering increasing the amount of design work it can perform by paying the staff to work evenings.

---

[2]This is the Kaldor criterion.

- One might wish to invest in the stock market, but the total cost of the investment is not fixed, and neither are the benefits.
- A car battery is needed. Batteries are available at different prices, and although each will provide the energy to start the vehicle, the useful lives of the various products are different.

What should the standard be in this category? Obviously, to be as economically efficient as possible, we must maximize the difference between the return from the investment (benefits) and the cost of the investment. Since the difference between the benefits and the costs is simply profit, a businessperson would define this standard as the **maximization of profit.**
For the three categories, the proper economic standards are:

Category	Economic Standard
Fixed input	Must maximize the benefits or other outputs.
Fixed output	Must minimize the costs or other inputs.
Neither input nor output fixed	Must maximize (benefits or other outputs minus costs or other inputs) or, stated another way, maximize profit.

6. Constructing the Model

At some point in the decision-making process, the various elements must be brought together. The *objective, relevant data, feasible alternatives,* and *selection standards* must be merged. For example, if one were considering borrowing money to pay for a car, there is a mathematical relationship between the following variables for the loan: amount, interest rate, duration, and monthly payment.

Constructing the interrelationships between the decision-making elements is frequently called **model building** or **constructing the model.** To an engineer, modelling may be a scaled *physical representation* of the real thing or system or a *mathematical equation,* or set of equations, describing the desired interrelationships. In a labouratory there may be a physical model, but in economic decision making, the model is usually mathematical.

In modelling, it is helpful to represent only that part of the real system that is important to the problem at hand. Thus, the mathematical model of the student capacity of a classroom might be

$$\text{Capacity} = \frac{lw}{k}$$

where l = length of classroom, in metres
w = width of classroom, in metres
k = classroom arrangement factor

The equation for student capacity of a classroom is a very simple model; yet it may be adequate for the problem being solved.

7. Predicting the Outcomes for Each Alternative

A model and the data are used to predict the outcomes for each feasible alternative. As was suggested earlier, each alternative might produce a variety of outcomes. Selecting a motorcycle, rather than a bicycle, for example, may make the fuel supplier happy, the neighbours unhappy, the environment more polluted, and one's savings account smaller. But, to avoid unnecessary complications, we assume that decision making is based on a single criterion for measuring the relative attractiveness of the various alternatives. If necessary, one could devise a single composite criterion that is the weighted average of several different choice criteria.

To choose the best alternative, the outcomes for each alternative must be stated in a *comparable* way. Usually the consequences of each alternative are stated in terms of money, that is, in the form of costs and benefits. This **resolution of consequences** is done with all monetary and non-monetary consequences. The consequences can also be categorized as follows:

Market consequences—where there are established market prices available

Extra-market consequences—no direct market prices, so priced indirectly

Intangible consequences—valued by judgment, not monetary prices.

In the initial problems we will examine, the costs and benefits occur over a short time period and can be considered as occurring at the same time. In other situations the various costs and benefits take place in a longer time period. The result may be costs at one point in time followed by periodic benefits. In the next chapter we will resolve these into a *cash flow diagram* to show the timing of the various costs and benefits.

For these longer-term problems, the most common error is to assume that the current situation will be unchanged for the do-nothing alternative. For example, current profits will shrink or vanish as a result of the actions of competitors and the expectations of customers; and traffic congestion normally increases over the years as the number of vehicles increases—doing nothing does not imply that the situation will not change.

8. Choosing the Best Alternative

Earlier we said that choosing the best alternative may be simply a matter of determining which alternative best meets the selection criterion. But the solutions to most problems in economics have market consequences, extra-market consequences, and intangible consequences. Since the intangible consequences of possible alternatives are left out of the numerical calculations, they should be introduced into the decision-making process at this point. The alternative to be chosen is the one that best meets the choice criterion after we have considered both the numerical consequences and the consequences not included in the monetary analysis.

During the decision-making process certain feasible alternatives are eliminated because they are dominated by other, better alternatives. For example, shopping for a computer on-line may allow you to buy a custom-configured computer for less money than a stock computer in a local store. Buying at the local store is feasible, but dominated. While eliminating dominated alternatives makes the decision-making process more efficient, there are dangers.

Having examined the structure of the decision-making process, we can ask: When is a decision made, and who makes it? If one person performs *all* the steps in decision making, then he is the decision maker. *When* he makes the decision is less clear. The selection of the feasible alternatives may be the key item, with the rest of the analysis a methodical process leading to the inevitable decision. We can see that the decision may be drastically affected, or even predetermined, by the way in which the decision-making process is carried out. This is illustrated by the following example.

Liz, a young engineer, was assigned to make an analysis of additional equipment needed for the machine shop. The single criterion for selection was that the equipment should be the most economical, considering both initial costs and future operating costs. A little investigation by Liz revealed three practical alternatives:

1. A new specialized lathe
2. A new general-purpose lathe
3. A rebuilt lathe available from a used-equipment dealer

A preliminary analysis indicated that the rebuilt lathe would be the most economical. Liz did not like the idea of buying a rebuilt lathe, so she decided to discard that alternative. She prepared a two-alternative analysis that showed that the general-purpose lathe was more economical than the specialized lathe. She presented this completed analysis to her manager. The manager assumed that the two alternatives presented were the best of all feasible alternatives, and he approved Liz's recommendation.

At this point we should ask: Who was the decision maker, Liz or her manager? Although the manager signed his name at the bottom of the economic analysis worksheets to authorize the purchase of the general-purpose lathe, he was merely authorizing what had already been made inevitable, and thus he was not the decision maker. Rather Liz had made the key decision when she decided to discard the most economical alternative from further consideration. The result was a decision to buy the better of the two *less economically desirable* alternatives.

9. Audit the Results

An audit of the results is a comparison of what happened against the predictions. Do the results of a decision analysis reasonably agree with its projections? If a new machine tool was purchased to save labour and improve quality, did it? If so, the economic analysis seems to be accurate. If the savings are not being obtained, what was overlooked? The audit may help ensure that projected operating advantages are ultimately obtained. On the other hand, the economic analysis projections may have been unduly optimistic. We want to know this, too, so that the mistakes that led to the inaccurate projection are not repeated. Finally, an effective way to promote *realistic* economic analysis calculations is for all people involved to know that there *will* be an audit of the results!

ENGINEERING DECISION MAKING FOR CURRENT COSTS

Some of the easiest forms of engineering decision making deal with problems of alternative *designs, methods,* or *materials.* If results of the decision occur in a very short period of time, one can quickly add up the costs and benefits for each alternative. Then, by using the suitable economic standards, one can identify the best alternative. Three example problems illustrate these situations.

EXAMPLE 1-2

A concrete aggregate mix must contain at least 31% sand by volume for proper batching. One source of material, which has 25% sand and 75% coarse aggregate, sells for $3 per cubic metre (m^3). Another source, which has 40% sand and 60% coarse aggregate, sells for $4.40/m^3. Determine the least cost per cubic metre of blended aggregates.

SOLUTION

The least cost of blended aggregates will result from maximum use of the lower-cost material. The higher-cost material will be used to increase the proportion of sand up to the minimum level (31%) specified.

Let x = Portion of blended aggregates from $3.00/m^3 source

$1 - x$ = Portion of blended aggregates from $4.40/m^3 source

Sand Balance

$$x(0.25) + (1 - x)(0.40) = 0.31$$

$$0.25x + 0.40 - 0.40x = 0.31$$

$$x = \frac{0.31 - 0.40}{0.25 - 0.40} = \frac{-0.09}{-0.15}$$

$$= 0.60$$

Thus the blended aggregates will contain

60% of $3.00/m^3 material

40% of $4.40/m^3 material

The least cost per cubic metre of blended aggregates is

$$0.60(\$3.00) + 0.40(\$4.40) = 1.80 + 1.76$$

$$= \$3.56/m^3$$

EXAMPLE 1-3

A machine part is manufactured at a unit cost of 40¢ for material and 15¢ for direct labour. An investment of $500,000 in tooling is required. The order calls for 3 million pieces. Halfway through the order, a new method of manufacture can be put into effect that will reduce the unit costs to 34¢ for material and 10¢ for direct labour—but it will require $100,000 for additional tooling. This tooling will not be useful for future orders. Other costs are allocated at 2.5 times the direct labour cost. What, if anything, should be done?

SOLUTION

Since there is only one way to handle first 1.5 million pieces, our problem concerns only the second half of the order.

Alternative A: Continue with Present Method

Material cost	1,500,000 pieces $\times$ 0.40 =	$ 600,000
Direct labour cost	1,500,000 pieces $\times$ 0.15 =	225,000
Other costs	2.50 $\times$ direct labour cost =	562,500
Cost for remaining 1,500,000 pieces		$1,387,500

Alternative B: Change the Manufacturing Method

Additional tooling cost		$ 100,000
Material cost	1,500,000 pieces $\times$ 0.34 =	510,000
Direct labour cost	1,500,000 pieces $\times$ 0.10 =	150,000
Other costs	2.50 $\times$ direct labour cost =	375,000
Cost for remaining 1,500,000 pieces		$1,135,000

Before making a final decision, one should closely examine the *Other costs* to see that they do, in fact, vary as the *Direct labour cost* varies. Assuming they do, the decision would be to change the manufacturing method.

EXAMPLE 1-4

In the design of a cold-storage warehouse, the specifications call for a maximum heat transfer through the warehouse walls of 30,000 joules per hour per square metre of wall when there is a 30°C temperature difference between the inside surface and the outside surface of the insulation. The two insulation materials being considered are as follows:

Insulation Material	Cost per Cubic Metre	Conductivity (J-m/m²-°C-hr)
Rock wool	$12.50	140
Foamed insulation	14.00	110

The basic equation for heat conduction through a wall is:

$$Q = \frac{K(\Delta T)}{L}$$

where Q = heat transfer, in J/hr/m^2 of wall
  K = conductivity in J-m/m^2-°C-hr
  ΔT = difference in temperature between the two surfaces, in °C
  L = thickness of insulating material, in metres

Which insulation material should be selected?

SOLUTION

Two steps are needed to solve the problem. First, the required thickness of each of the alternate materials must be calculated. Then, since the problem is one of providing a fixed output (heat transfer through the wall limited to a fixed maximum amount), the criterion is to minimize the input (cost).

Required Insulation Thickness

$$\text{Rock wool} \qquad 30{,}000 = \frac{140(30)}{L} \qquad L = 0.14 \text{ m}$$

$$\text{Foamed insulation} \quad 30{,}000 = \frac{110(30)}{L} \qquad L = 0.11 \text{ m}$$

Cost of Insulation per Square Metre of Wall

Unit cost $= \text{Cost/m}^3 \times$ Insulation thickness, in metres
Rock wool   Unit cost $= \$12.50 \times 0.14 \text{ m} = \$1.75/\text{m}^2$
Foamed insulation   Unit cost $= \$14.00 \times 0.11 \text{ m} = \$1.54/\text{m}^2$

The foamed insulation is the lesser-cost alternative. However, there is an intangible constraint that must be considered. How thick is the available wall space? Engineering economy and the time value of money are needed to decide what the maximum heat transfer should be. What is the cost of more insulation versus the cost of cooling the warehouse over its life?

SUMMARY

Classifying Problems

Many problems are simple and thus easy to solve. Others are of intermediate difficulty and need considerable thought and/or calculation to properly evaluate. These intermediate problems tend to have a substantial economic component and hence are good candidates for economic analysis. Complex problems, on the other hand, often contain people elements, along with political and economic components. Economic analysis is still very important, but the best alternative must be selected with all criteria in mind—not just economics.

The Decision-Making Process

Rational decision making uses a logical method of analysis to select the best alternative from among the feasible alternatives. The following nine steps can be followed sequentially, but decision makers often repeat some steps, undertake some simultaneously, and skip others altogether.

1. Recognize the problem.
2. Define the goal or objective: What is the task?
3. Assemble relevant data: What are the facts? Are more data needed, and is it worth more than the cost to obtain it?
4. Identify feasible alternatives.

5. Select the criterion for choosing the best alternative: possible criteria include the political, economic, environmental, and humanitarian. The single criterion may be a composite of several different criteria.
6. *Mathematically model* the various interrelationships.
7. Predict the outcomes for each alternative.
8. Choose the best alternative.
9. Audit the results.

Engineering decision making refers to solving substantial engineering problems in which economic aspects dominate and economic efficiency is the criterion for choosing from among possible alternatives. It is a particular case of the general decision-making process. Some of the unusual aspects of engineering decision making are as follows:

1. Cost-accounting systems, while an important source of cost data, contain allocations of indirect costs that may be unsuitable for use in economic analysis.
2. The various consequences—costs and benefits—of an alternative may be of three types:
    (a) Market consequences—there are established market prices
    (b) Extra-market consequences—there are no direct market prices, but prices can be assigned by indirect means
    (c) Intangible consequences—valued by judgment, not by monetary prices
3. The economic criteria for judging alternatives can be reduced to three cases:
    (a) For fixed input: maximize benefits or other outputs.
    (b) For fixed output: minimize costs or other inputs.
    (c) When neither input nor output is fixed: maximize the difference between benefits and costs or, more simply stated, maximize profit.
    The third case states the general rule from which both the first and second cases may be derived.
4. To choose among the alternatives, the market consequences and extra-market consequences are organized into a cash flow diagram. We will see in Chapter 3 that engineering economic calculations can be used to compare differing cash flows. These outcomes are compared against the selection criterion. From this comparison *plus* the consequences not included in the monetary analysis, the best alternative is selected.
5. An essential part of engineering decision making is the post-audit of results. This step helps to ensure that projected benefits are obtained and to encourage realistic estimates in analyses.

PROBLEMS

1-1 Think back to your first hour after awakening this morning. List 15 decision-making opportunities that existed during that hour. After you have done that, mark the decision-making opportunities that you actually recognized this morning and upon which you made a conscious decision.

1-2 Some of the following problems would be suitable for solution by engineering economic analysis. Which ones are they?
    (*a*) Would it be better to buy a car with a diesel engine or a gasoline engine?

(b) Should an automatic machine be purchased to replace three workers now doing a task by hand?

(c) Would it be wise to enroll for an early morning class so you could avoid travelling during the morning traffic rush hours?

(d) Would you be better off if you changed your major?

(e) One of the people you might marry has a job that pays very little money, while another one has a professional job with an excellent salary. Which one should you marry?

1-3 Which one of the following problems is *most* suitable for analysis by engineering economic analysis?

(a) Some 45¢ chocolate bars are on sale for 12 bars for $3. Sandy, who eats a couple of chocolate bars a week, must decide whether to buy a dozen at the lower price.

(b) A woman has $150,000 in a bank chequing account that pays no interest. She can either invest it immediately at a desirable interest rate or wait a week and know that she will be able to obtain an interest rate that is 0.15% higher.

(c) Joe backed his car into a tree, damaging the fender. He has car insurance that will pay for the fender repair. But if he files a claim for payment, they may change his "good driver" rating downward and charge him more for car insurance in the future.

1-4 If you have $300 and could make the right decisions, how long would it take you to become a millionaire? Explain briefly what you would do.

1-5 Many people write books explaining how to make money in the stock market. Apparently the authors plan to make *their* money selling books telling other people how to profit from the stock market. Why don't these authors forget about the books and make their money in the stock market?

1-6 The owner of a small machine shop has just lost one of his larger customers. The solution to his problem, he says, is to fire three machinists to balance his workforce with his current level of business. The owner says it is a simple problem with a simple solution. The three machinists disagree. Why?

1-7 Every university student had the problem of selecting the college or university to attend. Was this a simple, intermediate, or complex problem for you? Explain.

1-8 Toward the end of the twentieth century, the US government wanted to save money by closing a small portion of all its military installations throughout the United States. Though many people agreed that saving money was a desirable goal, areas that might be affected by a closing soon reacted negatively. Congress finally selected a panel of people whose task was to develop a list of installations to close, with the legislation specifying that Congress could not alter the list. Since the goal was to save money, why was this problem so hard to solve?

1-9 The university bookstore has put pads of engineering computation paper on sale at half price. What is the minimum and maximum number of pads you might buy during the sale? Explain.

1-10 Consider the following seven situations. Which one seems most suitable for solution by engineering economic analysis?

(a) Jane has met two university students that interest her. Bill is a music major who is lots of fun to be with. Alex, on the other hand, is a fellow engineering student, but he does not like to dance. Jane wonders what to do.

(b) You drive periodically to the post office to pick up your mail. The parking metres require 10¢ for six minutes—about twice the time it takes to get from your car to the post office and back. If parking fines cost $8, do you put money in the meter or not?

(c) At the local market, chocolate bars are 45¢ each or three for $1. Should you buy them three at a time?

(d) The cost of automobile insurance varies widely from insurance company to insurance company. Should you check with several companies when your insurance comes up for renewal?

(e) There is a special local sales tax ('sin tax') on a variety of things that the town council would like to remove from local distribution. As a result a store has opened up just outside the town and offers an abundance of these specific items at prices about 30% less than is charged in town. Should you shop there?

(f) Your mother reminds you that she wants you to attend the annual family picnic. That same Saturday you already have a date with a person you have been trying to date for months.

(g) One of your professors mentioned that you have a poor attendance record in her class. You wonder whether to drop the course now or wait to see how you do on the first mid-term exam. Unfortunately, the course is required for graduation.

1-11 A car manufacturer is considering locating a car assembly plant in your region. List two simple, two intermediate, and two complex problems associated with this proposal.

1-12 Consider the following situations. Which ones appear to represent rational decision making? Explain.
   (*a*) Joe's best friend has decided to become a civil engineer, and so Joe has decided that he, too, will become a civil engineer.
   (*b*) Jill needs to get to the university from her home. She bought a car and now drives to the university each day. When Jim asks her why she didn't buy a bicycle instead, she replies, 'Gee, I never thought of that.'
   (*c*) Don needed a wrench to replace the spark plugs in his car. He went to the local automobile supply store and bought the cheapest one they had. It broke before he had finished replacing all the spark plugs in his car.

1-13 Identify possible objectives for NASA. For your favourite of these, how should alternative plans to achieve the objective be evaluated?

1-14 Suppose you have just 2 hours to answer the question, How many people in your home town would be interested in buying a pair of left-handed scissors? Give a step-by-step outline of how you would seek to answer this question within two hours.

1-15 A university student determines that he will have only $250 per month available for his housing for the coming year. He is determined to continue in the university, so he has decided to list all feasible alternatives for his housing. To help him, list five feasible alternatives.

1-16 Describe a situation where a poor alternative was selected because there was a poor search for better alternatives.

1-17 If there are only two alternatives available and both are unpleasant and undesirable, what should you do?

1-18 The three economic criteria for choosing the best alternative are minimize input, maximize output, and maximize the difference between output and input. For each of the following situations, what is the right economic criterion?
   (*a*) A manufacturer of plastic drafting triangles can sell all the triangles he can produce at a fixed price. As he increases production, his unit costs increase as a result of overtime pay and so forth. The manufacturer's criterion should be _____.

   (*b*) An architectural and engineering firm has been awarded the contract to design a wharf for a petroleum company for a fixed sum of money. The engineering firm's criterion should be _____.
   (*c*) A book publisher is about to set the list price (retail price) on a textbook. The choice of a low list price would mean less advertising than would be used for a higher list price. The amount of advertising will affect the number of copies sold. The publisher's criterion should be _____.
   (*d*) At an auction of antiques, a bidder for a particular porcelain statue would be trying to _____.

1-19 As in Problem 1-18, state the right economic criterion for each of the following situations.
   (*a*) The engineering school held a raffle of a car with tickets selling for 50¢ each or three for $1. When the students were selling tickets, they noted that many people had trouble deciding whether to buy one or three tickets. This indicates the buyers' criterion was _____.
   (*b*) A student organization bought a soft-drink machine for use in a student area. There was considerable discussion over whether they should set the machine to charge 50¢, 75¢, or $1 per drink. The organization recognized that the number of soft drinks sold would depend on the price charged. Eventually the decision was made to charge 75¢. Their criterion was _____.
   (*c*) In many cities, grocery stores find that their sales are much greater on days when they have advertised their special bargains. However, the advertised special prices do not appear to increase the total physical volume of groceries sold by a store. This leads us to conclude that many shoppers' criterion is _____.
   (*d*) A recently graduated engineer has decided to return to school in the evenings to obtain a master's degree. He feels it should be accomplished in a manner that will allow him the maximum amount of time for his regular day job plus time for recreation. In working for the degree, he will _____.

1-20 Seven criteria are given in the chapter for judging which is the best alternative. After reviewing the list, devise three additional criteria that might be used.

1-21 Suppose you are assigned the task of determining the route of a new highway through an older section of town. The highway will require that many older homes be either relocated or torn down. Two possible criteria that might be used in deciding exactly where to locate the highway are:

(a) Ensure that there are benefits to those who gain from the decision and that no one is harmed by the decision.

(b) Ensure that the benefits to those who gain from the decision are greater than the losses of those who are harmed by the decision.

Which criterion will you select to use in determining the route of the highway? Explain.

1-22 Identify benefits and costs for Problem 1-21.

1-23 In the fall, Jay Thompson decided to live in a college residence. He signed a contract under which he was obligated to pay the room rent for the full college year. One clause stated that if he moved out during the year, he could sell his contract to another student, who would move into the residence as his replacement. The cost of residence was $2000 for the two semesters, which Jay had already paid.

A month after he moved into residence, he decided he would prefer to live in an apartment. That week, after some searching for a replacement to buy his contract, Jay had two offers. One student offered to move in immediately and to pay Jay $100 per month for the eight remaining months of the school year. A second student offered to move in the second semester and pay $700 to Jay.

Jay estimates his food cost per month is $300 if he lives in residence and $250 if he lives in an apartment with three other students. His share of the apartment rent and utilities will be $200 per month. Assume each semester is $4^1/2$ months long. Disregard the small differences in the timing of the disbursements or receipts.

(a) What are the three alternatives available to Jay?

(b) Evaluate the cost for each of the alternatives.

(c) What do you recommend that Jay do?

1-24 In decision making we talk about the construction of a model. What kind of model is meant?

1-25 An electric motor on a conveyor burned out. The foreman told the plant manager that the motor had to be replaced. The foreman said there were no alternatives and asked for authorization to order the replacement. In this situation, is any decision making taking place? If so, who is making the decision(s)?

1-26 Bill Jones's parents insisted that Bill buy himself a new sport shirt. Bill's father gave specific instructions, saying the shirt must be in 'good taste', that is,

neither too wildly coloured nor too extreme in tailoring. Bill found three types of sport shirts in the local department store:

• Rather sombre shirts that Bill's father would want him to buy
• Good-looking shirts that appealed to Bill
• Weird shirts that were even too much for Bill

He wanted a good-looking shirt but wondered how to persuade his father to let him keep it. The clerk suggested that Bill take home two shirts for his father to see and return the one he did not like. Bill selected a good-looking blue shirt he liked, and also a weird lavender shirt. His father took one look and insisted that Bill keep the blue shirt and return the lavender one. Bill did as his father instructed. What was the crucial decision in this decision process, and who made it?

1-27 A farmer must decide what combination of seed, water, fertilizer, and pest control will be most profitable for the coming year. The local agricultural college did a study of this farmer's situation and prepared the following table.

Plan	Cost/acre	Income/acre
A	$ 600	$ 800
B	1500	1900
C	1800	2250
D	2100	2500

The last page of the college's study was torn off, and hence the farmer is not sure which plan the agricultural college recommends. Which plan should the farmer adopt? Explain.

1-28 Identify the alternatives, outcomes, criteria, and process for the selection of your university program. Did you make the best choice for you?

1-29 Describe a major problem you must address in the next two years. Use the techniques of this chapter to structure the problem and recommend a decision.

1-30 One strategy for solving a complex problem is to break the problem into a group of less complex problems and then find solutions to the smaller problems. The result is the solution of the complex problem. Give an example in which this strategy will work. Then give another example in which this strategy will not work.

1-31 On her first engineering job, Joy Hayes was given the responsibility of determining the production rate for a new product. She has assembled data as indicated on the two graphs:

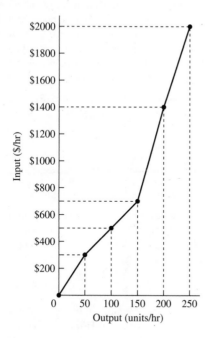

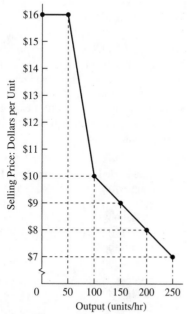

(*a*) Select a suitable economic criterion and estimate the production rate based upon it.

(*b*) Joy's boss told Joy: 'I want you to maximize output with minimum input.' Joy wonders if it

is possible to meet her boss's criterion. She asks your advice. What would you tell her?

1-32 Willie Lohmann travels from city to city in the conduct of his business. Every other year he buys a used car for about $12,000. The car dealer allows about $8000 as a trade-in allowance with the result that the salesman spends $4000 every other year for a car. Willie keeps accurate records, which show that all other expenses on his car amount to 14¢ per kilometre for each kilometre he drives. Willie's employer has two plans by which salesmen are reimbursed for their car expenses:

(1) Willie will receive all his operating expenses, and in addition will receive $2000 each year for the decline in value of the car.

(2) Willie will receive 20¢ per kilometre but no operating expenses and no depreciation allowance.

If Willie travels 29,000 kilometres per year, which method of computation gives him the larger reimbursement? At what annual distance do the two methods give the same reimbursement?

1-33 Maria, a university student, is getting ready for three final examinations. Between now and the start of exams, she has 15 hours of study time available. She would like to get as high a grade average as possible in her math, physics, and engineering economy classes. She feels she must study at least 2 hours for each course and, if necessary, will settle for the low grade that the limited study would yield. How much time should Maria devote to each class if she estimates her grade in each subject as follows:

Mathematics		Physics		Engineering Economy	
Study Hours	Grade	Study Hours	Grade	Study Hours	Grade
2	25	2	35	2	50
3	35	3	41	3	61
4	44	4	49	4	71
5	52	5	59	5	79
6	59	6	68	6	86
7	65	7	77	7	92
8	70	8	85	8	96

1-34 Two manufacturing companies, located in cities 150 kilometres apart, have discovered that they both send their trucks four times a week to the other city full of cargo and return empty. Each company pays its driver $185 a day (the round trip takes all day) and has truck operating costs (excluding the driver) of 37.5¢ a kilometre. How much could each company save each week if they shared the task, with each sending its truck twice a week and hauling the other company's cargo on the return trip?

1-35 A city needs to increase its rubbish disposal facilities. There is a choice of two rubbish disposal areas, as follows.

> *Area A:* A gravel pit with a capacity of 16 million cubic metres. Owing to the possibility of high ground water, however, the provincial environment ministry has restricted the lower 2 million cubic metres of fill to inert material only (earth, concrete, asphalt, paving, brick, etc.). The inert material, principally clean earth, must be purchased and hauled to this area for the bottom fill.
>
> *Area B:* Capacity is 14 million cubic metres. The entire capacity may be used for general rubbish disposal. This area will require an average increase in a round-trip haul of 8 km for 60% of the city, a decreased haul of 3 km for 20% of the city. For the remaining 20% of the city, the haul is the same distance as for Area A.

Assume the following conditions:

- Cost of inert material placed in Area A will be $2.35/m³.
- Average speed of trucks from last pickup to disposal site is 25 kilometres per hour.
- The rubbish truck and a two-man crew will cost $35 per hour.
- Truck capacity of $4\frac{1}{2}$ tons per load or 20 m³.
- Sufficient cover material is available at all areas; however, inert material for the bottom fill in Area A must be hauled in.

Which of the sites do you recommend? (*Answer:* Area B)

1-36 An oil company is considering adding an additional grade of fuel at its service stations. To do this, an additional 12,000 litre tank must be buried at each station. Discussions with tank fabricators indicate that the least expensive tank would be cylindrical with minimum surface area. What size of tank should be ordered?

1-37 The vegetable buyer for a group of grocery stores has decided to sell packages of sprouted grain in the vegetable section of the stores. The product is perishable, and any remaining unsold after one week in the store is discarded. The supplier will deliver the packages to the stores, arrange them in the display space, and remove and dispose of any old packages. The price the supplier will charge the stores depends on the size of the total weekly order for all the stores.

Weekly Order	Price per Package
Less than 1000 packages	35¢
1000–1499	28
1500–1999	25
2000 or more	20

The vegetable buyer estimates the quantity that can be sold per week, at various selling prices, as follows:

Selling Price	Packages Sold per Week
60¢	300
45	600
40	1200
33	1700
26	2300

The sprouted grain will be sold at the same price in all the grocery stores. How many packages should be purchased per week, and at which of the five prices listed above should they be sold?

1-38 Cathy Gwynn, a recently graduated engineer, decided to invest some of her money in a Quick Shop grocery store. The store emphasizes quick service, a limited assortment of grocery items, and rather high prices. Cathy wants to study the business to see if the store hours (currently 0600 to 0100) can be changed to make the store more profitable. Cathy assembled the following information.

Time Period	Daily Sales in the Time Period
0600–0700	$ 20
0700–0800	40
0800–0900	60
0900–1200	200
1200–1500	180
1500–1800	300
1800–2100	400
2100–2200	100
2200–2300	30
2300–2400	60
2400–0100	20

The cost of the groceries sold averages 70% of sales. The incremental cost to keep the store open, including the clerk's wage and other incremental operating costs, is $10 per hour. To maximize profit, when should the store be opened, and when should it be closed?

1-39 Jim Jones, a motel owner, noticed that just down the street the Motel 36 advertises a $36-per-night room rental rate on its sign. As a result, this competitor rents all 80 rooms every day by late afternoon. Jim, on the other hand, does not advertise his rate, which is $54 per night, and he averages only a 68% occupancy of his 50 rooms.

There are a lot of other motels nearby, but only Motel 36 advertises its rate on its sign. (Rates at the other motels vary from $48 to $80 per night.) Jim estimates that his actual incremental cost per night for each room rented, rather than remaining vacant, is $12. This $12 pays for all the cleaning, laundering, maintenance, utilities, and so on. Jim believes his eight alternatives are:

Alternative		Resulting Occupancy Rate
Advertise and Charge		
1	$35 per night	100%
2	42 per night	94
3	48 per night	80
4	54 per night	66
Do Not Advertise and Charge		
5	$48 per night	70%
6	54 per night	68
7	62 per night	66
8	68 per night	56

What should Jim do? Show how you reached your conclusion.

1-40 A firm is planning to manufacture a new product. The sales department estimates that the quantity that can be sold depends on the selling price. As the selling price is increased, the quantity that can be sold

decreases. Numerically they estimate:

$$P = \$35.00 - 0.02Q$$

where P = selling price per unit
 Q = quantity sold per year

On the other hand, the management estimates that the average cost of manufacturing and selling the product will decrease as the quantity sold increases. They estimate

$$C = \$4.00Q + \$8000$$

where C = cost to produce and sell Q per year

The firm's management wishes to produce and sell the product at the rate that will maximize profit, that is, where income minus cost will be a maximum. What quantity should the decision makers plan to produce and sell each year?   (*Answer:* 775 units)

1-41 A manufacturing firm has received a contract to assemble 1000 units of test equipment in the next year. The firm must decide how to organize its assembly operation. Skilled workers, at $22 per hour each, could be assigned to assemble the test equipment individually. Each worker would do all the assembly steps, and it would take 2.6 hours to complete one unit. An alternative approach would be to set up teams of four less skilled workers (at $13 per hour each) and organize the assembly tasks so that each worker does part of the assembly. The four-man team would be able to assemble a unit in one hour. Which approach would result in more economically assembly?

1-42 A grower estimates that if he picks his apple crop now, he will obtain 1000 boxes of apples, which he can sell at $3 per box. However, he thinks his crop will increase by 120 boxes of apples for each week he delays picking, but that the price will drop at a rate of 15¢ per box per week; in addition, he estimates approximately 20 boxes per week will spoil for each week he delays picking. When should he pick his crop to obtain the largest total cash return? How much will he receive for his crop at that time?

CHAPTER 2

Engineering Costs and Cost Estimating

Webvan Hits the Skids

Webvan, an on-line supermarket, aimed to revolutionize the humdrum business of selling groceries. Consumers could order their weekly provisions with a few clicks and have the goods delivered right to their door. It sounded like a great business plan, and the company had no trouble attracting capital during the dotcom boom of the late 1990s. Eager investors happily poured hundreds of millions into the company.

With that kind of money to spend, Webvan invested lavishly in building infrastructure, including large warehouses capable of filling 8000 orders a day. The firm rapidly expanded to serve multiple cities nationwide and even acquired a competing on-line company, Home Grocer.

But the hoped-for volume of customers never materialized. By early 2000, Internet grocers had managed to capture only a small part of the food sales market—far short of the 20% they had anticipated. When the dot-com boom went bust, Webvan suddenly looked much less attractive to investors, who quickly snapped their wallets shut.

Without new money coming in, Webvan suddenly had to face an uncomfortable fact: it was spending far more than it was earning. Finally, in 2001, Webvan went bankrupt. A rival on-line grocer, Peapod, narrowly escaped the same fate—but only because a Dutch retailer was willing to buy the company and continue pumping money into it.

Interestingly, at the same time Webvan was burning through millions in dot-com cash, a bricks-and-mortar supermarket chain in Britain called Tesco also decided to get into the on-line grocery business. Tesco invested around $56 million in a computerized processing system and, instead of building warehouses, had employees in each store walk the aisles filling orders. Unlike Webvan, Tesco made a profit.

After Completing This Chapter...

The student should be able to:

- Define various cost concepts.
- Provide specific examples of how and why these engineering cost concepts are important.
- Define engineering cost estimating.
- Explain the three types of engineering estimate, as well as common difficulties encountered in making engineering cost estimates.
- Use several common mathematical estimating models in cost estimating.
- Discuss the effect of the *learning curve* on cost estimates.
- State the relationship between cost estimating and estimating project benefits.
- Draw *cash flow diagrams* to show project costs and benefits.

QUESTIONS TO CONSIDER

1. By investing heavily in warehouses and other infrastructure, Webvan incurred large "fixed costs" that it would have to pay regardless of whether it attracted customers. By contrast, Tesco invested a more modest sum up front and hired employees only when customer orders increased enough to warrant it. How might these choices have affected the financial fates of the two companies?

2. In most cases, businesses that are seeking financing for start-up or expansion must develop detailed estimates of their likely costs and future earnings. But Webvan convinced investors that it was operating in a 'new world' of Internet commerce, to which the old rules did not apply. How did this affect investors' willingness to accept Webvan's estimates of its financial prospects?

3. Generally, businesses view cost considerations as a constraint. In this case, however, the dot-com boom of the 1990s stood that rule on its head: the more Webvan spent, the more money investors seemed willing to give the company—at least until the boom ran its course. Would Webvan have been better off with investors who asked more questions and imposed more limits? Why or why not?

This chapter defines fundamental cost concepts. These include fixed and variable costs, marginal and average costs, sunk and opportunity costs, recurring and non-recurring costs, incremental cash costs, book costs, and life-cycle costs. We then describe the various types of estimates and difficulties sometimes encountered. The models that are described include unit factor, segmenting, cost indexes, power sizing, triangulation, and learning curves. The chapter discusses estimating benefits, developing cash flow diagrams, and drawing these diagrams with spreadsheets.

An understanding of engineering costs is fundamental to the engineering economic analysis process, and therefore this chapter addresses an important question: where do the numbers come from?

ENGINEERING COSTS

Evaluating a set of feasible alternatives requires that many costs be analyzed. Some examples are costs for initial investment, new construction, facility modification, general labour, parts and materials, inspection and quality, contractor and subcontractor labour, training, computer hardware and software, material handling, fixtures and tooling, data management, and technical support, as well as general support costs (overhead). In this section we describe several concepts for classifying and understanding these costs.

Fixed, Variable, Marginal, and Average Costs

Fixed costs are constant or unchanging regardless of the level of output or activity. In contrast, **variable** costs depend on the level of output or activity. A **marginal** cost is the variable cost for one more unit, while the **average** cost is the total cost divided by the number of units.

For example, in a production environment fixed costs, such as those for factory floor space and equipment, remain the same even though production quantity, number of employees, and level of work-in-process may vary. Labour costs are classified as a *variable* cost because they depend on the number of employees in the factory. Thus *fixed* costs are level or constant regardless of output or activity, and *variable* costs are changing and related to the level of output or activity.

As another example, many universities charge full-time students a fixed cost for 12 to 18 hours and a cost per credit hour for each credit hour over 18. Thus for full-time students who are taking an overload (>18 hours), there is a variable cost that depends on the level of activity.

This example can also be used to distinguish between *marginal* and *average* costs. A marginal cost is the cost of one more unit. This will depend on how many credit hours the student is taking. If a student is currently enrolled for 12 to 17 hours, adding one more is free. The marginal cost of an additional credit hour is $0. However, if the student is taking 18 or more hours, then the marginal cost equals the variable cost of one more hour.

To illustrate average costs, the fixed and variable costs need to be specified. Suppose the cost of 12 to 18 hours is $1800 per term and overload credits are $120 an hour. If a student takes 12 hours, the *average* cost is $1800/12 = $150 per credit hour. If the student were to take 18 hours, the *average* cost would decrease to $1800/18 = $100 per credit hour. If the student takes 21 hours, the *average* cost is $102.86 per credit hour [$1800 + (3 × $120)/21].

Average cost is thus calculated by dividing the total cost for all units by the total number of units. Decision makers use **average** cost to attain an overall cost picture of the investment on a per unit basis.

Â Â Â Â **Marginal** cost is used to decide whether the additional unit should be made, purchased, or enrolled in. For the full-time student in our example, the marginal cost of another credit is $0 or $120 depending on how many credits the student has already signed up for.

EXAMPLE 2-1

An entrepreneur named DK was considering the money-making potential of chartering a bus to take people from his hometown to an event in a larger city. DK planned to provide transportation, tickets to the event, and refreshments on the bus for his customers. He gathered data and categorized the predicted expenses as either fixed or variable.

DK's Fixed Costs		DK's Variable Costs	
Bus rental	$80	Event ticket	$12.50 per person
Gas	75	Refreshments	7.50 per person
Other fuels	20		
Bus driver	50		

Develop an expression of DK's total fixed and total variable costs for operating this trip.

SOLUTION

DK's fixed costs will be incurred regardless of how many people sign up for the trip (even if only one person signs up!). These costs are bus rental, gas and fuel expense, and the cost of hiring a driver:

$$\text{Total fixed costs} = 80 + 75 + 20 + 50 = \$225$$

DK's variable costs depend on how many people sign up for the charter, which is the level of activity. Thus for event tickets and refreshments, we would write

$$\text{Total variable costs} = 12.50 + 7.50 = \$20 \text{ per person}$$

Â Â Â Â From Example 2-1 we see how it is possible to calculate total fixed and total variable costs. Furthermore, these values can be combined into a single **total** cost equation as follows:

$$\textbf{Total cost} = \textbf{Total fixed cost} + \textbf{Total variable cost} \qquad (2\text{-}1)$$

Â Â Â Â The relationship between total cost and fixed and variable costs is shown in Figure 2-1. The fixed-cost portion of $3000 is the same across the entire range of the output variable x. Often, the variable costs are *linear* (y equals a constant times x), but they can be non-linear. For example, employees are often paid at 150% of their normal hourly rate for overtime hours, so that production levels requiring overtime have higher variable costs.

FIGURE 2-1 Fixed, variable, and total costs.

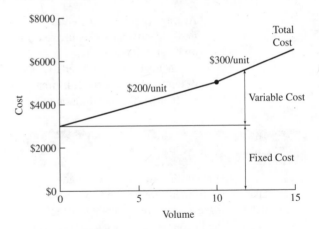

Total cost in Figure 2-1 is a fixed cost of $3000 plus a variable cost of $200 per unit for straight-time production of up to 10 units and $300 per unit for overtime production of up to 5 more units.

Figure 2-1 can also be used to illustrate marginal and average costs. At a volume of 5 units the marginal cost is $200 per unit, while at a volume of 12 units the marginal cost is $300 per unit. The respective average costs are $800 per unit, or $(3000 + 200 \times 5)/5$, and $467 per unit, or $(3000 + 200 \times 10 + 300 \times 2)/12$.

EXAMPLE 2-2

In Example 2-1, DK developed an overall total cost equation for his business expenses. Now he wants to evaluate the potential to make money from this chartered-bus trip.

SOLUTION

We use Equation 2-1 to find DK's total cost equation:

$$\text{Total cost} = \text{total fixed cost} + \text{total variable cost}$$

$$= \$225 + (\$20)(\text{number of people on the trip})$$

where number of people on the trip $= x$. Thus,

$$\text{Total cost} = 225 + 20x$$

Using this relationship, DK can calculate the total cost for any number of people—up to the capacity of the bus. What he lacks is a *revenue equation* to offset his costs. DK's total revenue from this trip can be expressed as:

$$\text{Total revenue} = (\text{Charter ticket price})(\text{Number of people on the trip}) = (\text{Ticket price})(x)$$

DK believes that he could attract 30 people at a charter ticket price of $35. Thus

$$\text{Total profit} = (\text{Total revenue}) - (\text{Total costs}) = (35x) - (225 + 20x) = 15x - 225$$

At $x = 30$,

$$\text{Total profit} = 35 \times 30 - (225 + 20 \times 30) = \$225$$

So, if 30 people take the charter, DK will net a profit of $225. This somewhat simplistic analysis ignores the value of DK's time—he would have to "pay himself" out of his $225 profit.

In Examples 2-1 and 2-2, DK developed *total cost* and *total revenue* equations to describe the charter-bus proposal. These equations can be used to create what is called a *profit-loss break-even chart* (see Figure 2-2). Both the *costs* and *revenues* associated with various levels of output (activity) are placed on the same set of *x-y* axes. This allows one to illustrate a *break-even point* (in terms of costs and revenue) and regions of *profit* and *loss* for some business activity. These terms can be defined as follows.

Break-even point: The level of business activity at which the total costs to provide the product, good, or service are *equal to* the revenue (or savings) generated by providing the service. This is the level at which one "just breaks even."

Profit region: The output level of the variable *x* greater than the break-even point, where total revenue is greater than total costs.

Loss region: The output level of the variable *x* less than the break-even point, where total costs are greater than total revenue.

Notice in Figure 2-2 that the *break-even* point for the number of persons on the charter trip is 15. For more than 15 people, DK will make a profit. If fewer than 15 sign up there

FIGURE 2-2 Profit-loss break-even chart for Examples 2-1 and 2-2.

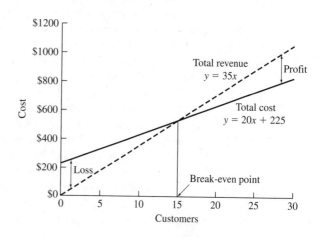

will be a net loss. At the break-even level the total cost of providing the charter equals the revenue received from the 15 passengers. We can solve for the *break-even point* by setting the *total costs* and *total revenue* expressions equal to each other and solving for the unknown value of x. From Examples 2-1 and 2-2:

$$\text{Total cost} = \text{Total revenue}$$

$$\$225 + 20x = 35x$$

$$x = 15 \text{ people}$$

Sunk Costs

A **sunk cost** is money already spent as a result of a *past* decision. Sunk costs should be disregarded in our engineering economic analysis because current decisions cannot change the past. For example, dollars spent last year to purchase new production machinery is money that is *sunk:* the money allocated to purchase the production machinery has already been spent—there is nothing that can be done now to change that action. As engineering economists we deal with present and future opportunities.

Many times it is difficult not to be influenced by sunk costs. Consider 100 shares of stock in XYZ, Inc., purchased for $15 per share last year. The share price has steadily declined over the past 12 months to a price of $10 per share today. Current decisions must focus on the $10 per share that could be attained today (as well as future price potential), not the $15 per share that was paid last year. The $15 per share paid last year is a *sunk cost* and has no influence on present opportunities.

As another example, when Ruth was a second year student, she purchased a newest-generation laptop from the university bookstore for $2000. By the time she graduated, the most anyone would pay her for the computer was $400 because the newest models were faster and cheaper and had more capabilities. For Ruth the original purchase price was a *sunk cost* that has no influence on her present opportunity to sell the laptop at its current market value ($400).

Opportunity Costs

An **opportunity cost** is associated with using a resource in one activity instead of another. Every time we use a business resource (equipment, dollars, manpower, etc.) in one activity, we give up the opportunity to use the same resources at that time in some other activity.

Every day, businesses use resources to accomplish various tasks—forklifts are used to transport materials, engineers are used to design products and processes, assembly lines are used to make a product, and parking lots are used to provide parking for employees' cars. Each of these resources costs the company money to maintain for those intended purposes. However, that cost is not just made up of the dollar cost; it also includes the opportunity cost. Each resource that a firm owns can feasibly be used in several alternative ways. For instance, the assembly line could produce a different product, and the parking lot could be rented out, used as a building site, or converted into a small airstrip. Each of these alternative uses would provide some benefit to the company.

A firm that chooses to use the resource in one way is giving up the benefits that would be derived from using it in those other ways. The benefit that would be derived by using the resource in this "other activity" is the **opportunity cost** of using it in the chosen activity. Opportunity cost may also be considered a **forgone opportunity cost** because we are forgoing the benefit that could have been realized. A formal definition of opportunity cost might be:

> An *opportunity cost* is the benefit that is *forgone* by engaging a business resource in a chosen activity instead of engaging that same resource in a forgone activity.

As an example, suppose that a university student is invited to travel through Europe over the summer break. In considering the offer, the student might calculate all the *out-of-pocket* cash costs that would be incurred. Cost estimates might be made for items such as air travel, lodging, meals, entertainment, and train passes. Suppose this amounts to $3000 for a 10-week period. After checking his bank account, the student reports that indeed he can afford the $3000 trip. However, the *true* cost to the student includes not only his *out-of-pocket* cash costs but also his *opportunity cost*. By taking the trip, the student is giving up the *opportunity* to earn $5000 as a summer intern at a local business. The student's total cost will comprise the $3000 cash cost as well as the $5000 opportunity cost (wages forgone)—thus the total cost to our traveller is $8000.

EXAMPLE 2-3

A distributor of electric pumps must decide what to do with a "lot" of old electric pumps purchased 3 years ago. Soon after the distributor purchased the lot, technology advances made the old pumps less desirable to customers. The pumps are becoming obsolescent as they sit in inventory. The pricing manager has the following information:

Distributor's purchase price 3 years ago	$ 7,000
Distributor's storage costs to date	1,000
Distributor's list price 3 years ago	9,500
Current list price of the same number of new pumps	12,000
Amount offered for the old pumps from a buyer 2 years ago	5,000
Current price the lot of old pumps would bring	3,000

Looking at the data, the pricing manager has concluded that the price should be set at $8000. This is the money that the firm has 'tied up' in the lot of old pumps ($7000 purchase and $1000 storage), and it was reasoned that the company should at least recover this cost. Furthermore, the pricing manager has argued that an $8000 price would be $1500 less than the list price from three years ago, and it would be $4000 less than what a lot of new pumps would cost ($12,000 − $8000). What would be your advice on price?

SOLUTION

Let's look more closely at each of the data items.

Distributor's purchase price three years ago: This is a sunk cost that should not be considered in setting the price today.

Distributor's storage costs to date: The storage costs for keeping the pumps in inventory are sunk costs; that is, they have been paid. Hence they should not influence the pricing decision.

Distributor's list price three years ago: If there have been no willing buyers in the past three years at this price, it is unlikely that a buyer will emerge in the future. This past list price should have no influence on the current pricing decision.

Current list price of newer pumps: Newer pumps now include technology and features that have made the older pumps less valuable. Directly comparing the older pumps to those with new technology is misleading. However, the price of the new pumps and the value of the new features help determine the market value of the old pumps.

Amount offered by a buyer two years ago: This is a forgone opportunity. At the time of the offer, the company chose to keep the lot, and thus the $5000 offered became an opportunity cost for keeping the pumps. This amount should not influence the current pricing decision.

Current price the lot could bring: The price a willing buyer in the marketplace offers is called the asset's *market value*. The lot of old pumps in question is believed to have a current market value of $3000.

From this analysis, it is easy to see the flaw in the pricing manager's reasoning. In an engineering economist analysis we deal only with *today's* and prospective *future* opportunities. It is impossible to go back in time and change decisions that have been made. Thus, the pricing manager should recommend to the distributor that the price be set at the current value that a buyer assigns to the item: $3000.

Recurring and Non-recurring Costs

Recurring costs refer to any expense that is known and anticipated and that occurs at regular intervals. **Non-recurring costs** are one-of-a-kind expenses that occur at irregular intervals and thus are sometimes difficult to plan for or anticipate from a budgeting perspective.

Examples of recurring costs would be those for resurfacing a highway and reshingling a roof. Annual expenses for maintenance and operation are also recurring expenses. Examples of non-recurring costs would be the cost of installing a new machine (including any facility modifications required), the cost of augmenting equipment based on older technology to restore its usefulness, emergency maintenance expenses, and the disposal or close-down expenses associated with ending operations.

In engineering economic analyses, *recurring costs* are modelled as cash flows that occur at regular intervals (such as every year or every five years.) Their magnitude can be estimated, and they can be included in the overall analysis. *Non-recurring costs* can be handled easily in our analysis if we are able to anticipate their timing and size. However, this is not always so easy to do.

Incremental Costs

One of the fundamental principles in engineering economic analysis is that when one makes a choice among a set of competing alternatives, the emphasis should be on the *differences* between those alternatives. This is the concept of **incremental costs.** For instance, one may be interested in comparing two options for leasing a vehicle for personal use. The two lease options may have several specifics for which costs are the same. However, there may be incremental costs associated with one option but not with the other. In a comparison of the two leases, the focus should be on the differences between the alternatives, not on the costs that are the same.

EXAMPLE 2-4

Philip is choosing between model A (a budget model) and model B (with more features and a higher purchase price). What *incremental costs* would Philip incur if he chose model B instead of the less expensive model A?

Cost Items	Model A	Model B
Purchase price	$10,000	$17,500
Installation costs	3,500	5,000
Annual maintenance costs	2,500	750
Annual utility expenses	1,200	2,000
Disposal costs after useful life	700	500

SOLUTION

We are interested in the incremental or *extra* costs that are associated with choosing model B instead of model A. To obtain these we subtract model A costs from model B costs for each category (cost item) with the following results.

Cost Items	(Model B Cost – A Cost)	Incremental Cost of B
Purchase price	17,500 – 10,000	$7500
Installation costs	5,000 – 3,500	1500
Annual maintenance costs	750 – 2,500	–1750/yr
Annual utility expenses	2,000 – 1,200	800/yr
Disposal costs after useful life	500 – 700	–200

Notice that for the cost categories given, the incremental costs of model B are both positive and negative. Positive incremental costs mean that model B costs more than model A, and negative incremental costs mean that there would be a *savings* (reduction in cost) if model B where chosen instead.

Because model B has more features, a decision would also have to take into account the incremental benefits offered by that model.

Cash Costs versus Book Costs

A *cash cost* requires the cash transaction of dollars 'out of one person's pocket' into 'the pocket of someone else'. When you buy dinner for your friends or make your monthly car payment, you are incurring a **cash cost** or **cash flow.** Cash costs and cash flows are the basis for engineering economic analysis.

Book costs do not require the transaction of dollars 'from one pocket to another'. Rather, **book costs** are cost effects from past decisions that are recorded 'in the books' (accounting books) of a firm. In one common book cost, asset depreciation (which we discuss in Chapter 11), the expense paid for a particular business asset is 'written off' on a company's accounting system over a number of periods. Book costs do not ordinarily represent cash flows and thus are not included in engineering economic analysis. One exception to this is the effect of asset depreciation on tax payments—which are cash flows and are included in after-tax analyses.

Life-Cycle Costs

The products, goods, and services designed by engineers all progress through a **life cycle** very much like the human life cycle. People are conceived, go through a growth phase, reach their peak during maturity, and then gradually decline and expire. The same general pattern holds for products, goods, and services. As with humans, the duration of the different phases, the height of the peak at maturity, and the time of the onset of decline and termination all vary depending on the individual product, good, or service. Figure 2-3 illustrates the typical phases that a product, good, or service progresses through over its life cycle.

   Life-cycle costing refers to the concept of designing products, goods, and services with a full and explicit recognition of the associated costs over the various phases of their life cycles. Two key concepts in life-cycle costing are that the later a design change is made, the higher the cost, and that decisions made early in the life cycle tend to 'lock in' costs that will be incurred later. Figure 2-4 illustrates how costs are committed early in the product

Beginning ──────────────────── Time ──────────────────→ End					
Needs Assessment and Justification Phase	Conceptual or Preliminary Design Phase	Detailed Design Phase	Production or Construction Phase	Operational Use Phase	Decline and Retirement Phase
Requirements	Impact analysis	Allocation of resources	Product, goods and services built	Operational use	Declining use
Overall feasibility	Proof of concept	Detailed specifications	All supporting facilities built	Use by ultimate customer	Phase-out
Conceptual design planning	Prototype or breadboard	Component and supplier selection	Operational use planning	Maintenance and support	Retirement
	Development and testing	Production or construction phase		Processes, materials, and methods use	Responsible disposal
	Detailed design planning			Decline and retirement planning	

FIGURE 2-3 Typical life cycle for products, goods, and services.

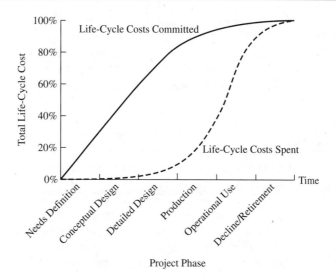

FIGURE 2-4 Cumulative life-cycle costs committed and dollars spent.

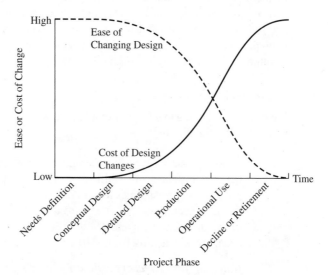

FIGURE 2-5 Life-cycle design change costs and ease of change.

life cycle—nearly 70%–90% of all costs are set during the design phases. At the same time, as the figure shows, only 10%–30% of cumulative life-cycle costs have been spent.

Figure 2-5 reinforces these concepts by illustrating that downstream product changes are more costly and that upstream changes are easier (and less costly) to make. When planners try to save money at an early design stage, the result is often a poor design that results in change orders during construction and prototype development. These changes, in turn, are more costly than working out a better design would have been.

From Figures 2-4 and 2-5 we see that the time to consider all life-cycle effects and make design changes is during the needs and conceptual or preliminary design phases—before a lot of money is committed. Some of the life-cycle effects that engineers should consider at design time are product costs for liability, production, material, testing and quality assurance, and maintenance and warranty. Other life-cycle effects are product features based on customer input and the effects of product disposal on the environment. The key point is that engineers who design products and the systems that produce them should consider all life-cycle costs.

COST ESTIMATING

Engineering economic analysis focuses on the future consequences of current decisions. Because these consequences are in the future, they usually must be estimated and cannot be known with certainty. The estimates that may be needed in engineering economic analysis include purchase costs, annual revenue, yearly maintenance, interest rates for investments, annual labour and insurance costs, equipment salvage values, and tax rates.

Estimating is the foundation of economic analysis. As in any analysis procedure, the outcome is only as good as the numbers used to reach the decision. For example, a person who wants to estimate her income taxes for a given year could do a very detailed analysis, including income assistance deductions, retirement savings deductions, itemized personal deductions, exemption calculations, and estimates of likely changes to the tax laws. However, this very technical and detailed analysis·will be grossly inaccurate if poor data are used to predict the next year's income. Thus, to ensure that an analysis is a reasonable evaluation of future events, it is very important to make careful estimates.

Types of Estimate

We can define three general types of estimate whose purposes, accuracies, and underlying methods are quite different.

Rough estimates: Order-of-magnitude estimates used for high-level planning, for determining macro-feasibility, and in a project's initial planning and evaluation phases. Rough estimates tend to involve back-of-the-envelope numbers with little detail or accuracy. The intent is to quantify and consider the order of magnitude of the numbers involved. These estimates can be made quickly and easily; their accuracy is generally from −30% to +60%.

Notice the non-symmetry in the estimating error. This is because decision makers tend to underestimate the magnitude of costs (negative economic effects). Also as Murphy's law predicts, there seem to be more ways for results to be worse than expected than there are for the results to be better than expected.

Budget estimates: Used for budgeting purposes at the conceptual or preliminary design stages of a project. These estimates are more detailed, and they require additional time and resources. Greater sophistication is used in developing budget estimates than the rough-order type, and their accuracy is generally −15% to +20%.

Detailed estimates: Used during a project's detailed design and contract bidding phases. These estimates are made from detailed quantitative models, blueprints, product specification sheets, and vendor quotes. Detailed estimates involve the most time and resources to develop and thus are much more accurate than rough or budget estimates. The accuracy of these estimates is generally −3 to +5%.

The upper limits of +60% for rough order, +20% for budget, and +5% for detailed estimates are based on construction data for plants and infrastructure. Final costs for software, research and development, and new military weapons often correspond to much higher percentages.

In considering the three types of estimate it is important to recognize that each has its unique purpose, place, and function in a project's life. Rough estimates are used for general feasibility activities, budget estimates support budgeting and preliminary design decisions, and detailed estimates are used for establishing design details and contracts. As one moves from rough to detailed design, one moves from less to much more accurate estimates.

However, this increased accuracy requires added time and resources. Figure 2-6 illustrates the trade-off between accuracy and cost. In engineering economic analysis, the resources spent must be justified by the need for detail in the estimate. As an illustration, during the project feasibility stages we would not want to use our people, time, and money to develop detailed estimates for unfeasible alternatives that will be quickly eliminated from further consideration. However, regardless of how accurate an estimate is assumed to be, it is only an estimate of what the future will be. There will be some error even if ample resources and sophisticated methods are used.

Difficulties in Estimation

Estimating is difficult because the future is unknown. With few exceptions (such as with legal contracts) it is difficult to foresee future economic consequences exactly. In this section we discuss several aspects of estimating that make it a difficult task.

One-of-a-Kind Estimates

Estimated parameters can be for one-of-a-kind or first-run projects. The first time something is done, it is difficult to estimate costs required to design, produce, and maintain a product

FIGURE 2-6 Accuracy versus cost trade-off in estimating.

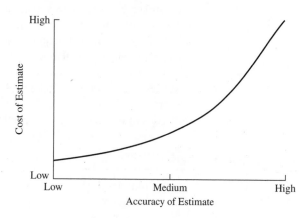

over its life cycle. Consider the projected cost estimates that were developed for the first NASA missions. The U.S. space program initially had no experience with human flight in outer space; thus the development of the cost estimates for design, production, launch, and recovery of the astronauts, flight hardware, and payloads was a first-time experience. The same is true for any endeavour lacking local or global historical cost data. New products or processes that are unique and fundamentally different make estimating costs difficult.

The good news is that there are very few one-of-a-kind estimates to be made in engineering design and analysis. Nearly all new technologies, products, and processes have 'close cousins' that have led to their development. The concept of **estimation by analogy** allows one to use knowledge about well-understood activities to anticipate costs for new activities. In the 1950s, at the start of the military missile program, aircraft companies drew on their in-depth knowledge of designing and producing aircraft when they bid on missile contracts. As another example, consider the problem of estimating the production labour requirements for a brand new product, X. A company may use its labour knowledge about Product Y, a similar type product, to build up the estimate for X. Thus, although first-run estimates are difficult to make, estimation by analogy can be an effective tool.

Time and Effort Available

Our ability to develop engineering estimates is constrained by time and the availability of person-power. In an ideal world, it would *cost nothing* to use *unlimited resources* over an *extended period of time*. However, reality requires the use of limited resources in fixed intervals of time. Thus for a rough estimate only limited effort is used.

Constraints on time and person-power can make the overall estimating task more difficult. If the estimate does not need as much detail (such as when a rough estimate is the goal), then time and personnel constraints may not be a factor. When detail is necessary and critical (such as in legal contracts), however, requirements must be anticipated and resource use planned.

Estimator Expertise

Consider two common expressions: *The past is our greatest teacher* and *knowledge is power*. These simple axioms hold true for much of what we encounter during life, and they are true in engineering estimating as well. The more experienced and knowledgeable the engineering estimator is, the less difficult the estimating process will be, the more accurate the estimate will be, the less likely it is that a major error will occur, and the more likely it is that the estimate will be of high quality.

How is experience acquired in industry? One approach is to assign inexperienced engineers relatively small jobs in order to create expertise and build familiarity with products and processes. Another strategy is to pair inexperienced engineers with mentors who have vast technical experience. Technical boards and review meetings conducted to 'justify the numbers' are also used to build knowledge and experience. Finally, many firms maintain databases of their past estimates and the costs that were actually incurred.

ESTIMATING MODELS

This section develops several estimating models that can be used at the rough, budget, or detailed design levels. For rough estimates the models are used with rough data; likewise for detailed design estimates they are used with detailed data. The level of detail will depend upon the accuracy of the data used.

Per-Unit Model

The **per-unit model** uses a 'per unit' factor, such as cost per square foot, to develop the estimate desired. This is a very simplistic yet useful technique, especially for developing estimates of the rough or order-of-magnitude type. The per unit model is commonly used in the construction industry. As an example, you may be interested in a new home that is constructed with a certain type of material and has a specific construction style. Using this information, a contractor may quote a cost of $65 per square foot for your home. If you are interested in a 2000 square foot floor plan, your cost would thus be: $2000 \times 65 = \$130,000$. Per unit factors are also used for the following estimates.

- Service cost per customer
- Safety cost per employee
- Gasoline cost per kilometre
- Cost of defects per batch
- Maintenance cost per window
- Mileage cost per vehicle
- Utility cost per square foot of floor space
- Housing cost per student

It is important to note that the per-unit model does not make allowances for economies of scale (the fact that higher quantities usually cost less on a per-unit basis). In most cases, however, the model can be effective at getting the decision maker 'in the ballpark' of likely costs, and it can be very accurate if accurate data are used.

EXAMPLE 2-5

Use the per-unit model to estimate the cost per student that you will incur for hosting 24 foreign exchange students at a local island campground for 10 days. During camp you are planning the following activities:

- 2 days of canoeing
- 3 campsite-sponsored day hikes
- 3 days at the lake beach (swimming, volleyball, etc.)
- Nightly entertainment

After calling the campground and collecting other information, you have accumulated the following data:

- Van rental from your city to the camp (one way) is $50 per 15-person van plus gas.
- Camp is 25 kilometres away, the van uses 20 litres per hundred kilometres and gas is $1 per litre.

- Each cabin at the camp holds 4 campers, and rent is $10 per day per cabin.
- Meals are $10 per day per camper; no outside food is allowed.
- Boat transportation to the island is $2 per camper (one way).
- Insurance, grounds fee, and overhead is $1 per day per camper.
- Canoe rentals are $5 per day per canoe, canoes hold 3 campers.
- Day hikes are $2.50 per camper (plus the cost of meals).
- Beach rental is $25 per group per half-day.
- Nightly entertainment is free.

SOLUTION

You are asked to use the per unit factor to estimate the cost per student on this trip. For planning purposes we assume that there will be 100% participation in all activities. We will break the total cost down into categories of transportation, living, and entertainment.

Transportation Costs

> *Van travel to and from camp:* 2 vans × 2 trips × ($50/van + 25 km × 20 L/100 km × $1/L)
> = $220

> *Boat travel to and from island:* 2 trips × $2/camper × 24 campers = $96

Transportation costs = $220 + $96 = $316

Living Costs

> *Meals for the 10-day period:* 24 campers × $10/camper/day × 10 days = $2400

> *Cabin rental for the 10-day period:* 24 campers × 1 cabin/4 campers × $10/day/cabin ×
> 10 days = $600

> *Insurance and overhead expense for the 10-day period:* 24 campers × $1/day/camper ×
> 10 days = $240

Living costs = $2400 + $600 + $240 = $3240

Entertainment Costs

> *Canoe rental costs:* 2 canoe days × 24 campers × 1 canoe/3 campers × $5/day/canoe
> = $80

> *Beach rental costs:* 3 days × 2 half-days/day × $25/half-day = $150

> *Day hike costs:* 24 campers × 3 day hikes × $2.50/camper/day hike = $180

> *Nightly entertainment:* This is free! Can you believe it?

Entertainment costs = $80 + $150 + $180 + 0 = $410

Total cost

Total cost for 10-day period = Transportation costs + Living costs + Entertainment costs

$$= \$316 + \$3240 + \$410 = \$3966$$

Thus, the cost per student would be $\$3966/24 = \$165.25$.

Thus, it would cost you $165.25 per student to host the students at the island campground for the 10-day period. In this case the per-unit model gives you a very detailed cost estimate (although its accuracy depends on the accuracy of your data and assumptions you've made).

Segmenting Model

The **segmenting model** can be described as 'divide and conquer'. An estimate is decomposed into its individual components, estimates are made at those lower levels, and then the estimates are aggregated (added) back together. It is much easier to estimate at the lower levels because they are more readily understood. This approach is common in engineering estimating in many applications and for any level of accuracy needed. In planning the camp trip of Example 2-5, the overall estimate was **segmented** into the costs for travel, living, and entertainment. The example illustrated the segmenting model (division of the overall estimate into the various categories) together with the unit factor model to make the estimate for each category. Example 2-6 provides another example of the segmenting approach.

EXAMPLE 2-6

Clean Lawn Corp. a manufacturer of yard equipment, is planning to introduce a new high-end industrial-use lawn mower called the Grass Grabber. The Grass Grabber is designed as a walk-behind self-propelled mower. Clean Lawn engineers have been asked by the accounting department to estimate the cost of the materials that will make up the new mower. The material cost estimate will be used, along with estimates for labour and overhead to evaluate the potential of this new model.

SOLUTION

The engineers decide to decompose the design specifications for the Grass Grabber into its components, estimate the material costs for each of the components, and then add these costs up to obtain their overall estimate. The engineers are using a segmenting approach to build up their estimate. After careful consideration, they have divided the mower into the following major subsystems: chassis, drive train, controls, and cutting and collection system. Each of these is further divided as appropriate, and unit material costs are estimated at this lowest of levels as follows:

Cost Item	Unit Material Cost Estimate	Cost Item	Unit Material Cost Estimate
A. Chassis		**C. Controls**	
A.1 Deck	$ 7.40	C.1 Handle assembly	$ 3.85
A.2 Wheels	10.20	C.2 Engine linkage	8.55
A.3 Axles	4.85	C.3 Blade linkage	4.70
	$22.45	C.4 Speed control linkage	21.50
		C.5 Drive control assembly	6.70
B. Drive train		C.6 Cutting height adjuster	7.40
B.1 Engine	$38.50		$52.70
B.2 Starter assembly	5.90		
B.3 Transmission	5.45	**D. Cutting and Collection system**	
B.4 Drive disc assembly	10.00	D.1 Blade assembly	$10.80
B.5 Clutch linkage	5.15	D.2 Side chute	7.05
B.6 Belt assemblies	7.70	D.3 Grass bag and adapter	7.75
	$72.70		$25.60

The total material cost estimate of $173.45 was calculated by adding up the estimates for each of the four major subsystem levels (chassis, drive train, controls, and cutting and collection system). It should be noted that this cost represents only the material portion of the overall cost to produce the mowers. There would also be costs for labour and overhead.

In Example 2-6 the engineers at Clean Lawn Corp. decomposed the cost estimation problem into logical elements. The scheme they used of decomposing cost items and numbering the material components (A.1, A.1, A.2, etc.) is known as a **work breakdown structure.** This technique is commonly used in engineering cost estimating and project management of large products, processes, or projects. A work breakdown structure decomposes a large 'work package' into its constituent parts, which can then be estimated or managed individually. In Example 2-6 the work breakdown structure of the Grass Grabber has three levels. At the top level is the product itself, at the second level are the four major subsystems, and at the third level are the individual cost items. Imagine what the product work breakdown structure for a Boeing 777 looks like. Then imagine trying to manage the 777's design, engineering, construction, and costing without a tool like the work breakdown structure.

Cost Indexes

Cost indexes are numerical values that record historical change in engineering (and other) costs. The cost index numbers are dimensionless, and reflect relative price change in either individual cost items (labour, material, utilities) or groups of costs (consumer prices, producer prices). Indexes can be used to update historical costs with the basic ratio relationship given in Equation 2-2.

$$\frac{\text{Cost at time } A}{\text{Cost at time } B} = \frac{\text{Index value at time } A}{\text{Index value at time } B} \qquad (2\text{-}2)$$

Equation 2-2 states that the ratio of the cost index numbers at two points in time (A and B) is equivalent to the dollar cost ratio of the item at the same times (see Example 2-7).

EXAMPLE 2-7

Miriam is interested in estimating the annual labour and material costs for a new production facility. She was able to obtain the following cost data:

Labour costs

- Labour cost index value was at 124 10 years ago and is 188 today.
- Annual labour costs for a similar facility were $575,500 10 years ago.

Material Costs

- Material cost index value was at 544 three years ago and is 715 today.
- Annual material costs for a similar facility were $2,455,000 three years ago.

SOLUTION

Miriam will use Equation 2-2 to develop her cost estimates for annual labour and material costs.

Labour

$$\frac{\text{Annual cost today}}{\text{Annual cost 10 years ago}} = \frac{\text{Index value today}}{\text{Index value 10 years ago}}$$

$$\text{Annual cost today} = \frac{188}{124} \times \$575,500 = \$871,800$$

Materials

$$\frac{\text{Annual cost today}}{\text{Annual cost 3 years ago}} = \frac{\text{Index value today}}{\text{Index value 3 years ago}}$$

$$\text{Annual cost today} = \frac{715}{544} \times \$2,455,000 = \$3,227,000$$

Cost index data are collected and published by several private and public sources in Canada and the United States. Canadian data are available from Statistics Canada. The US government publishes data through the Bureau of Labor Statistics of the Department of Commerce. The *Statistical Abstract of the United States* publishes cost indexes for labour, construction, and materials. Another useful source for engineering cost index data is the *Engineering News Record*.

Power-Sizing Model

The **power-sizing model** is used to estimate the costs of industrial plants and equipment. The model 'scales up' or 'scales down' known costs, thereby accounting for economies of scale that are common in industrial plant and equipment costs. Consider the cost of building a refinery. Would it cost twice as much to build the same facility with double the capacity? It is unlikely. The *power-sizing model* uses the exponent (x), called the *power-sizing exponent*, to represent economies of scale in the size or capacity:

$$\frac{\text{Cost of equipment A}}{\text{Cost of equipment B}} = \left(\frac{\text{Size(capacity) of equipment A}}{\text{Size(capacity) of B}} \right)^x \tag{2-3}$$

where x is the power-sizing exponent, costs of A and B are at the same point in time (same dollar basis), and size or capacity is in the same physical units for both A and B.

The power-sizing exponent (x) can be 1.0 (indicating a linear cost-versus-size or capacity relationship) or greater than 1.0 (indicating *dis*economies of scale), but it is usually less than 1.0 (indicating economies of scale). Generally the ratio should be less than 2, and it should never exceed 5. This model works best in a middle range—when the plants and equipment are not very small or very large.

Exponent values for plants and equipment of many types may be found in several sources, including industry reference books, research reports, and technical journals. Such exponent values may also be found in *Perry's Chemical Engineers' Handbook, Plant Design and Economics for Chemical Engineers,* and *Preliminary Plant Design in Chemical Engineering.* Table 2-1 gives power sizing exponent values for several types of industrial facilities and equipment. The exponent given applies only to equipment within the size range specified.

In Equation 2-3 equipment costs for both A and B occur at the same point in time. This equation is useful for scaling equipment costs but *not* for updating those costs. When the time of the desired cost estimate is different from the time in which the scaling occurs (per Equation 2-3), cost indexes accomplish the time updating. Thus, in cases like Example 2-8 involving both scaling and updating, we use the power sizing model together with cost indexes.

TABLE 2-1 Example Power-Sizing Exponent Values

Equipment or Facility	Size Range	Power-Sizing Exponent
Blower, centrifugal	10,000–100,000 ft^3/min	0.59
Compressor	200–2100 hp	0.32
Crystallizer, vacuum batch	500–7000 ft^2	0.37
Dryer, drum, single atmospheric	10–100 ft^2	0.40
Fan, centrifugal	20,000–70,000 ft^2/min	1.17
Filter, vacuum rotary drum	10–1500 ft^2	0.48
Lagoon, aerated	0.05–20 million gal/day	1.13
Motor	5–20 hp	0.69
Reactor, 300 psi	100–1000 gal	0.56
Tank, atmospheric, horizontal	100–40,000 gal	0.57

EXAMPLE 2-8

Because of her work in Example 2-7, Miriam has been asked to estimate the cost today of a 2500 ft^2 heat exchange system for the new plant being analyzed. She has the following data.

- Her company paid $50,000 for a 1000 ft^2 heat exchanger 5 years ago.
- Heat exchangers within this range of capacity have a power sizing exponent (x) of 0.55.
- Five years ago the Heat Exchanger Cost Index (HECI) was 1306; it is 1487 today.

SOLUTION

Miriam will first use Equation 2-3 and the 0.55 power-sizing exponent to scale up the cost of the 1000 ft^2 exchanger to one that is 2500 ft^2.

$$\frac{\text{Cost of 2500 ft}^2 \text{ equipment}}{\text{Cost of 1000 ft}^2 \text{ equipment}} = \left(\frac{2500 \text{ ft}^2 \text{ equipment}}{1000 \text{ ft}^2 \text{ equipment}}\right)^{0.55}$$

$$\text{Cost of 2500 ft}^2 \text{ equipment} = \left(\frac{2500}{1000}\right)^{0.55} \times 50,000 = \$82,800$$

Miriam knows that the $82,800 reflects only the scaling up of the cost of the 1000 ft^2 model to a 2500 ft^2 model. Now she will use Equation 2-2 and the HECI data to estimate the cost of a 2500 ft^2 exchanger today. Miriam's cost estimate would be:

$$\frac{\text{Equipment cost today}}{\text{Equipment cost 5 years ago}} = \frac{\text{Index value today}}{\text{Index value 5 years ago}}$$

$$\text{Equipment cost today} = \frac{1487}{1306} \times \$82,800 = \$94,300$$

Triangulation

Triangulation is used in engineering surveying. A geographical area is divided into triangles from which the surveyor is able to map points within that region by using three fixed points and horizontal angular distances to locate fixed points of interest (e.g., property line reference points). Since any point can be located with two lines, the third line represents an extra perspective and check. We will not use trigonometry to arrive at our cost estimates, but we can use the concept of triangulation. We should approach our economic estimate from different perspectives because such varied perspectives add richness, confidence, and quality to the estimate. **Triangulation** in cost estimating might involve using different sources of data or different quantitative models to arrive at the value being estimated. As decision makers we should always seek out varied perspectives.

Improvement and the Learning Curve

One common phenomenon observed, regardless of the task being performed, is that as the number of repetitions increases, performance becomes faster and more accurate. This is the concept of learning and improvement in the activities that people perform. From our own experience we all know that our fiftieth repetition is done in much less time than we needed to do the task the first time.

The **learning curve** captures the relationship between task performance and task repetition. In general, as output *doubles,* the unit production time will be reduced to some fixed percentage, the **learning curve percentage** or **learning curve rate.** For example, it may take 300 minutes to produce the third unit in a production run involving a task with a 95% learning time curve. In this case the sixth (2×3) unit will take $300(0.95) = 285$ minutes to produce. Sometimes the learning curve is also known as the progress curve, improvement curve, experience curve, or manufacturing progress function.

Equation 2-4 gives an expression that can be used for estimating time in repetitive tasks.

$$T_N = T_{initial} \times N^b \tag{2-4}$$

where T_N = time required for the Nth unit of production
$T_{initial}$ = time required for the first (initial) unit of production
N = number of completed units (cumulative production)
b = learning curve exponent (slope of the learning curve on a log–log plot)

As just given, a learning curve is often referred to by its percentage learning slope. Thus, a curve with $b = -0.074$ is a 95% learning curve because $2^{-0.074} = 0.95$. This equation uses 2 because the learning curve percentage applies for doubling cumulative production. The learning curve exponent is calculated with Equation 2-5.

$$b = \frac{\log (\text{learning curve expressed as a decimal})}{\log 2.0} \tag{2-5}$$

EXAMPLE 2-9

Calculate the time required to produce the hundredth unit of a production run if the first unit took 32.0 minutes to produce and the learning curve rate for production is 80%.

SOLUTION

$$T_{100} = T_1 \times 100^{\log 0.80/ \log 2.0}$$

$$T_{100} = 32.0 \times 100^{-0.3219}$$

$$T_{100} = 7.27 \text{ minutes}$$

It is particularly important to account for the learning-curve effect if the production run involves a small number of units instead of a large number. When thousands or even millions of units are being produced, early inefficiencies tend to be averaged out because of the larger batch sizes. However, in the short run, inefficiencies of the same magnitude can lead to rather poor estimates of production time requirements, and thus production cost estimates may be understated. Consider Example 2-10 and the results that might be observed if the learning-curve effect is ignored. Notice in this example that a 'steady state' time is given. Steady state is the time at which the physical constraints of performing the task prevent the achievement of any more learning or improvement.

EXAMPLE 2-10

Estimate the overall labour cost portion due to a task that has a learning-curve rate of 85% and reaches a steady state value after 16 units of 5.0 minutes per unit. Labour and benefits are $22 per hour, and the task requires two skilled workers. The overall production run is 20 units.

SOLUTION

Because we know the time required for the 16th unit, we can use Equation 2-4 to calculate the time required to produce the first unit.

$$T_{16} = T_1 \times 16^{\log 0.85 / \log 2.0}$$

$$5.0 = T_1 \times 16^{-0.2345}$$

$$T_1 = 9.6 \text{ minutes}$$

Now we use Equation 2-4 to calculate the time requirements for each unit in the production run as well as the total production time required.

$$T_N = 9.6 \times N^{-0.2345}$$

Unit Number, N	Time (min) to produce Nth Unit	Cumulative Time from 1 to N	Unit Number, N	Time (min) to produce Nth Unit	Cumulative Time from 1 to N
1	9.6	9.6	11	5.5	74.0
2	8.2	17.8	12	5.4	79.2
3	7.4	24.2	13	5.3	84.5
4	6.9	32.1	14	5.2	89.7
5	6.6	38.7	15	5.1	94.8
6	6.3	45.0	16	5.0	99.8
7	6.1	51.1	17	5.0	104.8
8	5.9	57.0	18	5.0	109.8
9	5.7	62.7	19	5.0	114.8
10	5.6	68.3	20	5.0	119.8

The total cumulative time of the production run is 119.8 minutes (2.0 hours). Thus the total labour cost estimate would be

2.0 hours × $22/hour per worker × 2 workers = $88

If we ignore the learning-curve effect and calculate the labour cost portion from only the steady state labour rate, the estimate would be

0.083 hours/unit × 20 units × $22/hour per worker × 2 workers = $73.04

This estimate is understated by about 20% from what the true cost would be.

ESTIMATING BENEFITS

This chapter has focused on cost terms and cost estimating. However, engineering economists must often also estimate benefits. Benefits would include sales of products, revenues from bridge tolls and electric power sales, cost reductions from reduced material or labour costs, less time spent in traffic jams, and reduced risk of flooding. Many engineering projects are undertaken precisely to secure these benefits.

The cost concepts and cost estimating models can also be applied to economic benefits. Fixed and variable benefits, recurring and non-recurring benefits, incremental benefits, and life-cycle benefits all have meaning. Also, issues regarding the type of estimate (rough, budget, and detailed), as well as difficulties in estimation (one of a kind, time and effort, and estimator expertise), all apply directly to estimating benefits. Last, per-unit, segmented, and indexed models are used to estimate benefits. The concept of triangulation is particularly important for estimating benefits.

The uncertainty in benefit estimates is also usually asymmetric, with a broader limit for negative outcomes. Benefits are more likely to be overestimated than underestimated, so an example set of limits might be (–50%, +20%). One difference between cost and benefit estimation is that many costs of engineering projects occur in the near future (for design and construction), but the benefits are further in the future. Because benefits are often further in the future, they are more difficult to estimate accurately, and more uncertainty is typical.

The estimation of economic benefits for inclusion in our analysis is an important step that should not be overlooked. Many of the models, concepts, and issues that apply in the estimation of costs also apply in the estimation of economic benefits.

CASH FLOW DIAGRAMS

The costs and benefits of engineering projects occur over time and are summarized on a cash flow diagram (CFD). Specifically, a CFD illustrates the size, sign, and timing of individual cash flows. In this way the CFD is the basis for engineering economic analysis.

A **cash flow diagram** is created by first drawing a segmented time-based horizontal line, divided into time units. The time units on the CFD can be years, months, quarters, or any other consistent time unit. Then at each time at which a cash flow will occur, a vertical arrow is added—pointing down for costs and up for revenues or benefits. These cash flows are drawn to scale.

The cash flows are **assumed** to occur at time 0 or at the **end** of each period. Consider Figure 2-7, the CFD for a specific investment opportunity whose cash flows are described as follows:

Timing of Cash Flow	Size of Cash Flow
At time zero (now or today)	A positive cash flow of $100
1 time period from today	A negative cash flow of $100
2 time periods from today	A positive cash flow of $100
3 time periods from today	A negative cash flow of $150
4 time periods from today	A negative cash flow of $150
5 time periods from today	A positive cash flow of $50

FIGURE 2-7 An example of a cash flow diagram (CFD).

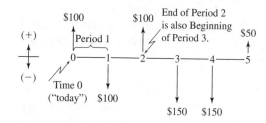

Categories of Cash Flows

The expenses and receipts due to engineering projects usually fall into one of the following categories.

First cost ≡ expense to build or to buy and install

Operations and maintenance (O&M) ≡ annual expense, such as electricity, labour, and minor repairs

Salvage value ≡ receipt at project termination for sale or transfer of the equipment (can be a salvage cost)

Revenues ≡ annual receipts due to sale of products or services

Overhaul ≡ major capital expenditure that occurs during the life of the asset

Individual projects will often have specific costs, revenues, or user benefits. For example, annual operations and maintenance expenses on an assembly line might be divided into direct labour, power, and other. Similarly, a public-sector dam project might have its annual benefits divided into flood control, agricultural irrigation, and recreation.

Drawing a Cash Flow Diagram

The cash flow diagram shows when all cash flows occur. Look at Figure 2-7 and the $100 positive cash flow at the end of period 2. From the time line one can see that this cash flow can also be described as occurring at the *beginning* of period 3. Thus, in a CFD the end of *period t* is the same time as the beginning of *period t* + 1. Beginning-of-period cash flows (such as rent, lease, and insurance payments) are thus easy to handle: just draw your CFD and put them in where they occur. Thus O&M, salvages, revenues, and overhauls are assumed to be end-of-period cash flows.

The choice of time 0 is arbitrary. For example, it can be when a project is analyzed, when funding is approved, or when construction begins. When construction periods are short, first costs are assumed to occur at time 0, and the first annual revenues and costs start at the end of the first period. When construction periods are long, for example, several years, time 0 is usually the date of commissioning—when the facility comes on stream.

Perspective is also important when one is drawing a CFD. Consider the simple transaction of paying $5000 for some equipment. To the firm buying the equipment, the cash flow is a cost and hence negative in sign. To the firm selling the equipment, the cash flow is a revenue and positive in sign. This simple example shows that a consistent perspective is required when one is using a CFD to model the cash flows of a problem. One person's cash outflow is another person's inflow.

Often two or more cash flows occur in the same year, such as an overhaul and an O&M expense or the salvage value and the last year's O&M expense. Combining these into one total cash flow per year would simplify the cash flow diagram. However, it is better to show each individually so as to ensure a clear connection from the problem statement to each cash flow in the diagram.

Drawing Cash Flow Diagrams with a Spreadsheet

One simple way to draw cash flow diagrams with 'arrows' proportional to the size of the cash flows is to use a spreadsheet to draw a stacked bar chart. The data for the cash flows is entered, as shown in the table part of Figure 2-8. To make a quick graph, select cells B1 to D8, which are the three columns of the cash flow. Then select the graph menu and choose column chart and select the stack option. Except for labelling the axes (using the cells for year 0 to year 6), choosing the scale for the y axis, and adding titles, the cash flow diagram is done. Refer to the appendix for a review of basic spreadsheet use. (*Note:* a bar chart labels periods rather than using an x axis with arrows at times 0, 1, 2 . . . .)

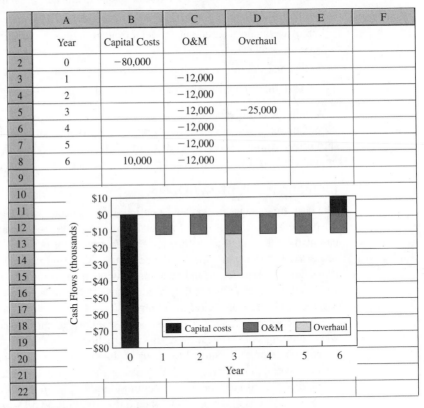

	A	B	C	D	E	F
1	Year	Capital Costs	O&M	Overhaul		
2	0	−80,000				
3	1		−12,000			
4	2		−12,000			
5	3		−12,000	−25,000		
6	4		−12,000			
7	5		−12,000			
8	6	10,000	−12,000			

FIGURE 2-8 Example of cash flow diagram in spreadsheets.

SUMMARY

This chapter has introduced the following cost concepts: fixed and variable, marginal and average, sunk, opportunity, recurring and non-recurring, incremental, cash and book, and life-cycle. **Fixed costs** are constant and unchanging as volumes change, while **variable**

costs change as output changes. Fixed and variable costs are used to find a break-even value between costs and revenues, as well as the regions of net profit and loss. A **marginal cost** is for one more unit, while the **average cost** is the total cost divided by the number of units.

Sunk costs result from past decisions and should not influence our attitude toward current and future opportunities. Remember, 'sunk costs are sunk.' **Opportunity costs** involve the benefit that is forgone when we choose to use a resource in one activity instead of another. **Recurring costs** can be planned and anticipated expenses; **non-recurring costs** are one-of-a-kind costs that are often more difficult to anticipate.

Incremental costs are economic consequences associated with the differences between two choices of action. **Cash costs** are also known as **out-of-pocket costs** that represent actual cash flows. **Book costs** do not result in the exchange of money, but rather are costs listed in a firm's accounting books. **Life-cycle costs** are all costs that are incurred over the life of a product, process, or service. Thus engineering designers must consider life-cycle costs when choosing materials and components, tolerances, processes, testing, safety, service and warranty, and disposal.

Cost estimating is the process of "developing the numbers" for engineering economic analysis. Unlike a textbook, the real world does not present its challenges with neat problem statements that provide all the data. **Rough estimates** give us order-of-magnitude numbers and are useful for high-level and initial planning as well as judging the feasibility of alternatives. **Budget estimates** are more accurate than rough-order estimates, thus requiring more resources (people, time, and money) to develop. These estimates are used in preliminary design and budgeting. **Detailed estimates** generally have an accuracy of ± 3–5%. They are used during the detailed design and contract bidding phases of a project.

Difficulties are common in developing estimates. **One-of-a-kind estimates** will have no basis in earlier work, but this disadvantage can be addressed through **estimation by analogy.** Lack of time available is best addressed by planning and by matching the estimate's detail to the purpose—one should not spend money developing a detailed estimate when only a rough estimate is needed. **Estimator expertise** must be developed through work experiences and mentors.

Several general models and techniques for developing cost estimates were discussed. The **per-unit** and **segmenting models** use different levels of detail and costs per square foot or other unit. **Cost index data** are useful for updating historical costs to formulate current estimates. The **power-sizing model** is useful for scaling up or down a known cost quantity to account for economies of scale, with different power-sizing exponents for industrial plants and equipment of different types. **Triangulation** suggests that one should seek varying perspectives when developing cost estimates. Different information sources, databases, and analytical models can all be used to create unique perspectives. As the number of task repetitions increases, efficiency improves because of learning or improvement. This is summarized in the **learning-curve percentage,** where doubling the cumulative production reduces the time to complete the task, which equals the learning-curve percentage times the current production time.

Cash flow estimation must include project benefits. These include labour cost savings, quality costs avoided, direct revenue from sales, reduced catastrophic risks, improved traffic flow, and cheaper power supplies. **Cash flow diagrams** are used to model the positive and negative cash flows of potential investment opportunities. These diagrams provide a consistent view of the problem (and the alternatives) to support economic analysis.

PROBLEMS

2-1 Bob Johnson decided to buy a new house. After looking at some new housing developments, he decided that a custom-built home was preferable. He hired an architect to prepare the drawings. In due time, the architect completed the drawings and submitted them. Bob liked the plans; he was less pleased that he had to pay the architect a fee of $4000 to design the house. Bob asked a building contractor for a bid to construct the home on a lot Bob already owned. While the contractor was working to assemble the bid, Bob came across a book of standard house plans. In the book was a house that he and his wife liked better than the one designed for them by the architect. Bob paid $75 and obtained a complete set of plans for this other house. Bob then asked the contractor for a bid to construct this 'stock plan' home. In this way Bob felt he could compare the costs and make a decision. The building contractor submitted the following bids:

Custom-designed home	$128,000
Stock-plan home	128,500

Bob was willing to pay the extra $500 for it. Bob's wife, however, felt they should go ahead with the custom-designed house, for, as she put it, 'We can't afford to throw away a set of plans that cost $4000.' Bob agreed, but he disliked the thought of building a house that is less desirable than the stock plan home. Then he asked your advice. Which house would you advise him to build? Explain.

2-2 Venus Computer can produce 23,000 personal computers a year on its daytime shift. The fixed manufacturing costs per year are $2 million and the total labour cost is $9,109,000. To increase its production to 46,000 computers per year, Venus is considering adding a second shift. The unit labour cost for the second shift would be 25% higher than for the day shift, but the total fixed manufacturing costs would increase only to $2.4 million from $2 million.

  (a) Compute the unit manufacturing cost for the daytime shift.

  (b) Would adding a second shift increase or decrease the unit manufacturing cost at the plant?

2-3 A small machine shop, with 30 hp of connected load, purchases electricity at the following monthly rates (assume any demand charge is included in this schedule):

First 50 kWh per hp of connected load at 8.6¢ per kWh

Next 50 kWh per hp of connected load at 6.6¢ per kWh

Next 150 kWh per hp of connected load at 4.0¢ per kWh

All electricity over 250 kWh per hp of connected load at 3.7¢ per kWh

The shop uses 2800 kWh per month.

  (a) Calculate the monthly bill for this shop. What are the marginal and average costs per kilowatt hour?

  (b) Suppose Jennifer, the proprietor of the shop, has the chance to secure additional business that will require her to operate her existing equipment more hours per day. This will use an extra 1200 kw-hr per month. What is the lowest figure that she might reasonably consider to be the "cost" of this additional energy? What is this per kilowatt hour?

  (c) She contemplates installing certain new machines that will reduce the labour time required on certain operations. These will increase the connected load by 10 hp, but since they will operate only on certain special jobs, will add only 100 kWh per month. In a study to determine the economy of installing these new machines, what should be considered as the "cost" of this energy? What is this per kilowatt hour?

2-4 Two automatic systems for dispensing maps are being compared by a provincial highway department. The accompanying break-even chart of the comparison of these systems (System I vs System II) shows total yearly costs for the number of maps dispensed per year for both alternatives. Answer the following questions.

  (a) What is the fixed cost for System I?

  (b) What is the fixed cost for System II?

  (c) What is the variable cost per map dispensed for System I?

  (d) What is the variable cost per map dispensed for System II?

  (e) What is the break-even point in terms of maps dispensed at which the two systems have equal annual costs?

  (f) For what range of annual number of maps dispensed is System I recommended?

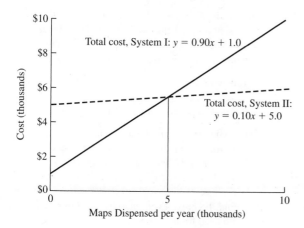

Total cost, System I: $y = 0.90x + 1.0$

Total cost, System II: $y = 0.10x + 5.0$

Cost (thousands)

Maps Dispensed per year (thousands)

(g) For what range of annual number of maps dispensed is System II recommended?

(h) At 3000 maps per year, what are the marginal and average map costs for each system?

2-5 Mr Sam Spade, the president of Ajax, recently read in a report that a competitor named Bendix has the following relationship between cost and production quantity:

$$C = \$3,000,000 - \$18,000Q + \$75Q^2$$

where C = total manufacturing cost per year and Q = number of units produced per year.

A newly hired employee, who previously worked for Bendix, tells Mr Spade that Bendix is now producing 110 units per year. If the selling price remains unchanged, Mr Spade wonders if Bendix is likely to increase the number of units produced per year, in the near future. He asks you to look at the information and tell him what you are able to deduce from it.

2-6 A privately owned summer camp for youngsters has the following data for a 12-week session:

Charge per camper	$120 per week
Fixed costs	$48,000 per session
Variable cost per camper	$80 per week
Capacity	200 campers

(a) Develop the mathematical relationships for total cost and total revenue.

(b) What is the total number of campers that will allow the camp to *just break even*?

(c) What is the profit or loss for the 12-week session if the camp operates at 80% capacity?

2-7 Two new rides are being compared by a local amusement park in terms of their annual operating costs. The two rides are assumed to be able to generate the same level of revenue (hence the focus on costs). The Tummy Tugger has fixed costs of $10,000 per year and variable costs of $2.50 per visitor. The Head Buzzer has fixed costs of $4000 per year, and variable costs of $4 per visitor. Provide answers to the following questions so the amusement park can make the needed comparison.

(a) Mathematically determine the break-even number of visitors per year for the two rides to have equal annual costs.

(b) Develop a graph that illustrates the following: (*Note:* Put visitors per year on the horizontal axis and costs on the vertical axis.)

- Accurate total cost lines for the two alternatives (show line, slopes, and equations).
- The break-even point for the two rides in terms of number of visitors.
- The ranges of visitors per year where each alternative is preferred.

2-8 Consider the accompanying break-even graph for an investment, and answer the following questions as they pertain to the graph.

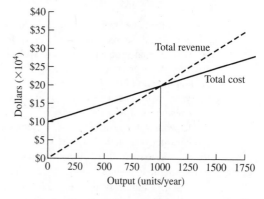

Total revenue

Total cost

Dollars ($\times 10^4$)

Output (units/year)

(a) Give the equation to describe total revenue for x units per year.

(b) Give the equation to describe total costs for x units per year.

(c) What is the 'break-even' level of x in terms of costs and revenues?

(d) If you sell 1500 units this year, will you have a profit or loss? How much?

2-9 Quatro Hermanas, Inc. is investigating the use of some new production machinery as part of its operations.

Three alternatives have been identified, and they have the following fixed and variable costs:

Alternative	Annual Fixed Costs	Annual Variable Costs per Unit
A	$100,000	$20.00
B	200,000	5.00
C	150,000	7.50

Determine the ranges of production (units produced per year) over which each alternative would be recommended for purchase by Quatro Hermanas. Be exact. (*Note:* Consider the range of production to be from 0–30,000 units per year.)

2-10 Three alternative designs have been created by Snakisco engineers for a new machine that spreads cheese between the crackers in a Snakisco snack. Each machine design has unique total costs (fixed and variable) based on the annual production rate of boxes of these crackers. The costs for the three designs are given (where x is the annual production rate of boxes of cheese crackers).

Design	Fixed Cost	Variable Cost ($/x$)
A	$100,000	$20.5x$
B	350,000	$10.5x$
C	600,000	$8.0x$

(a) Mathematically determine which of the machine designs would be recommended for different levels of annual production of boxes of snack crackers. Management is interested in the production interval of 0–150,000 boxes of crackers per year. Over what production volume would each design (A or B or C) be chosen?

(b) Depict your solution from part (a) graphically, putting x per year on the horizontal axis and $ on the vertical axis, so that management can more easily see the following:

  i. Accurate total cost lines for each alternative (show line, slopes, and line equations).

  ii. Any relevant break-even or crossover points in terms of costs between the alternatives.

  iii. Ranges of annual production where each alternative is preferred.

  iv. Clearly label your axes and include a *title* for the graph.

2-11 A small company manufactures a certain product. Variable costs are $20 per unit and fixed costs are $10,875. The price-demand relationship for this product is $P = -0.25D + 250$, where P is the unit sales price of the product and D is the annual demand. Use the data (and helpful hints) that follow to work out answers to parts (a)–(e).

- Total cost = Fixed cost + Variable cost
- Revenue = Demand × Price
- Profit = Revenue – Total cost

Set up your graph with dollars on the y axis (between 0 and $70,000) and, on the x axis, demand D: (units produced or sold), between 0 and 1000 units.

(a) Develop the equations for total cost and total revenue.

(b) Find the break-even quantity (in terms of profit and loss) for the product.

(c) What profit would the company obtain by maximizing its total revenue?

(d) What is the company's maximum possible profit?

(e) Neatly graph the solutions from parts (a), (b), (c), and (d).

2-12 A painting operation is performed by a production worker at a labour cost of $1.40 per unit. A robot spray-painting machine, costing $15,000, would reduce the labour cost to $0.20 per unit. If the device would be valueless at the end of 3 years, what is the minimum number of units that would have to be painted each year to justify the purchase of the robot machine?

2-13 Company A has fixed expenses of $15,000 per year, and each unit of product has a $0.002 variable cost. Company B has fixed expenses of $5000 per year and can produce the same product at a $0.05 variable cost. At what number of units of annual production will Company A have the same overall cost as Company B?

2-14 A firm believes the sales volume (S) of its product depends on its unit selling price (P) and can be determined from the equation $P = $100 - S$. The cost (C) of producing the product is $1000 + 10S$.

(a) Draw a graph with the sales volume (S) from 0 to 100 on the x axis, and total cost and total income from 0 to 2500 on the y axis. On the graph draw the line $C = $1000 + 10S$. Then plot the curve of total income [which is sales volume (S) × unit selling price ($100 - S$)]. Mark the break-even points on the graph.

(b) Determine the break-even point (lowest sales volume where total sales income just equals total production cost). (*Hint:* This may be done by trial and error or by using the quadratic equation to locate the point at which profit is zero.)

(c) Determine the sales volume (S) at which the firm's profit is a maximum. (*Hint:* Write an equation for profit and solve it by trial and error, or as a minima-maxima calculus problem.)

2-15 Consider the situation of owning rental properties that local university students rent from you for the academic year. Develop a set of costs that you could classify as recurring and others that could be classified as non-recurring.

2-16 Define the difference between a 'cash cost' and a 'book cost'. Is engineering economic analysis concerned with both types of cost? Give an example of each, and provide the context in which it is important.

2-17 In your own words, develop a statement of what the authors mean by 'life-cycle costs'. Is it important for a firm to be aware of life-cycle costs? Explain why.

2-18 In looking at Figures 2-4 and 2-5, restate in your own words what the authors are trying to get across with these figures. Do you agree that this is an important effect for companies? Explain.

2-19 In the text we describe three effects that complicate the process of making estimates to be used in engineering economy analyses. List these three effects and comment on which of these might be most influential.

2-20 Northern Tundra Telephone (NTT) has received a contract to install emergency phones along a new 100-kilometre section of the Snow-Moose Highway. Fifty emergency phone systems will be installed about 2 km apart. The material cost of a unit is $125. NTT will need to run underground communication lines that cost NTT $7500 per km (including labour) to install. There will also be a one-time cost of $10,000 to network these phones into NTT's current communication system. You are asked to develop a cost estimate of the project from NTT's perspective. If NTT adds a profit margin of 35% to its costs, how much will it cost the province to fund the project?

2-21 You and your spouse are planning a second honeymoon to the Cayman Islands this summer and would like to have your house painted while you are away. Estimate the total cost of the paint job from the information given below, where:

$$\text{Cost}_{\text{total}} = \text{Cost}_{\text{paint}} + \text{Cost}_{\text{labour}} + \text{Cost}_{\text{fixed}}$$

Paint information: Your house has a surface area of 6000 ft^2. One can of paint can cover 300 ft^2. You are estimating the cost to put on *two coats* of paint for the entire house, using the cost per can given. Note the incremental decrease in unit cost per can as you purchase more and more cans.

Number of Cans Purchased	Cost per Can
First 10 cans purchased	$15.00
Second 15 cans purchased	$10.00
Up to next 50 cans purchased	$7.50

Variable cost information: You plan to hire five painters who will paint for 10 hours per day each. You estimate that the job will require 4.5 days of their painting time. The painter's rate is $8.75 per hour.

Fixed cost information: There is a fixed cost of $200 per job that the painting company charges to cover travel expenses, clothing, cloths, thinner, administration, and so on.

2-22 You are interested in having a cottage built for weekend trips, vacations, and perhaps eventually to retire to. After discussing the project with a local contractor, you receive an estimate that the total construction cost of your 2000 ft^2 'cottage' will be $150,000. Costs within each category include labour, material, and overhead items. The percentage of costs for each of several items (categories) is broken down as follows:

Cost Items	Percentage of Total Costs
Construction permits, legal, and title fees	8%
Roadway, site clearing, preparation	15
Foundation, concrete, masonry	13
Wallboard, flooring, carpentry	12
Heating, ventilation, air conditioning	13
Electric, plumbing, communications	10
Roofing, flooring	12
Painting, finishing	17
	100

(a) What is the cost per square foot of the 2000 ft^2 cottage?

(b) If you are also considering a 4000 ft^2 layout option, estimate your construction costs if:

  i. All cost items (in the table) change proportionately to the size increase.

ii. The first two cost items do not change at all; all others are proportionate.

2-23 SungSam, Inc. is currently designing a new digital camcorder that is projected to have the following per unit costs to manufacture:

Cost Categories	Unit Costs
Materials costs	$112
Labour costs	85
Overhead costs	213
Total Unit Cost	$410

SungSam adds 30% to its manufacturing cost for corporate profit. Answer the following questions:
(a) What unit profit would SungSam realize on each camcorder?
(b) What is the overall cost to produce a batch of 10,000 camcorders?
(c) What would SungSam's profit be on the batch of 10,000 if historical data show that 1% of product will be scrapped in manufacturing, and 3% of finished product will go unsold, 2% of *sold* product will be returned for refund?
(d) How much can SungSam afford to pay for a contract that would lock in a 50% reduction in the unit material cost previously given? If SungSam does sign the contract, the sales price will remain the same as before.

2-24 Fifty years ago, Grandma Bell purchased a set of gold-plated dinnerware for $55, and last year you inherited it. Unfortunately a fire at your home destroyed the set. Your insurance company is at a loss to define the replacement cost and has asked your help. You do some research and find that the Aurum Flatware Cost Index (AFCI) for gold-plated dinnerware, which was 112 when Grandma Bell bought her set, is at 2050 today. Use the AFCI to update the cost of Bell's set to today's cost to show to the insurance company.

2-25 Your boss is the director of reporting for the Athens County Construction Agency (ACCA). It has been his job to track the cost of construction in Athens County. Twenty-five years ago he created the ACCA Cost Index to track these costs. Costs during the first year of the index were $12 per square foot of constructed space (the index value was set at 100 for that first year). This past year a survey of contractors revealed that costs were $72 per square foot. What index number will your boss publish in his report for this year? If the index value was 525 last year, what was the cost per square foot last year?

2-26 An refinisher of antiques named Constance has been so successful with her small business that she is planning to expand her shop and buy all new equipment. She is going to start enlarging her shop by purchasing the following equipment.

Equipment	Original Capacity	Cost of Original Equipment	Power-Sizing Exponent	Capacity of New Equipment
Varnish bath	50 gal	$3500	0.80	75 gal
Power scraper	3/4 hp	$250	0.22	1.5 hp
Paint booth	3 ft^3	$3000	0.6	12 ft^3

What would be the *net* cost to Constance to obtain this equipment—assume that she can trade the old equipment in for 15% of its original cost. Assume also that there has been no inflation in equipment prices.

2-27 Refer to Problem 2-26 and now assume the prices for the equipment that Constance wants to replace have not been constant. Use the cost index data for each piece of equipment to update the costs to the price that would be paid today. Develop the overall cost for Constance, again assuming the 15% trade-in allowance for the old equipment. Use any necessary data from Problem 2-26.

Original Equipment	Cost Index When Originally Purchased	Cost Index Today
Varnish bath	154	171
Power scraper	780	900
Paint booth	49	76

2-28 Five years ago, when the relevant cost index was 120, a nuclear centrifuge cost $40,000. The centrifuge had a capacity of separating 1500 gallons of ionized solution per hour. Today, a centrifuge with a capacity of 4500 gallons per hour is needed, but the cost index now is 300. Assuming a power-sizing exponent to reflect economies of scale, x, of 0.75, use the power-sizing model to determine the approximate cost (expressed in today's dollars) of the new reactor.

2-29 Pierre works for a trade magazine that publishes lists of Power-Sizing Exponents (PSE) that reflect economies of scale for developing engineering estimates of various types of equipment. Pierre has been unable to find any published data on the VMIC machine and wants to list its PSE value in his next issue. Given the following data (your staff was able to find data regarding costs and sizes of the VMIC machine), calculate the PSE value that Pierre should publish. (*Note:* The VMIC-100 can handle twice the volume of a VMIC-50.)

> Cost of VMIC-100 today: $100,000
> Cost of VMIC-50 five years ago: $45,000
> VMIC equipment index today $= 214$
> VMIC equipment index five years ago $= 151$

2-30 Develop an estimate for each of the following situations.

(*a*) The cost of an 800 kilometre trip by car if gasoline costs 75¢ per litre, vehicle wear and tear is $0.05 per kilometre, and our vehicle uses 11 L/100 km.

(*b*) The total number of hours in the average human life, if the average life is 75 years.

(*c*) The number of days it takes to travel around the equator in a hot air balloon, if the balloon averages 100 miles per day, the diameter of the earth is ~4000 miles. (*Note:* Circumference = π times diameter.)

2-31 If 200 worker hours were required to produce the 1st unit in a production run and 60 worker hours were required to produce the 7th unit, what was the *learning-curve rate* during production?

2-32 Rose is a project manager at the civil engineering consulting firm of Sands, Gravel, Concrete, and Waters, Inc. She has been collecting data on a project in which concrete pillars were being constructed; however not all the data are available. She has been able to find out that the 10th pillar required 260 person-hours to construct, and that a 75% learning curve applied. She is interested in calculating the time required to construct the 1st and 20th pillars. Compute the values for her.

2-33 Sally Statistics is implementing a system of statistical process control (SPC) charts in her factory in an effort to reduce the overall cost of scrapped product. The current cost of scrap is $X per month. If an 80% learning curve is expected in the use of the SPC charts to reduce the cost of scrap, what would the *percentage reduction* in monthly scrap cost be after the charts have been used for 12 months? (*Hint:* Model each month as a unit of production.)

2-34 Randy Duckout has been asked to develop an estimate of the *per-unit selling price* (the price that each unit will be sold for) of a new line of hand-crafted booklets that offer excuses for missed appointments. His assistant Doc Duckout has collected information

that Randy will need in developing his estimate:

Cost of direct labour	$20 per hour
Cost of materials	$43.75 per batch of 25 booklets
Cost of overhead items	50% of direct labour cost
Desired profit	20% of total manufacturing cost

Doc also finds out that (1) they should use a 75% learning curve for estimating the cost of direct labour, (2) the time to complete the 1st booklet is estimated at 0.60 hour, and (3) the estimated time to complete the 25th booklet should be used as their standard time for the purpose of determining the *unit selling price*. What would Randy and Doc's estimate be for the *unit selling price*?

2-35 Develop a statement that expresses the extent to which cost estimating topics also apply to estimating benefits. Provide examples to illustrate.

2-36 On December 1, Al Smith purchased a car for $18,500. He paid $5000 immediately and agreed to pay three additional payments of $6000 each (which includes principal and interest) at the end of 1, 2, and 3 years. Maintenance for the car is projected at $1000 at the end of the first year and $2000 at the end of each subsequent year. Al expects to sell the car at the end of the fourth year (after paying for the maintenance work) for $7000. Using these facts, prepare a table of cash flows.

2-37 Bonka Toys is considering a robot that will cost $20,000. After seven years its salvage value will be $2000. An overhaul costing $5000 will be needed in year 4. O&M costs will be $2500 per year. Draw the cash flow diagram.

2-38 Pine Village needs some additional recreation fields. Construction will cost $225,000, and annual O&M expenses are $85,000. The city council estimates that the value of added youth leagues is about $190,000 annually. In year 6 another $75,000 will be needed to refurbish the fields. The salvage value is estimated to be $100,000 after 10 years. Draw the cash flow diagram.

2-39 Identify your major cash flows for the current university term as first costs, O&M expenses, salvage values, revenues, overhauls, and so on. Using a week as the time period, draw the cash flow diagram.

CHAPTER 3

Interest and Equivalence

Going Up in Smoke

State governments throughout the United States were the beneficiaries of a large financial windfall in 1998. Tobacco companies agreed to pay in perpetuity to settle claims arising from the health effects of smoking. Payments over the first 25 years are estimated to be nearly $250 billion.

State officials announced they would earmark these funds for purposes such as health care, education, and, of course, anti-smoking campaigns. But several states involved in the settlement were chronically short of money and were desperate to plug budget deficits. They wondered how they could get their hands on the full value of that tobacco settlement now, instead of waiting for the payments to dribble in year by year.

That's when someone hit on the idea of raising instant money by issuing "tobacco bonds": the states would sell bonds to investors and pay them interest out of the tobacco settlement payments the states were receiving. State governments could get the whole sale price of the bond up front; investors would get ongoing income from the bond.

A growing number of states are now selling these bonds and pocketing quick billions. To attract buyers, however, the states have to pay a high rate of interest on their tobacco bonds, since investors view them as riskier than other government-backed securities. Investors reason that tobacco companies could go broke, after all—especially if the no-smoking laws being passed all over the country cause enough people to snuff out their cigarette habit for good.

After Completing This Chapter...

The student should be able to:

- Define and provide examples of the *time value of money*.
- Distinguish between *simple* and *compound interest,* and use compound interest in engineering economic analysis.
- Explain *equivalence* of cash flows.
- Solve problems using the single payment compound interest formulas.

QUESTIONS TO CONSIDER

1. Tobacco bonds often pay interest of as much as 7%, while conventional government bonds may pay only 5%. On a $1 million bond with a term of 25 years, how much does this interest differential amount to, in total?

2. Assuming a 2% interest differential, how much are governments losing over a 25-year period by opting to take the settlement money "up front" instead of collecting the long-term payout?

3. When the money they get from selling bonds runs out, governments will have to find a new way of closing their budget gaps. Already, some commentators are suggesting that governments may soon look to other business sectors as 'deep pockets.' The tobacco industry was particularly vulnerable because it sold a product that came to be seen as dangerous and socially undesirable. What other industries might face the same fate in years to come? How about fast-food restaurant chains or sellers of alcohol products? Should a company like Anheuser-Busch be worried? Should companies in potentially 'problematic' industries consider setting aside money for future settlement costs?

In the first chapter, we discussed the engineering economic decision process. In Chapter 2 we described models used to estimate the costs and benefits that are summarized in cash flow diagrams. For many of the situations we examined, the economic consequences of an alternative were immediate or took place in a very short period of time, as in Example 1-2 (the decision on the design of a concrete aggregate mix) or Example 1-3 (the change of manufacturing method). In such relatively simple situations, we total the various positive and negative aspects, compare our results, and quickly reach a decision. But can we do the same if the economic consequences occur over a considerable period of time?

No we cannot, because *money has value over time*. Would you rather (1) receive $1000 today or (2) receive $1000 10 years from today? Obviously, the $1000 today has more value. Money's value over time is expressed by an interest rate. In this chapter, we describe two introductory concepts involving the *time value of money:* interest and cash flow equivalence.

COMPUTING CASH FLOWS

Installing an expensive piece of machinery in a plant obviously has economic consequences that occur over an extended period of time. If the machinery were bought on credit, the simple process of paying for it may take several years. What about the usefulness of the machinery? Certainly it was purchased because it would be a beneficial addition to the plant. These favourable consequences may last as long as the equipment performs its useful function. In these circumstances, we do not add up the various consequences; instead, we describe each alternative as cash **receipts** or **disbursements** at different points in **time.** In this way, each alternative is resolved into a set of **cash flows.** This is illustrated by Examples 3-1 and 3-2.

EXAMPLE 3-1

The manager has decided to purchase a new $30,000 mixing machine. The machine may be paid for in one of two ways:

1. Pay the full price now *minus* a 3% discount.
2. Pay $5000 now; at the end of one year, pay $8000; at the end of each of the next four years, pay $6000.

List the alternatives in the form of a table of cash flows.

SOLUTION

In this problem the two alternatives represent different ways to pay for the mixing machine. While the first plan represents a lump sum of $29,100 now, the second one calls for payments continuing until the end of the fifth year. The problem is to convert an alternative into cash receipts or disbursements and show the timing of each receipt or disbursement. The result is called a **cash flow table** or, more simply, a set of *cash flows*.

The cash flows for both the alternatives in this problem are very simple. The cash flow table, with disbursements given negative signs, is as follows:

End of Year	Pay in Full Now	Pay over 5 Years
0 (now)	−$29,100	−$5000
1	0	−8000
2	0	−6000
3	0	−6000
4	0	−6000
5	0	−6000

EXAMPLE 3-2

A man borrowed $1000 from a bank at 8% interest. He agreed to repay the loan in two end-of-year payments. At the end of the first year, he will repay half of the $1000 principal amount plus the interest that is due. At the end of the second year, he will repay the remaining half of the principal amount plus the interest for the second year. Compute the borrower's cash flow.

SOLUTION

In engineering economic analysis, we normally refer to the beginning of the first year as 'time 0'. At this point the man receives $1000 from the bank. (A positive sign represents a receipt of money and a negative sign, a disbursement.) Thus, at time 0, the cash flow is +$1000.

At the end of the first year, the man pays 8% interest for the use of $1000 for one year. The interest is $0.08 \times \$1000 = \$80$. In addition, he repays half the $1000 loan, or $500. Therefore, the end-of-year-1 cash flow is −$580.

At the end of the second year, the payment is 8% for the use of the balance of the principal ($500) for the one-year period, or $0.08 \times 500 = \$40$. The $500 principal is also repaid for a total end-of-year-2 cash flow of −$540. The cash flow is

End of Year	Cash Flow
0 (now)	+$1000
1	−580
2	−540

In this chapter, we will demonstrate techniques for comparing the value of money at different dates, an ability that is essential to engineering economic analysis. We must be able to compare, for example, a low-cost motor with a higher-cost motor. If there were no other consequences, we would obviously prefer the low-cost one. But if the higher-cost motor were more efficient and thereby reduced the annual electric power cost, we would want to consider whether to spend more money now on the motor to reduce power costs in the future. This chapter will provide the methods for comparing the alternatives to determine which motor is preferrable.

TIME VALUE OF MONEY

We often find that the monetary consequences of any alternative occur over a substantial period of time—say, a year or more. When monetary consequences occur in a short period of time, we simply add up the various sums of money and obtain a net result. But can we treat money this way when the time span is greater?

Which would you prefer, $100 cash today or the assurance of receiving $100 a year from now? You might decide you would prefer the $100 now because that is one way to be certain of receiving it. But suppose you were convinced that you would receive the $100 one year hence. Now what would be your answer? A little thought should convince you that it *still* would be more desirable to receive the $100 now. If you had the money now, rather than a year hence, you would have the use of it for an extra year. And if you had no current use for $100, you could let someone else use it.

Money is quite a valuable asset—so valuable that people are willing to pay to have money available for their use. Money can be rented in roughly the same way one rents an apartment; only with money, the charge is called **interest** instead of rent. The importance of interest is demonstrated by banks and savings institutions continuously offering to pay for the use of people's money, that is, to pay interest.

If the current interest rate is 9% per year and you put $100 into the bank for one year, how much will you receive back at the end of the year? You will receive your original $100 together with $9 interest, for a total of $109. This example demonstrates the time preference for money: we would rather have $100 today than the assured promise of $100 one year hence; but we might well consider leaving the $100 in a bank if we knew it would be worth $109 one year hence. This is because there is a **time value of money** in the form of the willingness of banks, businesses, and people to pay interest for the use of various sums.

Simple Interest

Simple interest is interest that is computed only on the original sum and not on accrued interest. Thus if you were to loan a present sum of money P to someone at a simple annual interest rate i (stated as a decimal) for a period of n years, the amount of interest you would receive from the loan would be:

$$\text{Total interest earned} = P \times i \times n = Pin \qquad (3\text{-}1)$$

At the end of n years the amount of money due you, F, would equal the amount of the loan P plus the total interest earned. That is, the amount of money due at the end of the loan would be

$$F = P + Pin \qquad (3\text{-}2)$$

or $F = P(1 + in)$.

EXAMPLE 3-3

You have agreed to lend a friend $5000 for 5 years at a simple interest rate of 8% per year. How much interest will you receive from the loan? How much will your friend pay you at the end of 5 years?

SOLUTION

$$\text{Total interest earned} = Pin = (\$5000)(0.08)(5 \text{ yr}) = \$2000$$

$$\text{Amount due at end of loan} = P + Pin = 5000 + 2000 = \$7000$$

In Example 3-3 the interest earned at the end of the first year is $(5000)(0.08)(1) = \$400$, but this money is not paid to the lender until the end of the fifth year. As a result, the borrower has the use of the $400 for four years without paying any interest on it. This is how simple interest works, and it is easy to see why lenders seldom agree to make simple-interest loans.

Compound Interest

With simple interest, the amount earned (for invested money) or due (for borrowed money) in one period does not affect the principal for interest calculations in later periods. However, this is not how interest is normally calculated. In practice, interest is computed by the **compound interest** method. For a loan, any interest owed but not paid at the end of the year is added to the balance due. Thus, the next year's interest is calculated on the unpaid balance due, which includes the unpaid interest from the preceding period. In this way, compound interest can be thought of as *interest on top of interest*. This distinguishes compound interest from simple interest. In this section, the remainder of the book, and in practice you should assume that the rate is a compound interest rate. The few exceptions will clearly state use "simple interest."

EXAMPLE 3-4

To highlight the difference between simple and compound interest, rework Example 3-3 using an interest rate of 8% per year compound interest. How will this change affect the amount that your friend pays you at the end of 5 years?

SOLUTION

Original loan amount (original principal) = $5000

Loan term = 5 years

Interest rate charged = 8% per year compound interest

In the following table we calculate on a year-to-year basis the total dollar amount due at the end of each year. Notice that this amount becomes the principal upon which interest is calculated in the next year (this is the compounding effect).

Year	Total Principal (P) on Which Interest Is Calculated in Year n	Interest (I) Owed at the End of Year n from Year n's Unpaid Total Principal	Total Amount Due at the End of Year n, New Total Principal for Year $n+1$
1	$5000	$5000 × 0.08 = 400	$5000 + 400 = 5400
2	5400	5400 × 0.08 = 432	5400 + 432 = 5832
3	5832	5832 × 0.08 = 467	5832 + 467 = 6299
4	6299	6299 × 0.08 = 504	6299 + 504 = 6803
5	6803	6803 × 0.08 = 544	6803 + 544 = 7347

The total amount due at the end of the fifth year, $7347, is the amount that your friend will give you to repay the original loan. Notice that this amount is $347 more than the amount you received for lending the same amount, for the same period, at simple interest. This, of course, is because of the effect of interest being earned (by you) on top of interest.

Repaying a Debt

To understand better the mechanics of interest, let us say that $5000 is owed and is to be repaid in 5 years, together with 8% annual interest. There are a great many ways in which debts are repaid; for simplicity, we have selected four specific ways for our example. Table 3-1 tabulates the four plans.

In Plan 1, $1000 will be paid at the end of each year plus the interest due at the end of the year for the use of money to that point. Thus, at the end of the first year, we will have had the use of $5000. The interest owed is 8% × $5000 = $400. The end-of-year payment is, therefore, $1000 principal *plus* $400 interest, for a total payment of $1400. At the end of the second year, another $1000 principal plus interest will be repaid on the money owed during the year. This time the amount owed has declined from $5000 to $4000 because of the $1000 principal payment at the end of the first year. The interest payment is 8% × $4000 = $320, making the end-of-year payment a total of $1320. As indicated in Table 3-1, the series of payments continues each year until the loan is fully repaid at the end of the fifth year.

Plan 2 is another way to repay $5000 in 5 years with interest at 8%. This time the end-of-year payment is limited to the interest due, with no principal payment. Instead, the $5000 owed is repaid in a lump sum at the end of the fifth year. The end-of-year payment in each of the first four years of Plan 2 is 8% × $5000 = $400. The fifth year, the payment is $400 interest *plus* the $5000 principal, for a total of $5400.

Plan 3 calls for five equal end-of-year payments of $1252 each. At this point, we have not shown how the figure of $1252 was computed (see later: Example 4-3). However, it is

TABLE 3-1 Four Plans for Repayment of $5000 in 5 Years with Interest at 8%

(a) Year	(b) Amount Owed at Beginning of Year	(c) Interest Owed for That Year, 8% × (b)	(d) Total Owed at End of Year, (b) + (c)	(e) Principal Payment	(f) Total End-of-Year Payment
Plan 1: At end of each year pay $1000 principal *plus* interest due.					
1	$5000	$ 400	$5400	$1000	$1400
2	4000	320	4320	1000	1320
3	3000	240	3240	1000	1240
4	2000	160	2160	1000	1160
5	1000	80	1080	1000	1080
		$1200		$5000	$6200
Plan 2: Pay interest due at end of each year and principal at end of 5 years.					
1	$5000	$ 400	$5400	$ 0	$ 400
2	5000	400	5400	0	400
3	5000	400	5400	0	400
4	5000	400	5400	0	400
5	5000	400	5400	5000	5400
		$2000		$5000	$7000
Plan 3: Pay in five equal end-of-year payments.					
1	$5000	$ 400	$5400	$ 852	$1252*
2	4148	331	4479	921	1252
3	3227	258	3485	994	1252
4	2233	178	2411	1074	1252
5	1159	93	1252	1159	1252
		$1260		$5000	$6260
Plan 4: Pay principal and interest in one payment at end of 5 years.					
1	$5000	$ 400	$5400	$ 0	$ 0
2	5400	432	5832	0	0
3	5832	467	6299	0	0
4	6299	504	6803	0	0
5	6803	544	7347	5000	7347
		$2347		$5000	$7347

*The exact value is $1252.28, which has been rounded to an even dollar amount.

clear that there is some equal end-of-year amount that would repay the loan. By following the computations in Table 3-1, we see that this series of five payments of $1252 repays a $5000 debt in 5 years with interest at 8%.

Plan 4 is still another method of repaying the $5000 debt. In this plan, no payment is made until the end of the fifth year, when the loan is completely repaid. Note what happens at the end of the first year: the interest due for the first year—8% × $5000 = $400—is not paid; instead, it is added to the debt. At the second year, then, the debt has increased to $5400. The second year interest is thus 8% × $5400 = $432. This amount, again unpaid, is added to the debt, increasing it further to $5832. At the end of the fifth year, the total sum due has grown to $7347 and is paid at that time (see Example 3-4).

Note that when the $400 interest was not paid at the end of the first year, it was added to the debt and, in the second year, there was interest charged on this unpaid interest. That is, the $400 of unpaid interest resulted in 8% × $400 = $32 of additional interest charge in the second year. That $32, together with 8% × $5000 = $400 interest on the $5000 original debt, brought the total interest charge at the end of the second year to $432. Charging interest on unpaid interest is called **compound interest.** We will deal extensively with compound interest calculations later in this chapter.

With Table 3-1 we have illustrated four different ways of accomplishing the same task, that is, repaying a debt of $5000 in 5 years with interest at 8%. Having described the alternatives, we will now use them to present the important concept of *equivalence*.

EQUIVALENCE

When we are indifferent as to whether we have a quantity of money now or the assurance of some other sum of money in the future, or series of future sums of money, we say that the present sum of money is **equivalent** to the future sum or series of future sums.

If an industrial firm believed 8% was a reasonable interest rate, it would have no particular preference about whether it received $5000 now or was repaid by Plan 1 of Table 3-1. Thus $5000 today is equivalent to the series of five end-of-year payments. In the same fashion, the industrial firm would accept repayment Plan 2 as equivalent to $5000 now. Logic tells us that if Plan 1 is equivalent to $5000 now and Plan 2 is also equivalent to $5000 now, it must follow that Plan 1 is equivalent to Plan 2. In fact, *all four repayment plans must be equivalent to each other and to $5000 now*.

Equivalence is an essential factor in engineering economic analysis. In Chapter 2, we saw how an alternative could be represented by a cash flow table. How might two alternatives with different cash flows be compared? For example, consider the cash flows for Plans 1 and 2:

Year	Plan 1	Plan 2
1	−$1400	−$400
2	−1320	−400
3	−1240	−400
4	−1160	−400
5	−1080	−5400
	−$6200	−$7000

If you were given your choice between the two alternatives, which one would you choose? Obviously the two plans have cash flows that are different. Plan 1 requires that there be larger payments in the first four years, but the total payments are smaller than the sum of the Plan 2 payments. To make a decision, the cash flows must be altered so that they can be compared. The **technique of equivalence** is the way we accomplish this.

Using mathematical manipulation, we can determine an equivalent value at some point in time for Plan 1 and a **comparable equivalent value** for Plan 2, based on a selected interest rate. Then we can judge the relative attractiveness of the two alternatives, not from their cash flows, but from comparable equivalent values. Since Plan 1, like Plan 2, repays a *present* sum of $5000 with interest at 8%, the plans are equivalent to $5000 *now;* therefore, the alternatives are equally attractive. This cannot be deduced from the given cash flows alone. It is necessary to learn this by determining the equivalent values for each alternative at some point in time, which in this case is 'the present'.

Difference in Repayment Plans

The four plans computed in Table 3-1 are equivalent in nature but different in structure. Table 3-2 repeats the end-of-year payment schedules from Table 3-1 and also graphs each plan to show the debt still owed at any point in time. Since $5000 was borrowed at the beginning of the first year, all the graphs begin at that point. We see, however, that the four plans result in quite different amounts of money owed at any other point in time. In Plans 1 and 3, the money owed declines as time passes. With Plan 2 the debt remains constant, while Plan 4 increases the debt until the end of the fifth year. These graphs show an important difference among the repayment plans—the areas under the curves differ greatly. Since the axes are Money Owed and Time, the area is their product: Money owed × Time, in years.

In the discussion of the time value of money, we saw that the use of money over a time period was valuable, that people are willing to pay interest to have the use of money for periods of time. When people borrow money, they are acquiring the use of money as represented by the area under the curve for Money owed versus Time. It follows that, at a given interest rate, the amount of interest to be paid will be proportional to the area under the curve. Since in each case the $5000 loan is repaid, the interest for each plan is the total *minus* the $5000 principal:

Plan	Total Interest Paid
1	$1200
2	2000
3	1260
4	2347

We can use Table 3-2 and the data from Table 3-1 to compute the area under each of the four curves, that is, the area bounded by the abscissa, the ordinate, and the curve itself. We multiply the ordinate (Money owed) by the abscissa (1 year) for each of the five years, then *add:*

$$\text{Shaded area} = (\text{Money owed in Year 1})(1 \text{ year})$$

$$+ (\text{Money owed in Year 2})(1 \text{ year})$$

$$+ \cdots$$

$$+ (\text{Money owed in Year 5})(1 \text{ year})$$

or

$$\text{Shaded area } [(\text{Money owed})(\text{Time})] = \textbf{Dollar-Years}$$

TABLE 3-2 End-of-Year Payment Schedules and Their Graphs

Plan 1: At end of each year pay $1000 principal
plus interest due.

Year	End-of-Year Payment
1	$1400
2	1320
3	1240
4	1160
5	1080
	$6200

Plan 2: Pay interest due at end of each year and
principal at end of 5 years.

Year	End-of-Year Payment
1	$ 400
2	400
3	400
4	400
5	5400
	$7000

Plan 3: Pay in five equal end-of-year payments.

Year	End-of-Year Payment
1	$1252
2	1252
3	1252
4	1252
5	1252
	$6260

Plan 4: Pay principal and interest in one payment
at end of 5 years.

Year	End-of-Year Payment
1	$ 0
2	0
3	0
4	0
5	7347
	$7347

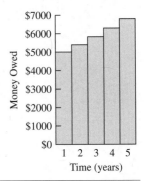

The dollar-years for the four plans would be as follows:

	Dollar-Years			
	Plan 1	Plan 2	Plan 3	Plan 4
(Money owed in Year 1)(1 year)	$ 5,000	$ 5,000	$ 5,000	$ 5,000
(Money owed in Year 2)(1 year)	4,000	5,000	4,148	5,400
(Money owed in Year 3)(1 year)	3,000	5,000	3,227	5,832
(Money owed in Year 4)(1 year)	2,000	5,000	2,233	6,299
(Money owed in Year 5)(1 year)	1,000	5,000	1,159	6,803
Total dollar-years	$15,000	$25,000	$15,767	$29,334

With the area under each curve computed in dollar-years, the ratio of total interest paid to area under the curve may be obtained:

Plan	Total Interest Paid	Area under Curve (dollar-years)	Ratio of Total Interest Paid to Area under Curve
1	$1200	15,000	0.08
2	2000	25,000	0.08
3	1260	15,767	0.08
4	2347	29,334	0.08

We see that the ratio of total interest paid to the area under the curve is constant and equal to 8%. Stated another way, the total interest paid equals the interest rate *times* the area under the curve.

From our calculations, we more easily see why the repayment plans require the payment of different total sums of money, yet are actually equivalent to each other. The key factor is that the four repayment plans provide the borrower with different quantities of dollar-years. Since dollar-years times interest rate equals the interest charge, the four plans result in different total interest charges.

Equivalence Is Dependent on Interest Rate

In the example of Plans 1 to 4, all calculations were made at an 8% interest rate. At this interest rate, it has been shown that all four plans are equivalent to a present sum of $5000. But what would happen if we were to change the problem by changing the interest rate?

If the interest rate were increased to 9%, we know that the interest payment for each plan would increase, and the calculated repayment schedules (Table 3-1, column f) could no longer repay the $5000 debt with the higher interest. Instead, each plan would repay a sum *less* than the principal of $5000, because more money would have to be used to repay the higher interest rate. By some calculations (to be explained later in this chapter and in Chapter 4), the equivalent present sum that each plan will repay at 9% interest is

Plan	Repay a Present Sum of
1	$4877
2	4806
3	4870
4	4775

As predicted, at the higher 9% interest each of the repayment plans of Table 3-1 repays a present sum less than $5000. But they do not repay the *same* present sum. Plan 1 would repay $4877 with 9% interest, while Plan 2 would repay $4806. Thus, with interest at 9%, Plans 1 and 2 are no longer equivalent, for they will not repay the same present sum. The two series of payments (Plan 1 and Plan 2) were equivalent at 8%, but not at 9%. This leads to the conclusion **that equivalence is dependent on the interest rate.** Changing the interest rate destroys the equivalence between two series of payments.

Could we create revised repayment schemes that would be equivalent to $5000 now with interest at 9%? Yes, of course we could: to revise Plan 1 of Table 3-1, we need to increase the total end-of-year payment in order to pay 9% interest on the outstanding debt.

Year	Amount Owed at Beginning of Year	9% Interest for Year	Total End-of-Year Payment ($1000 plus interest)
1	$5000	$450	$1450
2	4000	360	1360
3	3000	270	1270
4	2000	180	1180
5	1000	90	1090

Plan 2 of Table 3-1 is revised for 9% interest by increasing the first four payments to 9% × $5000 = $450 and the final payment to $5450. Two plans that repay $5000 in 5 years with interest at 9% are

Revised Year	End-of-Year Plan 1	Payments Plan 2
1	$1450	$ 450
2	1360	450
3	1270	450
4	1180	450
5	1090	5450

We have determined that Revised Plan 1 is equivalent to a present sum of $5000 and Revised Plan 2 is equivalent to $5000 now; it follows that at 9% interest, Revised Plan 1 is equivalent to Revised Plan 2.

Application of Equivalence Calculations

To understand the usefulness of equivalence calculations, consider the following:

Year	Alternative A: Lower Initial Cost, Higher Operating Cost	Alternative B: Higher Initial Cost, Lower Operating Cost
0 (now)	−$600	−$850
1	−115	−80
2	−115	−80
3	−115	−80
.	.	.
.	.	.
.	.	.
10	−115	−80

Is the least-cost alternative the one that has the lower initial cost and higher operating costs or the one with higher initial cost and lower continuing costs? Because of the time value of money, one cannot add up sums of money at different points in time directly. This means that alternatives cannot be compared in actual dollars at different points in time; instead comparisons must be made in some equivalent comparable sums of money.

It is not sufficient to compare the initial $600 against $850. Instead, we must compute a value that represents the entire stream of payments. In other words, we want to determine a sum that is equivalent to Alternative *A*'s cash flow; similarly, we need to compute the equivalent sum for Alternative *B*. By computing equivalent sums at the same point in time ('now'), we will have values that may be validly compared. The methods for accomplishing this will be presented later in this chapter and Chapter 4.

Thus far we have discussed computing equivalent present sums for a cash flow. But the technique of equivalence is not limited to a present computation. Instead, we could compute the equivalent sum for a cash flow at any point in time. We could compare alternatives in 'Equivalent Year 10' dollars rather than 'now' (Year 0) dollars. Further, the equivalence need not be a single sum; it could be a series of payments or receipts. In Plan 3 of Table 3-1, the series of equal payments was equivalent to $5000 now. But the equivalency works both ways. Suppose we ask: What is the equivalent equal annual payment continuing for 5 years, given a present sum of $5000 and interest at 8%? The answer is $1252.

SINGLE PAYMENT COMPOUND INTEREST FORMULAS

To facilitate equivalence computations, a series of **interest formulas** will be derived. To simplify the presentation, we'll use the following notation:

i = *interest rate per interest period.* In the equations the interest rate is stated as a decimal (that is, 9% interest is 0.09).

n = *number of interest periods.*

P = *a present sum of money.*

F = *a future sum of money.* The future sum F is an amount, n interest periods from the present, that is equivalent to P with interest rate i.

Suppose a present sum of money P is invested for one year[1] at interest rate i. At the end of the year, we should receive back our initial investment P, together with interest equal to iP, or a total amount $P + iP$. Factoring P, the sum at the end of one year is $P(1 + i)$.

Let us assume that, instead of removing our investment at the end of one year, we agree to let it remain for another year. How much would our investment be worth at the end of the second year? The end-of-first-year sum $P(1 + i)$ will draw interest in the second year of $iP(1 + i)$. This means that, at the end of the second year, the total investment will become

$$P(1 + i) + iP(1 + i)$$

---

[1] A more general statement is to specify 'one interest period' rather than 'one year'. Since, however, it is easier to visualize one year, the derivation will assume that one year is the interest period.

This may be rearranged by factoring $P(1 + i)$, which gives

$$P(1 + i)(1 + i)$$

or

$$P(1 + i)^2$$

If the process is continued for a third year, the end-of-the-third-year total amount will be $P(1 + i)^3$; at the end of n years, we will have $P(1 + i)^n$. The progression looks like this:

Year	Amount at Beginning of Interest Period	+ Interest for Period	= Amount at End of Interest Period
1	P	$+iP$	$= P(1 + i)$
2	$P(1 + i)$	$+iP(1 + i)$	$= P(1 + i)^2$
3	$P(1 + i)^2$	$+iP(1 + i)^2$	$= P(1 + i)^3$
n	$P(1 + i)^{n-1}$	$+iP(1 + i)^{n-1}$	$= P(1 + i)^n$

In other words, a present sum P increases in n periods to $P(1 + i)^n$. We therefore have a relationship between a present sum P and its equivalent future sum, F.

$$\text{Future sum} = (\text{Present sum})\,(1 + i)^n$$

$$F = P(1 + i)^n \tag{3-3}$$

This is the **single payment compound amount formula** and is written in functional notation as

$$F = P(F/P, i, n) \tag{3-4}$$

The notation in parentheses $(F/P, i, n)$ can be read as follows:

> To find a future sum F, given a present sum, P, at an interest rate i per interest period, and n interest periods hence.

Functional notation is designed so that the compound interest factors may be written in an equation in an algebraically correct form. In Equation 3-4, for example, the functional notation is interpreted as

$$F = P\left(\frac{F}{P}\right)$$

which is dimensionally correct. Without proceeding further, we can see that when we derive a compound interest factor to find a present sum P, given a future sum F, the factor will be $(P/F, i, n)$; so, the resulting equation would be

$$P = F(P/F, i, n)$$

which is dimensionally correct.

EXAMPLE 3-5

If $500 were deposited in a bank savings account, how much would be in the account three years hence if the bank paid 6% interest compounded annually?

We can draw a diagram of the problem. *Note:* To have a consistent notation, we will represent *receipts* by upward arrows (and positive signs), and *disbursements* (or payments) by downward arrows (and negative signs).

SOLUTION

From the viewpoint of the person depositing the $500, the diagram for 'today' (Time $= 0$) through Year 3 is as follows:

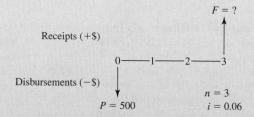

We need to identify the various elements of the equation. The present sum P is $500. The interest rate per interest period is 6%, and in 3 years there are three interest periods. The future sum F is to be computed from the formula

$$F = P(1+i)^n = 500(1+0.06)^3 = \$595.50$$

where $P = \$500$, $i = 0.06$, $n = 3$, and F is unknown.

Thus if we deposit $500 in the bank now at 6% interest, there will be $595.50 in the account in three years.

ALTERNATIVE SOLUTION

The equation $F = P(1+i)^n$ need not be solved with a hand calculator. Instead, *the single payment compound amount factor*, $(1+i)^n$, is readily determined from computed tables. The factor is written in convenient notation as

$$(1+i)^n = (F/P, i, n)$$

and in functional notation as

$$(F/P, 6\%, 3)$$

Knowing $n = 3$, locate the proper row in the 6% table.[2] To find F given P, look in the first column, which is headed 'Single Payment, Compound Amount Factor': for $n = 3$, we find 1.191.

---

[2]The appendix contains compound interest tables for rates between 1/4 and 60%.

Thus,

$$F = 500(F/P, 6\%, 3) = 500(1.191) = \$595.50$$

Before leaving this problem, let's draw another diagram of it, this time from the bank's point of view.

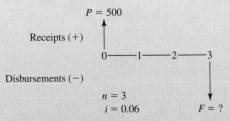

This indicates the bank receives $500 now and must make a disbursement of F at the end of 3 years. The computation, from the bank's point of view, is

$$F = 500(F/P, 6\%, 3) = 500(1.191) = \$595.50$$

This is exactly the same as what was computed from the depositor's viewpoint, since this is just the other side of the same transaction. The bank's future disbursement equals the depositor's future receipt.

If we take $F = P(1 + i)^n$ and solve for P, then

$$P = F\frac{1}{(1 + i)^n} = F(1 + i)^{-n}$$

This is the **single payment present worth formula.** The equation

$$P = F(1 + i)^{-n} \tag{3-5}$$

in our notation becomes

$$P = F(P/F, i, n) \tag{3-6}$$

EXAMPLE 3-6

If you wish to have $800 in a savings account at the end of 4 years, and 5% interest was paid annually, how much should you put into the savings account now?

SOLUTION

$$F = \$800 \qquad i = 0.05 \qquad n = 4 \qquad P = \text{unknown}$$

$$P = F(1 + i)^{-n} = 800(1 + 0.05)^{-4} = 800(0.8227) = \$658.16$$

Thus to have $800 in the savings account at the end of 4 years, we must deposit $658.16 now.

ALTERNATIVE SOLUTION

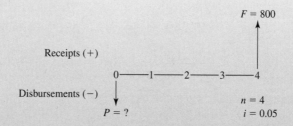

$$P = F(P/F, i, n) = \$800(P/F, 5\%, 4)$$

From the compound interest tables,

$$(P/F, 5\%, 4) = 0.8227$$

$$P = \$800(0.8227) = \$658.16$$

Here the problem has an exact answer. In many situations, however, the answer is rounded off since it can be only as accurate as the input information on which it is based.

EXAMPLE 3-7

Suppose the bank changed their interest policy in Example 3-5 to '6% interest, compounded quarterly'. For this situation, how much money would be in the account at the end of 3 years, assuming a $500 deposit now?

SOLUTION

First, we must be certain to understand the meaning of *6% interest, compounded quarterly*. There are two elements:

6% interest: Unless otherwise described, it is customary to assume that the stated interest is for a one-year period. *If the stated interest is for other than a one-year period, the time frame must be clearly stated.*

Compounded quarterly: This indicates there are four interest periods per year; that is, an interest period is 3 months long.

We know that the 6% interest is an annual rate because if it were for a different period, it would have been stated. Since we are dealing with four interest periods per year, it follows that the interest rate per interest period is $1\frac{1}{2}\%$. For the total 3-year duration, there are 12 interest periods. Thus

$$P = \$500 \qquad i = 0.015 \qquad n = (4 \times 3) = 12 \qquad F = \text{unknown}$$

$$F = P(1+i)^n = P(F/P, i, n)$$

$$= \$500(1 + 0.015)^{12} = \$500(F/P, 1\frac{1}{2}\%, 12)$$

$$= \$500(1.196) = \$598$$

A \$500 deposit now would yield \$598 in 3 years.

EXAMPLE 3-8

Consider the following situation:

Year	Cash Flow
0	+P
1	0
2	0
3	−400
4	0
5	−600

Solve for P assuming a 12% interest rate and using the compound interest tables. Recall that receipts have a plus sign and disbursements or payments have a negative sign. Thus, the diagram is:

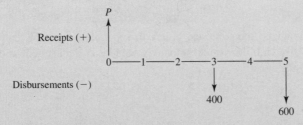

SOLUTION

$$P = 400(P/F, 12\%, 3) + 600(P/F, 12\%, 5)$$

$$= 400(0.7118) + 600(0.5674)$$

$$= \$625.16$$

It is important to understand just what the solution, \$625.16, represents. We can say that \$625.16 is the amount of money that would need to be invested at 12% annual interest to allow for the withdrawal of \$400 at the end of 3 years and \$600 at the end of 5 years.

Let's examine the computations further.

If $625.16 is invested for one year at 12% interest, it will increase to [625.16 + 0.12(625.16)] = $700.18. If for the second year the $700.18 is invested at 12%, it will increase to [700.18 + 0.12(700.18)] = $784.20. And if this is repeated for another year, [784.20 + 0.12(784.20)] = $878.30.

We are now at the end of Year 3. The original $625.16 has increased through the addition of interest to $878.30. It is at this point that the $400 is paid out. Deducting $400 from $878.30 leaves $478.30.

The $478.30 can be invested at 12% for the fourth year and will increase to [478.30 + 0.12(478.30)] = $535.70. And if left at interest for another year, it will increase to [535.70 + 0.12(535.70)] = $600. We are now at the end of Year 5; with a $600 payout; there is no money remaining in the account.

In other words, the $625.16 was just enough money, at a 12% interest rate, to provide exactly for a $400 disbursement at the end of Year 3 and also a $600 disbursement at the end of Year 5. We end up neither short of money nor with money left over: this is an illustration of equivalence. The initial $625.16 is *equivalent* to the combination of a $400 disbursement at the end of Year 3 and a $600 disbursement at the end of Year 5.

ALTERNATIVE FORMATION OF EXAMPLE 3-8

There is another way to see what the $625.16 value of P represents.

Suppose at Year 0 you were offered a piece of paper that guaranteed you would be paid $400 at the end of 3 years and $600 at the end of 5 years. How much would you be willing to pay for this piece of paper if you wanted your money to produce a 12% interest rate?

This alternative statement of the problem changes the signs in the cash flow and the diagram:

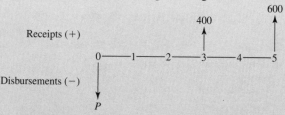

Year	Cash Flow
0	$-P$
1	0
2	0
3	+400
4	0
5	+600

Since the goal is to recover our initial investment P together with 12% interest per year, we can see that P must be *less* than the total amount to be received in the future (that is, $400 + 600 = 1000). We must calculate the present sum P that is *equivalent*, at 12% interest, to an aggregate of $400 in 3 years and $600 in 5 years.

Since we have already derived the relationship

$$P = (1+i)^{-n}F$$

we write

$$P = 400(1 + 0.12)^{-3} + 600(1 + 0.12)^{-5}$$

$$= \$625.17$$

This is virtually the same amount computed from the first statement of this example. [The slight difference is due to the rounding in the compound interest tables. For example, $(1 + 0.12)^{-5} = 0.567427$, but the compound interest table for 12% shows 0.5674.]

Both problems in Example 3-8 have been solved by computing the value of P that is equivalent to $400 at the end of Year 3 and $600 at the end of Year 5. In the first problem, we received $+P$ at Year 0 and were obligated to pay out the $400 and $600 in later years. In the second (alternative formation) problem, the reverse was true. We paid $-P$ at Year 0 and would receive the $400 and $600 sums in later years. In fact, the two problems could represent the buyer and seller of the same piece of paper. The seller would receive $+P$ at Year 0 while the buyer would pay $-P$. Thus, while the problems looked different, they could have been one situation examined first from the viewpoint of the seller and then from that of the buyer. Either way, the solution is based on an equivalence computation.

EXAMPLE 3-9

The second set of cash flows in Example 3-8 was:

Year	Cash Flow
0	$-P$
1	0
2	0
3	+400
4	0
5	+600

At a 12% interest rate, P was computed to be $625.17. Suppose the interest rate is increased to 15%. Will the value of P be larger or smaller?

SOLUTION

One can consider P as a sum of money invested at 15% from which one is to obtain $400 at the end of 3 years and $600 at the end of 5 years. At 12%, P is $625.17. At 15%, P will earn more interest each year, indicating that we can begin with a *smaller P* and still accumulate enough money for the subsequent cash flows. The computation is:

$$P = 400(P/F, 15\%, 3) + 600(P/F, 15\%, 5)$$

$$= 400(0.6575) + 600(0.4972)$$

$$= \$561.32$$

The value of P is smaller at 15% than at 12% interest.

SUMMARY

This chapter describes cash flow tables, the time value of money, and equivalence. The single payment compound interest formulas were derived. It is essential that these concepts and the use of the interest formulas be fully understood, since the remainder of this book and the practice of engineering economy are based on them.

> *Time value of money:* The continuing offer of banks to pay interest for the temporary use of other people's money is ample proof that there is a time value of money. Thus, we would always choose to receive $100 today rather than the promise of $100 to be paid at a future date.

> *Equivalence:* What sum would a person be willing to accept a year hence instead of $100 today? If a 9% interest rate is considered to be appropriate, he would require $109 a year hence. If $100 today and $109 a year hence are considered equally desirable, we say the two sums of money are equivalent. But, if on further consideration, we decided that a 12% interest rate is applicable, then $109 a year hence would no longer be equivalent to $100 today. This illustrates that equivalence is dependent on the interest rate.

Single Payment Formulas

These formulas are for compound interest, which is used in engineering economy.

Compound amount $F = P(1 + i)^n = P(F/P, i, n)$

Present worth $P = F(1 + i)^{-n} = F(P/F, i, n)$

where i = interest rate per interest period (stated as a decimal)
n = number of interest periods
P = a present sum of money
F = a future sum of money; the future sum F is an amount, n interest periods from the present, that is equivalent to P with interest rate i

This chapter also defined simple interest, where interest does not carry over and become part of the principal in subsequent periods. Unless otherwise specified, all interest rates in this text are compound rates.

PROBLEMS

3-1 In your own words explain the *time value of money*. From your own life (either now or in a situation that might occur in the future), provide an example in which the time value of money would be important.

3-2 Magdalen, Miriam, and Mary were asked to consider two different cash flows: $500 that they could receive today and $1000 that they would receive 3 years from today. Magdalen wanted the $500 dollars today, Miriam chose to collect $1000 in 3 years, and Mary was indifferent between these two options. Can you offer an explanation of the choice made by each woman?

3-3 A woman borrowed $2000 and agreed to repay it at the end of 3 years, together with 10% simple interest per year. How much will she pay 3 years hence?

3-4 A $5000 loan was to be repaid with 8% simple annual interest. A total of $5350 was paid. How long had the loan been outstanding?

3-5 Solve the diagram below for the unknown Q assuming a 10% interest rate.

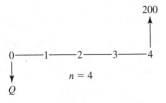

(*Answer: Q = $136.60*)

3-6 The following series of payments will repay a present sum of $5000 at an 8% interest rate. Using single payment factors, what present sum is equivalent to this series of payments at a 10% interest rate?

Year	End-of-Year Payment
1	$1400
2	1320
3	1240
4	1160
5	1080

3-7 A man went to his bank and borrowed $750. He agreed to repay the sum at the end of 3 years, together with the interest at 8% per year. How much will he owe the bank at the end of 3 years? (*Answer: $945*)

3-8 What sum of money now is equivalent to $8250 two years hence if interest is 4% per 6-month period? (*Answer: $7052*)

3-9 The local bank offers to pay 5% interest on savings deposits. In a nearby town, the bank pays 1.25% per 3-month period (quarterly). A man who has $3000 to put in a savings account wonders whether the higher interest paid in the nearby town justifies driving his car there to make the deposit. Assuming he will leave all money in the account for 2 years, how much additional interest would he obtain from the out-of-town bank over the local bank?

3-10 A sum of money invested at 2% per 6-month period (semi-annually) will double in amount in approximately how many years? (*Answer: $17\frac{1}{2}$ years*)

3-11 The Apex Company sold a water softener to Marty Smith. The price of the unit was $350. Marty asked for a deferred payment plan, and a contract was written. Under the contract, the buyer could delay paying for the water softener if he purchased the coarse salt for recharging the softener from Apex. At the end of 2 years, the buyer was to pay for the unit in a lump sum, with interest at a rate of 1.5% per quarter-year. According to the contract, if the customer ceased buying salt from Apex at any time prior to 2 years, the full payment due at the end of 2 years would automatically become due.

Six months later, Marty decided to buy salt elsewhere and stopped buying from Apex, whereupon Apex asked for the full payment that was to have been due 18 months hence. Marty was unhappy about this, so Apex offered as an alternative to accept the $350 with interest at 10% per semi-annual period for the 6 months that Marty had been buying salt from Apex. Which of these alternatives should Marty accept? Explain.

3-12 The United States recently purchased $1 billion in 30-year zero-coupon bonds from a struggling foreign nation. The bonds yield $4\frac{1}{2}$% per year interest. The zero-coupon bonds pay no interest during their 30-year life. Instead, at the end of 30 years, the U.S. government is to receive back its $1 billion together with interest at $4\frac{1}{2}$% per year. A U.S. senator objected to the purchase, claiming that the correct interest rate for bonds like this is $5\frac{1}{4}$%. The result, he said, was a multimillion dollar gift to the foreign country without the approval of Congress. Assuming the senator's math is correct, how much will the foreign country have saved in interest when it repays the bonds at $4\frac{1}{2}$% instead of $5\frac{1}{4}$% at the end of 30 years?

3-13 One thousand dollars is borrowed for one year at an interest rate of 1% per month. If the same sum of money could be borrowed for the same period at an interest rate of 12% per year, how much could be saved in interest charges?

3-14 A sum of money Q will be received 6 years from now. At 5% annual interest, the present worth now of Q is $60. At the same interest rate, what would be the value of Q in 10 years?

3-15 In 1995 an anonymous private collector purchased a painting by Picasso entitled *Angel Fernandez de Soto* for $29,152,000. The picture depicts Picasso's friend deSoto seated in a Barcelona cafe drinking absinthe. The painting was done in 1903 and valued then at $600. If the painting was owned by the same family until its sale in 1995, what rate of return did they receive on the $600 investment?

3-16 (*a*) If $100 at Time 0 will be worth $110 a year hence and was worth $90 a year ago, compute the interest rate for the past year and the interest rate next year.

(*b*) Assume that $90 invested a year ago will return $110 a year from now. What is the annual interest rate in this situation?

3-17 How much must you invest now at 7.9% interest to accumulate $175,000 in 63 years?

3-18 We know that a certain piece of equipment will cost $150,000 in 5 years. How much will it cost today if the interest rate is 10%?

3-19 The local garbage company charges $6 a month for garbage collection. It had been their practice to send out bills to their 100,000 customers at the end of each 2-month period. Thus, at the end of February it would send a bill to each customer for $12 for garbage collection during January and February.

Recently the firm changed its billing date: it now sends out the 2-month bills after one month's service has been performed. Bills for January and February, for example, are sent out at the end of January. The local newspaper points out that the firm is receiving half its money before the garbage collection. This unearned money, the newspaper says, could be temporarily invested for one month at 1% per month interest by the garbage company to earn extra income.

Compute how much extra income the garbage company could earn each year if it invests the money as described by the newspaper.   (*Answer:* $36,000)

3-20 Sally Stanford is buying a car that costs $12,000. She will pay $2000 immediately and the remaining $10,000 in four annual end-of-year principal payments of $2500 each. In addition to the $2500, she must pay 15% interest on the unpaid balance of the loan each year. Prepare a cash flow table to represent this situation.

CHAPTER 4

More Interest Formulas

Anne Scheiber's Bonanza

When Anne Scheiber died in 1995, aged 101, she left an estate worth more than $20 million. It all went to Yeshiva University in New York. The university's officials were grateful. But who, they asked, was Anne Scheiber?

She wasn't a mysterious heiress or a business tycoon, it turned out. She was a retired IRS auditor who had started investing in 1944, putting $5000—her life savings up to that point—into the stock market.

Scheiber's portfolio was not based on get-rich-quick companies. In fact, it contained mostly 'garden variety' stocks like Coca Cola and Exxon. Scheiber researched stock purchases carefully and then held onto her shares for years, rather than trading them. She also continually reinvested her dividends.

Despite her increasing wealth, Scheiber never indulged in a lavish life. Acquaintances described her as extremely frugal and reported that she lived as a near recluse in her small apartment.

What motivated Scheiber to accumulate so much money? In large part, it seems to have been a reaction to her life experiences. Scheiber worked for over 20 years as a tax auditor but failed to receive promotions. She attributed her lack of professional advancement to discrimination against her as a Jewish woman. At her request, the endowment she left to Yeshiva University was used to fund scholarships and interest-free loans for women students.

After Completing This Chapter...

The student should be able to:

- Solve problems modelled by the uniform series compound interest formulas.
- Use arithmetic and geometric gradients to solve appropriately modelled problems.
- Apply *nominal* and *effective interest rates*.
- Use *discrete* and *continuous* compounding in appropriate contexts.
- Use spreadsheets and financial functions to model and solve engineering economic analysis problems.

QUESTIONS TO CONSIDER

1. Set aside your calculator and test your intuition for a minute. What annual rate of return do you suppose it would take to turn $5000 into $20 million in 51 years? 100%? 200%?
2. Now get out the calculator. What was Anne Scheiber's actual average annual rate of return during the years she was investing?
3. How does that rate compare with the overall performance of the stock market during the period from 1944 to 1995?

Chapter 3 presented the fundamental components of engineering economic analysis, including formulas for computing equivalent single sums of money at different points in time. Most problems we will encounter are much more complex. Thus this chapter develops formulas for payments that are a uniform series or are increasing on an arithmetic or geometric gradient. Later in the chapter, nominal and effective interest are discussed. Finally, equations are derived for situations where interest is continuously compounded.

UNIFORM SERIES COMPOUND INTEREST FORMULAS

Many times we will find uniform series of receipts or disbursements. Automobile loans, house payments, and many other loans are based on a **uniform payment series.** It will often be convenient to use tables based on a uniform series of receipts or disbursements. The series A is defined as follows:

$A = $ An end-of-period[1] cash receipt or disbursement in a uniform series, continuing for n periods, the entire series equivalent to P or F at interest rate i

The horizontal line in Figure 4-1 is a representation of time with four interest periods illustrated. Uniform payments A have been placed at the end of each interest period, and there are as many A's as there are interest periods n. (Both these conditions are specified in the definition of A.) Figure 4-1 uses January 1 and December 31, but other 1-year periods could be used.

In the section in Chapter 3 on single payment formulas, we saw that a sum P at one point in time would increase to a sum F in n periods, according to the equation

$$F = P(1+i)^n$$

We will use this relationship in our uniform series derivation.

FIGURE 4-1 The general relationship between A and F.

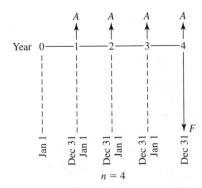

$n = 4$

---

[1] In textbooks on economic analysis, it is customary to define A as an end-of-period event rather than a beginning-of-period or, possibly, middle-of-period event. The derivations that follow are based on this end-of-period assumption. One could, of course, derive other equations based on beginning-of-period or mid-period assumptions.

Looking at Figure 4-1, we see that if an amount A is invested at the end of each year for 4 years, the total amount F at the end of 4 years will be the sum of the compound amounts of the individual investments.

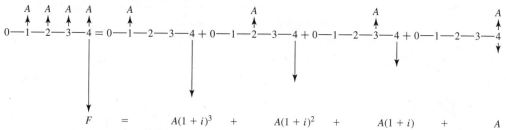

$$F \quad = \quad A(1+i)^3 \quad + \quad A(1+i)^2 \quad + \quad A(1+i) \quad + \quad A$$

In the general case for n years,

$$F = A(1+i)^{n-1} + \cdots + A(1+i)^3 + A(1+i)^2 + A(1+i) + A \qquad (4\text{-}1)$$

Multiplying Equation (4-1) by $(1+i)$, we have

$$(1+i)F = A(1+i)^n + \cdots + A(1+i)^4$$
$$+ A(1+i)^3 + A(1+i)^2 + A(1+i) \qquad (4\text{-}2)$$

Factoring out A and subtracting Equation 4-1 gives

$$(1+i)F = A[(1+i)^n + \cdots + (1+i)^4 + (1+i)^3 + (1+i)^2 + (1+i)] \qquad (4\text{-}3)$$
$$- \quad F = A[(1+i)^{n-1} + \cdots + (1+i)^3 + (1+i)^2 + (1+i) + 1]$$
$$\overline{iF = A[(1+i)^n - 1]} \qquad (4\text{-}4)$$

Solving Equation 4-4 for F gives

$$F = A\left[\frac{(1+i)^n - 1}{i}\right] = A(F/A, i\%, n) \qquad (4\text{-}5)$$

Thus we have an equation for F when A is known. The term inside the brackets

$$\left[\frac{(1+i)^n - 1}{i}\right]$$

is called the **uniform series compound amount factor** and has the notation $(F/A, i, n)$.

EXAMPLE 4-1

A man deposits $500 in a credit union at the end of each year for 5 years. The credit union pays 5% interest, compounded annually. At the end of 5 years, immediately after the fifth deposit, how much does the man have in his account?

SOLUTION

The diagram on the left shows the situation from the man's point of view; the one on the right, from the credit union's point of view. Either way, the diagram of the five deposits and the desired computation of the future sum F duplicates the situation for the uniform series compound amount formula

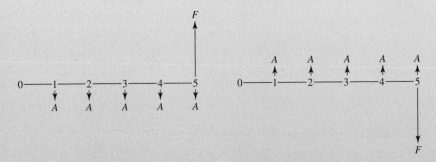

$$F = A \left[\frac{(1+i)^n - 1}{i} \right] = A(F/A, i\%, n)$$

where $A = \$500$, $n = 5$, $i = 0.05$, $F = $ unknown. Filling in the known variables gives

$$F = \$500(F/A, 5\%, 5) = \$500(5.526) = \$2763$$

There will be $2763 in the account following the fifth deposit.

If Equation 4-5 is solved for A, we have

$$A = F \left[\frac{i}{(1+i)^n - 1} \right]$$

$$= F(A/F, i\%, n) \tag{4-6}$$

where

$$\left[\frac{i}{(1+i)^n - 1} \right]$$

is called the **uniform series sinking fund**[2] factor and is written as $(A/F, i, n)$.

---

[2] A *sinking fund* is a separate fund into which one makes a uniform series of money deposits (A) with the goal of accumulating some desired future sum (F) at a given future point in time.

EXAMPLE 4-2

Jim Hayes read that out west, a parcel of land could be purchased for $1000 cash. Jim decided to save a uniform amount at the end of each month so that he would have the required $1000 at the end of one year. The local credit union pays 6% interest, compounded monthly. How much would Jim have to deposit each month?

SOLUTION

In this example,

$$F = \$1000 \qquad n = 12 \qquad i = \tfrac{1}{2}\% \qquad A = \text{unknown}$$

$$A = 1000(A/F, \tfrac{1}{2}\%, 12) = 1000(0.0811) = \$81.10$$

Jim would have to deposit $81.10 each month.

If we use the sinking fund formula (Equation 4-6) and substitute for F the single payment compound amount formula (Equation 3-3), we obtain

$$A = F \left[\frac{i}{(1+i)^n - 1} \right] = P(1+i)^n \left[\frac{i}{(1+i)^n - 1} \right]$$

$$A = P \left[\frac{i(1+i)^n}{(1+i)^n - 1} \right] = P(A/P, i\%, n) \tag{4-7}$$

We now have an equation for determining the value of a series of end-of-period payments—or disbursements—A when the present sum P is known.

The portion inside the brackets

$$\left[\frac{i(1+i)^n}{(1+i)^n - 1} \right]$$

is called the **uniform series capital recovery factor** and has the notation $(A/P, i, n)$.

EXAMPLE 4-3

Consider a situation in which you borrow $5000. You will repay the loan in five equal end-of-the-year payments. The first payment is due one year after you receive the loan. Interest on the loan is 8%. What is the size of each of the five payments?

SOLUTION

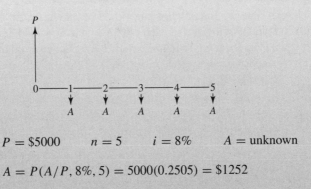

$$P = \$5000 \qquad n = 5 \qquad i = 8\% \qquad A = \text{unknown}$$

$$A = P(A/P, 8\%, 5) = 5000(0.2505) = \$1252$$

The annual loan payment is $1252.

In Example 4-3, with interest at 8%, a present sum of $5000 is equivalent to five equal end-of-period disbursements of $1252. This is another way of stating Plan 3 of Table 3-1. The method for determining the annual payment that would repay $5000 in 5 years with 8% interest has now been explained. The calculation is simply

$$A = 5000(A/P, 8\%, 5) = 5000(0.2505) = \$1252$$

If the capital recovery formula (Equation 4-7) is solved for the present sum P, we obtain the uniform series present worth formula

$$P = A \left[\frac{(1+i)^n - 1}{i(1+i)^n} \right] = A(P/A, i\%, n) \tag{4-8}$$

and

$$(P/A, i\%, n) = \left[\frac{(1+i)^n - 1}{i(1+i)^n} \right]$$

which is the **uniform series present worth factor.**

EXAMPLE 4-4

An investor holds a time payment purchase contract on some machine tools. The contract calls for the payment of $140 at the end of each month for a 5-year period. The first payment is due in one month. He offers to sell you the contract for $6800 cash today. If you otherwise can make 1% per month on your money, would you accept or reject the investor's offer?

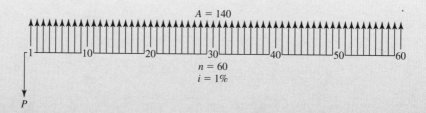

$$A = 140$$

$$n = 60$$
$$i = 1\%$$

P

SOLUTION

In this problem we are being offered a contract that will pay $140 per month for 60 months. We must determine whether the contract is worth $6800 if we consider 1% per month to be a suitable interest rate. Using the uniform series present worth formula, we will compute the present worth of the contract.

$$P = A(P/A, i, n) = 140(P/A, 1\%, 60)$$

$$= 140(44.955)$$

$$= \$6293.70$$

It is clear that if we pay the $6800 asking price for the contract, we will receive less than the 1% per month interest we desire. We will, therefore, reject the investor's offer.

EXAMPLE 4-5

Suppose we decided to pay the $6800 for the time purchase contract in Example 4-4. What monthly rate of return would we obtain on our investment?

SOLUTION

In this situation, we know P, A, and n, but we do not know i. The problem may be solved by using either the uniform series present worth formula

$$P = A(P/A, i, n)$$

or the uniform series capital recovery formula

$$A = P(A/P, i, n)$$

Either way, we have one equation with one unknown.

$$P = \$6800 \qquad A = \$140 \qquad n = 60 \qquad i = \text{unknown}$$

$$P = A(P/A, i, n)$$

$$\$6800 = \$140(P/A, i, 60)$$

$$(P/A, i, 60) = \frac{6800}{140} = 48.571$$

We know the value of the uniform series present worth factor, but we do not know the interest rate i. As a result, we need to look through several compound interest tables and then compute the rate of return i by interpolation. Entering values from the tables in the appendix, we find

Interest Rate	$(P/A, i, 60)$
$\frac{1}{2}\%$	51.726
i	48.571
$\frac{3}{4}\%$	48.174

The rate of return, which is between $\frac{1}{2}\%$ and $\frac{3}{4}\%$, may indeed be computed by a linear interpolation. The interest formulas are not linear, so a linear interpolation will not give an exact solution. To minimize the error, the interpolation should be computed with interest rates as close to the correct answer as possible. [Since $a/b = c/d$, $a = b(c/d)$], we write

$$\begin{aligned}
\text{Rate of return} \quad i &= 0.5\% + a \\
&= 0.5\% + b(c/d) \\
&= 0.50\% + 0.25\% \left(\frac{51.726 - 48.571}{51.726 - 48.174} \right) \\
&= 0.50\% + 0.25\% \left(\frac{3.155}{3.552} \right) = 0.50\% + 0.22\% \\
&= 0.72\% \text{ per month}
\end{aligned}$$

The monthly rate of return on our investment would be 0.72% per month.

EXAMPLE 4-6

Using a 15% interest rate, compute the value of F in the following cash flow:

Year	Cash Flow
1	+100
2	+100
3	+100
4	0
5	−F

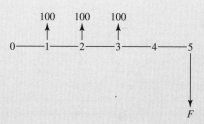

SOLUTION

We see that the cash flow diagram is not the same as the diagram shown in Example 4-1, which is a sinking fund factor diagram.

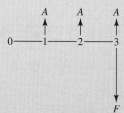

Since the diagrams do not agree, the problem is more difficult than those we've discussed so far. The general approach to use in this situation is to convert the cash flow from its present form into standard forms, for which we have compound interest factors and compound interest tables.

One way to solve this problem is to consider the cash flow as a series of single payments P and then to compute their sum F. In other words, the cash flow is broken into three parts, each one of which we can solve.

$$F = F_1 + F_2 + F_3 = 100(F/P, 15\%, 4) + 100(F/P, 15\%, 3) + 100(F/P, 15\%, 2)$$

$$= 100(1.749) + 100(1.521) + 100(1.322)$$

$$= \$459.20$$

The value of F in the illustrated cash flow is \$459.20.

ALTERNATE SOLUTION

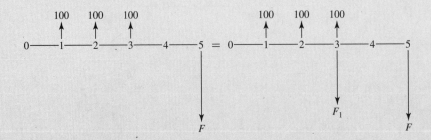

Looked at this way, we first solve for F_1.

$$F_1 = 100(F/A, 15\%, 3) = 100(3.472) = \$347.20$$

Now F_1 can be considered a present sum P in the diagram

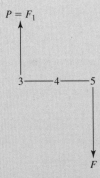

and so

$$F = F_1(F/P, 15\%, 2)$$

$$= 347.20(1.322)$$

$$= \$459.00$$

The slightly different value from the preceding computation is due to rounding in the compound interest tables.

This has been a two-step solution:

$$F_1 = 100(F/A, 15\%, 3)$$

$$F = F_1(F/P, 15\%, 2)$$

One could substitute the value of F_1 from the first equation into the second equation and solve for F, without computing F_1.

$$F = 100(F/A, 15\%, 3)(F/P, 15\%, 2)$$
$$= 100(3.472)(1.322)$$
$$= \$459.00$$

EXAMPLE 4-7

Consider the following situation:

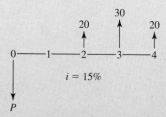

$$i = 15\%$$

The diagram is not in a standard form, indicating that there will be a multiple-step solution. There are at least three different ways of computing the answer. (It is important that you understand how the three computations are made, so please study all three solutions.)

SOLUTION 1

$$P = P_1 + P_2 + P_3$$
$$= 20(P/F, 15\%, 2) + 30(P/F, 15\%, 3) + 20(P/F, 15\%, 4)$$
$$= 20(0.7561) + 30(0.6575) + 20(0.5718)$$
$$= \$46.28$$

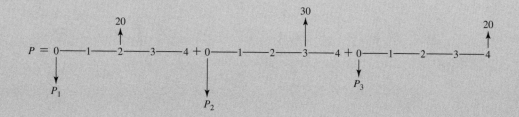

SOLUTION 2

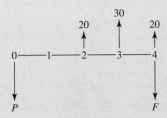

The relationship between P and F in the diagram is

$$P = F(P/F, 15\%, 4)$$

Next we compute the future sums of the three payments, as follows:

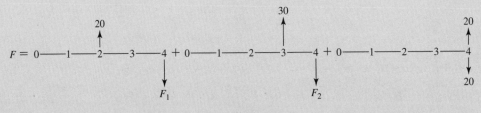

$$F = F_1 + F_2 + 20$$
$$= 20(F/P, 15\%, 2) + 30(F/P, 15\%, 1) + 20$$

Combining the two equations, we have

$$P = [F_1 + F_2 + 20](P/F, 15\%, 4)$$
$$= [20(F/P, 15\%, 2) + 30(F/P, 15\%, 1) + 20](P/F, 15\%, 4)$$
$$= [20(1.322) + 30(1.150) + 20](0.5718)$$
$$= \$46.28$$

SOLUTION 3

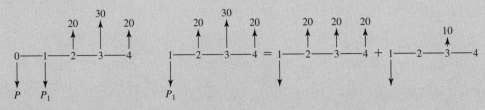

$$P = P_1(P/F, 15\%, 1)$$

$$P_1 = 20(P/A, 15\%, 3) + 10(P/F, 15\%, 2)$$

Combining, we have

$$P = [20(P/A, 15\%, 3) + 10(P/F, 15\%, 2)](P/F, 15\%, 1)$$
$$= [20(2.283) + 10(0.7561)](0.8696)$$
$$= \$46.28$$

RELATIONSHIPS BETWEEN COMPOUND INTEREST FACTORS

From the derivations, we see there are several simple relationships between the compound interest factors. They are summarized here.

Single Payment

$$\text{Compound amount factor} = \frac{1}{\text{Present worth factor}}$$

$$(F/P, i, n) = \frac{1}{(P/F, i, n)} \tag{4-9}$$

Uniform Series

$$\text{Capital recovery factor} = \frac{1}{\text{Present worth factor}}$$

$$(A/P, i, n) = \frac{1}{(P/A, i, n)} \tag{4-10}$$

$$\text{Compound amount factor} = \frac{1}{\text{Sinking fund factor}}$$

$$(F/A, i, n) = \frac{1}{(A/F, i, n)} \tag{4-11}$$

The uniform series present worth factor is simply the sum of the n terms of the single payment present worth factor

$$(P/A, i, n) = \sum_{J=1}^{n} (P/F, i, J) \tag{4-12}$$

For example:

$$(P/A, 5\%, 4) = (P/F, 5\%, 1) + (P/F, 5\%, 2) + (P/F, 5\%, 3) + (P/F, 5\%, 4)$$
$$3.546 = 0.9524 + 0.9070 + 0.8638 + 0.8227$$

The uniform series compound amount factor equals 1 *plus* the sum of $(n - 1)$ terms of the single payment compound amount factor

$$(F/A, i, n) = 1 + \sum_{J=1}^{n-1} (F/P, i, J) \tag{4-13}$$

For example,

$$(F/A, 5\%, 4) = 1 + (F/P, 5\%, 1) + (F/P, 5\%, 2) + (F/P, 5\%, 3)$$

$$4.310 = 1 + 1.050 + 1.102 + 1.158$$

The uniform series capital recovery factor equals the uniform series sinking fund factor *plus i*:

$$(A/P, i, n) = (A/F, i, n) + i \tag{4-14}$$

For example,

$$(A/P, 5\%, 4) = (A/F, 5\%, 4) + 0.05$$

$$0.2820 = 0.2320 + 0.05$$

This may be proved as follows:

$$(A/P, i, n) = (A/F, i, n) + i$$

$$\left[\frac{i(1+i)^n}{(1+i)^n - 1} \right] = \left[\frac{1}{(1+i)^n - 1} \right] + i$$

Multiply by $(1 + i)^n - 1$ to get

$$i(1+i)^n = i + i(1+i)^n - i = i(1+i)^n$$

ARITHMETIC GRADIENT

It frequently happens that the cash flow series is not of constant amount A. Instead, there is a uniformly increasing series as shown:

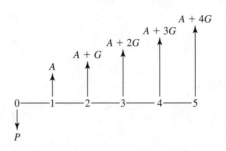

Cash flows of this form may be resolved into two components:

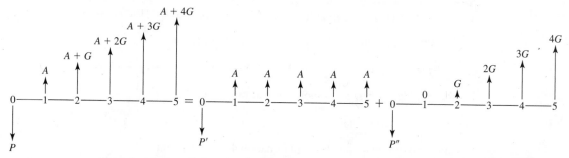

Note that by resolving the problem in this manner, the first cash flow in the arithmetic gradient series becomes zero. This is done so that G is the change from period to period, and because the gradient (G) series normally is used along with a uniform series (A). We already have an equation for P', and we need to derive an equation for P''. In this way, we will be able to write

$$P = P' + P'' = A(P/A, i, n) + G(P/G, i, n)$$

Derivation of Arithmetic Gradient Factors

The arithmetic gradient is a series of increasing cash flows as follows:

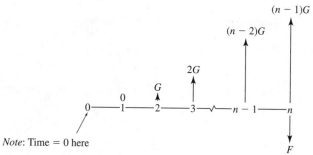

The arithmetic gradient series may be thought of as a series of individual cash flows:

The value of F for the sum of the cash flows $= F^{\mathrm{I}} + F^{\mathrm{II}} + \cdots + F^{\mathrm{III}} + F^{\mathrm{IV}}$, or

$$F = G(1+i)^{n-2} + 2G(1+i)^{n-3} + \cdots + (n-2)(G)(1+i)^1 + (n-1)G \qquad (4\text{-}15)$$

Multiply Equation 4-15 by $(1+i)$ and factor out G, or

$$(1+i)F = G[(1+i)^{n-1} + 2(1+i)^{n-2} + \cdots + (n-2)(1+i)^2 + (n-1)(1+i)^1] \qquad (4\text{-}16)$$

Rewrite Equation 4-15 to show other terms in the series,

$$F = G[(1+i)^{n-2} + \cdots + (n-3)(1+i)^2 + (n-2)(1+i)^1 + n - 1] \qquad (4\text{-}17)$$

Subtracting Equation 4-17 from Equation 4-16, we obtain

$$F + iF - F = G[(1+i)^{n-1} + (1+i)^{n-2} + \cdots + (1+i)^2 + (1+i)^1 + 1] - nG \qquad (4\text{-}18)$$

In the derivation of Equation 4-5, the terms inside the brackets of Equation 4-18 were shown to equal the series compound amount factor:

$$[(1+i)^{n-1} + (1+i)^{n-2} + \cdots + (1+i)^2 + (1+i)^1 + 1] = \frac{(1+i)^n - 1}{i}$$

Thus, Equation 4-18 becomes

$$iF = G\left[\frac{(1+i)^n - 1}{i}\right] - nG$$

Rearranging and solving for F, we write

$$F = \frac{G}{i}\left[\frac{(1+i)^n - 1}{i} - n\right] \qquad (4\text{-}19)$$

Multiplying Equation 4-19 by the single payment present worth factor gives

$$P = \frac{G}{i}\left[\frac{(1+i)^n - 1}{i} - n\right]\left[\frac{1}{(1+i)^n}\right]$$

$$= G\left[\frac{(1+i)^n - in - 1}{i^2(1+i)^n}\right]$$

$$(P/G, i, n) = \left[\frac{(1+i)^n - in - 1}{i^2(1+i)^n}\right] \qquad (4\text{-}20)$$

Equation 4-20 is the **arithmetic gradient present worth factor.** Multiplying Equation 4-19 by the sinking fund factor, we have

$$A = \frac{G}{i}\left[\frac{(1+i)^n - 1}{i} - n\right]\left[\frac{i}{(1+i)^n - 1}\right] = G\left[\frac{(1+i)^n - in - 1}{i(1+i)^n - i}\right]$$

$$(A/G, i, n) = \left[\frac{(1+i)^n - in - 1}{i(1+i)^n - i}\right] = \left[\frac{1}{i} - \frac{n}{(1+i)^n - 1}\right] \qquad (4\text{-}21)$$

Equation 4-21 is the **arithmetic gradient uniform series factor.**

EXAMPLE 4-8

A man has purchased a new car. He wishes to set aside enough money in a bank account to pay the maintenance on the car for the first 5 years. It has been estimated that the maintenance cost of a car is as follows:

Year	Maintenance Cost
1	$120
2	150
3	180
4	210
5	240

Assume the maintenance costs occur at the end of each year and that the bank pays 5% interest. How much should the car owner deposit in the bank now?

SOLUTION

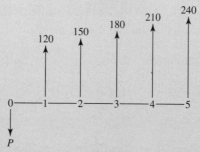

The cash flow may be broken into its two components:

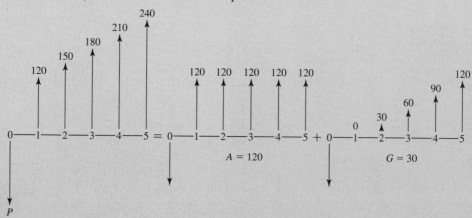

Both components represent cash flows for which compound interest factors have been derived. The first is a uniform series present worth, and the second is an arithmetic gradient series present worth:

$$P = A(P/A, 5\%, 5) + G(P/G, 5\%, 5)$$

Note that the value of n in the gradient factor is 5, not 4. In deriving the gradient factor, we had $(n - 1)$ terms containing G. Here there are four terms containing G.

Thus, $(n - 1) = 4$, so $n = 5$.

$$P = 120(P/A, 5\%, 5) + 30(P/G, 5\%, 5)$$

$$= 120(4.329) + 30(8.237)$$

$$= 519 + 247$$

$$= \$766$$

He should deposit $766 in the bank now.

EXAMPLE 4-9

On a certain piece of machinery, it is estimated that the maintenance expense will be as follows:

Year	Maintenance
1	$100
2	200
3	300
4	400

What is the equivalent uniform annual maintenance cost for the machinery if 6% interest is used?

SOLUTION

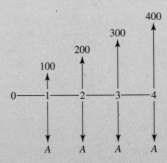

The first cash flow in the arithmetic gradient series is zero; hence the diagram is *not* in proper form for the arithmetic gradient equation. As in Example 4-8, the cash flow must be resolved into two components:

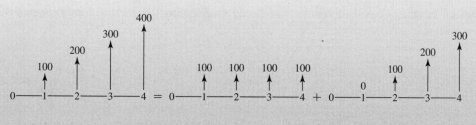

$$A = 100 + 100(A/G, 6\%, 4) = 100 + 100(1.427) = \$242.70$$

The equivalent uniform annual maintenance cost is \$242.70.

EXAMPLE 4-10

A textile mill in India installed a number of new looms. It is expected that initial maintenance costs and expenses for repairs will be high but will then decline for several years. The projected cost is:

Year	Maintenance and Repair Costs (rupees)
1	24,000
2	18,000
3	12,000
4	6,000

What is the projected equivalent annual maintenance and repair cost if interest is 10%?

SOLUTION

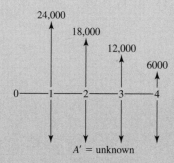

The projected cash flow is not in the form of the arithmetic gradient factors. Both factors were derived for an increasing gradient over time. The factors cannot be used directly for a declining gradient. Instead, we will subtract an increasing gradient from an assumed uniform series of payments.

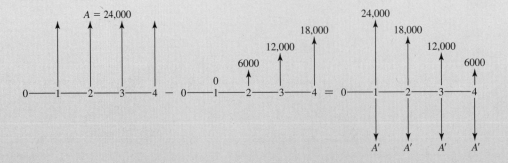

$$A' = 24,000 - 6000(A/G, 10\%, 4)$$
$$= 24,000 - 6000(1.381)$$
$$= 15,714 \text{ rupees}$$

The projected equivalent uniform maintenance and repair cost is 15,714 rupees per year.

EXAMPLE 4-11

Compute the value of P in the diagram. Use a 10% interest rate.

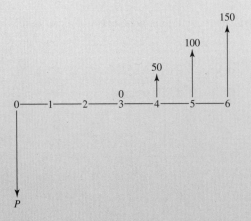

SOLUTION

With the arithmetic gradient series present worth factor, we can compute a present sum J.

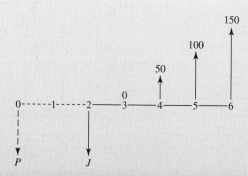

It is important that you closely examine the location of J. Based on the way the factor was derived, there will be one zero value in the gradient series to the right of J. (If this seems strange or incorrect, review the beginning of this section on arithmetic gradients.)

$$J = G(P/G, i, n)$$

$$= 50(P/G, 10\%, 4) \qquad (\textit{Note:} \text{ 3 would be incorrect.})$$

$$= 50(4.378) = 218.90$$

Then

$$P = J(P/F, 10\%, 2)$$

To obtain the present worth of the future sum J, use the $(P/F, i, n)$ factor. Combining, we have

$$P = 50(P/G, 10\%, 4)(P/F, 10\%, 2)$$

$$= 50(4.378)(0.8264) = \$180.90$$

The value of P is $180.90.

GEOMETRIC GRADIENT

In the preceding section, we saw that the arithmetic gradient is applicable where the period-by-period change in a cash receipt or payment is a uniform amount. There are other situations where the period-by-period change is a **uniform rate,** g. For example, if the maintenance costs for an automobile are $100 the first year and they increase at a uniform rate, g, of 10% per year, the cash flow for the first 5 years would be as follows:

Year			Cash Flow
1	100.00	$=$	$100.00
2	$100.00 + 10\%(100.00) = 100(1 + 0.10)^1$	$=$	110.00
3	$110.00 + 10\%(110.00) = 100(1 + 0.10)^2$	$=$	121.00
4	$121.00 + 10\%(121.00) = 100(1 + 0.10)^3$	$=$	133.10
5	$133.10 + 10\%(133.10) = 100(1 + 0.10)^4$	$=$	146.41

From the table, we can see that the maintenance cost in any year is

$$\$100(1 + g)^{n-1}$$

Stated in a more general form,

$$A_n = A_1(1 + g)^{n-1} \tag{4-22}$$

where g = uniform *rate* of cash flow increase or decrease from period
to period, that is, the geometric gradient
A_1 = value of cash flow at Year 1 ($100 in the example)
A_n = value of cash flow at any Year n

Since the present worth P_n of any cash flow A_n at interest rate i is

$$P_n = A_n(1 + i)^{-n} \tag{4-23}$$

we can substitute Equation 4-22 into Equation 4-23 to get

$$P_n = A_1(1 + g)^{n-1}(1 + i)^{-n}$$

This may be rewritten as

$$P = A_1(1 + i)^{-1} \sum_{x=1}^{n} \left(\frac{1 + g}{1 + i} \right)^{x-1} \tag{4-24}$$

The present worth of the entire gradient series of cash flows may be obtained by expanding Equation 4-24:

$$P = A_1(1 + i)^{-1} \sum_{x=1}^{n} \left(\frac{1 + g}{1 + i} \right)^{x-1} \tag{4-25}$$

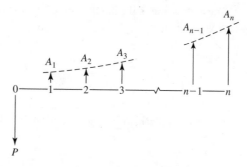

In the general case, where $i \neq g$, Equation 4-24 may be written out as follows:

$$P = A_1(1+i)^{n-1} + A_1(1+i)^{-1}\left(\frac{1+g}{1+i}\right) + A_1(1+i)^{-1}\left(\frac{1+g}{1+i}\right)^2$$

$$+ \cdots + A_1(1+i)^{-1}\left(\frac{1+g}{1+i}\right)^{n-1} \tag{4-26}$$

Let $a = A_1(1+i)^{-1}$ and $b = (1+g)/(1+i)$. Equation 4-26 becomes

$$P = a + ab + ab^2 + \cdots + ab^{n-1} \tag{4-27}$$

Multiply Equation 4-27 by b:

$$bP = ab + ab^2 + ab^3 + \cdots + ab^{n-1} + ab^n \tag{4-28}$$

Subtract Equation 4-28 from Equation 4-27:

$$P - bP = a - ab^n$$

$$P(1-b) = a(1-b^n)$$

$$P = \frac{a(1-b^n)}{1-b}$$

Replacing the original values for a and b, we obtain:

$$P = A_1(1+i)^{-1}\left[\frac{1-\left(\dfrac{1+g}{1+i}\right)^n}{1-\left(\dfrac{1+g}{1+i}\right)}\right] = A_1\left[\frac{1-\left(\dfrac{1+g}{1+i}\right)^n}{(1+i)-\left(\dfrac{1+g}{1+i}\right)(1+i)}\right]$$

$$= A_1\left[\frac{1-(1+g)^n(1+i)^{-n}}{1+i-1-g}\right]$$

$$P = A_1\left[\frac{1-(1+g)^n(1+i)^{-n}}{i-g}\right] \tag{4-29}$$

where $i \neq g$.

The expression in the brackets of Equation 4-29 is the **geometric series present worth factor** where $i \neq g$.

$$(P/A, g, i, n) = \left[\frac{1 - (1+g)^n(1+i)^{-n}}{i - g} \right] \qquad \text{where } i \neq g \qquad (4\text{-}30)$$

In the special case of $i = g$, Equation 4-29 becomes

$$P = A_1 n (1+i)^{-1}$$

$$(P/A, g, i, n) = [n(1+i)^{-1}] \qquad \text{where } i = g \qquad (4\text{-}31)$$

EXAMPLE 4-12

The first-year maintenance cost for a new car is estimated to be $100, and it increases at a uniform rate of 10% per year. Using an 8% interest rate, calculate the present worth of cost of the first 5 years of maintenance.

STEP-BY-STEP SOLUTION

Year n		Maintenance Cost		$(P/F, 8\%, n)$	PW of Maintenance
1	100.00	= 100.00	×	0.9259 =	$ 92.59
2	100.00 + 10%(100.00)	= 110.00	×	0.8573 =	94.30
3	110.00 + 10%(110.00)	= 121.00	×	0.7938 =	96.05
4	121.00 + 10%(121.00)	= 133.10	×	0.7350 =	97.83
5	133.10 + 10%(133.10)	= 146.41	×	0.6806 =	99.65
					$480.42

SOLUTION USING GEOMETRIC SERIES PRESENT WORTH FACTOR

$$P = A_1 \left[\frac{1 - (1+g)^n(1+i)^{-n}}{i - g} \right] \qquad \text{where } i \neq g$$

$$= 100.00 \left[\frac{1 - (1.10)^5(1.08)^{-5}}{-0.02} \right] = \$480.42$$

The present worth of cost of maintenance for the first 5 years is $480.42.

NOMINAL AND EFFECTIVE INTEREST

EXAMPLE 4-13

Consider the situation of a person depositing $100 into a bank that pays 5% interest, compounded semi-annually. How much would be in the savings account at the end of one year?

SOLUTION

Five per cent interest, compounded semi-annually, means that the bank pays $2\frac{1}{2}\%$ every 6 months. Thus, the initial amount $P = \$100$ would be credited with $0.025(100) = \$2.50$ interest at the end of 6 months, or

$$P \rightarrow P + Pi = 100 + 100(0.025) = 100 + 2.50 = \$102.50$$

The $102.50 is left in the savings account; at the end of the second 6-month period, the interest earned is $0.025(102.50) = \$2.56$, for a total in the account at the end of one year of $102.50 + 2.56 = \$105.06$, or

$$(P + Pi) \rightarrow (P + Pi) + i(P + Pi) = P(1 + i)^2 = 100(1 + 0.025)^2$$
$$= \$105.06$$

Nominal interest rate per year, r, is the annual interest rate without considering the effect of any compounding.

In the example, the bank pays $2\frac{1}{2}\%$ interest every 6 months. The nominal interest rate per year, r, therefore, is $2 \times 2\frac{1}{2}\% = 5\%$.

Effective interest rate per year, i_a, is the annual interest rate taking into account the effect of any compounding during the year.

In Example 4-13 we saw that $100 left in the savings account for one year increased to $105.06, so the interest paid was $5.06. The effective interest rate per year, i_a, is $5.06/$100.00 = 0.0506 = 5.06\%$.

r = Nominal interest rate per interest period (usually one year)

i = Effective interest rate per interest period

i_a = Effective interest rate per year

m = Number of compounding sub-periods per time period

Using the method presented in Example 4-13, we can derive the equation for the effective interest rate. If a $1 deposit were made to an account that compounded interest m times per year and paid a nominal interest rate per year, r, the *interest rate per compounding sub-period* would be r/m, and the total in the account at the end of one year would be

$$\$1\left(1 + \frac{r}{m}\right)^m \quad \text{or simply} \quad \left(1 + \frac{r}{m}\right)^m$$

If we deduct the $1 principal sum, the expression would be

$$\left(1 + \frac{r}{m}\right)^m - 1$$

Therefore,

Effective interest rate per year $\qquad i_a = \left(1 + \frac{r}{m}\right)^m - 1$ $\qquad$ (4-32)

where $\quad r =$ nominal interest rate per year
$\qquad m =$ number of compounding sub-periods per year

Or, substituting the effective interest rate per compounding sub-period, $i = (r/m)$,

Effective interest rate per year $\qquad i_a = (1 + i)^m - 1$ $\qquad$ (4-33)

where $\quad i =$ effective interest rate per compounding sub-period
$\qquad m =$ number of compounding sub-periods per year

Either Equation 4-32 or 4-33 may be used to compute an effective interest rate per year.

One should note that i was described in Chapter 3 simply as the interest rate per interest period. We were describing the effective interest rate without making any fuss about it. A more precise definition, we now know, is that i is the *effective* interest rate per interest period. Although it seems more complicated, we are describing exactly the same situation, but with more care.

The nominal interest rate r is often given for a one-year period (but it could be given for either a shorter or a longer time period). In the special case of a nominal interest rate that is given per compounding sub-period, the effective interest rate per compounding sub-period, i, equals the nominal interest rate per sub-period, r.

In the typical effective interest computation, there are multiple compounding sub-periods ($m > 1$). The resulting effective interest rate is either the solution to the problem or an intermediate solution, which allows us to use standard compound interest factors to proceed to solve the problem.

For **continuous compounding** (which is described in the next section),

Effective interest rate per year $\qquad i_a = e^r - 1$ $\qquad$ (4-34)

EXAMPLE 4-14

If a savings bank pays $1\frac{1}{2}\%$ interest every 3 months, what are the nominal and effective interest rates per year?

SOLUTION

$$\text{Nominal interest rate per year} \quad r = 4 \times 1\frac{1}{2}\% = 6\%$$

$$\text{Effective interest rate per year} \quad i_a = \left(1 + \frac{r}{m}\right)^m - 1$$

$$= \left(1 + \frac{0.06}{4}\right)^4 - 1 = 0.061$$

$$= 6.1\%$$

Alternately,

$$\text{Effective interest rate per year} \quad i_a = (1 + i)^m - 1$$

$$= (1 + 0.015)^4 - 1 = 0.061$$

$$= 6.1\%$$

Table 4-1 tabulates the effective interest rate for a range of compounding frequencies and nominal interest rates. It should be noted that when a nominal interest rate is compounded

TABLE 4-1 Nominal and Effective Interest

Nominal Interest Rate per Year	Effective Interest Rate per Year, i_a When Nominal Rate Is Compounded				
r (%)	Yearly	Semi-annually	Monthly	Daily	Continuously
1	1.0000%	1.0025%	1.0046%	1.0050%	1.0050%
2	2.0000	2.0100	2.0184	2.0201	2.0201
3	3.0000	3.0225	3.0416	3.0453	3.0455
4	4.0000	4.0400	4.0742	4.0809	4.0811
5	5.0000	5.0625	5.1162	5.1268	5.1271
6	6.0000	6.0900	6.1678	6.1831	6.1837
8	8.0000	8.1600	8.3000	8.3278	8.3287
10	10.0000	10.2500	10.4713	10.5156	10.5171
15	15.0000	15.5625	16.0755	16.1798	16.1834
25	25.0000	26.5625	28.0732	28.3916	28.4025

annually, the nominal interest rate equals the effective interest rate. Also, it will be noted that increasing the frequency of compounding (for example, from monthly to continuously) has only a small effect on the effective interest rate. But if the amount of money is large, even small differences in the effective interest rate can be significant.

EXAMPLE 4-15

A loan shark lends money on the following terms: 'If I give you $50 on Monday, you owe me $60 on the following Monday.'

   (a) What nominal interest rate per year (r) is the loan shark charging?
   (b) What effective interest rate per year (i_a) is he charging?
   (c) If the loan shark started with $50 and was able to keep it, as well as all the money he received, out in loans at all times, how much money would he have at the end of one year?

SOLUTION TO PART a

$$F = P(F/P, i, n)$$

$$60 = 50(F/P, i, 1)$$

$$(F/P, i, 1) = 1.2$$

Therefore, $i = 20\%$ per week.

$$\text{Nominal interest rate per year} = 52 \text{ weeks} \times 0.20 = 10.40 = 1040\%$$

SOLUTION TO PART b

$$\text{Effective interest rate per year} \quad i_a = \left(1 + \frac{r}{m}\right)^m - 1$$

$$= \left(1 + \frac{10.40}{52}\right)^{52} - 1 = 13,105 - 1$$

$$= 13,104 = 1,310,400\%$$

Or

$$\text{Effective interest rate per year} \quad i_a = (1 + i)^m - 1$$

$$= (1 + 0.20)^{52} - 1 = 13,104$$

$$= 1,310,400\%$$

SOLUTION TO PART c

$$F = P(1+i)^n = 50(1+0.20)^{52}$$
$$= \$655,200$$

With a nominal interest rate of 1040% per year and effective interest rate of 1,310,400% per year, if he started with $50, the loan shark would have $655,200 at the end of one year.

When the various time periods in a problem match, we generally can solve the problem by simple calculations. Thus in Example 4-3, where we had $5000 in an account paying 8% interest, compounded annually, the five equal end-of-year withdrawals are simply computed as follows:

$$A = P(A/P, 8\%, 5) = 5000(0.2505) = \$1252$$

Consider how this simple problem becomes more difficult if the compounding period is changed so that it no longer matches the annual withdrawals.

EXAMPLE 4-16

On January 1, a woman deposits $5000 in a credit union that pays 8% nominal annual interest, compounded quarterly. She wishes to withdraw all the money in five equal yearly sums, beginning December 31 of the first year. How much should she withdraw each year?

SOLUTION

Since the 8% nominal annual interest rate r is compounded quarterly, we know that the effective interest rate per interest period, i, is 2%; and there are a total of $4 \times 5 = 20$ interest periods in 5 years. For the equation $A = P(A/P, i, n)$ to be used, there must be as many periodic withdrawals as there are interest periods, n. In this example we have 5 withdrawals and 20 interest periods.

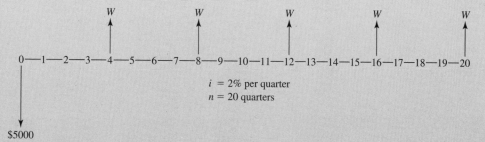

To solve the problem, we must adjust it so that it is in one of the standard forms for which we have compound interest factors. This means we must first compute either an equivalent A for each

3-month interest period or an effective i for each time period between withdrawals. Let's solve the problem both ways.

SOLUTION 1

Compute an equivalent A for each 3-month time period.
If we had been required to compute the amount that could be withdrawn quarterly, the diagram would be as follows:

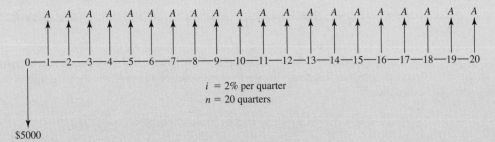

$$A = P(A/P, i, n) = 5000(A/P, 2\%, 20) = 5000(0.0612) = \$306$$

Now, since we know A, we can construct the diagram that relates it to our desired equivalent annual withdrawal, W:

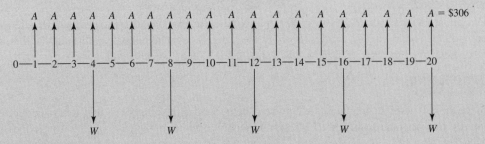

Looking at a one-year period,

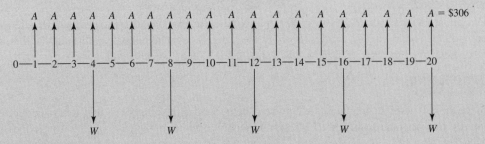

$A = \$306$

$i = 2\%$ per quarter
$n = 4$ quarters

$$W = A(F/A, i, n) = 306(F/A, 2\%, 4) = 306(4.122)$$

$$= \$1260$$

SOLUTION 2

Compute an effective i for the time period between withdrawals.

Between withdrawals, W, there are four interest periods; hence $m = 4$ compounding sub-periods per year. Since the nominal interest rate per year, r, is 8%, we can proceed to compute the effective interest rate per year.

$$\text{Effective interest rate per year} \quad i_a = \left(1 + \frac{r}{m}\right)^m - 1 = \left(1 + \frac{0.08}{4}\right)^4 - 1$$

$$= 0.0824 \doteq 8.24\% \text{ per year}$$

Now the problem may be redrawn as follows:

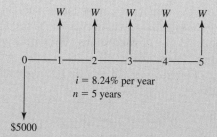

$i = 8.24\%$ per year
$n = 5$ years

This diagram may be solved directly to determine the annual withdrawal W with the capital recovery factor:

$$W = P(A/P, i, n) = 5000(A/P, 8.24\%, 5)$$

$$= P\left[\frac{i(1+i)^n}{(1+i)^n - 1}\right] = 5000\left[\frac{0.0824(1 + 0.0824)^5}{(1 + 0.0824)^5 - 1}\right]$$

$$= 5000(0.2520) = \$1260$$

The depositor should withdraw $1260 per year.

CONTINUOUS COMPOUNDING

Two variables we have introduced are:

r = Nominal interest rate per interest period

m = Number of compounding sub-periods per time period

Since the interest period is normally one year, the definitions become:

r = Nominal interest rate per year

m = Number of compounding sub-periods per year

$$\frac{r}{m} = \text{Interest rate per interest period}$$

$mn = $ Number of compounding sub-periods in n years

Single Payment Interest Factors: Continuous Compounding

The single payment compound amount formula (Equation 3-3)

$$F = P(1+i)^n$$

may be rewritten as

$$F = P\left(1 + \frac{r}{m}\right)^{mn}$$

If we increase m, the number of compounding sub-periods per year, without limit, m becomes very large and approaches infinity, and r/m becomes very small and approaches zero.

This is the condition of **continuous compounding,** that is, where the duration of the interest period decreases from some finite duration Δt to an infinitely small duration dt, and the number of interest periods per year becomes infinite. In this situation of continuous compounding:

$$F = P \lim_{m \to \infty} \left(1 + \frac{r}{m}\right)^{mn} \tag{4-35}$$

An important limit in calculus is:

$$\lim_{x \to 0}(1+x)^{1/x} = 2.71828 = e \tag{4-36}$$

If we set $x = r/m$, then mn may be written as $(1/x)(rn)$. As m becomes infinite, x becomes 0. Equation 4-35 becomes

$$F = P[\lim_{x \to 0}(1+x)^{1/x}]^{rn}$$

Equation 4-36 tells us the quantity inside the brackets equals e. So returning to Equation 3-3, we find that

$$F = P(1+i)^n \quad \text{becomes} \quad F = Pe^{rn} \tag{4-37}$$

and

$$P = F(1+i)^{-n} \quad \text{becomes} \quad P = Fe^{-rn} \tag{4-38}$$

We see that for continuous compounding,

$$(1+i) = e^r$$

or, as shown earlier,

Effective interest rate per year $\quad i_a = e^r - 1 \tag{4-34}$

To find compound amount and present worth for continuous compounding and a single payment, we write:

Compound amount $F = P(e^{rn}) = P[F/P, r, n]$ (4-39)

Present worth $P = F(e^{-rn}) = F[P/F, r, n]$ (4-40)

Square brackets around the factors distinguish continuous compounding. If your hand calculator does not have e^x, use the table of e^{rn} and e^{-rn}, provided at the end of the appendix containing the compound interest tables.

EXAMPLE 4-17

If you were to deposit $2000 in a bank that pays 5% nominal interest, compounded continuously, how much would be in the account at the end of 2 years?

SOLUTION

The single payment compound amount equation for continuous compounding is

$$F = Pe^{rn}$$

where $r =$ nominal interest rate $= 0.05$
$n =$ number of years $= 2$

$$F = 2000e^{(0.05)(2)} = 2000(1.1052) = \$2210.40$$

There would be $2210.40 in the account at the end of 2 years.

EXAMPLE 4-18

A bank offers to sell savings certificates that will pay the purchaser $5000 at the end of 10 years but will pay nothing to the purchaser in the meantime. If interest is computed at 6%, compounded continuously, at what price is the bank selling the certificates?

SOLUTION

$$P = Fe^{-rn} \qquad F = \$5000 \qquad r = 0.06 \qquad n = 10 \text{ years}$$

$$P = 5000e^{-0.06 \times 10} = 5000(0.5488) = \$2744$$

Therefore, the bank is selling the $5000 certificates for $2744.

EXAMPLE 4-19

How long will it take for money to double at 10% nominal interest, compounded continuously?

$$F = Pe^{rn}$$

$$2 = 1e^{0.10n}$$

$$e^{0.10n} = 2$$

or

$$0.10n = \ln 2 = 0.693$$

$$n = 6.93 \text{ years}$$

It will take 6.93 years for money to double at 10% nominal interest, compounded continuously.

EXAMPLE 4-20

If the savings bank in Example 4-14 changes its interest policy to 6% interest, compounded continuously, what are the nominal and the effective interest rates?

SOLUTION

The nominal interest rate remains at 6% per year.

$$\text{Effective interest rate} = e^r - 1$$

$$= e^{0.06} - 1 = 0.0618$$

$$= 6.18\%$$

Uniform Payment Series: Continuous Compounding at Nominal Rate r per Period

Let us now substitute the equation $i = e^r - 1$ into the equations for end-of-period compounding.

Continuous Compounding Sinking Fund

$$[A/F, r, n] = \frac{e^r - 1}{e^{rn} - 1} \tag{4-41}$$

Continuous Compounding Capital Recovery

$$[A/P, r, n] = \frac{e^{rn}(e^r - 1)}{e^{rn} - 1} \tag{4-42}$$

Continuous Compounding Series Compound Amount

$$[F/A, r, n] = \frac{e^{rn} - 1}{e^r - 1} \tag{4-43}$$

Continuous Compounding Series Present Worth

$$[P/A, r, n] = \frac{e^{rn} - 1}{e^{rn}(e^r - 1)} \tag{4-44}$$

EXAMPLE 4-21

In Example 4-1, a man deposited $500 per year into a credit union that paid 5% interest, compounded annually. At the end of 5 years, he had $2763 in the credit union. How much would he have if the institution paid 5% nominal interest, compounded continuously?

SOLUTION

$$A = \$500 \qquad r = 0.05 \qquad n = 5 \text{ years}$$

$$F = A[F/A, r, n] = A\left(\frac{e^{rn} - 1}{e^r - 1}\right) = 500\left(\frac{e^{0.05(5)} - 1}{e^{0.05} - 1}\right)$$

$$= \$2769.84$$

He would have $2769.84.

EXAMPLE 4-22

In Example 4-2, Jim Hayes wished to save a uniform amount each month so he would have $1000 at the end of a year. Using 6% nominal interest, compounded monthly, he calculated that he had to deposit $81.10 per month. How much would he have to deposit if his credit union paid 6% nominal interest, compounded continuously?

SOLUTION

The deposits are made monthly; hence, there are 12 compounding sub-periods in the one-year time period.

$$F = \$1000 \qquad r = \text{nominal interest rate/interest period} = \frac{0.06}{12} = 0.005$$

$$n = 12 \text{ compounding sub-periods in the one-year period of the problem}$$

$$A = F[A/F, r, n] = F\left(\frac{e^r - 1}{e^{rn} - 1}\right) = 1000\left(\frac{e^{0.005} - 1}{e^{0.005(12)} - 1}\right)$$

$$= 1000\left(\frac{0.005013}{0.061837}\right) = \$81.07$$

He would have to deposit \$81.07 per month. Note that the difference between monthly and continuous compounding is just 3 cents per month.

Continuous, Uniform Cash Flow (One Period) with Continuous Compounding at Nominal Interest Rate r

Equations for a continuous, uniform cash flow during one period only, with continuous compounding, can be derived as follows. Let the continuous, uniform cash flow totaling $\bar{P}$ be distributed over m sub-periods within one period ($n = 1$). Thus $\bar{P}/m$ is the cash flow at the end of each sub-period. Since the nominal interest rate per period is r, the effective interest rate per sub-period is r/m. Substituting these values into the uniform series compound amount equation (Equation 4-5) gives

$$F = \frac{\bar{P}}{m}\left[\frac{[1 + (r/m)]^m - 1}{r/m}\right] \tag{4-45}$$

Setting $x = r/m$, we obtain

$$F = \frac{\bar{P}}{m}\left[\frac{[1 + x]^{r/x} - 1}{r/m}\right] = \bar{P}\left[\frac{[(1 + x)^{1/x}]^r - 1}{r}\right] \tag{4-46}$$

As m increases, x approaches zero. Equation 4-36 says

$$\lim_{x \to 0}(1 + x)^{1/x} = e$$

hence Equation 4-46 for one period becomes

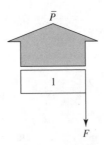

$$F = \bar{P}\left(\frac{e^r - 1}{r}\right) \tag{4-47}$$

Multiplying Equation 4-43 by the single payment, continuous compounding factors, we can find F for any future time and P for any present time.

For Any Future Time

$$F = \bar{P}\left(\frac{e^r - 1}{r}\right)(e^{r(n-1)}) = \bar{P}\left[\frac{(e^r - 1)(e^{rn})}{re^r}\right]$$

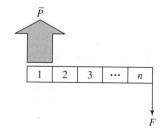

Compound amount $[F/\bar{P}, r, n] = \left[\dfrac{(e^r - 1)(e^{rn})}{re^r}\right]$ (4-48)

For Any Present Time

$$P = \bar{F}\left(\frac{e^r - 1}{r}\right)(e^{-rn}) = \bar{F}\left[\frac{e^r - 1}{re^{rn}}\right]$$

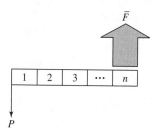

Present worth $[P/\bar{F}, r, n] = \left[\dfrac{e^r - 1}{re^{rn}}\right]$ (4-49)

EXAMPLE 4-23

A self-service gasoline station has been equipped with an automatic teller machine (ATM). Customers may obtain gasoline simply by inserting their ATM bank card into the machine and filling their tank with gasoline. When they have finished, the ATM unit automatically deducts the gasoline purchase from the customer's bank account and credits it to the gas station's bank account. The gas station receives $40,000 per month in this manner, with the cash flowing uniformly throughout the month. If the bank pays 9% nominal interest, compounded continuously, how much will be in the gasoline station bank account at the end of the month?

SOLUTION

The problem may be solved by either Equation 4-47 or 4-48. Here the latter, the general equation, is used.

$$r = \text{Nominal interest rate per month} = \frac{0.09}{12} = 0.0075$$

$$F = \bar{P}\left[\frac{(e^r - 1)(e^{rn})}{re^r}\right] = 40{,}000\left[\frac{(e^{0.0075} - 1)(e^{0.0075(1)})}{0.0075e^{0.0075}}\right]$$

$$= 40{,}000\left[\frac{(0.0075282)(1.0075282)}{0.00755646}\right] = \$40{,}150$$

There will be $40,150 in the station's bank account.

SPREADSHEETS FOR ECONOMIC ANALYSIS

Spreadsheets are used in most real world applications of engineering economy. Common tasks include the following:

1. Constructing tables of cash flows.
2. Using annuity functions to calculate a P, F, A, n, or i.
3. Using a block function to find the present worth or internal rate of return for a table of cash flows.
4. Making graphs for analysis and convincing presentations.
5. Calculating 'what-if' for different assumed values of problem variables.

Constructing tables of cash flows relies mainly on the spreadsheet basics that are covered in Appendix A. These basics include using and naming spreadsheet variables, understanding the difference between absolute and relative addresses when copying a formula, and formatting a cell. Appendix A uses the example of the amortization schedule shown in Table 3-1, Plan 3. This amortization schedule divides each scheduled loan payment into principal and interest and includes the outstanding balance for each period.

Because spreadsheet functions can be found through pointing and clicking on menus, those steps are not detailed. In Excel the starting point is the f_x button. Excel functions are used here, but there are only minor syntax differences in most other spreadsheet programs.

Spreadsheet Annuity Functions

In tables of engineering economy factors, i is the table, n is the row, and two of P, F, A, and G define a column. The spreadsheet annuity functions list four arguments chosen from n, A, P, F, and i, and solve for the fifth argument. The *Type* argument is optional. If it is omitted or 0, then the A value is assumed to be the end-of-period cash flow. If the A value represents the beginning-of-period cash flow, then a value of 1 can be entered for the Type variable.

To find the equivalent P	$-$PV$(i,n,A,F,$Type$)$
To find the equivalent A	$-$PMT$(i,n,P,F,$Type$)$
To find the equivalent F	$-$FV$(i,n,A,P,$Type$)$
To find n	NPER$(i,A,P,F,$Type$)$
To find i	RATE$(n,A,P,F,$Type,guess$)$

The sign convention for the first three functions seems odd to some students. The PV of \$200 per period for 10 periods is negative, and the PV of $-$\$200 per period is positive. So a minus sign is inserted to find the equivalent P, A, or F. Without this minus sign, the calculated value is not equivalent to the four given values.

EXAMPLE 4-24

A new engineer wants to save money for down payment on a house. The initial deposit is \$685, and \$375 is deposited at the end of each month. The savings account earns interest at an annual nominal rate of 6% with monthly compounding. How much is on deposit after 48 months?

SOLUTION

Because deposits are made monthly, the nominal annual interest rate of 6% must be converted to $1/2$% per month for the 48 months. Thus we must find F if $i = 0.5\% = 0.005$, $n = 48$, $A = 375$, and $P = 685$. Note both the initial and periodic deposits are positive cash flows for the savings account. The Excel function is multiplied by -1 or $-$FV$(0.005,48,375,685,0)$, and the result is \$21,156.97.

EXAMPLE 4-25

A new engineer buys a car with 0% down financing from the dealer. The cost with all taxes, registration, and licence fees is \$15,732. If each of the 48 monthly payments is \$398, what is the monthly interest rate? What is the effective annual interest rate?

SOLUTION

The RATE function can be used to find the monthly interest rate, given that $n = 48$, $A = -398$, $P = 15,732$, and $F = 0$. The Excel function is RATE$(48,-398,15732,0)$ and the result is 0.822%. The effective annual interest rate is $1.00822^{12} - 1 = 10.33\%$.

Spreadsheet Block Functions

Cash flows can be specified period by period as a block of values. These cash flows are analyzed by **block functions** that identify the row or column entries for which a present worth or an internal rate of return should be calculated. In Excel the two functions are NPV$(i,$values$)$ and IRR$($values,guess$)$.

Economic Criteria	**Excel Function**	**Values for Periods**
Net present value	NPV(i,values)	1 to n
Internal rate of return	IRR(values,guess);	0 to n
	guess argument is optional	

Excel's IRR function can be used to find the interest rate for a loan with irregular payments (other applications are covered in Chapter 7).

These block functions make different assumptions about the range of years included. For example, NPV(i,values) assumes that Year 0 is NOT included, while IRR(values,guess) assumes that Year 0 is included. These functions require that a cash flow be identified for each period. You cannot leave cells blank even if the cash flow is $0. The cash flows for 1 to n are assumed to be end-of-period flows. All periods are assumed to be the same length of time.

Also the NPV functions returns the present worth equivalent to the cash flows, unlike the PV annuity function, which returns the negative of the equivalent value.

For cash flows involving only constant values of P, F, and A, this block approach seems to be inferior to the annuity functions. However, this is a conceptually easy approach for more complicated cash flows, such as arithmetic gradients. Suppose the years (row 1) and the cash flows (row 2) are specified in columns B through E.

	A	B	C	D	E	F
1	Year	0	1	2	3	4
2	Cash flow	−25,000	6000	8000	10,000	12,000

If an interest rate of 8% is assumed, then the present worth of the cash flows can be calculated as =B2+NPV(.08,C2:F2), which equals $4172.95. This is the present worth equivalent to the five cash flows, rather than the negative of the present worth equivalent returned by the PV annuity function. The internal rate of return calculated using IRR(B2:F2) is 14.5%. Notice how the NPV function does not include the Year 0 cash flow in B2, whereas the IRR function does.

- PW = B2+NPV(.08,C2:F2)   NPV range without Year 0
- IRR = IRR(B2:F2)   IRR range with Year 0

Using Spreadsheets for Basic Graphing

Often we are interested in the relationship between two variables. Examples include the number and size of payments to repay a loan, the present worth of an M.S. degree and how long until we retire, and the interest rate and present worth for a new machine. This kind of two-variable relationship is best shown with a graph.

As we will show in Example 4-26, the goal is to place one variable on each axis of the graph and then plot the relationship. Modern spreadsheets automate most steps of drawing a graph, so that it is quite easy. However, there are two very similar types of chart, and we must be careful to choose the *xy* chart and not the *line* chart. Both charts measure the *y* variable, but they treat the *x* variable differently. The *xy* chart measures the *x* variable; thus its *x* value is measured along the *x* axis. For the *line* chart, each *x* value is placed an equal distance along the *x* axis. Thus *x* values of 1, 2, 4, 8 would be spaced evenly, rather than doubling each distance. The *line* chart is really designed to plot *y* values for different categories, such as prices for models of cars or enrolments for different universities.

Drawing an *xy* plot with Excel is easiest if the table of data lists the *x* values before the *y* values. This convention makes it easy for Excel to specify one set of *x* values and several sets of *y* values. The block of *xy* values is selected, and then the chart tool is selected. Then the spreadsheet guides the user through the rest of the steps.

EXAMPLE 4-26

Graph the loan payment as a function of the number of payments for a possible auto loan. Let the number of monthly payments vary between 36 and 60. The nominal annual interest rate is 12%, and the amount borrowed is $18,000.

SOLUTION

The spreadsheet table shown in Figure 4-2 is constructed first. Cells A5:B10 are selected, and then the Chartwizard icon is selected. The first step is to select an *xy (scatter)* plot with smoothed

	A	B	C	D	E	F
1	12%	nominal annual interest rate				
2	1%	monthly interest rate				
3	$18,000	amount borrowed				
4						
5	# Payments	Monthly Payment				
6	36	$597.86	=PMT($A$2,A6,-$A$3)			
7	42	$526.96				
8	48	$474.01				
9	54	$433.02				
10	60	$400.40				
11						
12						
13						
14						
15						
16						
17						
18						
19						
20						
21						
22						

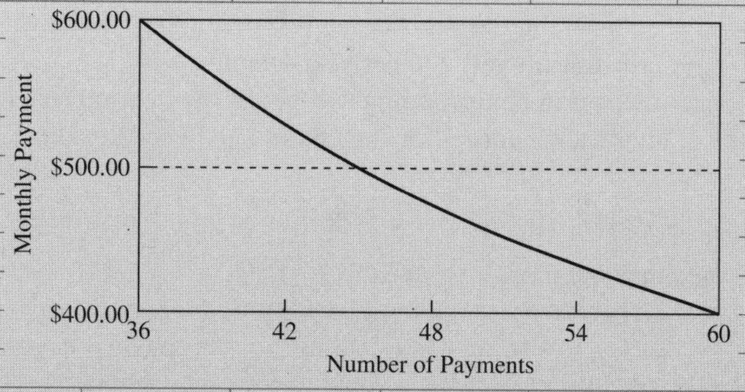

FIGURE 4-2 Example of a spreadsheet graph.

lines and without markers as the chart type. (The other choices are no lines, straight lines, and adding markers for the data points.) The second step shows us the graph and allows the option of changing the data cells selected. The third step is for chart options. Here we add titles for the two axes and turn off 'showing the legend'. (Since we have only one line in our graph, the legend is not needed. Deleting it leaves more room for the graph.) In the fourth step we choose where the chart is placed.

Because xy plots are normally graphed with the origin set to (0, 0), an attractive plot is best obtained if the minimum and maximum values are changed for each axis. This is done by placing the mouse cursor over the axis and left-clicking. Handles or small black boxes should appear on the axis to show that it has been selected. Right-clicking brings up a menu to select format axis. The scale tab allows us to change the minimum and maximum values.

SUMMARY

The compound interest formulas described in this chapter, along with those in Chapter 3, will be referred to throughout the rest of the book. It is very important that the reader understand the concepts presented and how these formulas are used. The following notation is used consistently.

i = effective interest rate per interest period[3] (stated as a decimal)

n = number of interest periods

P = a present sum of money

F = a future sum of money: the future sum F is an amount, n interest periods from the present, that is equivalent to P with interest rate i

A = an end-of-period cash receipt or disbursement in a uniform series continuing for n periods; the entire series equivalent to P or F at interest rate i

G = uniform period-by-period increase or decrease in cash receipts or disbursements; the arithmetic gradient

g = uniform rate of cash flow increase or decrease from period to period; the geometric gradient

r = nominal interest rate per interest period (see footnote 3)

i_a = effective interest rate per year (annum)

m = number of compounding sub-periods per period (see footnote 3)

$\overline{P}, \overline{F}$ = amount of money flowing continuously and uniformly during a given period

Single Payment Formulas (derived in Chapter 3)

Compound amount:

$$F = P(1+i)^n = P(F/P, i, n)$$

---

[3]Normally the interest period is one year, but it could be some other period (e.g., quarter, month, half year).

Present worth:

$$P = F(1 + i)^{-n} = F(P/F, i, n)$$

Uniform Series Formulas

Compound amount:

$$F = A\left[\frac{(1 + i)^n - 1}{i}\right] = A(F/A, i, n)$$

Sinking fund:

$$A = F\left[\frac{i}{(1 + i)^n - 1}\right] = F(A/F, i, n)$$

Capital recovery:

$$A = P\left[\frac{i(1 + i)^n}{(1 + i)^n - 1}\right] = P(A/P, i, n)$$

Present worth:

$$P = A\left[\frac{(1 + i)^n - 1}{i(1 + i)^n}\right] = A(P/A, i, n)$$

Arithmetic Gradient Formulas

Arithmetic gradient present worth:

$$P = G\left[\frac{(1 + i)^n - in - 1}{i^2(1 + i)^n}\right] = G(P/G, i, n)$$

Arithmetic gradient uniform series:

$$A = G\left[\frac{(1 + i)^n - in - 1}{i(1 + i)^n - i}\right] = G\left[\frac{1}{i} - \frac{n}{(1 + i)^n - 1}\right] = G(A/G, i, n)$$

Geometric Gradient Formulas

Geometric series present worth, where $i \neq g$:

$$P = A_1\left[\frac{1 - (1 + g)^n(1 + i)^{-n}}{i - g}\right] = A_1(P/A, g, i, n)$$

Geometric series present worth, where $i = g$:

$$P = A_1[n(1 + i)^{-1}] = A_1(P/A, g, i, n) = A_1(P/A, i, i, n)$$

Single Payment Formulas: Continuous Compounding at Nominal Rate r per Period

Compound amount:

$$F = P(e^{rn}) = P[F/P, r, n]$$

Present worth:

$$P = F(e^{-rn}) = F[P/F, r, n]$$

Note that square brackets around the factors are used to distinguish continuous compounding.

Uniform Payment Series: Continuous Compounding at Nominal Rate r per Period

Continuous compounding sinking fund:

$$A = F\left[\frac{e^r - 1}{e^{rn} - 1}\right] = F[A/F, r, n]$$

Continuous compounding capital recovery:

$$A = P\left[\frac{e^{rn}(e^r - 1)}{e^{rn} - 1}\right] = P[A/P, r, n]$$

Continuous compounding series compound amount:

$$F = A\left[\frac{e^{rn} - 1}{e^r - 1}\right] = A[F/A, r, n]$$

Continuous compounding series present worth:

$$P = A\left[\frac{e^{rn} - 1}{e^{rn}(e^r - 1)}\right] = A[P/A, r, n]$$

Continuous, Uniform Cash Flow (One Period) with Continuous Compounding at Nominal Interest Rate r

Compound amount

$$F = \bar{P}\left[\frac{(e^r - 1)(e^{rn})}{re^r}\right] = \bar{P}[F/\bar{P}, r, n]$$

Present worth

$$P = \overline{F}\left[\frac{e^r - 1}{re^{rn}}\right] = \overline{F}[P/\overline{F}, r, n]$$

Nominal Interest Rate per Year, r

The annual interest rate without considering the effect of any compounding.

Effective Interest Rate per Year, i_a

The annual interest rate taking into account the effect of any compounding during the year. Effective interest rate per year (periodic compounding):

$$i_a = \left(1 + \frac{r}{m}\right)^m - 1$$

or

$$i_a = (1 + i)^m - 1$$

Effective interest rate per year (continuous compounding):

$$i_a = e^r - 1$$

PROBLEMS

4-1 Solve diagrams (a)–(c) for the unknowns R, S, and T, assuming a 10% interest rate.

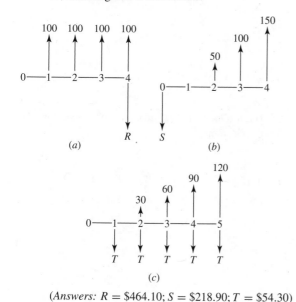

(Answers: $R = \$464.10$; $S = \$218.90$; $T = \$54.30$)

4-2 For diagrams (a)–(d), compute the unknown values— B, C, V, x, respectively—using the minimum number of compound interest factors.

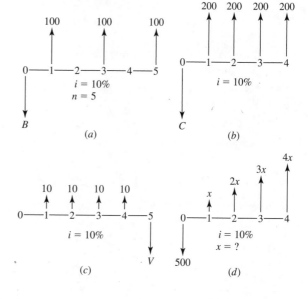

(*Answers:* $B = \$228.13; C = \$634; V = \$51.05;$ $x = \$66.24$)

4-3 For diagrams (*a*)–(*d*), compute the unknown values C, A, F, A, respectively.

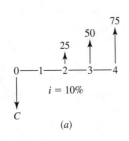

(*a*)

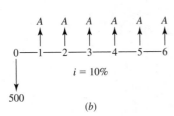

(*b*)

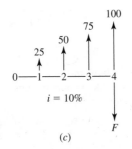

(*c*)

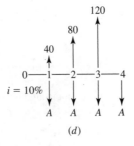

(*d*)

(*Answers:* $C = \$109.45; A = \$115; F = \$276.37;$ $A = \$60.78$)

4-4 For diagrams (*a*)–(*d*), compute the unknown values W, X, Y, Z, respectively.

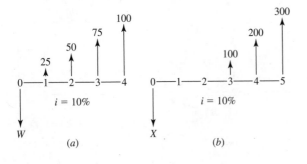

(*a*)   (*b*)

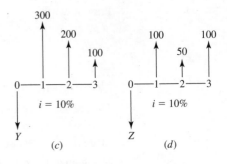

(*c*)   (*d*)

4-5 Compute the value of P in the diagram.

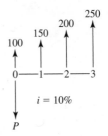

(*Answer:* $P = \$589.50$)

4-6 Use a 10% interest rate to compute the value of X in the diagram.

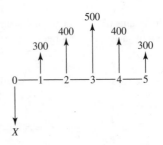

4-7 Use a 15% interest rate to compute the value of P in the diagram.

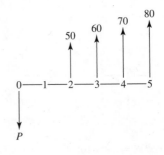

4-8 If $i = 12\%$, what is the value of B in the diagram?

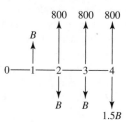

4-9 Compute the value of n for the diagram, assuming a 10% interest rate.

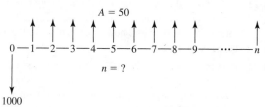

$A = 1$

$n = ?$

4-10 What is the value of n, for this diagram, based on a $3^1/2\%$ interest rate?

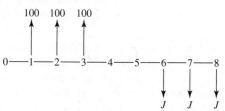

$A = 50$

$n = ?$

1000

4-11 Compute the value of J for the diagram, assuming a 10% interest rate.

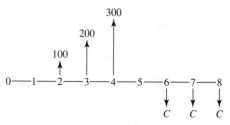

100 100 100

J J J

4-12 Compute the value of C for the diagram, assuming a 10% interest rate.

300

200

100

0—1—2—3—4—5—6—7—8

C C C

4-13 If $i = 12\%$, compute G in the diagram.

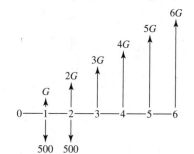

$6G$

$5G$

$4G$

$3G$

$2G$

G

500 500

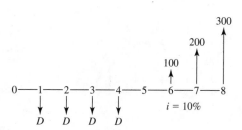

$F = 35.95$

4-14 Compute the value of D in the diagram.

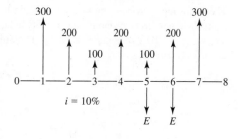

300

200

100

0—1—2—3—4—5—6—7—8

$i = 10\%$

D D D D

4-15 Compute E for the diagram.

300   300

200   200   200

100   100

0—1—2—3—4—5—6—7—8

$i = 10\%$

E E

4-16   Using a 10% interest rate, compute *B* in the diagram.

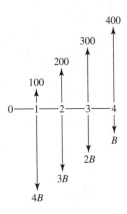

4-17   Find the value of *P* for the following cash flow diagram.

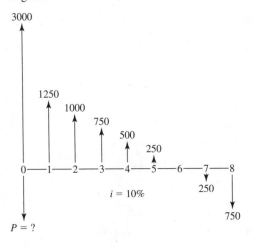

4-18   The following cash flow transactions are said to be equivalent in terms of economic desirability at an interest rate of 12% compounded annually. Determine the unknown value *A*.

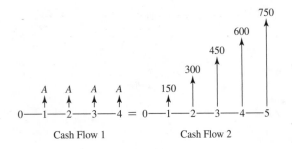

4-19   Upon the birth of his first child, Dick Jones decided to establish a savings account to partly pay for his son's education. He plans to deposit $20 per month in the account, beginning when the boy is 13 months old. The savings and loan association has a current interest policy of 6% per annum, compounded monthly, paid quarterly. Assuming no change in the interest rate, how much will be in the savings account when Dick's son becomes 16 years old?

4-20   Mary Lavor plans to save money at her bank for use in December. She will deposit $30 a month, beginning on March 1 and continuing until November 1. She will withdraw all the money on December 1. If the bank pays $1/2\%$ interest each month, how much money will she receive on December 1?

4-21   Derive an equation to find the end-of-year future sum *F* that is equivalent to a series of *n* beginning-of-year payments *B* at interest rate *i*. Then use the equation to determine the future sum *F* equivalent to six *B* payments of $100 at 8% interest.
(*Answer: F = $792.28*)

4-22   If $200 is deposited in a savings account at the beginning of each of 15 years, and the account draws interest at 7% per year, how much will be in the account at the end of 15 years?

4-23   A company deposits $2000 in a bank at the end of every year for 10 years. The company makes no deposits during the subsequent 5 years. If the bank pays 8% interest, how much would be in the account at end of 15 years?

4-24   How much must be deposited now at 5.25% interest to produce $300 at the end of every year for 10 years?

4-25   A student wants to have $30,000 at graduation 4 years from now to buy a new car. His grandfather gave him $10,000 as a high school graduation present. How much must the student save each year if he deposits the $10,000 today and can earn 12% on both the $10,000 and his earnings in a mutual fund his grandfather recommends?

4-26   A city engineer knows that she will need $25 million in 3 years to replace toll booths on a toll road in the city. Traffic on the road is estimated to be 20 million vehicles per year. How much per vehicle should the toll be to cover the cost of the toll booth replacement project? Interest is 10%. (Simplify your analysis by assuming that the toll receipts are received at the end of each year in a lump sum.)

4-27 Using linear interpolation, determine the value of $(P/A, 6^{1}/_{2}\%, 10)$ from the compound interest tables. Compute this same value using the equation. Why do the values differ?

4-28 A man wants to help provide a university education for his young daughter. He can afford to invest $600/yr for the next 4 years, beginning on the girl's fourth birthday. He wishes to give his daughter $4000 on her 18th, 19th, 20th, and 21st birthdays, for a total of $16,000. Assuming 5% interest, what uniform annual investment will he have to make on the girl's 8th to 17th birthdays?   (*Answer:* $792.73)

4-29 How many months will it take to pay off a $525 debt, with monthly payments of $15 at the end of each month, if the interest rate is 18%, compounded monthly?   (*Answer:* 50 months)

4-30 For how many months, at an interest rate of 1% per month, does money have to be invested before it will double in value?

4-31 A bank recently announced an 'instant cash' plan for holders of its bank credit cards. A cardholder may receive cash from the bank up to a preset limit (about $500). There is a special charge of 4% made at the time the 'instant cash' is sent to the cardholders. The debt may be repaid in monthly instalments. Each month the bank charges $1^{1}/_{2}\%$ on the unpaid balance. The monthly payment, including interest, may be as little as $10. Thus, for $150 of 'instant cash', an initial charge of $6 is made and added to the balance due. Assume the cardholder makes a monthly payment of $10 (this includes both principal and interest). How many months are needed to repay the debt? If your answer includes a fraction of a month, round up to the next month.

4-32 A man borrowed $500 from a bank on October 15th. He must repay the loan in 16 equal monthly payments, due on the 15th of each month, beginning November 15th. If interest is computed at 1% per month, how much must he pay each month?   (*Answer:* $33.95)

4-33 In Table 3-1 in the text, four plans were presented for the repayment of $5000 in 5 years with interest at 8%. Still another way to repay the $5000 would be to make four annual end-of-year payments of $1000 each, followed by a final payment at the end of the fifth year. How much would the final payment be?

4-34 An engineer borrowed $3000 from the bank, payable in six equal end-of-year payments at 8%. The bank agreed to reduce the interest on the loan if interest rates declined in Canada before the loan was fully repaid. At the end of 3 years, at the time of the third payment, the bank agreed to reduce the interest rate from 8% to 7% on the remaining debt. What was the amount of the equal annual end-of-year payments for each of the first 3 years? What was the amount of the equal annual end-of-year payments for each of the last 3 years?

4-35 A $150 bicycle was purchased on December 1 with a $15 down payment. The balance is to be paid at the rate of $10 at the end of each month, with the first payment due on December 31. The last payment may be some amount less than $10. If interest on the unpaid balance is computed at $1^{1}/_{2}\%$ per month, how many payments will there be, and what is the amount of the final payment?   (*Answers:* 16 payments; final payment: $1.99)

4-36 A company buys a machine for $12,000, which it agrees to pay for in five equal annual payments, beginning one year after the date of purchase, at an interest rate of 4% per annum. Immediately after the second payment, the terms of the agreement are changed to allow the balance due to be paid off in a single payment the next year. What is the final single payment? (*Answer:* $7778)

4-37 An engineering student bought a car at a local used car lot. Including tax and insurance, the total price was $3000. He is to pay for the car in 12 equal monthly payments, beginning with the first payment immediately (in other words, the first payment was the down payment). Nominal interest on the loan is 12%, compounded monthly. After six payments (the down payment plus five additional payments), he decides to sell the car. A buyer agrees to pay a cash amount to pay off the loan in full at the time the next payment is due and also to pay the engineering student $1000. If there are no penalty charges for this early payment of the loan, how much will the car cost the new buyer?

4-38 A realtor sold a house on August 31, 1997, for $150,000 to a buyer in which a 20% down payment was made. The buyer took a 15-year loan on the property with an effective interest rate of 8% per annum. The buyer intends to pay off the loan owed in yearly payments starting on August 31, 1998.

(*a*) How much of the loan will still be owed after the payment due on August 31, 2004, has been made?

(*b*) Solve the same problem by separating the interest and the principal amounts.

4-39 To provide for a university education for his daughter, a man opened an escrow account in which equal deposits were made. The first deposit was made on January 1, 1981, and the last deposit was made on January 1, 1998. The yearly university expenses including tuition were estimated to be $8000 for each of the 4 years. Assuming the interest rate to be 5.75%, how much did the father have to deposit each year in the escrow account for his daughter to draw $8000 a year for 4 years beginning January 1, 1998?

4-40 Develop a complete amortization table for a loan of $4500, to be paid back in 24 uniform monthly instalments, based on an interest rate of 6%. The amortization table must include the following column headings:

>Payment Number, Principal Owed (beginning of period), Interest Owed in Each Period, Total Owed (end of each period), Principal Paid in Each Payment, Uniform Monthly Payment Amount

You must also show the equations used to calculate each column of the table. You are encouraged to use spreadsheets. *The entire table must be shown.*

4-41 Using the loan and payment plan developed in Problem 4-40, determine the month that the final payment is due, and the amount of the final payment, if $500 is paid for payment 8 and $280 is paid for payment 10. This problem requires a separate amortization table giving the balance due, principal payment, and interest payment for each period of the loan.

4-42 A couple borrowed $80,000 at 7% to purchase a house. The loan is to be repaid in equal monthly payments over a 30-year period. The first payment is paid exactly at the end of the first month. Calculate the interest and principal in the second payment if the second payment is made 33 days after the first payment.

4-43 Jim Duggan made an investment of $10,000 in a savings account 10 years ago. This account paid interest of $5\frac{1}{2}\%$ for the first 4 years and $6\frac{1}{2}\%$ interest for the remaining 6 years. The interest charges were compounded quarterly. How much is this investment worth now?

4-44 Consider the cash flow:

Year	Cash Flow
0	−$100
1	+50
2	+60
3	+70
4	+80
5	+140

Which one of the following is correct for this cash flow?

(1) $100 = 50 + 10(A/G, i, 5) + 50(P/F, i, 5)$

(2) $\dfrac{50(P/A, i, 5) + 10(P/G, i, 5) + 50(P/F, i, 5)}{100} = 1$

(3) $100(A/P, i, 5) = 50 + 10(A/G, i, 5)$

(4) None of the equations are correct.

4-45 Consider the following cash flow:

Year	Cash Flow
0	−$P
1	+1000
2	+850
3	+700
4	+550
5	+400
6	+400
7	+400
8	+400

Alice was asked to compute the value of P for the cash flow at 8% interest. She wrote three equations:

(1) $P = 1000(P/A, 8\%, 8) − 150(P/G, 8\%, 8) + 150(P/G, 8\%, 4)(P/F, 8\%, 4)$

(2) $P = 400(P/A, 8\%, 8) + 600(P/A, 8\%, 5) − 150(P/G, 8\%, 4)$

(3) $P = 150(P/G, 8\%, 4) + 850(P/A, 8\%, 4) + 400(P/A, 8\%, 4)(P/F, 8\%, 4)$

Which of the equations is correct?

4-46 It is estimated that the maintenance cost on a new car will be $40 the first year. Each subsequent year, this cost is expected to increase by $10. How much would you need to set aside when you bought a new car to pay all future maintenance costs if you planned to keep the vehicle for 7 years? Assume interest is 5% per annum. (*Answer:* $393.76)

4-47 A young engineer wishes to become a millionaire by the time he is 60 years old. He believes that by careful investment he can obtain a 15% rate of return. He plans to add a uniform sum of money to his investment program each year, beginning on his 20th birthday and continuing through his 59th birthday. How much money must the engineer set aside in this project each year?

4-48   The council members of a small town have decided that the breakwater that protects the town from a nearby river should be rebuilt and strengthened. The town engineer estimates that the cost of the work at the end of the first year will be $85,000. He estimates that in subsequent years the annual repair costs will decline by $10,000, making the second-year cost $75,000; the third-year $65,000, and so forth. The council members want to know what the equivalent present cost is for the first 5 years of repair work if interest is 4%.   (*Answer:* $292,870)

4-49   A company expects to install smog control equipment on the exhaust of a gasoline engine. The local smog control district has agreed to pay to the firm a lump sum of money to provide for the first cost of the equipment and maintenance during its 10-year useful life. At the end of 10 years the equipment, which initially cost $10,000, is valueless. The company and smog control district have agreed that the following are reasonable estimates of the end-of-year maintenance costs:

Year 1	$500	Year 6	$200
2	100	7	225
3	125	8	250
4	150	9	275
5	175	10	300

Assuming interest at 6% per year, how much should the smog control district pay to the company now to provide for the first cost of the equipment and its maintenance for 10 years?   (*Answer:* $11,693)

4-50   A debt of $5000 can be repaid, with interest at 8%, by the following payments.

Year	Payment
1	$ 500
2	1000
3	1500
4	2000
5	X

The payment at the end of the fifth year is shown as X. How much is X?

4-51   A man is purchasing a small garden tractor. There will be no maintenance cost during the first 2 years

because the tractor is sold with 2 years free maintenance. For the third year, the maintenance is estimated at $20. In subsequent years the maintenance cost will increase by $20 per year (i.e., fourth-year maintenance will be $40, fifth-year $60, etc.). How much would need to be set aside now at 8% interest to pay the maintenance costs on the tractor for the first 6 years of ownership?

4-52   Mark Johnson saves a fixed percentage of his salary at the end of each year. This year he saved $1500. For the next 5 years, he expects his salary to increase at an 8% annual rate, and he plans to increase his savings at the same 8% annual rate. He invests his money in the stock market. Thus there will be six end-of-year investments (the initial $1500 plus five more). Solve the problem by using the geometric gradient factor.

(*a*)   How much will the investments be worth at the end of 6 years if they increase in the stock market at a 10% annual rate?

(*b*)   How much will Mark have at the end of 6 years if his stock market investments increase only at 8% annually?

4-53   The Macintosh Company has an employee savings plan that allows every employee to invest up to 5% of his or her annual salary. The money is invested in company common stock with the company guaranteeing that the annual return will never be less than 8%. Jill was hired at an annual salary of $52,000. She immediately joined the savings plan, investing the full 5% of her salary each year. If Jill's salary increases at an 8% uniform rate, and she continues to invest 5% of it each year, what amount of money is she guaranteed to have at the end of 20 years?

4-54   The football coach at a university was given a 5-year employment contract that paid $225,000 the first year, and increased at an 8% uniform rate in each subsequent year. At the end of the first year's football season, the alumni demanded that the coach be fired. The alumni agreed to buy his remaining years on the contract by paying him the equivalent present sum, computed with a 12% interest rate. How much will the coach receive?

4-55   Traffic at a certain intersection is currently 2000 cars per day. A consultant has told the city that traffic is expected to grow at a continuous rate of 5% a year for the next 4 years. How much traffic will be expected at the end of 2 years?

4-56 A local bank will lend a customer $1000 on a 2-year car loan as follows:

Money to pay for car	$1000
Two years' interest at 7%: $2 \times 0.07 \times 1000$	140
	$1140

$$24 \text{ monthly payments} = \frac{1140}{24} = \$47.50$$

The first payment must be made in 30 days. What is the nominal annual interest rate the bank is receiving?

4-57 A local lending institution advertises the '51–50 Club'. A person may borrow $2000 and repay $51 for the next 50 months, beginning 30 days after receiving the money. Compute the nominal annual interest rate for this loan. What is the effective interest rate?

4-58 A loan company has been advertising on television a plan that allows people to borrow $1000 and make a payment of $10.87 per month. This payment is for interest only and includes no payment on the principal. What is the nominal annual interest rate that they are charging?

4-59 What effective interest rate per annum corresponds to a nominal rate of 12% compounded monthly? (*Answer:* 12.7%)

4-60 A woman opened an account in a local store. In the charge account agreement, the store indicated it charges $1^{1}/_{2}\%$ each month on the unpaid balance. What nominal annual interest rate is being charged? What is the effective interest rate?

4-61 The *Bawl Street Journal* costs $206, payable now, for a 2-year subscription. The newspaper is published 252 days a year (5 days a week, except holidays). If a 10% nominal annual interest rate, compounded quarterly, is used:
   (*a*) What is the effective annual interest rate in this problem?
   (*b*) Compute the equivalent interest rate per $1/252$ of a year.
   (*c*) What is a subscriber's cost per copy of the newspaper, taking interest into account?

4-62 A new graduate borrows $10,000 to purchase a car. He must repay the loan in 48 equal end-of-period monthly payments. Interest is calculated at 1.25% per month. Determine the following:
   (*a*) The nominal annual interest rate
   (*b*) The effective annual interest rate
   (*c*) The amount of the monthly payment

4-63 Penelope borrows $1000. To repay the amount she makes 12 equal monthly payments of $90.30. Determine the following:
   (*a*) The effective monthly interest rate
   (*b*) The nominal annual interest rate
   (*c*) The effective annual interest rate

4-64 At the Central Furniture Company, customers who buy on credit pay an effective annual interest rate of 16.1%, based on monthly compounding. What is the nominal annual interest rate that they pay?

4-65 What monthly interest rate is equivalent to an effective annual interest rate of 18%?

4-66 A bank advertises it pays 7% annual interest, compounded daily, on savings accounts, provided the money is left in the account for 4 years. What effective annual interest rate do they pay?

4-67 To repay a $1000 loan, a man paid $91.70 at the end of each month for 12 months. Compute the nominal interest rate he paid.

4-68 A student bought a $75 used guitar and agreed to pay for it with a single $85 payment at the end of 6 months. Assuming semi-annual (every 6 months) compounding, what is the nominal annual interest rate? What is the effective interest rate?

4-69 A firm charges its credit customers $1^{3}/_{4}\%$ interest per month. What is the effective interest rate?

4-70 A thousand dollars is invested for 7 months at an interest rate of 1% per month. What is the nominal interest rate? What is the effective interest rate? (*Answers:* 12%; 12.7%)

4-71 What interest rate, compounded quarterly, is equivalent to a 9.31% effective interest rate?

4-72 If the nominal annual interest rate is 12% compounded quarterly, what is the effective annual interest rate?

4-73 A contractor wishes to set up a special fund by making uniform semi-annual end-of-period deposits for 20 years. The fund is to provide $10,000 at the end of each of the last 5 years of the 20-year period. If interest is 8%, compounded semi-annually, what is the required semi-annual deposit?

4-74 What amount will be required to purchase, on an engineer's 40th birthday, an annuity to provide him with 30 equal semiannual payments of $1000 each, the first to be received on his 50th birthday, if nominal interest is 4% compounded semi-annually?

4-75 A student decides to deposit $50 in the bank today and to make 10 additional deposits every 6 months beginning 6 months from now, the first of which will be $50 and increasing $10 per deposit after that. A few minutes after making the last deposit, she decides to withdraw all the money deposited. If the bank pays 6% nominal interest compounded semi-annually, how much money will she receive?

4-76 A man makes an investment every 3 months at a nominal annual interest rate of 28%, compounded quarterly. His first investment was $100, followed by investments *increasing* $20 each 3 months. Thus, the second investment was $120, the third investment $140, and so on. If he continues to make this series of investments for a total of 20 years, what will be the value of the investments at the end of that time?

4-77 A 25-year old engineer is opening an individual retirement account (RRSP) at a bank. Her goal is to accumulate $1 million in the account by the time she retires from work in 40 years. The bank manager estimates she may expect to receive 8% nominal annual interest, compounded quarterly, throughout the 40 years. The engineer believes her income will increase at a 7% annual rate during her career. She wishes to start with as low a deposit as possible to her RRSP now and increase it at a 7% rate each year. Assuming end-of-year deposits, how much should she deposit the first year?

4-78 What single amount on April 1, 1998, is equivalent to a series of equal, semi-annual cash flows of $1000 that starts with a cash flow on January 1, 1996, and ends with a cash flow on January 1, 2005? The interest rate is 14% and compounding is quarterly.

4-79 Paco's saving account earns 13% compounded weekly and receives quarterly deposits of $38,000. His first deposit was made on October 1, 1996, and the last deposit is scheduled for April 1, 2012. Tisha's account earns 13% compounded weekly. Semi-annual deposits of $18,000 are made into her account, with the first one being made on July 1, 2006, and the last one on January 1, 2015. What single amount on January 1, 2007, is equivalent to the sum of both cash flow series?

4-80 The first of a series of equal, monthly cash flows of $2000 occurred on April 1, 1998, and the last of the monthly cash flows was made on February 1, 2000. This series of monthly cash flows is equivalent to a series of semi-annual cash flows. The first semi-annual

cash flow was made on July 1, 2001, and the last semi-annual cash flow will be made on January 1, 2010. What is the amount of each semi-annual cash flow? Use a nominal interest rate of 12% with monthly compounding on all accounts.

4-81 A series of monthly cash flows is deposited into an account that earns 12% nominal interest compounded monthly. Each monthly deposit is equal to $2100. The first monthly deposit was made on June 1, 1998, and the last monthly deposit will be on January 1, 2005. The account (the series of monthly deposits, 12% nominal interest, and monthly compounding) also has equivalent quarterly withdrawals from it. The first quarterly withdrawal is equal to $5000 and was made on October 1, 1998. The last $5000 withdrawal will be made on January 1, 2005. How much remains in the account after the last withdrawal?

4-82 Ann deposits $100 at the end of each month into her bank savings account. The bank paid 6% nominal interest, compounded and paid quarterly. No interest was paid on money not in the account for the full 3-month period. How much was in Ann's account at the end of 3 years? (*Answer:* $3912.30)

4-83 What is the present worth of a series of equal quarterly payments of $3000 that extends over a period of 8 years if the interest rate is 10% compounded monthly?

4-84 The first of a series of equal semi-annual cash flows occurs on July 1, 1997, and the last occurs on January 1, 2010. Each cash flow is equal to $128,000. The nominal interest rate is 12% compounded semi-annually. What single amount on July 1, 2001 is equivalent to this cash flow system?

4-85 A man buys a car for $3000 with no money down. He pays for the car in 30 equal monthly payments with interest at 12% per annum, compounded monthly. What is his monthly loan payment? (*Answer:* $116.10)

4-86 On January 1, Frank Jenson bought a used car for $4200 and agreed to pay for it as follows: $1/3$ down payment; the balance to be paid in 36 equal monthly payments; the first payment due February 1; an annual interest rate of 9%, compounded monthly.
   (*a*) What is the amount of Frank's monthly payment?
   (*b*) During the summer, Frank made enough money that he decided to pay off the entire balance due on the car as of October 1. How much did Frank owe on October 1?

4-87 On January 1, Laura Brown borrowed $1000 from the Friendly Finance Company. The loan is to be repaid by four equal payments, which are due at the end of March, June, September, and December. If the finance company charges 18% interest, compounded quarterly, what is the amount of each payment? What is the effective annual interest rate? (*Answers:* $278.70; 19.3%)

4-88 The **Rule of 78's** is a commonly used method of computing the amount of interest when the balance of a loan is repaid in advance.

Adding the numbers representing 12 months gives

$$1 + 2 + 3 + 4 + 5 + \cdots + 11 + 12 = 78$$

If a 12-month loan is repaid at the end of one month, for example, the interest the borrower would be charged is 12/78 of the year's interest. If the loan is repaid at the end of 2 months, the total interest charged would be $(12 + 11)/78$, or 23/78 of the year's interest. After 11 months the interest charge would therefore be 77/78 of the total year's interest.

Shannon borrowed $10,000 on January 1 at 9% annual interest, compounded monthly. The loan was to be repaid in 12 equal end-of-period payments. Helen made the first two payments and then decided to repay the balance of the loan when she pays the third payment. Thus she will pay the third payment plus an additional sum.

Calculate the amount of this additional sum
(a) Using the rule of 78s.
(b) Using exact economic analysis methods.

4-89 A bank is offering a loan of $25,000 with a nominal interest rate of 18% compounded monthly, payable in 60 months. (*Hint:* The loan origination fee of 2% will be taken out of the loan amount.)
(a) What is the monthly payment?
(b) If a loan origination fee of 2% is charged at the time of the loan, what is the effective interest rate?

4-90 Our cat, Fred, wants to buy a new litter box. The cost is $100 and he'll finance it over 2 years at an annual rate of 18% compounded monthly and to be repaid in 24 monthly payments.
(a) What is his monthly payment?
(b) At the time of the thirteenth payment, Fred decides to pay off the remainder of the loan.

Using regular compound interest factors, determine the amount of this last payment.

4-91 Our cat, Fred, has convinced me that I should set up an account that will ensure that he gets his Meow Mix for the next 4 years. I will deposit an amount P today that will permit Fred to make end-of-the-month withdrawals of $10 for the next 48 months. Consider an interest rate of 6% compounded monthly and that the account will be emptied with the last withdrawal.
(a) What is the value of P that I must deposit today?
(b) What is the account balance immediately after the 24th withdrawal has been made?

4-92 When Jerry Garcia was alive he bought a house for $500,000 and made a $100,000 down payment. He obtained a 30-year loan for the remaining amount. Payments were made monthly. The nominal annual interest rate was 9%. After 10 years (120 payments) he decided to pay the remaining balance on the loan.
(a) What was his monthly loan payment?
(b) What must he have paid (in addition to his regular 120th monthly payment) to pay the remaining balance of his loan?
(c) Recompute part (a) using 6% compounded continuously.

4-93 A car may be purchased with a $3000 down payment now and 60 monthly payments of $280. If the interest rate is 12% compounded monthly, what is the price of the car?

4-94 A man has $5000 on deposit in a bank that pays 5% interest compounded annually. He wonders how much more advantageous it would be to transfer his funds to another bank whose dividend policy is 5% interest, compounded continuously. Compute how much he would have in his savings account at the end of 3 years under each of these situations.

4-95 A friend was left $50,000 by his uncle. He has decided to put it into a savings account for the next year or so. He finds there are varying interest rates at savings institutions: $4\frac{3}{8}\%$ compounded annually, $4\frac{1}{4}\%$ compounded quarterly, and $4\frac{1}{8}\%$ compounded continuously. He wishes to select the savings institution that will give him the highest return on his money. What interest rate should he select?

4-96 One of the local banks says that it computes the interest it pays on savings accounts by the continuous compounding method. Suppose you deposited $100 in the bank and they pay 4% per annum, compounded continuously. After 5 years, how much money will there be in the account?

4-97 A university professor won $85,000 in a lottery; income taxes will take about half the amount. She plans to spend her sabbatical year on leave from the university on an around-the-world trip with her husband, but she must continue to teach 3 more years first. She estimates the trip will cost $40,000 and they will spend the money as a continuous flow of funds during their year of travel. She will put enough of her lottery winnings in a bank account now to pay for the trip. The bank pays 7% nominal interest, compounded continuously. She asks you to compute how much she should set aside in the account for the trip.

4-98 Michael Jacks deposited $500,000 into a bank for 6 months. At the end of that time, he withdrew the money and received $520,000. If the bank paid interest based on continuous compounding:
(a) What was the effective annual interest rate?
(b) What was the nominal annual interest rate?

4-99 How long will it take for $10,000, invested at 5% per year, compounded continuously, to triple in value?

4-100 A bank pays 10% nominal annual interest on special three-year certificates. What is the effective annual interest rate if interest is compounded:
(a) Every three months?
(b) Daily?
(c) Continuously?

4-101 Bart wishes to tour the country with his friends. To do this, he is saving money for a bus.
(a) How much money must Bart deposit in a savings account paying 8% nominal annual interest, compounded continuously, in order to have $8000 in $4^1/2$ years?
(b) A friend offers to repay Bart $8000 in $4^1/2$ years if Bart gives him $5000 now. Assuming continuous compounding, what is the nominal annual interest rate of this offer?

4-102 Select the best of the following five alternatives. Assume the investment is for a period of 4 years and $P = \$10,000$.
(1) 11.98% interest rate compounded continuously
(2) 12.00% interest rate compounded daily
(3) 12.01% interest rate compounded monthly
(4) 12.02% interest rate compounded quarterly
(5) 12.03% interest rate compounded yearly

4-103 What single amount on October 1, 1997, is equal to a series of $1000 quarterly deposits made into an account? The first deposit is made on October 1, 1997 and the last deposit occurs on January 1, 2011. The account earns 13% compounded continuously.

4-104 You are taking a $2000 loan. You will pay it back in four equal amounts, paid every 6 months starting 3 years from now. The interest rate is 6% compounded semi-annually. Calculate:
(a) The effective interest rate, based on both semi-annual and continuous compounding
(b) The amount of each semi-annual payment
(c) The total interest paid

4-105 If you want a 12% rate of return, continuously compounded, on a project that will yield $6000 at the end of $2^1/2$ years, how much must you be willing to invest now? (*Answer:* $4444.80)

4-106 A department store charges $1^3/4$% interest per month, compounded continuously, on its customer's charge accounts. What is the nominal annual interest rate? What is the effective interest rate? (*Answers:* 21%; 23.4%)

4-107 A bank is offering to sell 6-month certificates of deposit for $9500. At the end of 6 months, the bank will pay $10,000 to the certificate owner. Using a 6-month interest period, compute the nominal annual interest rate and the effective annual interest rate.

4-108 Two savings banks are located across the street from each other. The West Bank put a sign in the window saying, "We pay 6.50%, compounded daily." The East Bank put up a sign saying, "We pay 6.50%, compounded continuously."

Jean Silva has $10,000 which she will put in a bank for one year. How much additional interest will Jean receive by placing her money in the East Bank rather than the West Bank?

4-109 Sally Struthers wants to have $10,000 in a savings account at the end of 6 months. The bank pays 8%, compounded continuously. How much should Sally deposit now? (*Answer:* $9608)

4-110 The I've Been Moved Corporation receives a constant flow of funds from its worldwide operations. This money (in the form of cheques) is continuously deposited in many banks with the goal of earning as much interest as possible for 'IBM'. One billion dollars is deposited each month, and the money earns an average of $1/2$% interest per month, compounded continuously. Assume all the money remains in the accounts until the end of the month.

(a) How much interest does IBM earn each month?

(b) How much interest would IBM earn each month if it held the cheques and made deposits to its bank accounts just four times a month?

4-111 A group of 10 public-spirited citizens have agreed that they will support the local school hot lunch program. Each year one of the group is to pay the $15,000 cost that occurs continuously and uniformly during the year. Each member of the group is to underwrite the cost for one year. Slips of paper numbered year 1 through year 10 are put in a hat. As one of the group, you draw the slip marked year 6. Assuming an 8% nominal interest rate per year, how much do you need to set aside now to meet your obligation in year 6?

4-112 A forklift truck costs $29,000. A company agrees to purchase such a truck with the understanding that it will make a single payment for the balance due in 3 years. The vendor agrees to the deal and offers two different interest schedules. The first schedule uses an annual effective interest rate of 13%. The second schedule uses 12.75% compounded continuously.

(a) Which schedule should the company accept?

(b) What would be size of the single payment?

4-113 PARC Company has money to invest in an employee benefit plan, and you have been chosen as the plan's trustee. As an employee yourself, you want to maximize the interest earned on this investment and have found an account that pays 14% compounded continuously. PARC is providing you with $1200 per month to put into your account for 7 years. What will be the balance in this account at the end of the 7-year period?

4-114 Barry, a recent engineering graduate, never took engineering economics. When he graduated, he was hired by a prominent architectural firm. The earnings from this job allowed him to deposit $750 each quarter into a savings account. There were two banks offering a savings account in his town (a small town!). The first bank was offering 4.5% interest compounded continuously. The second bank offered 4.6% compounded monthly. Barry decided to deposit in the first bank since it offered continuous compounding. Did he make the right decision?

4-115 A local finance company will lend $10,000 to a homeowner. It is to be repaid in 24 monthly payments of $499 each. The first payment is due 30 days after the $10,000 is received. What interest rate per month are they charging?   (*Answer*: $1^1/_2\%$)

4-116 Mr Sansome withdrew $1000 from a savings account and invested it in common stock. At the end of 5 years, he sold the stock and received a check for $1307. If Mr Sansome had left his $1000 in the savings account, he would have received an interest rate of 5%, compounded quarterly. Mr Sansome would like to compute a comparable interest rate on his common stock investment. If compounding is quarterly, what nominal annual interest rate did Mr Sansome receive on his investment in stock? What effective annual interest rate did he receive?

4-117 The treasurer of a firm noted that many invoices were received with the following terms of payment: '2%– 10 days, net 30 days'. Thus, if he were to pay the bill within 10 days of its date, he could deduct 2%. On the other hand, if he did not pay the bill promptly, the full amount would be due 30 days from the date of the invoice. Assuming a 20-day compounding period, the 2% deduction for prompt payment is equivalent to what effective interest rate per year?

4-118 In 1555, King Henry borrowed money from his bankers on the condition that he pay 5% of the loan at each fair (there were four fairs a year) until he had made 40 payments. At that time the loan would be considered repaid. What effective annual interest did King Henry pay?

4-119 One of the largest car dealers in the city advertises a 3-year-old car for sale as follows:

> Cash price $3575, or a down payment of $375 with 45 monthly payments of $93.41.

Susan DeVaux bought the car and made a down payment of $800. The dealer charged her the same interest rate used in his advertised offer. How much will Susan pay each month for 45 months? What effective interest rate is being charged?   (*Answers*: $81.03; 16.1%)

4-120 In 1990 Mrs John Hay Whitney sold her painting by Renoir, *Au Moulin de la Galette,* depicting an open-air Parisian dance hall, for $71 million. The buyer also had to pay the auction house commission of 10%, or a total of $78.1 million. Mrs Whitney had bought the painting in 1929 for $165,000.

(a) What rate of return did she receive on her investment?

(b) Was the rate of return really as high as you computed in (a)? Explain.

4-121 A woman made 10 annual end-of-year purchases of $1000 worth of common shares. The shares paid no dividends. Then for 4 years she held the shares. At the end of the 4 years she sold all the shares for $28,000. What interest rate did she obtain on her investment?

4-122 For some interest rate i and some number of interest periods n, the uniform series capital recovery factor is 0.1728 and the sinking fund factor is 0.0378. What is the interest rate?

4-123 The following beginning-of-month (BOM) and end of month (EOM) amounts are to be deposited in a savings account that pays interest at 9%, compounded monthly:

Today (BOM 1)	$400
EOM 2	270
EOM 6	100
EOM 7	180
BOM 10	200

Set up a spreadsheet to calculate the account balance at the end of the first year (EOM12). The spreadsheet must include the following column headings: Month Number, Deposit BOM, Account Balance at BOM, Interest Earned in Each Month, Deposit EOM, Account Balance at EOM. Also, use the compound interest tables to draw a cash flow diagram of this problem and solve for the account balance at the EOM 12.

4-124 Net revenues at an older manufacturing plant will be $2 million for this year. The net revenue will decrease 15% a year for 5 years, when the assembly plant will be closed (at the end of Year 6). If the firm's interest rate is 10%, calculate the PW of the revenue stream.

4-125 What is the present worth of cash flows that begin at $10,000 and increase at 8% per year for 4 years? The interest rate is 6%.

4-126 What is the present worth of cash flows that begin at $30,000 and decrease at 15% a year for 6 years? The interest rate is 10%.

4-127 Calculate and print out the balance due, principal payment, and interest payment for each period of a used-car loan. The nominal interest is 12% per year, compounded monthly. Payments are made monthly for 3 years. The original loan is for $11,000.

4-128 Calculate and print out the balance due, principal payment, and interest payment for each period of a new car loan. The nominal interest is 9% per year,

compounded monthly. Payments are made monthly for 5 years. The original loan is for $17,000.

4-129 For the used car loan of Problem 4-127, graph the monthly payment.
(a) As a function of the interest rate (5%–15%).
(b) As a function of the number of payments (24–48).

4-130 For the new car loan of Problem 4-128, graph the monthly payment.
(a) As a function of the interest rate (4–14%).
(b) As a function of the number of payments (36–84).

4-131 Your beginning salary is $50,000. You deposit 10% at the end of each year in a savings account that earns 6% interest. Your salary increases by 5% per year. What value does your savings book show after 40 years?

4-132 The market volume for widgets is increasing by 15% a year from current profits of $200,000. Investing in a design change will allow the profit per widget to stay steady; otherwise they will drop 3% a year. What is the present worth of the savings over the next 5 years? Ten years? The interest rate is 10%.

4-133 A 30-year mortgage for $120,000 has been issued. The interest rate is 10% and payments are made monthly. Print out the balance due, principal payment, and interest payment for each period.

4-134 A homeowner may upgrade a furnace that runs on fuel oil to a natural gas unit. The investment will be $2500 installed. The cost of the natural gas will average $60 per month over the year, instead of the $145 per month that the fuel oil costs. If the interest rate is 9% per year, how long will it take to recover the initial investment?

4-135 Develop a general-purpose spreadsheet to calculate the balance due, principal payment, and interest payment for each period of a loan. The user inputs to the spreadsheet will be the loan amount, the number of payments per year, the number of years payments are made, and the nominal interest rate. Submit printouts of your analysis of a loan in the amount of $15,000 at 8.9% nominal rate for 36 months and for 60 months of payments.

4-136 Use the spreadsheet developed for Problem 4-135 to analyze 180-month and 360-month house loan payments. Analyze a $100,000 mortgage loan at a nominal interest rate of 7.5% and submit a graph of the interest and principal paid over time. You need not submit the printout of the 360 payments because it will not fit on one page.

CHAPTER 5

Present Worth Analysis

Boeing versus Airbus

For several years, Boeing has been working on development of the Sonic Cruiser, an innovative lightweight plane that can fly near the speed of sound. This is about 15% faster than a conventional jet, so flying the Sonic Cruiser would allow airlines to cut long-haul flight times.

Meanwhile, Airbus, Boeing's rival in the passenger aircraft business, has also been working on its version of a lighter plane. But Airbus is betting on size rather than speed. Its new model, the A380, is known as the super-jumbo and will be the largest passenger plane in the world.

Airbus is hoping to use several recently developed composite materials, with the aim of reducing by 20% both the weight and the manufacturing cost. Airbus estimates that flying the composite plane will allow airlines to cut their operating costs by about 8%.

After Completing This Chapter...

The student should be able to:

- Define and apply the *present worth criteria*.
- Using present worth (PW), compare two competing choices.
- Apply the PW model in cases with equal, unequal, and infinite project lives.
- Compare multiple alternatives using the PW criteria.
- Develop and use spreadsheets to make *present worth* calculations.

QUESTIONS TO CONSIDER

1. So far, Boeing has not been able to sell the Sonic Cruiser idea to airlines. Fast is good, they say, but cheap is better. Why isn't increased speed more attractive to airlines, especially over long-haul flights? Can't airlines charge more for faster flights?

2. In many cases, air travellers actually spend as much time on the ground as in the air. Going through security, changing planes, clearing customs, and picking up baggage can add hours to their travel time. How might this affect a traveller's decision about whether to pay more to fly on a faster plane like the Sonic Cruiser?

3. Airlines seem to be more interested in the new Airbus model. Can you state some reasons for this apparent preference?

In Chapters 3 and 4 we accomplished two important tasks. First, we presented the concept of equivalence. We are powerless to compare series of cash flows unless we can resolve them into some equivalent arrangement. Second, equivalence, with alteration of cash flows from one series to an equivalent sum or series of cash flows, created the need for compound interest factors. So, we derived a whole series of compound interest factors—some for periodic compounding and some for continuous compounding. This background sets the stage for the chapters that follow.

ASSUMPTIONS IN SOLVING ECONOMIC ANALYSIS PROBLEMS

One of the difficulties of solving problems is that most problems tend to be very complicated. It becomes apparent that *some* simplifying assumptions are needed to make such problems manageable. The trick, of course, is to solve the simplified problem and still be satisfied that the solution is applicable to the *real* problem! In the subsections that follow, we will consider six different items and explain the customary assumptions that are made. These assumptions apply to all problems and examples unless other assumptions are given.

End-of-Year Convention

As we said in Chapter 4, economic analysis textbooks and practice follow the end-of-period convention. This makes 'A' a series of end-of-period receipts or disbursements. (We generally assume in problems that all series of receipts or disbursements occur at the *end* of the interest period. This allows us to use values from our compound interest tables without any adjustments.)

A cash flow diagram of P, A, and F for the end-of-period convention is as follows:

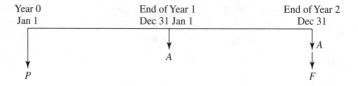

If one were to adopt a middle-of-period convention, the diagram would be:

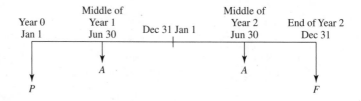

As the diagrams illustrate, only A shifts; P remains at the beginning-of-period and F at the end-of-period, regardless of the convention. The compound interest tables in the Appendix are based on the end-of-period convention.

Viewpoint of Economic Analysis Studies

When we make economic analysis calculations, we must proceed from a point of reference. Generally, we will want to take the point of view of a total firm when doing industrial economic analyses. Example 1-1 vividly illustrated the problem: a firm's shipping department decided it could save money by having its printing work done outside rather than by the in-house printing department. An analysis from the viewpoint of the shipping department supported this, for it could get for $688.50 the same printing it was paying $793.50 for in-house. Further analysis showed, however, that its printing department costs would decline *less* than using the commercial printer would save. From the viewpoint of the firm, the net result would be an increase in total cost.

From Example 1-1 we see it *is* important that the **viewpoint of the study** be carefully considered. Selecting a narrow viewpoint, like that of the shipping department, may result in a suboptimal decision from the viewpoint of the firm. For this reason, the viewpoint of the total firm is used in industrial economic analyses.

Sunk Costs

We know that it is the **differences between alternatives** that are relevant in economic analysis. Events that have occurred in the past really have no bearing on what we should do in the future. When the judge says, "$200 fine or 3 days in jail," the events that led to these unhappy alternatives really are unimportant. It is the current and future differences between the alternatives that *are* important. Past costs, like past events, have no bearing on deciding between alternatives unless the past costs somehow affect the present or future costs. In general, past costs do not affect the present or the future, so we refer to them as *sunk costs* and disregard them.

Borrowed Money Viewpoint

In most economic analyses, the proposed alternatives inevitably require money to be spent, and so it is natural to ask the source of that money. The source will vary from situation to situation. In fact, there are *two* aspects of money to determine: one is the **financing**—the obtaining of money—problem; the other is the **investment**—the spending of money—problem. Experience has shown that these two concerns should be distinguished. When separated, the problems of obtaining money and of spending it are both logical and straightforward. Failure to separate them sometimes produces confusing results and poor decision making.

The conventional assumption in economic analysis is that the money required to finance alternative solutions in problem solving is considered to be **borrowed at interest rate *i*.**

Effect of Inflation and Deflation

For the present we will assume that prices are stable. This means that a machine that costs $5000 today can be expected to cost the same amount several years hence. Inflation and deflation can be serious problems for after-tax analysis and for cost and revenues whose inflation rates differ from the economy's inflation rates, but we assume stable prices for now.

Income Taxes

Income taxes, like inflation and deflation, must be considered in order to find the real payoff of a project. However, taxes will often affect alternatives in the same way, allowing us to compare our choices without considering income taxes. So, we will defer our introduction of income taxes into economic analyses until later.

ECONOMIC RULES

We have shown how to manipulate cash flows in a variety of ways, and in so doing we can now solve many kinds of compound interest problems. But engineering economic analysis is more than simply solving interest problems. The decision process (see Figure 1-1) requires that the outcomes of feasible alternatives be arranged so that they may be judged for **economic efficiency** in terms of the selection rule. The economic rule will be one of the following, depending on the situation:[1]

Situation	**Rule**
For fixed input	Maximize output
For fixed output	Minimize input
Neither input nor output fixed	Maximize (output−input)

We will now examine ways to solve engineering problems, so that rules for economic efficiency can be applied.

Equivalence provides the logic by which we may adjust the cash flow for a given alternative into some equivalent sum or series. To apply the selection rule to the outcomes of the feasible alternatives, we must first resolve them into comparable units. The question is, How should they be compared? In this chapter we'll learn how analysis can resolve alternatives into *equivalent present consequences,* referred to simply as **present worth analysis.** Chapter 6 will show how given alternatives are converted into an *equivalent uniform annual cash flow,* and Chapter 7 solves for the interest rate at which favourable consequences—that is, *benefits*—are equivalent to unfavourable consequences—or *costs*.

As a general rule, any economic analysis problem may be solved by the methods presented in this and in the two following chapters. This is true because *present worth, annual cash flow,* and *rate of return* are exact methods that will always yield the same solution when one is selecting the best alternative from among a set of mutually exclusive alternatives.[2] Some problems, however, may be more easily solved by one method than by another. For this reason, we now focus on the kinds of problems that are most readily solved by present worth analysis.

---

[1] The short table summarizes the discussion on selection of criteria in Chapter 1 (see page 11, item 5, 'Select the rule for determining the best alternative').

[2] 'Mutually exclusive' means that selecting one alternative precludes selecting any other alternative. For example, constructing a gas station and constructing a drive-in restaurant on a particular piece of vacant land are mutually exclusive alternatives.

APPLYING PRESENT WORTH TECHNIQUES

One of the easiest ways to compare mutually exclusive alternatives is to resolve their consequences to the present time. The three rules for economic efficiency are restated in terms of present worth analysis in Table 5-1.

TABLE 5-1 Present Worth Analysis

	Situation	Rule
Fixed input	Amount of money or other input resources are fixed	Maximize present worth of benefits or other outputs
Fixed output	There is a fixed task, benefit, or other output to be accomplished	Minimize present worth of costs or other inputs
Neither input nor output is fixed	Neither amount of money, or other inputs, nor amount of benefits, or other output, is fixed	Maximize (present worth of benefits *minus* present worth of costs), that is, maximize net present worth

Present worth analysis is most frequently used to determine the present value of future money receipts and disbursements. It would help us, for example, to determine a present worth of income-producing property, like an oil well or an apartment building. If the future income and costs are known, then we can use a suitable interest rate to calculate the present worth of the property. This should provide a good estimate of the price at which the property could be bought or sold. Another application might be determining the valuation of stocks or bonds according to the anticipated future benefits from owning them.

In present worth analysis, careful consideration must be given to the time period covered by the analysis. Usually the task to be accomplished has a time period associated with it. In that case, the consequences of each alternative must be considered for this period of time, which is usually called the **analysis period,** or sometimes the **planning horizon.**

The analysis period for an economy study should be determined from the situation. In some industries with rapidly changing technologies, a rather short analysis period or planning horizon might be in order. Industries with more stable technologies (like steelmaking) might use a longer period (say, 10–20 years), while government agencies frequently use analysis periods extending to 50 years or more.

Three different analysis period situations are encountered in economic analysis problems:

1. The useful life of each alternative equals the analysis period.
2. The alternatives have useful lives different from the analysis period.
3. There is an infinite analysis period, $n = \infty$.

1. Useful Lives Equal to the Analysis Period

Since different lives and an infinite analysis period present some complications, we will begin with four examples in which the useful life of each alternative equals the analysis period.

EXAMPLE 5-1

A firm is considering which of two mechanical devices to install to reduce costs in a particular situation. Both devices cost $1000 and have useful lives of 5 years and no salvage value. Device A can be expected to result in $300 savings annually. Device B will provide cost savings of $400 the first year, but those will decline $50 annually to second-year savings of $350, third-year savings of $300, and so forth. With interest at 7%, which device should the firm purchase?

SOLUTION

The analysis period can conveniently be selected as the useful life of the devices, or 5 years. Since both devices cost $1000, there is a fixed input (cost) of $1000 regardless of whether A or B is chosen. The right decision rule is to choose the alternative that maximizes the present worth of benefits.

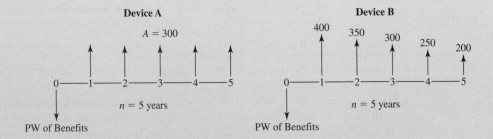

$$\text{PW of benefits A} = 300(P/A, 7\%, 5) = 300(4.100) = \$1230$$

$$\text{PW of benefits B} = 400(P/A, 7\%, 5) - 50(P/G, 7\%, 5)$$

$$= 400(4.100) - 50(7.647) = \$1257.65$$

Device B has the larger present worth of benefits and is, therefore, the preferred alternative. It is worth noting that, if we ignore the time value of money, both alternatives provide $1500 worth of benefits over the 5-year period. Device B provides greater benefits in the first 2 years and smaller benefits in the last 2 years. This more rapid flow of benefits from B, although the total magnitude equals that of A, results in a greater present worth of benefits.

EXAMPLE 5-2

A city plans to build an aqueduct to bring water in from the mountains in the west. The aqueduct can be built now at a reduced size for $300 million and enlarged 25 years hence for an additional $350 million. An alternative is to construct the full-sized aqueduct now for $400 million.

Both alternatives would provide the needed capacity for the 50-year analysis period. Maintenance costs are small and may be ignored. At 6% interest, which alternative should be selected?

SOLUTION

This problem illustrates staged construction. The aqueduct may be built in a single stage or in a smaller first stage followed many years later by a second stage to provide the additional capacity when needed.

For the Two-Stage Construction

$$\text{PW of cost} = \$300 \text{ million} + 350 \text{ million}(P/F, 6\%, 25)$$

$$= \$300 \text{ million} + 81.6 \text{ million}$$

$$= \$381.6 \text{ million}$$

For the Single-Stage Construction

$$\text{PW of cost} = \$400 \text{ million}$$

The two-stage construction has a smaller present worth of cost and is the preferred construction plan.

EXAMPLE 5-3

A purchasing agent is considering the purchase of some new equipment for the mailroom. Two different manufacturers have provided quotations. An analysis of the quotations shows the following:

Manufacturer	Cost	Useful Life (years)	End-of-Useful-Life Salvage Value
Speedy	$1500	5	$200
Allied	1600	5	325

The equipment of both manufacturers is expected to perform at the desired level of (fixed) output. For a 5-year analysis period, which manufacturer's equipment should be selected? Assume 7% interest and equal maintenance costs.

SOLUTION

For fixed output, the criterion is to minimize the present worth of cost.

Speedy

$$\text{PW of cost} = 1500 - 200(P/F, 7\%, 5)$$

$$= 1500 - 200(0.7130)$$

$$= 1500 - 143 = \$1357$$

Allied

$$PW \text{ of cost} = 1600 - 325(P/F, 7\%, 5) = 1600 - 325(0.7130)$$

$$= 1600 - 232 = \$1368$$

Since it is only the *differences between alternatives* that are relevant, maintenance costs may be left out of the economic analysis. Although the PWs of cost for all the alternatives are nearly identical, we would, nevertheless, choose the one with minimum present worth of cost unless there were other tangible or intangible differences that would change the decision. Buy the Speedy equipment.

EXAMPLE 5-4

A firm is trying to decide which of two weighing scales it should install to check a package-filling operation in the plant. The ideal scale would allow better control of the filling operation and result in less overfilling. If both scales have lives equal to the 6-year analysis period, which one should be selected? Assume an 8% interest rate.

Alternatives	Cost	Uniform Annual Benefit	End-of-Useful-Life Salvage Value
Atlas scale	$2000	$450	$100
Tom Thumb scale	3000	600	700

SOLUTION

Atlas Scale

$$PW \text{ of benefits} - PW \text{ of cost} = 450(P/A, 8\%, 6) + 100(P/F, 8\%, 6) - 2000$$

$$= 450(4.623) + 100(0.6302) - 2000$$

$$= 2080 + 63 - 2000 = \$143$$

Tom Thumb Scale

$$PW \text{ of benefits} - PW \text{ of cost} = 600(P/A, 8\%, 6) + 700(P/F, 8\%, 6) - 3000$$

$$= 600(4.623) + 700(0.6302) - 3000$$

$$= 2774 + 441 - 3000 = \$215$$

The salvage value of the scale, it should be noted, is simply treated as another benefit of the alternative. Since the criterion is to maximize the present worth of benefits minus the present worth of cost, the preferable alternative is the Tom Thumb scale.

In Example 5-4, we compared two alternatives and selected the one in which present worth of benefits *minus* present worth of cost was a maximum. The criterion is called the **net present worth criterion** and is written simply as **NPW:**

$$\text{Net present worth} = \text{Present worth of benefits} - \text{Present worth of cost}$$

$$\text{NPW} = \text{PW of benefits} - \text{PW of cost} \qquad (5\text{-}1)$$

2. Useful Lives Different from the Analysis Period

In present worth analysis, there always must be an identified analysis period. It follows, then, that each alternative must be considered for the entire period. In Examples 5-1 to 5-4, the useful life of each alternative was equal to the analysis period. Often we can arrange it this way, but in many situations the alternatives will have useful lives different from the analysis period. This section describes one way to evaluate alternatives with lives different from the study period.

In Example 5-3, suppose that the Allied equipment was expected to have a 10-year useful life, or twice that of the Speedy equipment. Assuming the Allied salvage value would still be $325 in 10 years, which equipment should now be purchased? We can recompute the present worth of cost of the Allied equipment, starting as follows:

$$\text{PW of cost} = 1600 - 325(P/F, 7\%, 10)$$

$$= 1600 - 325(0.5083)$$

$$= 1600 - 165 = \$1435$$

The present worth of cost has increased. This is due, of course, to the more distant recovery of the salvage value. More importantly, we now find ourselves attempting to compare Speedy equipment, with its 5-year life, against the Allied equipment with a 10-year life. Because of the variation in the useful life of the equipment, we no longer have a situation of *fixed output*. Speedy equipment in the mailroom for 5 years is certainly not the same as 10 years of service with Allied equipment.

For present worth calculations, it is important that we select an analysis period and judge the consequences of each of the alternatives during that period. Therefore, it is not fair to compare the NPW of the Allied equipment over its 10-year life with the NPW of the Speedy equipment over its 5-year life.

Not only the firm and its economic environment are important when we choose an analysis period, but so too is the specific situation being analyzed. If the Allied equipment (Example 5-3) has a useful life of 10 years, and the Speedy equipment will last 5 years, one method is to select an analysis period that is the **least common multiple** of their useful lives. Thus we would compare the 10-year life of Allied equipment against an initial purchase of Speedy equipment *plus* its replacement with new Speedy equipment in 5 years. The result is to judge the alternatives on the basis of a 10-year requirement in the mailroom. On this basis the economic analysis is as follows.

Assuming the replacement Speedy equipment 5 years hence will also cost $1500,

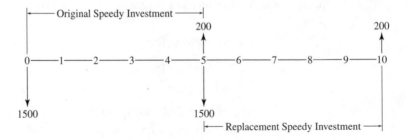

$$\text{PW of cost} = 1500 + (1500 - 200)(P/F, 7\%, 5) - 200(P/F, 7\%, 10)$$
$$= 1500 + 1300(0.7130) - 200(0.5083)$$
$$= 1500 + 927 - 102 = \$2325$$

For the Allied equipment, on the other hand, we have the following results:

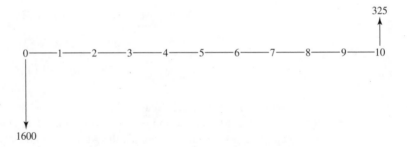

$$\text{PW of cost} = 1600 - 325(P/F, 7\%, 10) = 1600 - 325(0.5083) = \$1435$$

For the fixed output of 10 years of service in the mailroom, the Allied equipment, with its smaller present worth of cost, is preferable.

We have seen that setting the analysis period equal to the least common multiple of the lives of the two alternatives seems reasonable in the revised Example 5-3. What would one do, however, if the alternatives had useful lives of 7 and 13 years respectively? Here the least common multiple of lives is 91 years. An analysis period of 91 years hardly seems realistic. Instead, a suitable analysis period should be based on how long the equipment is likely to be needed. This may require that terminal values be estimated for the alternatives at some point before the end of their useful lives.

As Figure 5-1 shows, it is not necessary for the analysis period to equal the useful life of an alternative or some multiple of the useful life. To properly reflect the situation at the end of the analysis period, an estimate is required of the market value of the equipment at that time. The calculations might be easier if everything came out even, but it is not essential.

Alternative 1

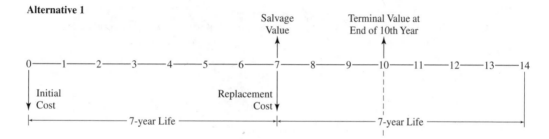

Alternative 2

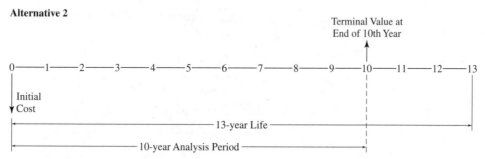

FIGURE 5-1 Superimposing a 10-year analysis period on 7- and 13-year alternatives.

EXAMPLE 5-5

A diesel manufacturer is considering the two alternative production machines graphically depicted in Figure 5-1. Specific data are as follows:

	Alt. 1	**Alt. 2**
Initial cost	$50,000	$75,000
Estimated salvage value at end of useful life	$10,000	$12,000
Useful life of equipment, in years	7	13

The manufacturer uses an interest rate of 8% and wants to use the PW method to compare these alternatives over an analysis period of 10 years.

	Alt. 1	**Alt. 2**
Estimated market value, end of 10-year analysis period	$20,000	$15,000

SOLUTION

In this case, the decision maker is setting the analysis period at 10 years rather than accepting a common multiple of the lives of the alternatives or assuming that the period of needed service is infinite (to be discussed in the next section). This is a legitimate approach—perhaps the diesel manufacturer will be phasing out this model at the end of the 10-year period. In any event, we need to compare the alternatives over the 10 years.

As illustrated in Figure 5-1, we may assume that Alternative 1 will be replaced by an identical machine after its 7-year useful life. Alternative 2 has a 13-year useful life. The diesel manufacturer has provided an estimated market value of the equipment at the time of the analysis period. So we can now compare the two choices over 10 years as follows:

$$\text{PW (Alt. 1)} = -50{,}000 + (10{,}000 - 50{,}000)(P/F, 8\%, 7) + 20{,}000(P/F, 8\%, 10)$$

$$= -50{,}000 - 40{,}000(0.5835) + 20{,}000(0.4632)$$

$$= -\$64{,}076$$

$$\text{PW (Alt. 2)} = -75{,}000 + 15{,}000(P/F, 8\%, 10)$$

$$= -75{,}000 + 15{,}000(0.4632)$$

$$= -\$69{,}442$$

To minimize PW of costs the diesel manufacturer should select Alternative 1.

3. Infinite Analysis Period: Capitalized Cost

Another difficulty in present worth analysis arises when we encounter an infinite analysis period ($n = \infty$). In governmental analyses, a service or condition sometimes must be maintained for an infinite period. The need for roads, dams, pipelines, and so on, is sometimes considered to be permanent. In these situations a present worth of cost analysis would have an infinite analysis period. We call this particular analysis **capitalized cost.**

Capitalized cost is the present sum of money that would need to be set aside now, at some interest rate, to yield the funds required to provide the service (or whatever) indefinitely. To accomplish this, the money set aside for future expenditures must not decline. The interest received on the money set aside can be spent, but not the principal. When one stops to think about an infinite analysis period (as opposed to something relatively short, like a hundred years), we see that an undiminished principal sum is essential; otherwise one will of necessity run out of money before infinity.

In Chapter 4 we saw that

$$\text{principal sum} + \text{interest for the period} = \text{amount at end of period, or}$$

$$P \quad + \quad iP \quad = \quad P + iP$$

If we spend iP, then in the next interest period the principal sum P will again increase to $P + iP$. Thus, we can again spend iP.

This concept may be illustrated by a numerical example. Suppose you deposited $200 in a bank that paid 4% interest annually. How much money could be withdrawn each year without reducing the balance in the account below the initial $200? At the end of the first year, the $200 would have earned 4%($200) = $8 interest. If this interest were withdrawn, the $200 would remain in the account. At the end of the second year, the $200 balance would again earn 4%($200) = $8. This $8 could also be withdrawn and the account would still have $200. This procedure could be continued indefinitely and the bank account would always contain $200.

The year-by-year situation would be depicted like this:

Year 1: $200 initial P → 200 + 8 =   208
                Withdrawal $i\,P$ =   − 8

                    Year 2: $200 → 200 + 8   =   208
                          Withdrawal $i\,P$ =   − 8
                                              $200

                                        and so on

Thus, for any initial present sum P, there can be an end-of-period withdrawal of A equal to $i\,P$ each period, and these withdrawals can continue forever without diminishing the initial sum P. This gives us the basic relationship:

$$\text{For} \quad n = \infty, \quad A = Pi$$

This relationship is the key to capitalized cost calculations. Earlier we defined capitalized cost as the present sum of money that would need to be set aside at some interest rate to yield the funds to provide the desired task or service forever. Capitalized cost is therefore the P in the equation $A = i\,P$. It follows that

$$\textbf{Capitalized cost} \qquad P = \frac{A}{i} \qquad\qquad (5\text{-}2)$$

If we can resolve the desired task or service into an equivalent A, the capitalized cost can be computed. The following examples illustrate such computations.

EXAMPLE 5-6

How much should one set aside to pay $50 per year for maintenance on a gravesite if interest is assumed to be 4%? For perpetual maintenance, the principal sum must remain undiminished after the annual disbursement is made.

SOLUTION

$$\text{Capitalized cost } P = \frac{\text{Annual disbursement } A}{\text{Interest rate } i}$$

$$P = \frac{50}{0.04} = \$1250$$

One should set aside $1250.

EXAMPLE 5-7

A city plans a pipeline to transport water from a distant area to the city. The pipeline will cost $8 million and will have an expected life of 70 years. The city expects it will need to keep the water line in service indefinitely. Compute the capitalized cost, assuming 7% interest.

SOLUTION

The capitalized cost equation

$$P = \frac{A}{i}$$

is simple to apply when there are end-of-period disbursements A. Here we have renewals of the pipeline every 70 years. To compute the capitalized cost, it is first necessary to compute an end-of-period disbursement A that is equivalent to $8 million every 70 years.

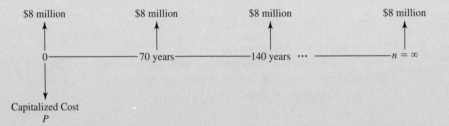

The $8 million disbursement at the end of 70 years may be resolved into an equivalent A.

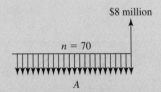

$$A = F(A/F, i, n) = \$8 \text{ million}(A/F, 7\%, 70)$$

$$= \$8 \text{ million}(0.00062)$$

$$= \$4960$$

Each 70-year period is identical to this one and the infinite series is shown in Figure 5-2.

$$\text{Capitalized cost } P = \$8 \text{ million} + \frac{A}{i} = \$8 \text{ million} + \frac{4960}{0.07}$$

$$= \$8,071,000$$

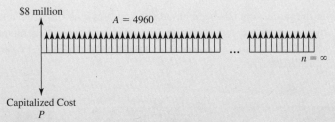

FIGURE 5-2 Using the sinking fund factor to compute an infinite series.

ALTERNATIVE SOLUTION 1

Instead of solving for an equivalent end-of-period payment A based on a *future* $8 million disbursement, we could find A, given a *present* $8 million disbursement.

$$A = P(A/P, i, n) = \$8 \text{ million}(A/P, 7\%, 70)$$

$$= \$8 \text{ million}(0.0706) = \$565,000$$

On this basis, the infinite series is shown in Figure 5-3. Carefully note the difference between this and Figure 5-2. Now:

$$\text{Capitalized cost } P = \frac{A}{i} = \frac{565,000}{0.07} = \$8,071,000$$

FIGURE 5-3 Using the capital recovery factor to compute an infinite series.

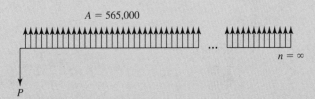

ALTERNATIVE SOLUTION 2

Another way of solving the problem is to assume the interest is 70 years. Compute an equivalent interest rate for the 70-year period. Then the capitalized cost may be computed by using Equation 4-33 for $m = 70$

$$i_{70\,\text{yr}} = (1 + i_{1\,\text{yr}})^{70} - 1 = (1 + 0.07)^{70} - 1 = 112.989$$

$$\text{Capitalized cost} = \$8 \text{ million} + \frac{\$8 \text{ million}}{112.989} = \$8,071,000$$

Multiple Alternatives

So far the discussion has been based on examples with only two alternatives. But multiple-alternative problems may be solved by exactly the same methods. (The only reason for avoiding multiple alternatives was to simplify the examples.) Examples 5-8 and 5-9 have multiple alternatives.

EXAMPLE 5-8

A contractor has been awarded the contract to construct a 6 km tunnel in the mountains. During the 5-year construction period, the contractor will need water from a nearby stream. He will construct a pipeline to carry the water to the main construction yard. An analysis of costs for various pipe sizes is as follows:

	Pipe Sizes (in.)			
	2	**3**	**4**	**6**
Installed cost of pipeline and pump	$22,000	$23,000	$25,000	$30,000
Cost per hour for pumping	$1.20	$0.65	$0.50	$0.40

The pipe and pump will have a salvage value at the end of 5 years equal to the cost of removing them. The pump will operate 2000 hours a year. The lowest interest rate at which the contractor is willing to invest money is 7%. (The minimum required interest rate for invested money is called the **minimum attractive rate of return,** or MARR.) Select the alternative with the least present worth of cost.

SOLUTION

We can compute the present worth of cost for each alternative. For each size of pipe, the present worth of cost is equal to the installed cost of the pipeline and pump plus the present worth of 5 years of pumping costs.

	Pipe Size (in.)			
	2	**3**	**4**	**6**
Installed cost of pipeline and pump	$22,000	$23,000	$25,000	$30,000
1.20×2000 hr $\times (P/A, 7\%, 5)$	9,840			
0.65×2000 hr $\times 4.100$		5,330		
0.50×2000 hr $\times 4.100$			4,100	
0.40×2000 hr $\times 4.100$				3,280
Present worth of cost	$31,840	$28,330	$29,100	$33,280

Select the 3 in. pipe.

EXAMPLE 5-9

An investor paid $8000 to a consulting firm to analyze what he might do with a small parcel of land on the edge of town that can be bought for $30,000. In their report, the consultants suggested four alternatives:

Alternatives		Total Investment Including Land*	Uniform Net Annual Benefit	Terminal Value at End of 20 Years
A	Do nothing	$ 0	$ 0	$ 0
B	Vegetable market	50,000	5,100	30,000
C	Gas station	95,000	10,500	30,000
D	Small motel	350,000	36,000	150,000

*Includes the land and structures but does not include the $8000 fee to the consulting firm.

Assuming 10% is the minimum attractive rate of return, what should the investor do?

SOLUTION

Alternative A is the 'do-nothing' alternative. Generally, one feasible alternative in any situation is to remain in the present status and do nothing. In this problem, the investor could decide that the most attractive alternative is not to purchase the property and develop it. This is clearly a do-nothing decision.

We note, however, that if he does nothing, the total venture would not be a very satisfactory one. This is because the investor spent $8000 for professional advice on the possible uses of the property. But because the $8000 is a past cost, it is a **sunk cost.** The only relevant costs in an economic analysis are *present* and *future* costs; past events and past, or sunk, costs are gone and cannot be allowed to affect future planning. (Past costs may be relevant in computing depreciation charges and income taxes, but nowhere else.) The past should not deter the investor from making the best decision now, regardless of the costs that brought him to this situation and point of time.

This problem is one of neither fixed input nor fixed output, so our criterion will be to maximize the present worth of benefits *minus* the present worth of cost or, simply stated, maximize net present worth.

Alternative A, Do Nothing

$$NPW = 0$$

Alternative B, Vegetable Market

$$NPW = -50,000 + 5100(P/A, 10\%, 20) + 30,000(P/F, 10\%, 20)$$

$$= -50,000 + 5100(8.514) + 30,000(0.1486)$$

$$= -50,000 + 43,420 + 4460$$

$$= -\$2120$$

Alternative C, Gas Station

$$NPW = -95,000 + 10,500(P/A, 10\%, 20) + 30,000(P/F, 10\%, 20)$$

$$= -95,000 + 89,400 + 4460$$

$$= -\$1140$$

Alternative D, Small Motel

$$NPW = -350,000 + 36,000(P/A, 10\%, 20) + 150,000(P/F, 10\%, 20)$$

$$= -350,000 + 306,500 + 22,290$$

$$= -\$21,210$$

The criterion is to maximize net present worth. In this situation, one alternative has NPW equal to zero, and three alternatives have negative values for NPW. We will select the best of the four alternatives, namely, the do-nothing Alt. A, with NPW equal to zero.

EXAMPLE 5-10

A piece of land may be bought for $610,000 to be strip-mined for coal. Annual net income will be $200,000 a year for 10 years. At the end of the 10 years, the surface of the land will be restored as required by a federal law on strip mining. The reclamation will cost $1.5 million more than the resale value of the land after it is restored. Using a 10% interest rate, determine whether the project is desirable.

SOLUTION

The investment opportunity may be described by the following cash flow:

Year	Cash Flow (thousands)
0	−$610
1–10	+200 (per year)
10	−1500

$$NPW = -610 + 200(P/A, 10\%, 10) - 1500(P/F, 10\%, 10)$$

$$= -610 + 200(6.145) - 1500(0.3855)$$

$$= -610 + 1229 - 578$$

$$= +\$41$$

Since NPW is positive, the project is desirable. (See Appendix 7A for a more complete analysis of this type of problem. At interest rates of 4.07% and 18.29% the NPW = 0.)

EXAMPLE 5-11

Two pieces of construction equipment are being analyzed:

Year	Alt. A	Alt. B
0	−$2000	−$1500
1	+1000	+700
2	+850	+300
3	+700	+300
4	+550	+300
5	+400	+300
6	+400	+400
7	+400	+500
8	+400	+600

At an 8% interest rate, which alternative should be selected?

SOLUTION

Alternative A

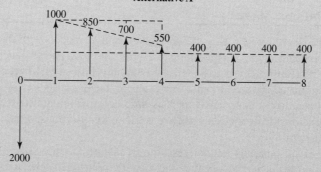

$$\text{PW of benefits} = 400(P/A, 8\%, 8) + 600(P/A, 8\%, 4) - 150(P/G, 8\%, 4)$$

$$= 400(5.747) + 600(3.312) - 150(4.650) = 3588.50$$

PW of cost = 2000

Net present worth = 3588.50 − 2000 = +$1588.50

Alternative B

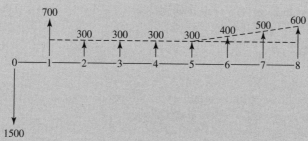

$$\text{PW of benefits} = 300(P/A, 8\%, 8) + (700 - 300)(P/F, 8\%, 1)$$

$$+ 100(P/G, 8\%, 4)(P/F, 8\%, 4)$$

$$= 300(5.747) + 400(0.9259) + 100(4.650)(0.7350)$$

$$= 2436.24$$

$$\text{PW of cost} = 1500$$

$$\text{Net present worth} = 2436.24 - 1500$$

$$= +\$936.24$$

To maximize NPW, choose Alternative A.

SPREADSHEETS AND PRESENT WORTH

Spreadsheets make it easy to build more accurate models with shorter time periods. When one is using factors, it is common to assume that costs and revenues are uniform for n years. With spreadsheets it is easy to use 120 months instead of 10 years, and the cash flows can be estimated for each month. For example, energy costs for air conditioning peak in the summer, and in many areas there is little construction during the winter. Cash flows that depend on population, such as for electric power and transportation costs, often increase at $x\%$ a year.

In spreadsheets any interest rate is entered exactly—so no interpolation is needed. This makes it easy to calculate the monthly repayment schedule for a car loan or a house mortgage. Examples 5-12 and 5-13 illustrate the use of spreadsheets to calculate PWs.

EXAMPLE 5-12

NLE Construction is bidding on a project whose costs are divided into $30,000 for start-up and $240,000 for the first year. If the interest rate is 1% per month, or 12.68% per year, what is the present worth with monthly compounding?

SOLUTION

Figure 5-4 illustrates the spreadsheet solution with the assumption that costs are distributed evenly throughout the year ($-20,000 = -240,000/12$).

FIGURE 5-4 Spreadsheet with monthly cash flows.

	A	B	C	D
1	1%	i		
2	−30,000	initial cash flow		
3	−240,000	annual amount		
4				
5	Month	Cash Flow		
6	0	−30000		
7	1	−20000		
8	2	−20000		
9	3	−20000		
10	4	−20000		
11	5	−20000		
12	6	−20000		
13	7	−20000		
14	8	−20000		
15	9	−20000		
16	10	−20000		
17	11	−20000		
18	12	−20000		
19	NPV	−$255,102	=NPV(A1,B7:B18)+B6	

Since the costs are uniform, the factor solution is:

$$PW_{mon} = -30,000 - 20,000(P/A, 1\%, 12) = -\$255,102$$

The value of monthly periods can be illustrated by computing the PW assuming an annual period. The results differ by more than $12,000, because $20,000 at the end of Months 1 to 12 is not the same as $240,000 at the end of Month 12. The timing of the cash flows makes the difference, even though the effective interest rates are the same.

$$PW_{annual} = -30,000 - 240,000(P/F, 12.68\%, 1) = -30,000 - 240,000/1.1268 = -\$242,993$$

EXAMPLE 5-13

Regina Industries has a new product whose sales are expected to be 1.2, 3.5, 7, 5, and 3 million units per year over the next 5 years. Production, distribution, and overhead costs are stable at $120 per unit. The price will be $200 per unit for the first 2 years, and then $180, $160, and $140 for the next 3 years. The remaining R&D and production costs are $300 million. If i is 15%, what is the present worth of the new product?

SOLUTION

It is easiest to calculate the yearly net revenue per unit before building the spreadsheet shown in Figure 5-5. Those values are the yearly price minus the $120 of costs, which equals $80, $80, $60, $40, and $20.

	A	B	C	D	E
1	12%	*i*			
2			Net	Cash	
3	Year	Sales (M)	Revenue per Unit	Flow ($M)	
4	0			−300	
5	1	1.2	80	96	
6	2	3.5	80	280	
7	3	7	60	420	
8	4	5	40	200	
9	5	3	20	60	
10		D4+NPV(A1,D5:D9) =		$469	Million

FIGURE 5-5 Present worth of a new product.

SUMMARY

Present worth analysis is suitable for almost any economic analysis problem. But it is particularly desirable when we wish to know the present worth of future costs and benefits. And we frequently want to know the value today of such things as income-producing assets, stocks, and bonds.

For present worth analysis, the proper economic criteria are:

Fixed input	Maximize the PW of benefits
Fixed output	Minimize the PW of costs
Neither input nor output is fixed	Maximize (PW of benefits − PW of costs) or, more simply stated: Maximize NPW

To make valid comparisons, we need to analyze each alternative in a problem over the same **analysis period** or **planning horizon.** If the alternatives do not have equal lives, some technique must be used to achieve a common analysis period. One method is to select an analysis period equal to the least common multiple of the alternative lives. Another method is to select an analysis period and then compute end-of-analysis-period salvage values for the alternatives.

Capitalized cost is the present worth of cost for an infinite analysis period ($n = \infty$). When $n = \infty$, the fundamental relationship is $A = iP$. Some form of this equation is used whenever there is a problem with an infinite analysis period.

The numerous assumptions routinely made in solving economic analysis problems include the following.

1. Present sums P are beginning of period and all series receipts or disbursements A and future sums F occur at the end of the interest period. The compound interest tables were derived on this basis.

2. In industrial economic analyses, the point of reference from which to compute the consequences of alternatives is the total firm. Taking a narrower view of the consequences can result in suboptimal solutions.

3. Only the differences between the alternatives are relevant. Past costs are sunk costs and generally do not affect present or future costs. For this reason they are ignored.

4. The investment problem is isolated from the financing problem. We generally assume that all money required is borrowed at interest rate i.

5. For now, stable prices are assumed. The inflation-deflation problem is deferred to Chapter 14. Similarly, our discussion of income taxes is deferred to Chapter 12.

6. Often uniform cash flows or arithmetic gradients are reasonable assumptions. However, spreadsheets simplify the finding of PW in more complicated problems.

PROBLEMS

Most problems could be solved with a spreadsheet, but calculators and tabulated factors are often easier. The icon indicates that a spreadsheet is recommended.

5-1 Compute P for the following diagram.

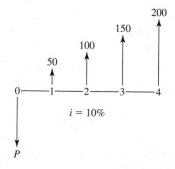

5-2 Compute the value of P that is equivalent to the four cash flows in the following diagram.

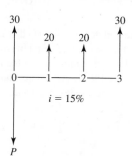

5-3 What is the value of P for the situation diagrammed?

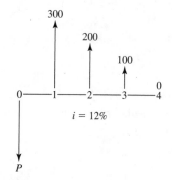

(*Answer: P* = $498.50)

5-4 Compute the value of Q in the following diagram.

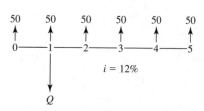

5-5   Compute the value of P for the following diagram.

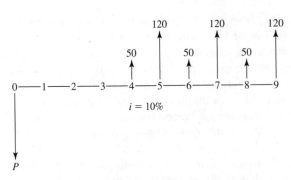

5-6   Compute the value of P for the following diagram.

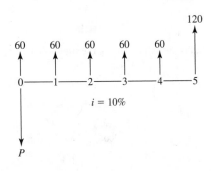

(*Answer: P* = $324.71)

5-7   Use a geometric gradient factor to solve the following diagram for P.

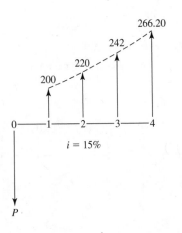

5-8   If $i = 10\%$, what is the value of P?

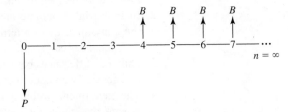

5-9   A stonecutter assigned to carve the headstone for a well-known engineering economist began with the following design.

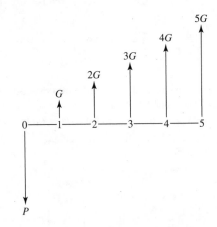

He then started the equation as follows:

$$P = G(P/G, i, 6)$$

He realized he had made a mistake. The equation should have been

$$P = G(P/G, i, 5) + G(P/A, i, 5)$$

The stonecutter does not want to discard the stone and start over. He asks you to help him with his problem. The right-hand side must be, for the first equation to be correct for the carved figure, multiplied by one compound interest factor, taking the form:

$$P = G(P/G, i, 6)(\quad, i,)$$

Write the complete equation.

5-10 Using 5% nominal interest, compounded continuously, solve for P.

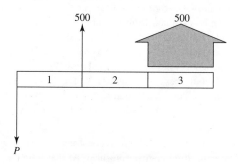

5-11 Find P for the following cash flow diagram.

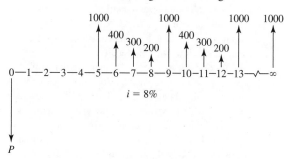

$$i = 8\%$$

5-12 The annual income from a rented house is $12,000. The annual expenses are $3000. If the house can be sold for $145,000 at the end of 10 years, how much could you afford to pay for it now, if you considered 18% to be a suitable interest rate?
(*Answer:* $68,155)

5-13 Consider the following cash flow. At a 6% interest rate, what value of P at the end of Year 1 is equivalent to the benefits at the end of Years 2 to 7?

Year	Cash Flow
1	$-P$
2	$+100$
3	$+200$
4	$+300$
5	$+400$
6	$+500$
7	$+600$

5-14 How much would the owner of a building be justified in paying for a sprinkler system that will save $750 a year in insurance premiums if the system has to be replaced every 20 years and has a salvage value equal to 10% of its initial cost? Assume money is worth 7%.
(*Answer:* $8156)

5-15 A manufacturer is considering purchasing equipment which will have the following financial effects:

Year	Disbursements	Receipts
0	$4400	$ 0
1	660.	880
2	660	1980
3	440	2420
4	220	1760

If money is worth 6%, should he invest in the equipment?

5-16 Jerry Stans, a young industrial engineer, prepared an economic analysis for some equipment to replace one production worker. The analysis showed that the present worth of benefits (of employing one less production worker) just equalled the present worth of the equipment costs, assuming a 10-year useful life for the equipment. It was decided not to buy the equipment.

A short time later, the production workers won a new 3-year union contract that granted them an immediate 40¢-per-hour wage increase, plus an additional 25¢-per-hour wage increase in each of the two subsequent years. Assume that in each and every future year, a 25¢-per-hour wage increase will be granted.

Jerry Stans has been asked to revise his earlier economic analysis. The present worth of benefits of replacing one production employee will now increase. Assuming an interest rate of 8%, by how much will the justifiable cost of the automation equipment (with a 10-year useful life) increase? Assume the plant operates a single 8-hour shift, 250 days a year.

5-17 In a present worth analysis of certain equipment, one alternative has a net present worth of +$420, based on a 6-year analysis period that equals the useful life of the alternative. A 10% interest rate was used in the computations.

The alternative device is to be replaced at the end of the 6 years by an identical item with the same cost, benefits, and useful life. Using a 10% interest rate, compute the net present worth of the alternative equipment for the 12-year analysis period.
(*Answer:* NPW = +$657.09)

5-18 A project has a net present worth of –$140 as of January 1, 2000. If a 10% interest rate is used, what is the project NPW as of December 31, 1997?

5-19 On February 1, the Miro Company needs to purchase some office equipment. The company is short of cash and expects to be short for several months. The company treasurer has said that he could pay for the equipment as follows:

Date	Payment
April 1	$150
June 1	300
Aug. 1	450
Oct. 1	600
Dec. 1	750

A local office supply firm will agree to sell the equipment to the firm now and accept payment according to the treasurer's schedule. If interest is charged at 3% every 2 months, with compounding once every 2 months, how much office equipment can the Miro Company buy now?   (*Answer*: $2020)

5-20 By installing some elaborate inspection equipment on its assembly line, the Robot Corp. can avoid hiring an extra worker who would have earned $26,000 a year in wages and an additional $7500 a year in employee benefits. The inspection equipment has a 6-year useful life and no salvage value. Use a nominal 18% interest rate in your calculations. How much can Robot afford to pay for the equipment if the wages and worker benefits would have been paid

(*a*) At the end of each year

(*b*) Monthly

(*c*) Continuously

(*d*) Explain why the answers in (*b*) and (*c*) are larger than in (*a*).

Assume the compounding matches the way the wages and benefits would have been paid (that is, annually, monthly, and continuously, respectively).

5-21 Annual maintenance costs for a particular section of highway pavement are $2000. The placement of a new surface would reduce the annual maintenance cost to $500 a year for the first 5 years and to $1000 a year for the next 5 years. After 10 years the annual maintenance would again be $2000. If maintenance costs are the only saving, what investment can be justified for the new surface? Assume interest at 4%.

5-22 An investor is considering buying a 20-year corporate bond. The bond has a face value of $1000 and pays 6% interest per year in two semi-annual payments. Thus the purchaser of the bond will receive $30 every 6 months in addition to $1000 at the end of 20 years, along with the last $30 interest payment. If the investor wants to receive 8% interest, compounded semi-annually, how much would he or she be willing to pay for the bond?

5-23 A road building contractor has received a major highway construction contract that will require 50,000 m³ of crushed stone each year for 5 years. The stone can be obtained from a quarry for $5.80/m³. As an alternative, the contractor has decided to try to buy the quarry. He believes that if he owned the quarry, the stone would cost him only $4.30/m³. He thinks he could resell the quarry at the end of 5 years for $40,000. If the contractor uses a 10% interest rate, how much would he be willing to pay for the quarry?

5-24 A new office building was constructed 5 years ago by a consulting engineering firm. At that time the firm obtained a bank loan for $100,000 with a 12% annual interest rate, compounded quarterly. The terms of the loan call for equal quarterly payments to repay the loan in 10 years. The loan also allows for its prepayment at any time without penalty.

As a result of internal changes in the firm, it is now proposed to refinance the loan through an insurance company. The new loan would be for a 20-year term with an interest rate of 8% per year, compounded quarterly. The new equal quarterly payments would repay the loan in the 20-year period. The insurance company requires the payment of a 5% loan initiation charge (often described as a '5-point loan fee'), which will be added to the new loan.

(*a*) What is the balance due on the original mortgage if 20 payments have been made in the last five years?

(*b*) What is the difference between the equal quarterly payments on the present bank loan and the proposed insurance company loan?

5-25 IBP Inc. is considering establishing a new machine to automate a meat packing process. The machine will save $50,000 in labour annually. The machine can be purchased for $200,000 today and will be used for 10 years. It is has a salvage value of $10,000 at the end of its useful life. The new machine will require an annual maintenance cost of $9000. The corporation has a minimum rate of return of 10%. Do you recommend automating the process?

5-26 Argentina is considering constructing a bridge across the Rio de la Plata to connect its northern coast to the southern coast of Uruguay. If this bridge is constructed, it will reduce the travel time from Buenos Aires, Argentina, to São Paulo, Brazil, by over 10 hours and has the potential to improve significantly the flow of manufactured goods between the two countries. The cost of the new bridge, which will be the longest bridge in the world, spanning over 50 miles, will be $700 million. The bridge will require an annual maintenance of $10 million for repairs and upgrades and is estimated to last 80 years. It is estimated that 550,000 vehicles will use the bridge during the first year of operation, and an additional 50,000 vehicles a year until the tenth year. These data are based on a toll charge of $90 per vehicle. The annual traffic for the remainder of the life of the bridge life will be 1,000,000 vehicles a year. The Argentine government requires a minimum rate of return of 9% to proceed with the project.

(a) Does this project provide sufficient revenues to offset its costs?

(b) What considerations are there besides economic factors in deciding whether to construct the bridge?

5-27 A student has a job that leaves her with $250 a month in disposable income. She decides that she will use the money to buy a car. Before looking for a car, she arranges a 100% loan whose terms are $250 per month for 36 months at 18% annual interest. What is the maximum car purchase price that she can afford with her loan?

5-28 The student in Problem 5-27 finds a car she likes and the dealer offers to arrange financing. His terms are 12% interest for 60 months and no down payment. The car's sticker price is $12,000. Can she afford to buy this car with her $250 monthly disposable income?

5-29 The student in Problem 5-28 really wants this particular car. She decides to try and negotiate a different interest rate. What is the highest interest rate that she can accept, given a 60-month term and $250 per month payments?

5-30 We know that a car costs 60 monthly payments of $199. The car dealer has set us a nominal interest rate of 4.5% compounded daily. What is the purchase price of the car?

5-31 A machine costs $980,000 to purchase and will provide $200,000 a year in benefits. The company plans to use the machine for 13 years and then will sell the machine for scrap, receiving $20,000. The company interest rate is 12%. Should the machine be purchased?

5-32 A corporate bond has a face value of $1000 with maturity date 20 years from today. The bond pays interest semi-annually at a rate of 8% a year based on the face value. The interest rate paid on similar corporate bonds has decreased to a current rate of 6%. Determine the market value of the bond.

5-33 Calculate the present worth of a 4.5%, $5000 bond with interest paid semi-annually. The bond matures in 10 years, and the investor wants to make 8% a year compounded quarterly on the investment.

5-34 A rather wealthy man decided he would like to arrange for his descendants to be well educated. He would like each child to have $60,000 for his or her education. He plans to set up a perpetual trust fund so that six children will receive this assistance in each generation. He estimates that there will be four generations per century, spaced 25 years apart. He expects the trust to be able to obtain a 4% rate of return, and the first recipients to receive the money 10 years hence. How much money should he now set aside in the trust? (*Answer:* $389,150)

5-35 The president of the E. L. Echo Corporation thought it would be appropriate for his firm to endow a chair in the Department of Industrial Engineering of the local university; that is, he was considering giving the university enough money to pay the salary of one professor forever. That professor, who would be designated the E. L. Echo Professor of Industrial Engineering, would be paid from the fund established by the Echo Corporation. If the professor holding that chair receives $67,000 a year, and the interest received on the endowment fund is expected to remain at 8%, what lump sum of money will the Echo Corporation need to provide to establish the endowment fund? (*Answer:* $837,500)

5-36 A man who likes cherry blossoms very much would like to have an urn full of them put on his grave once each year forever after he dies. In his will, he intends to leave a certain sum of money in the trust of a local bank to pay the florist's annual bill. How much money should be left for this purpose? Make whatever assumptions you feel are justified by the facts presented. State your assumptions, and compute a solution.

5-37 A home builder must construct a sewage treatment plant and deposit sufficient money in a perpetual trust fund to pay the $5000 per year operating cost and to replace the treatment plant every 40 years. The plant will cost $150,000, and future replacement plants will also cost $150,000 each. If the trust fund earns

8% interest, what is the builder's capitalized cost to construct the plant and future replacements, and to pay the operating costs?

5-38 The local botanical society wants to ensure that the gardens in the town park are properly cared for. They recently spent $100,000 to plant the gardens. They would like to set up a perpetual fund to provide $100,000 for future replantings of the gardens every 10 years. If interest is 5%, how much money would be needed to forever pay the cost of replanting?

5-39 An elderly lady decided to donate most of her considerable wealth to charity and to keep for herself only enough money to provide for her living. She feels that $1000 a month will amply provide for her needs. She will establish a trust fund at a bank that pays 6% interest, compounded monthly. At the end of each month she will withdraw $1000. She has arranged that, upon her death, the balance in the account is to be paid to her niece, Susan. If she opens the trust fund and deposits enough money to pay herself $1000 a month in interest as long as she lives, how much will Susan receive when her aunt dies?

5-40 What amount of money deposited 50 years ago at 8% interest would provide a perpetual payment of $10,000 a year beginning this year?

5-41 A small dam was constructed for $2 million. The annual maintenance cost is $15,000. If interest is 5%, compute the capitalized cost of the dam, including maintenance.

5-42 A depositor puts $25,000 in a savings account that pays 5% interest, compounded semi-annually. Equal annual withdrawals are to be made from the account, beginning one year from now and continuing forever. What is the maximum annual withdrawal?

5-43 A trust fund is to be established for three purposes: (1) to provide $750,000 for the construction and $250,000 for the initial equipment of a small engineering laboratory; (2) to pay the $150,000 a year laboratory operating cost; and (3) to pay for $100,000 of replacement equipment every 4 years, beginning 4 years from now.

At 6% interest, how much money is required in the trust fund to provide for the laboratory and equipment and its perpetual operation and equipment replacement?

5-44 The local Audubon Society has just put a new bird feeder in the park at a cost of $500. The feeder has a useful life of 5 years and an annual maintenance cost

of $50. Our cat, Fred, was very impressed with the project. He wants to establish a fund that will maintain the feeder in perpetuity (that's forever!). Replacement feeders cost $500 every 5 years. If the fund earns 5% interest, what amount must he raise for its establishment? Note that both maintenance and replacement costs following the intial investment must be covered.

5-45 We want to donate a marble birdbath to the city park as a memorial to our cat, Fred, while he can still enjoy it. We also want to set up a perpetual care fund to cover future expenses forever. The initial cost of the bath is $5000. Routine annual operating costs are $200 a year, but every fifth year the cost will be $500 to cover major cleaning and maintenance as well as operation.
(*a*) What is the capitalized cost of this project if the interest rate is 8%?
(*b*) How much is the present worth of this project if it is to be demolished after 75 years? The final $500 payment in the 75th year will cover the year's operating cost and the site reclamation.

5-46 Dr. Fog E. Professor is retiring and wants to endow a chair of engineering economics at his university. It is expected that he will need to cover an annual cost of $100,000 forever. What lump sum must he donate to the university today if the endowment will earn 10% interest?

5-47 A man had to have the muffler replaced on his 2-year-old car. The repairman offered two alternatives. For $50 he would install a muffler guaranteed for 2 years. But for $65 he would install a muffler guaranteed "for as long as you own the car." Assuming the present owner expects to keep the car for about 3 more years, which muffler would you advise him to have installed if you thought 20% was a suitable interest rate and the less expensive muffler would last only 2 years?

5-48 A consulting engineer has been engaged to advise a town how best to proceed with the construction of a 200,000 m^3 water supply reservoir. Since only 120,000 m^3 of storage will be required for the next 25 years, an alternative to building the full capacity now is to build the reservoir in two stages. Initially, the reservoir could be built with 120,000 m^3 of capacity, and then, 25 years hence, the additional 80,000 m^3 of capacity could be added by increasing the height of the reservoir. Estimated costs are as follows:

	Construction Cost	Annual Maintenance Cost
Build in two stages		
First stage: 120,000 m^3 reservoir	$14,200,000	$75,000
Second stage: Add 80,000 m^3 of capacity, additional construction and maintenance costs	$12,600,000	$25,000
Build full capacity now 200,000 m^3 reservoir	$22,400,000	$100,000

If interest is computed at 4%, which construction plan is preferred?

5-49 A weekly business magazine offers a 1-year subscription for $58 and a 3-year subscription for $116. If you thought you would read the magazine for at least the next 3 years, and you considered 20% to be a minimum rate of return, which way would you purchase the magazine: with three 1-year subscriptions or a single 3-year subscription? (*Answer:* Choose the 3-year subscription.)

5-50 Use an 8-year analysis period and a 10% interest rate to determine which alternative should be selected:

	A	B
First cost	$5300	$10,700
Uniform annual benefit	$1800	$2,100
Useful life, in years	4	8

5-51 Two alternative courses of action have the following schedules of disbursements:

Year	A	B
0	−$1300	
1	0	−$100
2	0	−200
3	0	−300
4	0	−400
5	0	−500
	−$1300	−1500

If the interest rate is 6%, which alternative should be selected?

5-52 Telefono Mexico is expanding its facilities to serve a new manufacturing plant. The new plant will require 2000 telephone lines this year, and another 2000 lines

after expansion in 10 years. The plant will be in operation for 30 years. The telephone company is evaluating two options to serve the demand.

Option 1 Provide one cable now with capacity to serve 4000 lines. The cable cost will be $200,000, and annual maintenance costs will be $15,000.

Option 2 Provide a cable with capacity to serve 2000 lines now and a second cable to serve the other 2000 lines in 10 years. The cost of each cable will be $150,000, and each cable will have an annual maintenance of $10,000.

The telephone cables will last at least 30 years, and the cost of removing the cables is offset by their salvage value.

(*a*) Which alternative should be selected assuming a 10% interest rate?

(*b*) Will your answer to (*a*) change if the demand for additional lines occurs in 5 years instead of 10 years?

5-53 Dick Dickerson Construction Inc. has asked to you help them select a new backhoe. You have a choice between a wheel-mounted version, which costs $50,000 and has an expected life of 5 years, and a salvage value of $2000, and a track-mounted one, which costs $80,000, with a 5-year life and an expected salvage value of $10,000. Both machines will achieve the same productivity. Interest is 8%. Which one will you recommend? Use a present worth analysis.

5-54 Walt Wallace Construction Enterprises is investigating the purchase of a new dump truck. Interest is 9%. The cash flows for two likely models are as follows:

Model	First Cost	Annual Operating Cost	Annual Income	Salvage Value	Life
A	$50,000	$2000	$9,000	$10,000	10 yr
B	$80,000	$1000	$12,000	$30,000	10 yr

(*a*) Using present worth analysis, decide which truck should the firm buy, and explain why.

(*b*) Before the construction company can close the deal, the dealer sells out of Model B and cannot get any more. What should the firm do now and why?

5-55 Two different companies are offering a punch press for sale. Company A charges $250,000 to deliver and install the device. Company A has estimated that the machine will have maintenance and operating costs of $4000 a year and will provide an annual benefit of $89,000. Company B charges $205,000 to deliver and install the device. Company B has estimated maintenance and operating costs of the press at $4300 a year, with an annual benefit of $86,000. Both machines will last 5 years and can be sold for $15,000 for the scrap metal. Use an interest rate of 12%. Which machine should your company buy?

5-56 A battery manufacturing plant has been ordered to cease discharging acidic waste liquids containing mercury into the city sewer system. As a result, the firm must now adjust the pH and remove the mercury from its waste liquids. Three firms have provided quotations on the necessary equipment. An analysis of the quotations provided the following table of costs.

Bidder	Installed Cost	Annual Operating Cost	Annual Income from Mercury Recovery	Salvage Value
Foxhill Instrument	$ 35,000	$8000	$2000	$20,000
Quicksilver	40,000	7000	2200	0
Almaden	100,000	2000	3500	0

If the installation can be expected to last 20 years and money is worth 7%, which equipment should be purchased? (*Answer:* Almaden)

5-57 A firm is considering three mutually exclusive alternatives as part of a production improvement program. The alternatives are:

	A	B	C
Installed cost	$10,000	$15,000	$20,000
Uniform annual benefit	$1,625	$1,530	$1,890
Useful life, in years	10	20	20

The salvage value at the end of the useful life of each alternative is zero. At the end of 10 years, Alternative A could be replaced with another A with identical cost and benefits. The maximum attractive rate of return is 6%. Which alternative should be selected?

5-58 A steam boiler is needed as part of the design of a new plant. The boiler can be fired by natural gas, fuel oil, or coal. A decision must be made on which fuel to use. An analysis of the costs shows that the installed cost, with all controls, would be least for natural gas at $30,000; for fuel oil it would be $55,000; and for coal it would be $180,000. If natural gas is used rather than fuel oil, the annual fuel cost will increase by $7500. If coal is used rather than fuel oil, the annual fuel cost will be $15,000 per year less. Assuming 8% interest, a 20-year analysis period, and no salvage value, which is the most economical installation?

5-59 The General Hospital is evaluating new office equipment offered by three companies. In each case the interest rate is 15% and the useful life of the equipment is 4 years. Use NPW analysis to determine the company from which you should purchase the equipment.

	Company A	Company B	Company C
First cost	$15,000	$25,000	$20,000
Maintenance and operating costs	1,600	400	900
Annual benefit	8,000	13,000	11,000
Salvage value	3,000	6,000	4,500

5-60 The following costs are associated with three tomato-peeling machines being considered for use in a canning plant.

	Machine A	Machine B	Machine C
First cost	$52,000	$63,000	$67,000
Maintenance and operating costs	$15,000	$9,000	$12,000
Annual benefit	$38,000	$31,000	$37,000
Salvage value	$13,000	$19,000	$22,000
Useful life, in years	4	6	12

If the canning company uses an interest rate of 12%, which is the best alternative? Use NPW to make your decision. (*Note:* Consider the least common multiple as the study period.)

5-61 A railway branch line to a landfill site is to be constructed. It is expected that the railway line will be used for 15 years, after which the land will be returned to agricultural use. The railway track and ties will be removed at that time.

In building the railway line, either treated or untreated wood ties may be used. Treated ties have an installed cost of $6 and a 10-year life; untreated ties are $4.50 with a 6-year life. If at the end of 15 years the ties then in place have a remaining useful life of

4 years or more, they will be used by the railway else-where and have an estimated salvage value of $3 each. Any ties that are removed at the end of their service life, or too close to the end of their service life to be used elsewhere, can be sold for $0.50 each.

Determine the most economical plan for the initial railway ties and their replacement for the 15-year period. Make a present worth analysis assuming 8% interest.

5-62 Consider A–E, five mutually exclusive alternatives:

	A	**B**	**C**	**D**	**E**
Initial cost	$600	$600	$600	$600	$600
Uniform annual benefits for first 5 years	100	100	100	150	150
For last 5 years	50	100	110	0	50

The interest rate is 10%. If all the alternatives have a 10-year useful life, and no salvage value, which alternative should be selected?

5-63 An investor has carefully studied a number of companies and their common shares. From his analysis, he has decided that the shares of six firms are the best of the many he has examined. They represent about the same amount of risk, and so he would like to choose one single share in which to invest. He plans to keep the shares for 4 years, and he requires a 10% minimum attractive rate of return.

Which share from Table P5-63, if any, should the investor consider buying? (*Answer:* Spartan Products)

5-64 Six mutually exclusive alternatives, A–F (see Table P5-64), are being examined. For an 8% interest

rate, which alternative should be selected? Each alternative has a six-year useful life.

5-65 The management of an electronics manufacturing firm believes it is desirable to automate its production facility. The automated equipment would have a 10-year life with no salvage value at the end of 10 years. The plant engineering department has surveyed the plant and suggested there are eight mutually exclusive alternatives available.

Plan	Initial Cost (thousands)	Net Annual Benefit (thousands)
1	$265	$51
2	220	39
3	180	26
4	100	15
5	305	57
6	130	23
7	245	47
8	165	33

If the firm expects a 10% rate of return, which plan, if any, should it adopt? (*Answer:* Plan 1)

5-66 A local symphony association offers memberships as follows:

Continuing membership, per year	$ 15
Patron lifetime membership	375

The patron membership has been based on the symphony association's belief that it can obtain a 4% rate of return on its investment. If you believed 4% to be an appropriate rate of return, would you be willing to purchase the patron membership? Explain why or why not.

TABLE P5-63

Common Stock	Price per Share	Annual End-of-Year Dividend per Share	Estimated Price at End of 4 Years
Western House	$23³/₄	$1.25	$32
Fine Foods	45	4.50	45
Mobile Motors	30⁵/₈	0	42
Spartan Products	12	0	20
U.S. Tire	33³/₈	2.00	40
Wine Products	52¹/₂	3.00	60

TABLE P5-64

	A	**B**	**C**	**D**	**E**	**F**
Initial cost	$20.00	$35.00	$55.00	$60.00	$80.00	$100.00
Uniform annual benefit	6.00	9.25	13.38	13.78	24.32	24.32

5-67 Using capitalized cost, determine which type of road surface is preferred on a particular section of highway. Use 12% interest rate.

	A	B
Initial cost	$500,000	$700,000
Annual maintenance	35,000	25,000
Periodic resurfacing	350,000	450,000
Resurfacing interval	10 years	15 years

5-68 A city has developed a plan to provide for future municipal water needs. The plan proposes an aqueduct that passes through 500 feet of tunnel in a nearby mountain. Two alternatives are being considered. The first proposes to build a full-capacity tunnel now for $556,000. The second proposes to build a half-capacity tunnel now (cost = $402,000), which should be adequate for 20 years, and then to build a second parallel half-capacity tunnel. The maintenance cost of the tunnel lining for the full-capacity tunnel is $40,000 every 10 years, and for each half-capacity tunnel it is $32,000 every 10 years.

The friction losses in the half-capacity tunnel will be greater than if the full-capacity tunnel were built. The estimated additional pumping costs in the single half-capacity tunnel will be $2000 per year, and for the two half-capacity tunnels it will be $4000 per year. On the basis of capitalized cost and a 7% interest rate, which alternative should be selected?

5-69 An engineer has received two bids for an elevator to be installed in a new building. The bids, plus his evaluation of the elevators, are shown in Table P5-69.

The engineer will make a present worth analysis using a 10% interest rate. Prepare the analysis and determine which bid should be accepted.

5-70 A building contractor obtained bids for some asphalt paving, based on a specification. Three paving subcontractors quoted the following prices and terms of payment:

Paving Co.	Price	Payment Schedule
Quick	$85,000	50% payable immediately 25% payable in 6 months 25% payable at the end of one year
Tartan	$82,000	Payable immediately
Faultless	$84,000	25% payable immediately 75% payable in 6 months

The building contractor uses a 12% nominal interest rate, compounded monthly, in this type of bid analysis. Which paving subcontractor should be awarded the paving job?

5-71 Given the following data, use present worth analysis to find the best alternative, A, B, or C.

	A	B	C
Initial cost	$10,000	$15,000	$12,000
Annual benefit	6,000	10,000	5,000
Salvage value	1,000	−2,000	3,000
Useful life	2 years	3 years	4 years

Use an analysis period of 12 years and 10% interest.

5-72 Consider the following four alternatives. Three are "do something" and one is "do nothing."

	A	B	C	D
Cost	$0	$50	$30	$40
Net annual benefit	0	12	4.5	6
Useful life, in years		5	10	10

At the end of the 5-year useful life of B, a replacement is not made. If a 10-year analysis period and a 10% interest rate are selected, which is the preferred alternative?

TABLE P5-69

			Engineer's Estimates	
Alternatives	Bids: Installed Cost	Service Life (years)	Annual Operating Cost, Including Repairs	Salvage Value at End of Service Life
Westinghome	$45,000	10	$2700 a year	$3000
Itis	54,000	15	2850 a year	4500

5-73 A cost analysis is to be made to determine what, if anything, should be done in a situation offering three 'do-something' and one 'do-nothing' alternatives. Estimates of the cost and benefits are as follows.

Alternatives	Cost	Uniform Annual Benefit	End-of-Useful-Life Salvage Value	Useful Life (years)
1	$500	$135	$ 0	5
2	600	100	250	5
3	700	100	180	10
4	0	0	0	0

Use a 10-year analysis period for the four mutually exclusive alternatives. At the end of 5 years, Alternatives 1 and 2 may be replaced with identical alternatives (with the same cost, benefits, salvage value, and useful life).

(*a*) If an 8% interest rate is used, which alternative should be selected?

(*b*) If a 12% interest rate is used, which alternative should be selected?

5-74 Assume monthly car payments of $500 a month for 4 years and an interest rate of 1/2% a month. What initial principal or PW will this repay?

5-75 Assume annual car payments of $6000 for 4 years and an interest rate of 6% a year. What initial principal or PW will this repay?

5-76 Assume annual car payments of $6000 for 4 years and an interest rate of 6.168% a year. What initial principal or PW will this repay?

5-77 Why do the values in Problems 5-74, 5-75, and 5-76 differ?

5-78 Assume mortgage payments of $1000 per month for 30 years and an interest rate of 1/2% per month. What initial principal or PW will this repay?

5-79 Assume annual mortgage payments of $12,000 for 30 years and an interest rate of 6% per year. What initial principal or PW will this repay?

5-80 Assume annual mortgage payments of $12,000 for 30 years and an interest rate of 6.168% per year. What initial principal or PW will this repay?

5-81 Why do the values in Problems 5-78, 5-79, and 5-80 differ?

5-82 Ding Bell Imports requires a return of 15% on all projects. If Ding is planning an overseas development project with the cash flows shown in Table P5-82, what is the project's net present value?

5-83 Maverick Enterprises is planning a new product. Annual sales, unit costs, and unit revenues are as tabulated; the first cost of R&D and setting up the assembly line is $42,000. If *i* is 10%, what is the PW?

Year	Annual Sales	Cost/unit	Price/unit
1	$ 5000	$3.50	$6
2	6000	3.25	5.75
3	9000	3.00	5.50
4	10,000	2.75	5.25
5	8000	2.5	4.5
6	4000	2.25	3

5-84 Northern Engineering is analyzing a mining project. Annual production, unit costs, and unit revenues are in the table. The first cost of the mine setup is $8 million. If *i* is 15%, what is the PW?

Year	Annual Production (tons)	Cost per ton	Price per ton
1	70,000	$25	$35
2	90,000	20	34
3	120,000	22	33
4	100,000	24	34
5	80,000	26	35
6	60,000	28	36
7	40,000	30	37

TABLE P5-82

Year	0	1	2	3	4	5	6	7
Net cash ($)	0	−120,000	−60,000	20,000	40,000	80,000	100,000	60,000

CHAPTER 6

Annual Cash Flow Analysis

Lowest Prices on the Net! Buy Now!

Next time you review your e-mail and scroll through the spam, take a look at how many unsolicited messages are offering cut-rate ink cartridges for your printer.

Strange, isn't it? Why would spam pests imagine they can capture your attention with ads for cheap printer ink, especially when competing spammers are advertising far more (shall we say) intriguing products?

The answer becomes clear when you look at the price of an ink cartridge for a typical inkjet printer. Cartridges often cost $30 or more, and they need to be replaced frequently. By contrast, a good quality inkjet printer can often be purchased for under $100.

After Completing This Chapter...

The student should be able to:

- Define *equivalent uniform annual cost (EUAC)* and *equivalent uniform annual benefits (EUAB)*.
- Resolve an engineering economic analysis problem into its annual cash flow equivalent.
- Conduct an *annual worth analysis* for a single investment.
- Use EUAC and EUAB to compare alternatives with equal, common multiple, or continuous lives, or over some fixed study period.
- Develop and use spreadsheets to analyze loans for purposes of building an amortization table, calculating interest versus principal, finding the balance due, and determining whether to pay off a loan early.

QUESTIONS TO CONSIDER

1. Retailers frequently attract buyers by advertising low sale prices for printers, but they almost never mention the cost of the ink cartridges. In analyzing the likely cost of operating a printer over its useful lifetime, how much weight should the buyer give to the price of the printer itself, and how much to the cost of the ink cartridges?
2. King Camp Gillette, inventor of the safety razor, designed his product to work with a specially crafted blade that did not need sharpening and could be disposed of after a few uses. Gillette eventually announced, to everyone's astonishment, that he would be giving his razors away free. Many people couldn't imagine how he could possibly make money this way. In fact, Gillette revolutionized marketing, and his business revenue soared. Can you explain why?
3. Would Gillette's strategy work with a product such as, say, a car? Why or why not?

This chapter is devoted to annual cash flow analysis—the second of the three major analysis techniques. As we've said, alternatives must be resolved into a form that allows them to be compared. This means we must use the equivalence concept to convert from a cash flow representing the alternative into some equivalent sum or equivalent cash flow.

With present worth analysis, we resolved an alternative into an equivalent cash sum. This might have been an equivalent present worth of cost, an equivalent present worth of benefit, or an equivalent net present worth. Here we compare alternatives on the basis of their equivalent annual cash flows. Depending on the particular situation, we may wish to compute the equivalent uniform annual cost (EUAC), the equivalent uniform annual benefit (EUAB), or their difference (EUAB − EUAC).

To prepare for a discussion of annual cash flow analysis, we will review some annual cash flow calculations, then examine annual cash flow criteria. Following this, we will proceed with annual cash flow analysis.

ANNUAL CASH FLOW CALCULATIONS

Resolving a Present Cost to an Annual Cost

Equivalence techniques were used in prior chapters to convert money, at one point in time, to some equivalent sum or series. In annual cash flow analysis, the goal is to convert money to an equivalent uniform annual cost or benefit. The simplest case is to convert a present sum P to a series of equivalent uniform end-of-period cash flows. This is illustrated in Example 6-1.

EXAMPLE 6-1

A student bought $1000 worth of home furniture. If it is expected to last 10 years, what will the equivalent uniform annual cost be if interest is 7%?

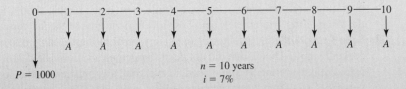

$$n = 10 \text{ years}$$
$$i = 7\%$$
$$P = 1000$$

SOLUTION

$$\text{Equivalent uniform annual cost} = P(A/P, i, n)$$
$$= 1000(A/P, 7\%, 10)$$
$$= \$142.40$$

Equivalent uniform annual cost is $142.40.

Treatment of Salvage Value

When there is a salvage value, or future value at the end of the useful life of an asset, the result is to decrease the equivalent uniform annual cost.

EXAMPLE 6-2

The student in Example 6-1 now believes the furniture can be sold at the end of 10 years for $200. Under these circumstances, what is the equivalent uniform annual cost?

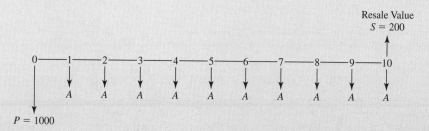

SOLUTION

For this situation, the problem may be solved by means of three different calculations.

SOLUTION 1

$$EUAC = P(A/P, i, n) - S(A/F, i, n) \tag{6-1}$$

$$= 1000(A/P, 7\%, 10) - 200(A/F, 7\%, 10)$$

$$= 1000(0.1424) - 200(0.0724)$$

$$= 142.40 - 14.48 = \$127.92$$

This method reflects the annual cost of the cash disbursement minus the annual benefit of the future resale value.

SOLUTION 2

Equation 6-1 describes a relationship that may be modified by an identity presented in Chapter 4:

$$(A/P, i, n) = (A/F, i, n) + i \tag{6-2}$$

Substituting this into Equation 6-1 gives:

$$EUAC = P(A/F, i, n) + Pi - S(A/F, i, n)$$

$$= (P - S)(A/F, i, n) + Pi \tag{6-3}$$

$$= (1000 - 200)(A/F, 7\%, 10) + 1000(0.07)$$

$$= 800(0.0724) + 70 = 57.92 + 70$$

$$= \$127.92$$

This method computes the equivalent annual cost due to the unrecovered $800 when the furniture is sold, and adds annual interest on the $1000 investment.

SOLUTION 3

If the value for $(A/F, i, n)$ from Equation 6-2 is substituted into Equation 6-1, we obtain:

$$\text{EUAC} = P(A/P, i, n) - S(A/P, i, n) + Si$$

$$= (P - S)(A/P, i, n) + Si \qquad (6\text{-}4)$$

$$= (1000 - 200)(A/P, 7\%, 10) + 200(0.07)$$

$$= 800(0.1424) + 14 = 113.92 + 14 = \$127.92$$

This method computes the annual cost of the $800 decline in value during the 10 years, plus interest on the $200 tied up in the furniture as the salvage value.

When there is an initial disbursement P followed by a salvage value S, the annual cost may be computed in the three different ways introduced in Example 6-2:

$$\textbf{EUAC} = P(A/P, i, n) - S(A/F, i, n) \qquad (6\text{-}1)$$

$$\textbf{EUAC} = (P - S)(A/F, i, n) + Pi \qquad (6\text{-}3)$$

$$\textbf{EUAC} = (P - S)(A/P, i, n) + Si \qquad (6\text{-}4)$$

Each of the three calculations gives the same results. In practice, the first and third methods are most commonly used. The EUAC calculated in Equations 6-1, 6-3, and 6-4 is also known as the *capital recovery cost* of a project.

EXAMPLE 6-3

Bill owned a car for 5 years. One day he wondered what his uniform annual cost for maintenance and repairs had been. He assembled the following data:

Year	Maintenance and Repair Cost for Year
1	$ 45
2	90
3	180
4	135
5	225

Compute the equivalent uniform annual cost (EUAC) assuming 7% interest and end-of-year disbursements.

SOLUTION

The EUAC may be computed for this irregular series of payments in two steps:

1. Use single payment present worth factors to compute the present worth of cost for the 5 years.
2. With the PW of cost known, use the capital recovery factor to compute EUAC.

$$\text{PW of cost} = 45(P/F, 7\%, 1) + 90(P/F, 7\%, 2) + 180(P/F, 7\%, 3)$$
$$+ 135(P/F, 7\%, 4) + 225(P/F, 7\%, 5)$$
$$= 45(0.9346) + 90(0.8734) + 180(0.8163) + 135(0.7629) + 225(0.7130)$$
$$= \$531$$
$$\text{EUAC} = 531(A/P, 7\%, 5) = 531(0.2439) = \$130$$

EXAMPLE 6-4

Bill re-examined his calculations and found that in his table he had reversed the maintenance and repair costs for years 3 and 4. This is the correct table.

Year	Maintenance and Repair Cost for Year
1	$ 45
2	90
3	135
4	180
5	225

Recompute the EUAC.

SOLUTION

This time the schedule of disbursements is an arithmetic gradient series plus a uniform annual cost, as follows:

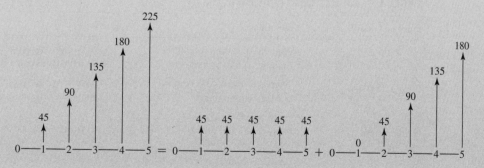

$$\text{EUAC} = 45 + 45(A/G, 7\%, 5)$$

$$= 45 + 45(1.865)$$

$$= \$129$$

Since the timing of the expenditures is different in Examples 6-3 and 6-4, we would not expect to obtain the same EUAC.

The examples have shown four essential points concerning cash flow calculations:

1. There is a direct relationship between the present worth of cost and the equivalent uniform annual cost. It is

$$\text{EUAC} = (\text{PW of cost})(A/P, i, n)$$

2. In a problem, an expenditure of money increases the EUAC, whereas a receipt of money (obtained, for example, from selling an item for its salvage value) decreases EUAC.

3. When there are irregular cash disbursements over the analysis period, a convenient method of solution is to first determine the PW of cost; then, using the equation in Item 1 above, the EUAC may be calculated.

4. Where there is an arithmetic gradient, EUAC may be rapidly computed by using the arithmetic gradient uniform series factor, $(A/G, i, n)$.

ANNUAL CASH FLOW ANALYSIS

The rules for economic efficiency are presented in Table 6-1. One notices immediately that the table is quite similar to Table 5-1. In the case of fixed input, for example, the present worth rule is *maximize PW of benefits,* and the annual cost rule is *maximize equivalent uniform annual benefits.* It is apparent that, if you are maximizing the present worth of benefits, simultaneously you must be maximizing the equivalent uniform annual benefits. This is illustrated in Example 6-5.

TABLE 6-1 Annual Cash Flow Analysis

Input or Output	Situation	Rule
Fixed input	Amount of money or other input resources is fixed	Maximize equivalent uniform benefits (maximize EUAB)
Fixed output	There is a fixed task, benefit, or other output to be accomplished	Minimize equivalent uniform annual cost (minimize EUAC)
Neither input nor output is fixed	Neither amount of money or other inputs, nor amount of benefits or other outputs is fixed	Maximize (EUAB − EUAC)

EXAMPLE 6-5

A firm is considering which of two devices to install to reduce costs in a particular situation. Both devices cost $1000 and have useful lives of 5 years with no salvage value. Device A can be expected to result in $300 savings annually. Device B will provide cost savings of $400 the first year but will decline $50 annually, making the second year savings $350, the third year savings $300, and so forth. With interest at 7%, which device should the firm purchase?

SOLUTION

Device A

$$EUAB = \$300$$

Device B

$$EUAB = 400 - 50(A/G, 7\%, 5)$$
$$= 400 - 50(1.865)$$
$$= \$306.75$$

To maximize EUAB, select Device B.

Example 6-5 was presented earlier, as Example 5-1, where we found:

$$PW \text{ of benefits } A = 300(P/A, 7\%, 5)$$
$$= 300(4.100) = \$1230$$

This is converted to EUAB by multiplying by the capital recovery factor:

$$EUAB_A = 1230(A/P, 7\%, 5) = 1230(0.2439) = \$300$$
$$PW \text{ of benefits } B = 400(P/A, 7\%, 5) - 50(P/G, 7\%, 5)$$
$$= 400(4.100) - 50(7.647) = \$1257.65$$

and, hence,

$$EUAB_B = 1257.65(A/P, 7\%, 5) = 1257.65(0.2439)$$
$$= \$306.75$$

We see, therefore, that it is easy to convert the present worth analysis results into the annual cash flow analysis results. We could go from annual cash flow to present worth just as easily, by using the series present worth factor. And, of course, both methods show that Device B is the preferred alternative.

EXAMPLE 6-6

Three alternatives are being considered for improving an operation on the assembly line along with the "do nothing" alternative. Equipment costs vary, as do the annual benefits of each in comparison to the present situation. Each of Plans A, B, and C has a 10-year life and a scrap value equal to 10% of its original cost.

	Plan A	Plan B	Plan C
Installed cost of equipment	$15,000	$25,000	$33,000
Material and labour savings per year	14,000	9,000	14,000
Annual operating expenses	8,000	6,000	6,000
End-of-useful life scrap value	1,500	2,500	3,300

If interest is 8%, which plan, if any, should be adopted?

SOLUTION

Since neither installed cost nor output benefits are fixed, the economic criterion is to maximize EUAB − EUAC.

	Plan A	Plan B	Plan C	Do Nothing
Equivalent uniform annual benefit (EUAB)				
Material and labour per year	$14,000	$9,000	$14,000	$0
Scrap value (A/F, 8%, 10)	104	172	228	0
EUAB =	$14,104	$9,172	$14,228	$0
Equivalent uniform annual cost (EUAC)				
Installed cost (A/P, 8%, 10)	$ 2,235	$3,725	$ 4,917	$0
Annual operating expenses	8,000	6,000	6,000	0
EUAC =	$10,235	$9,725	$10,917	$0
(EUAB − EUAC) =	$ 3,869	−$ 553	$ 3,311	$0

Based on our rule to maximize EUAB − EUAC, Plan A is the best of the four alternatives. We note, however, that since the do-nothing alternative has EUAB − EUAC = 0, it is a more desirable alternative than Plan B.

ANALYSIS PERIOD

In Chapter 5, we saw that the analysis period is an important consideration in computing present worth comparisons. In such problems, a common analysis period must be used for each alternative. In annual cash flow comparisons, we again have the analysis period question. Example 6-7 will help in examining the problem.

EXAMPLE 6-7

Two pumps are being considered for purchase. If interest is 7%, which pump should be bought?

	Pump A	**Pump B**
Initial cost	$7000	$5000
End-of-useful-life salvage value	1500	1000
Useful life, in years	12	6

SOLUTION

The annual cost for 12 years of Pump A can be found by using Equation 6-4:

$$\text{EUAC} = (P - S)(A/P, i, n) + Si$$

$$= (7000 - 1500)(A/P, 7\%, 12) + 1500(0.07)$$

$$= 5500(0.1259) + 105 = \$797$$

Now compute the annual cost for 6 years of Pump B:

$$\text{EUAC} = (5000 - 1000)(A/P, 7\%, 6) + 1000(0.07)$$

$$= 4000(0.2098) + 70 = \$909$$

For a common analysis period of 12 years, we need to replace Pump B at the end of its 6-year useful life. If we assume that another pump B′ can be obtained, having the same $5000 initial cost, $1000 salvage value, and 6-year life, the cash flow will be as follows:

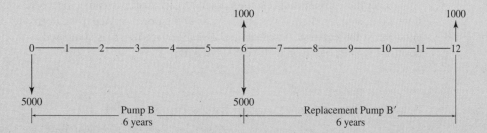

For the 12-year analysis period, the annual cost for Pump B is

$$\text{EUAC} = [5000 - 1000(P/F, 7\%, 6) + 5000(P/F, 7\%, 6)$$

$$- 1000(P/F, 7\%, 12)] \times (A/P, 7\%, 12)$$

$$= [5000 - 1000(0.6663) + 5000(0.6663) - 1000(0.4440)] \times (0.1259)$$

$$= (5000 - 666 + 3331 - 444)(0.1259)$$

$$= (7211)(0.1259) = \$909$$

The annual cost of B for the 6-year analysis period is the same as the annual cost for the 12-year analysis period. This is not a surprising conclusion when one recognizes that the annual cost of the first 6-year period is repeated in the second 6-year period. Thus the lengthy calculation of EUAC for 12 years of Pump B and B′ was not needed. By assuming that the shorter-life equipment is replaced by equipment with identical economic consequences, we have avoided a lot of calculations. Select Pump A.

Analysis Period Equal to Alternative Lives

In the ideal situation, the analysis period for an economy study coincides with the useful life for each alternative. The economy study is based on this analysis period.

Analysis Period a Common Multiple of Alternative Lives

When the analysis period is a common multiple of the alternative lives (for example, in Example 6-7, the analysis period was 12 years with 6- and 12-year alternative lives), a 'replacement with an identical item with the same costs, performance, and so forth' is frequently assumed. This means that when an alternative has reached the end of its useful life, we assume that it will be replaced with an identical item. As shown in Example 6-7, the result is that the EUAC for Pump B with a 6-year useful life is equal to the EUAC for the entire analysis period because we assume Pump B is replaced by Pump B′.

Under these circumstances of identical replacement, we can compare the annual cash flows computed for alternatives on the basis of their own service lives. In Example 6-7, the annual cost for Pump A, based on its 12-year service life, was compared with the annual cost for Pump B, based on its 6-year service life.

Analysis Period for a Continuing Requirement

Many times an economic analysis is undertaken to determine how to provide for a more or less continuing requirement. One might need to pump water from a well as a continuing requirement. There is no distinct analysis period. In this situation, the analysis period is assumed to be long but undefined.

If, for example, we had a continuing requirement to pump water and if alternative Pumps A and B had useful lives of 7 and 11 years respectively, what should we do? The customary assumption is that Pump A's annual cash flow (based on a 7-year life) may be compared to Pump B's annual cash flow (based on an 11-year life). This is done without much concern that the least common multiple of the 7- and 11-year lives is 77 years. This comparison of "different-life" alternatives assumes identical replacement (with identical costs, performance, etc.) when an alternative reaches the end of its useful life. Example 6-8 illustrates the situation.

EXAMPLE 6-8

Pump B in Example 6-7 is now believed to have a 9-year useful life. Assuming the same initial cost and salvage value, compare it with Pump A, using the same 7% interest rate.

SOLUTION

If we assume that the need for A or B will exist for some continuing period, the comparison of annual costs for the unequal lives is an acceptable technique. For 12 years of Pump A:

$$EUAC = (7000 - 1500)(A/P, 7\%, 12) + 1500(0.07) = \$797$$

For 9 years of Pump B:

$$EUAC = (5000 - 1000)(A/P, 7\%, 9) + 1000(0.07) = \$684$$

For minimum EUAC, select Pump B.

Infinite Analysis Period

At times we have an alternative with a limited (finite) useful life in an infinite analysis period situation. The equivalent uniform annual cost may be computed for the limited life. The assumption of identical replacement (replacements have identical costs, performance, etc.) is often correct. Thus, the same EUAC occurs for each replacement of the limited-life alternative. The EUAC for the infinite analysis period is therefore equal to the EUAC computed for the limited life. With identical replacement,

$$EUAC_{\text{infinite analysis period}} = EUAC_{\text{for limited life } n}$$

A somewhat different situation occurs when there is an alternative with an infinite life in a problem with an infinite analysis period:

$$EUAC_{\text{infinite analysis period}} = P(A/P, i, \infty) + \text{Any other annual costs}$$

When $n = \infty$, we have $A = Pi$ and, hence, $(A/P, i, \infty)$ equals i.

$$EUAC_{\text{infinite analysis period}} = Pi + \text{Any other annual costs}$$

EXAMPLE 6-9

In the construction of an aqueduct to expand the water supply of a city, there are two alternatives for a particular portion of the aqueduct. Either a tunnel can be constructed through a mountain, or a pipeline can be laid to go around the mountain. If there is a permanent need for the aqueduct, should the tunnel or the pipeline be chosen for this particular portion of the aqueduct? Assume a 6% interest rate.

	Tunnel through Mountain	**Pipeline around Mountain**
Initial cost	$5.5 million	$5 million
Maintenance	0	0
Useful life	Permanent	50 years
Salvage value	0	0

Tunnel

For the tunnel, with its permanent life, we want $(A/P, 6\%, \infty)$. For an infinite life, the capital recovery is simply the interest on the invested capital. So $(A/P, 6\%, \infty) = i$, and we write

$$\text{EUAC} = Pi = \$5.5 \text{ million}(0.06)$$

$$= \$330,000$$

Pipeline

$$\text{EUAC} = \$5 \text{ million}(A/P, 6\%, 50)$$

$$= \$5 \text{ million}(0.0634) = \$317,000$$

For fixed output, minimize EUAC. Select the pipeline.

The difference in annual cost between a long life and an infinite life is small unless an unusually low interest rate is used. In Example 6-9 the tunnel is assumed to be permanent. For purposes of comparison, compute the annual cost if an 85-year life is assumed for the tunnel.

$$\text{EUAC} = \$5.5 \text{ million}(A/P, 6\%, 85)$$

$$= \$5.5 \text{ million}(0.0604)$$

$$= \$332,000$$

The difference in time between 85 years and infinity is great indeed; yet the difference in annual costs in Example 6-9 is very small.

Some Other Analysis Period

The analysis period in a particular problem may be something other than one of the four we have so far described. It may be equal to the life of the shorter-life alternative, the longer-life alternative, or something entirely different. One must carefully examine the consequences of each alternative throughout the analysis period and, in addition, see what differences there might be in salvage values, and so forth, at the end of the analysis period.

EXAMPLE 6-10

Suppose that Alternative 1 has a 7-year life and a salvage value at the end of that time. The replacement cost at the end of 7 years may be more or less than the original cost. If the replacement is retired in less than 7 years, it will have a terminal value that exceeds the end-of-life salvage value. Alternative 2 has a 13-year life and a terminal value whenever it is retired. If the situation indicates that 10 years is the proper analysis period, set up the equations to compute the EUAC for each alternative.

SOLUTION

Alternative 1

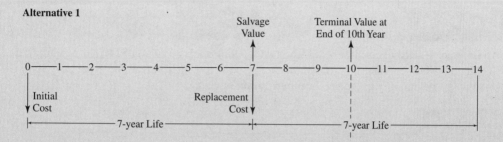

Alternative 2

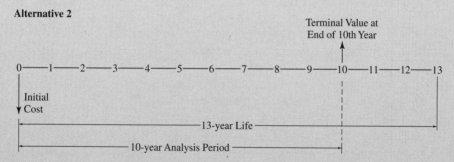

Alternative 1

$$\text{EUAC}_1 = [\text{Initial cost} + (\text{Replacement cost} - \text{Salvage value})(P/F, i, 7)$$

$$- (\text{Terminal value})(P/F, i, 10)](A/P, i, 10)$$

Alternative 2

$$\text{EUAC}_2 = [\text{Initial cost} - (\text{Terminal value})(P/F, i, 10)](A/P, i, 10)$$

USING SPREADSHEETS TO ANALYZE LOANS

Loan and bond payments are made by firms, agencies, and individual engineers. Usually, the payments in each period are constant. Spreadsheets make it easy to:

- Calculate the loan's amortization schedule
- Decide how a payment is to be split between principal and interest

- Find the balance due on a loan
- Calculate the number of payments remaining on a loan.

Building an Amortization Schedule

As illustrated in Chapter 4 and Appendix 1, an amortization schedule lists for each payment period: the loan payment, interest paid, principal paid, and remaining balance. For each period the interest paid equals the interest rate times the balance remaining from the period before. Then the principal payment equals the payment minus the interest paid. Finally, this principal payment is applied to the balance remaining from the preceding period to calculate the new remaining balance. As a basis for comparison with spreadsheet loan functions, Figure 6-1 shows this calculation for Example 6-11.

EXAMPLE 6-11

An engineer wanted to celebrate graduating and getting a job by buying $2400 of new furniture. Luckily the store was offering 6-month financing at the low interest rate of 6% per year nominal (really $1/2\%$ per month). Calculate the amortization schedule.

SOLUTION

	A	B	C	D	E
1	2400	Initial balance			
2	0.50%	i			
3	6	N			
4	$407.03	Payment	$= -\text{PMT(A2,A3,A1)}$		
5					
6			Principal	Ending	
7	Month	Interest	Payment	Balance	
8	0			2400.00	=A1
9	1	12.00	395.03	2004.97	=D8−C9
10	2	10.02	397.00	1607.97	
11	3	8.04	398.99	1208.98	
12	4	6.04	400.98	807.99	
13	5	4.04	402.99	405.00	
14	6	2.03	405.00	0.00	
15				=$A$4−B14	
16				=Payment−Interest	
17			=$A$2*D13		
18			=rate*previous balance		

FIGURE 6-1 Amortization schedule for furniture loan.

The first step is to calculate the monthly payment:

$$A = 2400(A/P, 1/2\%, 6) = 2400(0.1696) = \$407.0$$

With this information the engineer can use the spreadsheet of Figure 6-1 to obtain the amortization schedule.

How Much to Interest? How Much to Principal?

For a loan with constant payments, we can answer these questions for any period without the full amortization schedule. For a loan with constant payments, the functions IPMT and PPMT directly answer these questions. For simple problems, both functions have four arguments $(i, t, n, -P)$, where t is the time period being calculated. Both functions have optional arguments that permit us to add a balloon payment (an F) and change from end-of-period payments to beginning-of-period payments.

For example, consider Period 4 of Example 6-11. The spreadsheet formulas give the same answer as shown in Figure 6-1.

$$\text{Interest}_4 = \text{IPMT}(0.5\%, 4, 6, -2400) = \$6.04$$

$$\text{Principal payment}_4 = \text{PPMT}(0.5\%, 4, 6, -2400) = \$400.98$$

Finding the Balance Due on a Loan

An amortization schedule is one way to calculate the balance due on a loan. A second, easier way is to remember that the balance due equals the present worth of the remaining payments. Interest is paid in full after each payment, so later payments are simply based on the balance due.

EXAMPLE 6-12

A car is purchased with a 48-month, 9% nominal loan with an initial balance of $15,000. What is the balance due halfway through the 4 years?

SOLUTION

The first step is to calculate the monthly payment, at a monthly interest rate of $\frac{3}{4}\%$. This equals

$$\text{Payment} = 15,000(A/P, 0.75\%, 48) \qquad \text{or} \quad = \text{PMT}(0.75\%, 48, -15,000)$$

$$= (15,000)(0.0249) = \$373.50 \quad \text{or} \quad = \$373.28$$

The next step will use the spreadsheet answer, because it is more accurate (there are only three significant digits in the tabulated factor).

After 24 payments and with 24 left, the remaining balance equals $(P/A, i, N_{\text{remaining}})$ payment

$$\text{Balance} = (P/A, 0.75\%, 24)\$373.28 \qquad \text{or} \quad = \text{PV}(0.75\%, 24, 373.28)$$

$$= (21.889)(373.28) = \$8170.73 \quad \text{or} \quad = \$8170.78$$

Thus halfway through the repayment schedule, 54.5% of the original balance is still owed.

Pay Off Debt Sooner by Increasing Payments

Paying off debt can be a good investment because the investment earns the rate of interest on the loan. For example, this could be 8% for a mortgage, 10% for a car loan, or 19% for a credit card. When making extra payments on a loan, the common question is: how much sooner will the debt be paid off? Until the debt is paid off, any early payments are essentially locked up, since the same payment amount is owed each month.

The first reason that spreadsheets are convenient is fractional interest rates. For example, a car loan might be at a nominal rate of 13% with monthly compounding or 1.08333% per month. The second reason is that the function NPER calculates the number of periods remaining on a loan.

NPER can be used to calculate how much difference is made by one extra payment or by an increase of $x\%$ to all payments. Extra payments are applied entirely to principal, so the interest rate, remaining balance, and payment amounts are all known. $N_{remaining}$ equals NPER(i, payment, remaining balance) with optional arguments for beginning-of-period cash flows and balloon payments. The signs of the payment and the remaining balance must be different.

EXAMPLE 6-13

Maria has a 7.5% mortgage with monthly payments for 30 years. Her original balance was $100,000, and she just made her twelfth payment. Each month she also pays into a reserve account, which the bank uses to pay her fire and liability insurance ($900 annually) and property taxes ($1500 annually). By how much does she shorten the term of the loan if she makes an extra *loan* payment today? If she makes an extra *total* payment? If she increases each total payment to 110% of her current total payment?

SOLUTION

The first step is to calculate Maria's *loan* payment for the 360 months. Rather than calculating a six-significant digit monthly interest rate, it is easier to use 0.075/12 in the spreadsheet formulas.

$$\text{Payment} = \text{PMT}(0.075/12, 360, -100000) = \$699.21$$

The remaining balance after 12 such payments is the present worth of the remaining 348 payments.

$$\text{Balance}_{12} = \text{PV}(0.075/12, 348, 699.21) = \$99,077.53$$
(after 12 payments, she has paid off $922!)

If she pays an extra $699.21, then the number of periods remaining is

$$\text{NPER}(0.075/12, -699.21, 99077.53 - 699.21) = 339.5$$

This is 8.5 payments less than the 348 periods left before the extra payment. If she makes an extra total payment, then

$$\text{Total payment} = 699.21 + 900/12 + 1500/12 = \$899.21/\text{month}$$
$$\text{NPER}(0.075/12, -699.21, 99244 - 899.21) = 337.1$$

or 2.4 more payments saved. If she makes an extra 10% payment on the total payment of $899.21, then

$$\text{NPER}([0.075/12, -(1.1)(899.21 - 200)], 99077.53) = 246.5 \text{ payments}$$

or 101.5 payments saved.

Note that $200 of the total payment goes to pay for insurance and taxes.

MORTGAGES IN CANADA

Although technically a mortgage is a legal document, most people use the word to mean a long-term amortized loan that is used for buying real property such as a house or land. If the mortgage payments are not made, the lender can take the property and sell it to recover the outstanding debt.

A mortgage document outlines the terms and conditions for repaying the money borrowed: the amount borrowed, the interest rate, the first and last payment dates, the amortization period, and the date the balance is due (the renewal date or term). Prepayment options and penalties may also be included.

Amortization is the length of time it would take to pay off the mortgage—assuming that the interest rate never changed, all payments were made on time, and no additional payments were made. In Canada the shortest amortization is usually 5 years, and the longest is 40 years, although the norm is 20 to 25 years. The amortization of a mortgage is made up of smaller periods called 'terms'. A term can be anywhere from 3 months to 25 years. The term is the period of time during which the interest rate is fixed. At the end of the term, the mortgage can be renewed for a new term at the prevailing rates of interest.

Types of Mortgages Available

Before the advent of computers and imaginative interest calculations, most mortgages in Canada were either 'conventional mortgages' or 'high-ratio mortgages'. Conventional mortgages are for 75% (or less) of the appraised value of the property, and consequently they require the purchaser to make a down payment of at least 25%. High-ratio mortgages are those where the lender is providing an amount greater than 75% of the appraised value of the property. Generally, lenders will do this only if an outside agency, usually Central Mortgage and Housing Corporation (CMHC), will insure the mortgage. CMHC charges the borrower a fee for this insurance.

Today the following types of mortgages are also common: open mortgage, variable rate mortgage, ARM (adjustable rate mortgage), capped rate mortgage, closed mortgage, convertible rate mortgage, second mortgage, reverse mortgage, and CHIP mortgage—Canadian Home Income Plan.

Equity

Equity is the value remaining in your property after all mortgages and loans registered against the title are subtracted from the appraised value.

For example:

Appraised value $210,000
Minus mortgage $150,000
Minus second mortgage $25,000
equals Equity $35,000

Mortgage Interest Rate

In Canada the amount of interest charged to a borrower for residential first mortgages is stated as a percentage, for example, 8.25% 'compounded semi-annually, not in advance'. 'Compounded semi-annually' is an old English term, from the days when ledgers were kept by hand. The interest was calculated twice a year and added on to the principal owing, and then payments were subtracted. Even though we now use computers, the same method exists. Thus for a conventional Canadian mortgage the interest is stated as nominal, calculated semi-annually with a monthly payment period.

This means that a 6% mortgage is actually 3% semi-annually. To determine the actual effective monthly rate, one has to solve for i in the following equation:

$$(1 + i)^6 = 1.03$$
$$i = 0.49\% \text{ per month}$$

SUMMARY

Annual cash flow analysis is the second of the three major methods of resolving alternatives into comparable values. When an alternative has an initial cost P and salvage value S, there are three ways of computing the equivalent uniform annual cost:

- EUAC $= P(A/P, i, n) - S(A/F, i, n)$ (6-1)
- EUAC $= (P - S)(A/F, i, n) + Pi$ (6-3)
- EUAC $= (P - S)(A/P, i, n) + Si$ (6-4)

All three equations give the same answer. This quantity is also known as the *capital recovery cost* of the project.

The relationship between the present worth of cost and the equivalent uniform annual cost is:

- EUAC $= (\text{PW of cost})(A/P, i, n)$

The three annual cash flow criteria are:

For fixed input	Maximize EUAB
For fixed output	Minimize EUAC
Neither input nor output fixed	Maximize EUAB − EUAC

In present worth analysis there must be a common analysis period. Annual cash flow analysis, however, allows some flexibility provided the necessary assumptions are suitable in the situation being studied. The analysis period may be different from the lives of the alternatives, and provided the following criteria are met, a valid cash flow analysis may be made.

1. It is assumed that when an alternative has reached the end of its useful life, it will be replaced by an identical replacement (with the same costs, performance, etc.).
2. The analysis period is a common multiple of the useful lives of the alternatives, or there is a continuing or perpetual requirement for the selected alternative.

If neither condition applies, it is necessary to make a detailed study of the consequences of the various alternatives over the entire analysis period with particular attention to the difference between the alternatives at the end of the analysis period.

There is very little numerical difference between a long-life alternative and a perpetual alternative. As the value of n increases, the capital recovery factor approaches i. At the limit, $(A/P, i, \infty) = i$.

One of the most common uniform payment series is the repayment of loans. Spreadsheets are useful in analyzing loans (balance due, interest paid, etc.) for several reasons: they have specialized functions, many periods are easy, and any interest rate can be used.

PROBLEMS

6-1   Using a 10% interest rate, compute the value of C for the following diagram.

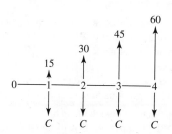

(*Answer: C = \$35.72*)

6-2   Compute the value of B for the following diagram:

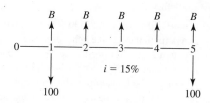

6-3   Compute the value of E:

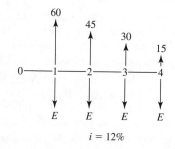

6-4   If $i = 6\%$, compute the value of D that is equivalent to the two disbursements shown.

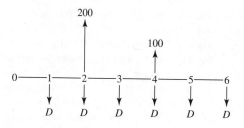

(*Answer: D = \$52.31*)

6-5   For the diagram, compute the value of D:

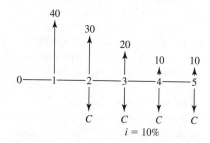

6-6   What is C in the accompanying figure?

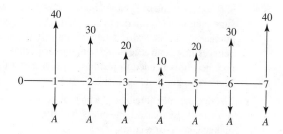

6-7   If interest is 10%, what is A?

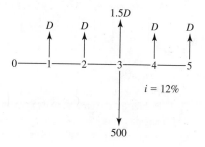

6-8   As shown in the cash flow diagram, there is an annual disbursement of money that varies from year to year from \$100 to \$300 in a fixed pattern that is repeated forever. If interest is 10%, compute the value of A, also continuing forever, that is equivalent to the fluctuating disbursements.

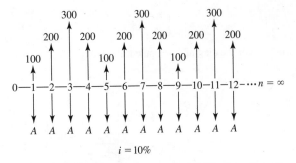

$i = 10\%$

6-9 On April 1 a man borrows $100. The loan is to be repaid in three equal semi-annual payments. If the annual interest rate is 7% compounded semi-annually, how much is each payment? (*Answer:* $35.69)

6-10 An electronics firm invested $60,000 in a precision inspection device. The device cost $4000 to operate and maintain in the first year and $3000 in each of the subsequent years. At the end of 4 years, the firm changed its inspection procedure, eliminating the need for the device. The purchasing agent was very fortunate in being able to sell the inspection device for $60,000, the original price. The plant manager asks you to compute the equivalent uniform annual cost of the device during the 4 years it was used. Assume interest at 10% per year. (*Answer:* $9287)

6-11 A firm is about to begin pilot plant operation on a process it has developed. One item of optional equipment that could be obtained is a heat exchanger unit. The company finds that a unit now available for $30,000 could be used in other company operations. It is estimated that the heat exchanger unit will be worth $35,000 at the end of 8 years. This seemingly high salvage value is due primarily to the fact that the $30,000 purchase price is really a rare bargain. If the firm believes 15% is an acceptable rate of return, what annual benefit is needed to justify the purchase of the heat exchanger unit? (*Answer:* $4135)

6-12 The maintenance foreman of a plant in reviewing his records found that a large press had the following maintenance cost record:

5 years ago	$ 600
4 years ago	700
3 years ago	800
2 years ago	900
Last year	1000

After consulting with a lubrication specialist, he changed the preventive maintenance schedule. He

believes that this year maintenance will be $900 and will decrease by $100 a year in each of the following 4 years. If his estimate of the future is correct, what will be the equivalent uniform annual maintenance cost for the 10-year period? Assume interest at 8%. (*Answer:* $756)

6-13 A firm purchased some equipment at a very favourable price of $30,000. The equipment resulted in an annual net saving of $1000 per year during the 8 years it was used. At the end of 8 years, the equipment was sold for $40,000. Assuming interest is 8%, did the equipment purchase prove to be desirable?

6-14 When he started work on his twenty-second birthday, D. B. Cooper decided to invest money each month with the objective of becoming a millionaire by the time he reaches age 65. If he expects his investments to yield 18% per annum, compounded monthly, how much should he invest each month? (*Answer:* $6.92 a month.)

6-15 Linda O'Shay deposited $30,000 in a savings account as a perpetual trust. She believes the account will earn 7% annual interest during the first 10 years and 5% interest thereafter. The trust is to provide a uniform end-of-year scholarship at the university. What uniform amount could be used for the student scholarship each year, beginning at the end of the first year and continuing forever?

6-16 A machine costs $20,000 and has a 5-year useful life. At the end of the 5 years, it can be sold for $4000. If annual interest is 8%, compounded semi-annually, what is the equivalent uniform annual cost of the machine? (An *exact* solution is expected.)

6-17 The average age of engineering students at graduation is a little over 23 years. This means that the working career of most engineers is almost exactly 500 months. How much would an engineer need to save each month to become a millionaire by the end of his working career? Assume a 15% interest rate, compounded monthly.

6-18 The Johnson Company pays $200 a month to a trucker to haul waste paper and cardboard to the city dump. The material could be recycled if the company were to buy a $6000 hydraulic press bailer and spend $3000 a year for labour to operate the bailer. The bailer has an estimated useful life of 30 years and no salvage value. Strapping material would cost $200 per year for the estimated 500 bales a year that would be produced. A waste paper company will pick up the bales at the plant and pay Johnson $2.30 per bale for them. Use an annual cash flow analysis in working this problem.

(a) If interest is 8%, is it economical to install and operate the bailer?

(b) Would you recommend that the bailer be installed?

6-19 An engineer has a fluctuating future budget for the maintenance of a particular machine. During each of the first 5 years, $1000 per year will be budgeted. During the second 5 years, the annual budget will be $1500 per year. In addition, $3500 will be budgeted for an overhaul of the machine at the end of the fourth year, and another $3500 for an overhaul at the end of the eighth year.

The engineer asks you to compute the uniform annual expenditure that would be equivalent to these fluctuating amounts, assuming interest is 6% per year.

6-20 An engineer wishes to have $5 million by the time he retires in 40 years. Assuming a nominal interest rate of 15%, compounded continuously, what annual sum must he set aside?   (*Answer:* $2011)

6-21 Art Arfons, an engineer, has made a considerable fortune. He wishes to start a perpetual scholarship for engineering students at his alma mater. The scholarship will provide a student with an annual stipend of $2500 for each of 4 years, plus an additional $5000 during the fourth year to cover entertainment expenses. Assume that students graduate in 4 years, a new award is given every 4 years, and the money is paid at the beginning of each year with the first award at the beginning of year one. The interest rate is 8%.

(a) Determine the equivalent uniform annual cost (EUAC) of providing the scholarship.

(b) How much money must Art donate to the university?

6-22 Jenny McCarthy is an engineer for a municipal power plant. The plant uses natural gas, which is currently obtained from an existing pipeline at an annual cost of $10,000 per year. Jenny is considering a project to construct a new pipeline. The initial cost of the new pipeline would be $35,000, but it would reduce the annual cost to $5000 per year. Assume an analysis period of 20 years and no salvage value for either the existing or new pipeline. The interest rate is 6%.

(a) Determine the equivalent uniform annual cost (EUAC) for the new pipeline.

(b) Should the new pipeline be built?

6-23 A machine has a first cost of $150,000, an annual operation and maintenance cost of $2500, a life of 10 years, and a salvage value of $30,000. At the end of years 4 and 8, it requires major servicing, which costs $20,000 and $10,000, respectively. At the end of year 5, it will need to be overhauled at a cost of $45,000. What is the equivalent uniform annual cost of owning and operating this particular machine?

6-24 Mr. Wiggley wants to buy a new house. It will cost $178,000. The bank will lend 90% of the purchase price at a nominal interest rate of 10.75% compounded weekly, and Mr. Wiggley will make monthly payments. What is the amount of the monthly payments if he intends to pay the house off in 25 years?

6-25 Steve Lowe must pay his property taxes in two equal instalments on December 1 and April 1. The two payments are for taxes for the fiscal year that begins on July 1 and ends the following June 30. Steve purchased a home on September 1. He estimates the annual property taxes will be $850 a year. Assuming the annual property taxes remain at $850 a year for the next several years, Steve plans to open a savings account and to make uniform monthly deposits the first of each month. The account is to be used to pay the taxes when they are due.

To begin the account, Steve deposits a lump sum equivalent to the monthly payments that will not have been made for the first year's taxes. The savings account pays 9% interest, compounded monthly and payable quarterly (March 31, June 30, September 30, and December 31). How much money should Steve put into the account when he opens it on September 1? What uniform monthly deposit should he make from that time on? (A careful *exact* solution is expected.) (*Answers:* Initial deposit $350.28; monthly deposit $69.02)

6-26 Claude Garneau, a salesman, needs a new car for use in his business. He expects to be promoted to a supervisory job at the end of 3 years, and so he needs to have a car for the 3 years he expects to be 'on the road'. The company reimburses salesmen each month at the rate of 25¢ per mile driven. Claude has decided to drive a low-priced car. He finds, however, that there are three different ways of obtaining the car:

(1) Purchase for cash; the price is $13,000.

(2) Lease the car; the monthly charge is $350 on a 36-month lease, payable at the end of each month; at the end of the 3-year period, the car is returned to the leasing company.

(3) Lease the car with an option to buy it at the end of the lease; pay $360 a month for 36 months; at

the end of that time, Claude could buy the car, if he chooses, for $3500.

Claude believes he should use a 12% interest rate in determining which alternative to choose. If the car could be sold for $4000 at the end of 3 years, which method should he use to obtain it?

6-27 A motorcycle is for sale for $2600. The motorcycle dealer is willing to sell it on the following terms:

> No down payment; pay $44 at the end of each of the first 4 months; pay $84 at the end of each month after that until the loan has been paid in full.

At a 12% annual interest rate compounded monthly, how many $84 payments will be required?

6-28 When he purchased his home, Al Silva borrowed $80,000 at 10% interest to be repaid in 25 equal annual end-of-year payments. After making 10 payments, Al found he could refinance the balance due on his loan at 9% interest for the remaining 15 years.

To refinance the loan, Al must pay the original lender the balance due on the loan, plus a penalty charge of 2% of the balance due; to the new lender he also must pay a $1000 service charge to obtain the loan. The new loan would be made equal to the balance due on the old loan, plus the 2% penalty charge, and the $1000 service charge. Should Al refinance the loan, assuming that he will keep the house for the next 15 years? Use an annual cash flow analysis in working this problem.

6-29 A company must decide whether to provide their salespeople with company-owned cars or pay them a mileage allowance and have them drive their own cars. New cars would cost about $18,000 each and could be resold 4 years later for about $7000 each. Annual operating costs would be $600 a year plus 7.5¢ per kilometre. If the salespeople drove their own cars, the company would probably pay them 18.75¢ per kilometre. Calculate the number of kilometres each salesperson would have to drive each year for it to be economically practical for the company to provide the cars. Assume a 10% annual interest rate. Use an annual cash flow analysis.

6-30 Your company must make a $500,000 balloon payment on a lease 2 years and 9 months from today. You have been directed to deposit an amount of money quarterly, beginning today, to provide for the $500,000 payment. The account pays 4% a year, com-

pounded quarterly. What is the required quarterly deposit? *Note:* Lease payments are due at beginning of the quarter.

6-31 A manufacturer is considering replacing a production machine tool. The new machine, costing $3700, would have a life of 4 years and no salvage value, but it would save the firm $500 a year in direct labour costs and $200 a year in indirect labour costs. The existing machine tool was purchased 4 years ago at a cost of $4000. It will last 4 more years and will have no salvage value at the end of that time. It could be sold now for $1000 cash. Assume that money is worth 8% and that the difference in taxes, insurance, and so forth, for the two alternatives is negligible. Use an annual cash flow analysis to determine whether the new machine should be purchased.

6-32 Two possible routes for a power line are under study. Data on the routes are as follows:

	Around the Lake	Under the Lake
Length	15 km	5 km
First cost	$5000/km	$25,000/km
Maintenance	$200/km/yr	$400/km/yr
Useful life, in years	15	15
Salvage value	$3000/km	$5000/km
Yearly power loss	$500/km	$500/km
Annual property taxes	2% of first cost	2% of first cost

If 7% interest is used, should the power line be routed around the lake or under the lake? (*Answer:* Around the lake.)

6-33 An oil refinery finds that it is now necessary to subject its waste liquids to a costly treating process before discharging them into a nearby stream. The engineering department estimates that the waste liquid processing will have cost $30,000 by the end of the first year. By making process and plant alterations, it is estimated that the waste treatment cost will decline by $3000 each year. As an alternative, a specialized firm, Hydro-Clean, has offered a contract to process the waste liquids for 10 years for a fixed price of $15,000 a year, payable at the end of each year. Either way, there should be no need for waste treatment after 10 years. The refinery manager considers 8% to be a suitable interest rate. Use an annual cash flow analysis to determine whether he should accept the Hydro-Clean offer.

6-34 Bill Anderson buys a car every 2 years as follows: initially he makes a down payment of $6000 on a

$15,000 car. The balance is paid in 24 equal monthly payments with annual interest at 12%. When he has made the last payment on the loan, he trades in the 2-year-old car for $6000 on a new $15,000 car, and the cycle begins over again.

Doug Jones decided on a different purchase plan. He thought he would be better off if he paid $15,000 cash for a new car. Then he would make a monthly deposit in a savings account so that, at the end of 2 years, he would have $9000 in the account. The $9000 plus the $6000 trade-in value of the car will allow Doug to replace his 2-year-old car by paying $9000 for a new one. The bank pays 6% interest, compounded quarterly.

(a) What is Bill Anderson's monthly payment to pay off the loan on the car?

(b) After he purchased the new car for cash, how much per month should Doug Jones deposit in his savings account to have enough money for the next car two years hence?

(c) Why is Doug's monthly savings account deposit smaller than Bill's payment?

6-35 Alice White has arranged to buy some home recording equipment. She estimates that it will have a 5-year useful life and no salvage value. The dealer, who is a friend, has offered Alice two alternative ways to pay for the equipment:

(1) Pay $2000 immediately and $500 at the end of one year.

(2) Pay nothing until the end of 4 years, when a single payment of $3000 must be made.

Alice believes 12% is a suitable interest rate. Use an annual cash flow analysis to determine which method of payment she should select. (*Answer:* Select (1))

6-36 Two mutually exclusive alternatives are being considered.

Year	A	B
0	−$3000	−$5000
1	+845	+1400
2	+845	+1400
3	+845	+1400
4	+845	+1400
5	+845	+1400

One of the alternatives must be selected. Using a 15% nominal interest rate, compounded continuously, determine which one. Solve by annual cash flow analysis.

6-37 A certain industrial firm desires an economic analysis to determine which of two different machines should be purchased. Each machine is capable of performing the same task in a given amount of time. Assume the minimum attractive return is 8%. The following data are to be used in this analysis:

	Machine X	Machine Y
First cost	$5000	$8000
Estimated life, in years	5	12
Salvage value	0	$2000
Annual maintenance cost	0	150

Which machine would you choose? Base your answer on annual cost. (*Answers:* X = $1252; Y = $1106)

6-38 A suburban taxi company is considering buying taxis with diesel engines instead of gasoline engines. The cars average 50,000 km a year, with a useful life of 3 years for the taxi with the gas engine and 4 years for the diesel taxi. Other comparative information is as follows:

	Diesel	Gasoline
Vehicle cost	$13,000	$12,000
Fuel cost per litre	48¢	51¢
Mileage, in km/litre	35	28
Annual repairs	$ 300	$ 200
Annual insurance premium	500	500
End-of-useful-life resale value	2,000	3,000

Use an annual cash flow analysis to determine the more economical choice if interest is 6%.

6-39 A company must decide whether to buy Machine A or Machine B:

	Machine A	Machine B
Initial cost	$10,000	$20,000
Useful life, in years	4	10
End-of-useful-life salvage value	$10,000	$10,000
Annual maintenance	1,000	0

At a 10% interest rate, which machine should be installed? Use an annual cash flow analysis in working this problem. (*Answer:* Machine A)

6-40 Consider the following alternatives:

	A	B
Cost	$50	$180
Uniform annual benefit	15	60
Useful life, in years	10	5

The analysis period is 10 years, but there will be no replacement for Alternative B at the end of 5 years. Based on a 15% interest rate, determine which alternative should be selected. Use an annual cash flow analysis in working this problem.

6-41 Consider the following two mutually exclusive alternatives:

	A	B
Cost	$100	$150
Uniform annual benefit	16	24
Useful life, in years	∞	20

Alternative B may be replaced with an identical item every 20 years at the same $150 cost and will have the same $24 uniform annual benefit. Using a 10% interest rate, and an annual cash flow analysis, determine which alternative should be selected.

6-42 Some equipment will be installed in a warehouse that a firm has leased for 7 years. There are two alternatives:

	A	B
Cost	$100	$150
Uniform annual benefit	55	61
Useful life, in years	3	4

At any time after the equipment is installed, it has no salvage value. Assume that Alternatives A and B will be replaced at the end of their useful lives by identical equipment with the same costs and benefits. For a 7-year analysis period and a 10% interest rate, use an annual cash flow analysis to determine which alternative should be selected.

6-43 A pump is needed for 10 years at a remote location. The pump can be driven by an electric motor if a power line is extended to the site. Otherwise, a gasoline engine will be used. Use an annual cash flow analysis to determine, using the following data and a 10% interest rate, how the pump should be powered.

	Gasoline	Electric
First cost	$2400	$6000
Annual operating cost	1200	750
Annual maintenance	300	50
Salvage value	300	600
Life, in years	5	10

6-44 The town of Oak Hills needs an additional supply of water from Pine Creek. The town engineer has selected two plans for comparison: a *gravity plan* (divert water at a point 10 kilometres up Pine Creek and carry it through a pipeline by gravity to the town) and a *pumping plan* (divert water at a point closer to the town and pump it to the town). The pumping plant would be built in two stages, with half-capacity installed initially and the other half installed 10 years later.

An analysis will assume a 40-year life, 10% interest, and no salvage value. Costs are as follows:

	Gravity	Pumping
Initial investment	$2,800,000	$1,400,000
Additional investment in 10th year	None	200,000
Operation and maintenance	10,000/yr	25,000/yr
Power cost		
Average first 10 years	None	50,000/yr
Average next 30 years	None	100,000/yr

Use an annual cash flow analysis to determine the more economical plan.

6-45 Uncle Elmo needs to replace the family privy. The local sanitary engineering firm has submitted two alternative structural proposals with respective cost estimates as shown. Which construction should Uncle Elmo choose if his minimum attractive rate of return is 6%? Use both a present worth and annual cost approach in your comparison.

	Masonite	Brick
First cost	$250	$1000
Annual maintenance	20	10
Salvage value	10	100
Service life, in years	4	20

6-46 The manager in a canned food processing plant is trying to decide between two labelling machines. Their respective costs and benefits are as follows:

	Machine A	Machine B
First cost	$15,000	$25,000
Maintenance and operating costs	1,600	400
Annual benefit	8,000	13,000
Salvage value	3,000	6,000
Useful life, in years	7	10

Assume an interest rate of 12%. Use annual cash flow analysis to determine which machine should be chosen.

6-47 Carp, Inc. wants to evaluate two methods of shipping their products. The following cash flows are associated with the alternatives:

	A	**B**
First cost	$700,000	$1,700,000
Maintenance and operating costs	18,000	29,000
+ Cost gradient (begin Year 1)	+900/yr	+750/yr
Annual benefit	154,000	303,000
Salvage value	142,000	210,000
Useful life, in years	10	20

Use an interest rate of 15% and annual cash flow analysis to decide which is the most desirable alternative.

6-48 A university student has been looking for a new tire for his car and has found the following alternatives:

Tire Warranty (months)	Price per Tire
12	$39.95
24	59.95
36	69.95
48	90.00

The student feels that the warranty period is a good estimate of the tire life and that a 10% interest rate is reasonable. Use an annual cash flow analysis to determine which tire he should buy.

6-49 Consider the following three mutually exclusive alternatives:

	A	**B**	**C**
Cost	$100	$150.00	$200.00
Uniform annual benefit	10	17.62	55.48
Useful life, in years	∞	20	5

Assuming that Alternatives B and C are replaced with identical replacements at the end of their useful lives, and an 8% interest rate, which alternative should be selected? Use an annual cash flow analysis in working this problem.   (*Answer:* Select C)

6-50 A new car is purchased for $12,000 with a 0% down, 9% loan. The loan is for 4 years. After making 30 payments, the owner wants to pay off the loan's remaining balance. How much is owed?

6-51 A year after buying her car, Annick has been offered a job in Europe. Her car loan is for $15,000 at a 9%

nominal interest rate for 60 months. If she can sell the car for $12,000, how much will she be able to keep after paying off the loan?

6-52 A $78,000 conventional Canadian mortgage has a 30-year term and a 9% nominal interest rate.
(*a*) What is the monthly payment?
(*b*) After the first year of payments, what is the outstanding balance?
(*c*) How much interest is paid in month 13? How much principal?

6-53 A $92,000 conventional Canadian mortgage has a 30-year term and a 9% nominal interest rate.
(*a*) What is the monthly payment?
(*b*) After the first year of payments, what fraction of the loan has been repaid?
(*c*) After the first 10 years of payments, what is the outstanding balance?
(*d*) How much interest is paid in Month 25? How much principal?

6-54 A 30-year conventional Canadian mortgage for $95,000 is issued at a 9% nominal interest rate.
(*a*) What is the monthly payment?
(*b*) How long does it take to pay off the mortgage if $1000 per month is paid?
(*c*) How long does it take to pay off the mortgage if double payments are made?

6-55 A 30-year conventional Canadian mortgage for $145,000 is issued at a 6% nominal interest rate.
(*a*) What is the monthly payment?
(*b*) How long does it take to pay off the mortgage if $1000 per month is paid?
(*c*) How long does it take to pay off the mortgage if 20% extra is paid each month?

6-56 Solve Problem 6-32 for the break-even first cost per kilometre of going under the lake.

6-57 Redo Problem 6-38 to calculate the EUAW of the alternatives as a function of kilometres driven per year to see if there is a crossover point in the decision process. Graph your results.

6-58 Set up Problem 6-18 on a spreadsheet to make all the input data variable and determine various scenarios which would make the bailer economical.

6-59 Develop a spreadsheet to solve Problem 6-44. What is the break-even cost of the additional pumping investment in year 10?

CHAPTER 7

Rate of Return Analysis

Big Brother Is Tracking You

Do you ever get the feeling that your closet is spying on you? Well, you may be right.

Manufacturers of products ranging from jeans to shampoo are trying out a new technology called radio-frequency identification (RFID), which embeds tiny tracking chips in merchandise. The chips can follow goods from the time they leave the plant until you buy them at your local shopping mall—and even after you take them home.

RFID chips offer a wealth of advantages to product makers and retailers. They can tell shippers when trucks are delayed, and they can alert store security when products are being stolen. They can also tell managers when stocks of popular products are running low.

The chips still suffer from some technical drawbacks, including vulnerability to interference from cell phones and other electronic devices. A bigger drawback, however, is their cost. As of early 2003, RFID chips cost about 30 cents each, which makes them too expensive to embed in inexpensive products like soap and shaving cream. Moreover, the scanners required to read the chips cost about $1000 each. And companies that want to use the technology must also invest in the electronic equipment necessary to keep track of the chips.

The cost hurdle may be falling soon, though. A major manufacturer of RFID chips says it can produce them for 5 cents each if the customer orders a batch of at least a billion. Even more substantial cost reductions are likely when the technology becomes more widely adopted.

After Completing This Chapter...

The student should be able to:

- Evaluate project cash flows by the *internal rate of return* (IRR) method.
- Plot the present worth (PW) of a project against the interest rate.
- Use an *incremental* rate of return analysis to evaluate competing alternatives.
- Develop and use spreadsheets to make IRR and incremental rate of return calculations.

QUESTIONS TO CONSIDER

1. Imagine that you are a manufacturer who is considering using RFID chips to track your products. What cost and benefit factors would you consider in deciding whether to embed chips in your products and how much to invest in the necessary technology?

2. Now assume that you are a retailer who is also considering the RFID technology. How would your cost and benefit considerations differ from those of a product manufacturer? What specific factors would you consider? What if the manufacturers you buy products from have already adopted the technology? How would this affect your decision?

3. Since October 2000, the U.S. military has been using RFID technology in its Joint Total Asset Visibility (JTAV) program, which tracks equipment and supplies. JTAV has reduced duplicate orders, helped speed up troop deployment, and allowed military planners to accomplish missions with fewer ships, planes, and ground vehicles. How is the JTAV program likely to affect the pace of RFID's adoption by private industry?

4. Some consumers have already expressed concern about RFID chips, calling them an invasion of privacy. What ethical and social concerns are raised by RFID chips?

5. In response to consumer worries, RFID chip manufacturers are developing 'kill' technologies that can allow consumers to disable RFID chips after they have purchased goods containing these miniature tracking devices. How might this affect consumer attitudes?

In this chapter we will examine three aspects of rate of return, the third of the three major analysis methods. First, the meaning of 'rate of return' is explained; then, the calculation of rate of return is illustrated; finally, rate of return analysis problems are presented. In an appendix to this chapter, we describe difficulties sometimes encountered in attempting to compute an interest rate for cash flows of certain kinds.

INTERNAL RATE OF RETURN

In Chapter 3 we examined four plans to repay $5000 in 5 years with interest at 8% (Table 3-1). In each of the four plans the amount loaned ($5000) and the loan duration (5 years) were the same. Yet the total interest paid to the lender varied from $1200 to $2347, depending on the loan repayment plan. We saw, however, that the lender received 8% interest each year on the amount of money actually owed. And, at the end of 5 years, the principal and interest payments exactly repaid the $5000 debt with interest at 8%. We say the lender received an '8% rate of return'.

> **Internal rate of return** is defined as the interest rate paid on the unpaid balance of a *loan* such that the payment schedule makes the unpaid loan balance equal to zero when the final payment is made.

Instead of lending money, we might invest $5000 in a machine tool with a 5-year useful life and an equivalent uniform annual benefit of $1252. A reasonable question is what rate of return we would receive on this investment. The cash flow would be as follows:

Year	Cash Flow
0	−$5000
1	+1252
2	+1252
3	+1252
4	+1252
5	+1252

We recognize the cash flow as Plan 3 of Table 3-1. We know that five payments of $1252 are equivalent to a present sum of $5000 when interest is 8%. Therefore, the rate of return on this investment is 8%. Stated in terms of an investment, we may define internal rate of return as follows:

> **Internal rate of return** is the interest rate earned on the unrecovered *investment* such that the payment schedule makes the unrecovered investment equal to zero at the end of the life of the investment.

It must be understood that the 8% rate of return does not mean an annual return of 8% on the $5000 investment, or $400 in each of the 5 years with $5000 returned at the end of Year 5. Instead, each $1252 payment represents an 8% return on the unrecovered investment *plus* the partial return of the investment. This may be tabulated as follows:

Year	Cash Flow	Unrecovered Investment at Beginning of Year	8% Return on Unrecovered Investment	Investment Repayment at End of Year	Unrecovered Investment at End of Year
0	−$5000				
1	+1252	$5000	$ 400	$ 852	$4148
2	+1252	4148	331	921	3227
3	+1252	3227	258	994	2233
4	+1252	2233	178	1074	1159
5	+1252	1159	93	1159	0
			$1260	$5000	

This cash flow represents a $5000 investment with benefits that produce an 8% rate of return on the unrecovered investment.

Although the two definitions of internal rate of return are stated differently, one in terms of a loan and the other in terms of an investment, there is only one fundamental concept being described. It is that **the internal rate of return is the interest rate at which the benefits are equivalent to the costs**, or at which the present worth (PW) is 0. Since we are describing situations of funds that remain within the investment throughout its life, the resulting rate of return is described as the internal rate of return, i.

CALCULATING RATE OF RETURN

To calculate a rate of return on an investment, we must convert the various consequences of the investment into a cash flow. Then we will solve the cash flow for the unknown value of the internal rate of return (IRR). Five forms of the cash flow equation are as follows:

$$\text{PW of benefits} - \text{PW of costs} = 0 \qquad (7\text{-}1)$$

$$\frac{\text{PW of benefits}}{\text{PW of costs}} = 1 \qquad (7\text{-}2)$$

$$\text{Net present worth} = 0 \qquad (7\text{-}3)$$

$$\text{EUAB} - \text{EUAC} = 0 \qquad (7\text{-}4)$$

$$\text{PW of costs} = \text{PW of benefits} \qquad (7\text{-}5)$$

The five equations represent the same concept in different forms. They can relate costs and benefits with the IRR as the only unknown. The calculation of rate of return is illustrated by the following examples.

EXAMPLE 7-1

An $8200 investment returned $2000 a year over a 5-year useful life. What was the rate of return on the investment?

SOLUTION

Using Equation 7-2, we write

$$\frac{\text{PW of benefits}}{\text{PW of costs}} = 1 \qquad \frac{2000(P/A, i, 5)}{8200} = 1$$

Rewriting, we see that

$$(P/A, i, 5) = \frac{8200}{2000} = 4.1$$

Then look at the compound interest tables for the value of i where $(P/A, i, 5) = 4.1$; if no tabulated value of i gives this value, we will then find values on either side of the desired value (4.1) and interpolate to find the IRR.

From interest tables we find:

i	$(P/A, i, 5)$
6%	4.212
7%	4.100
8%	3.993

In this example, no interpolation is needed because the internal rate of return is exactly 7%.

EXAMPLE 7-2

An investment resulted in the following cash flow. Compute the rate of return.

Year	Cash Flow
0	−$700
1	+100
2	+175
3	+250
4	+325

SOLUTION

$$\text{EUAB} - \text{EUAC} = 0$$

$$100 + 75(A/G, i, 4) - 700(A/P, i, 4) = 0$$

In this situation, we have two different interest factors in the equation. We will not be able to solve it as easily as Example 7-1. Since there is no convenient direct method of solution, we will solve the equation by trial and error. Try $i = 5\%$ first:

$$EUAB - EUAC = 0$$

$$100 + 75(A/G, 5\%, 4) - 700(A/P, 5\%, 4) = 0$$

$$100 + 75(1.439) - 700(0.2820) = 0$$

At $i = 5\%$,

$$EUAB - EUAC = 208 - 197 = +11$$

The EUAC is too low. If the interest rate is increased, EUAC will increase. Try $i = 8\%$:

$$EUAB - EUAC = 0$$

$$100 + 75(A/G, 8\%, 4) - 700(A/P, 8\%, 4) = 0$$

$$100 + 75(1.404) - 700(0.3019) = 0$$

At $i = 8\%$,

$$EUAB - EUAC = 205 - 211 = -6$$

This time the EUAC is too large. We see that the true rate of return is between 5% and 8%. Try $i = 7\%$:

$$EUAB - EUAC = 0$$

$$100 + 75(A/G, 7\%, 4) - 700(A/P, 7\%, 4) = 0$$

$$100 + 75(1.416) - 700(0.2952) = 0$$

At $i = 7\%$,

$$EUAB - EUAC = 206 - 206 = 0$$

The IRR is 7%.

EXAMPLE 7-3

Calculate the rate of return on the investment on the following cash flow.

Year	Cash Flow
0	−$100
1	+20
2	+30
3	+20
4	+40
5	+40

SOLUTION

Using a net present worth of 0, try $i = 10\%$:

$$NPW = -100 + 20(P/F, 10\%, 1) + 30(P/F, 10\%, 2) + 20(P/F, 10\%, 3)$$

$$+ 40(P/F, 10\%, 4) + 40(P/F, 10\%, 5)$$

$$= -100 + 20(0.9091) + 30(0.8264) + 20(0.7513) + 40(0.6830)$$

$$+ 40(0.6209)$$

$$= -100 + 18.18 + 24.79 + 15.03 + 27.32 + 24.84$$

$$= -100 + 110.16 = +10.16$$

The trial interest rate i is too low. Select a second trial, $i = 15\%$:

$$NPW = -100 + 20(0.8696) + 30(0.7561) + 20(0.6575) + 40(0.5718)$$

$$+ 40(0.4972)$$

$$= -100 + 17.39 + 22.68 + 13.15 + 22.87 + 19.89$$

$$= -100 + 95.98$$

$$= -4.02$$

FIGURE 7-1 Plot of NPW versus interest rate i.

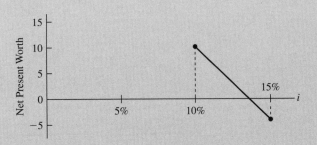

These 10% and 15% points are plotted in Figure 7-1. By linear interpolation we compute the rate of return as follows:

$$IRR = 10\% + (15\% - 10\%)\left(\frac{10.16}{10.16 + 4.02}\right) = 13.5\%$$

We can prove that the rate of return is very close to $13\frac{1}{2}\%$ by showing that the unrecovered investment is very close to zero at the end of the life of the investment.

Year	Cash Flow	Unrecovered Investment at Beginning of Year	$13\frac{1}{2}\%$ Return on Unrecovered Investment	Investment Repayment at End of Year	Unrecovered Investment at End of Year
0	−$100				
1	+20	$100.0	$13.5	$ 6.5	$93.5
2	+30	93.5	12.6	17.4	76.1
3	+20	76.1	10.3	9.7	66.4
4	+40	66.4	8.9	31.1	35.3
5	+40	35.3	4.8	35.2	0.1*

*This small unrecovered investment indicates that the rate of return is slightly less than $13\frac{1}{2}\%$.

If in Figure 7-1 net present worth (NPW) had been computed for a broader range of values of i, Figure 7-2 would have been obtained. From this figure it is apparent that the error resulting from linear interpolation increases as the interpolation width increases.

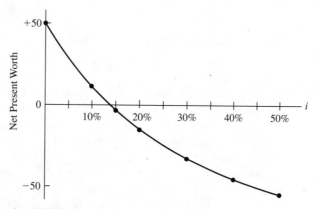

FIGURE 7-2 Replot of NPW versus interest rate i over a larger range of values.

Plot of NPW versus Interest Rate i

Figure 7-2—the second plot of NPW versus interest rate i—is an important source of information. For a cash flow representing an investment followed by benefits from the investment, the plot of NPW versus i (we will call it an **NPW plot** for convenience) would have the form of Figure 7-3.

Year	Cash Flow
0	−P
1	+Benefit A
2	+A
3	+A
4	+A
.	.
.	.
.	.

FIGURE 7-3 NPW plot for a typical investment.

If, on the other hand, borrowed money was involved, the NPW plot would appear as in Figure 7-4. This form of cash flow is common when one is a borrower of money. In such a case, the usual pattern is receipt of borrowed money early in the time period with a later repayment of an equal sum, plus payment of interest on the borrowed money. In all cases in which interest is charged, the NPW at 0% will be negative.

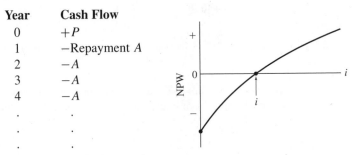

Year	Cash Flow
0	$+P$
1	$-$Repayment A
2	$-A$
3	$-A$
4	$-A$
.	.
.	.
.	.

FIGURE 7-4 Typical NPW plot for borrowed money.

How do we determine the interest rate paid by the borrower in this situation? Usually we would write an equation, such as PW of income $=$ PW of disbursements, and solve for the unknown IRR. Is the resulting IRR positive or negative from the borrower's point of view? If the lender said he was receiving, say, $+11\%$ on the debt, it might seem reasonable to state that the borrower is faced with -11% interest. Yet this is not the way interest is calculated. A banker says he pays 5% interest on savings accounts and charges 11% on personal loans. Both rates are positive.

Thus, we implicitly recognize interest as a charge for the use of someone else's money and a receipt for letting others use our money. In determining the interest rate in a particular situation, we solve for a single unsigned value of it. We then view this value of i in the customary way, that is, as either a charge for borrowing money or a receipt for lending money.

EXAMPLE 7-4

A new corporate bond was initially sold by a stockbroker to an investor for $1000. The issuing corporation promised to pay the bondholder $40 interest on the $1000 face value of the bond every 6 months, and to repay the $1000 at the end of 10 years. After one year the bond was sold by the original buyer for $950.

(a) What rate of return did the original buyer receive on his investment?

(b) What rate of return can the new buyer (paying $950) expect to receive if he keeps the bond for its remaining 9-year life?

SOLUTION TO a

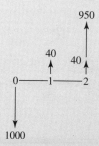

Since $40 is received each 6 months, we will use a 6-month interest period to solve the problem. Let PW of cost = PW of benefits, and write

$$1000 = 40(P/A, i, 2) + 950(P/F, i, 2)$$

Try $i = 1\frac{1}{2}\%$:

$$1000 \neq 40(1.956) + 950(0.9707) = 78.24 + 922.17 \neq 1000.41$$

The interest rate per 6 months, $IRR_{6\,mon}$, is very close to $1\frac{1}{2}\%$. This means the nominal (annual) interest rate is $2 \times 1.5\% = 3\%$. The effective (annual) interest rate $= IRR = (1 + 0.015)^2 - 1 = 3.02\%$.

SOLUTION TO b

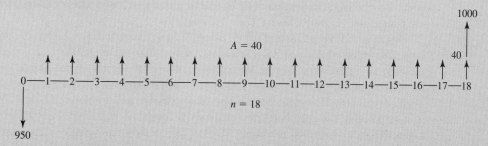

Given the same $40 semi-annual interest payments, for 6-month interest periods we write

$$950 = 40(P/A, i, 18) + 1000(P/F, i, 18)$$

Try $i = 5\%$:

$$950 \neq 40(11.690) + 1000(0.4155) = 467.60 + 415.50 \neq 883.10$$

The PW of benefits is too low. Try a lower interest rate, say, $i = 4\%$:

$$950 \neq 40(12.659) + 1000(0.4936) = 506.36 + 493.60$$

$$\neq 999.96$$

The value of the IRR is between 4% and 5%. By interpolation,

$$\text{IRR} = 4\% + (1\%)\left(\frac{999.96 - 950.00}{999.96 - 883.10}\right) = 4.43\%$$

The nominal interest rate is $2 \times 4.43\% = 8.86\%$. The effective interest rate is $(1 + 0.0443)^2 - 1 = 9.05\%$.

RATE OF RETURN ANALYSIS

Rate of return analysis is probably the most frequently used exact analysis technique in industry. Although problems in computing rate of return sometimes occur, its major advantage outweighs the occasional difficulty. The advantage is that we can compute a single figure of merit that is readily understood.

Consider these statements:

- The net present worth on the project is $32,000.
- The equivalent uniform annual net benefit is $2800.
- The project will produce a 23% rate of return.

While none of these statements tells the complete story, the third one gives a measure of desirability of the project in terms that are widely understood. It is this acceptance by engineers and business leaders alike of rate of return that has promoted its more frequent use than present worth or annual cash flow methods.

There is another advantage to rate of return analysis. In both present worth and annual cash flow calculations, one must select an interest rate for use in the calculations—and this may be a difficult and controversial item. In rate of return analysis, no interest rate is introduced into the calculations (except as described in Appendix 7A). Instead, we compute a rate of return (more accurately called *internal rate of return*) from the cash flow. To decide how to proceed, the calculated rate of return is compared with a pre-selected **minimum attractive rate of return,** or simply MARR. This is the same value of i used for present worth and annual cash flow analysis.

When there are two alternatives, rate of return analysis is performed by computing the *incremental rate of return*—ΔIRR—on the difference between the alternatives. Since we want to look at increments of investment, the cash flow for the difference between the alternatives is computed by taking the higher initial-cost alternative *minus* the lower initial-cost alternative. If ΔIRR is the same or greater than the MARR, choose the higher-cost alternative. If ΔIRR is less than the MARR, choose the lower-cost alternative.

Two-Alternative Situation	Decision
ΔIRR $\geq$ MARR	Choose the higher-cost alternative
ΔIRR $<$ MARR	Choose the lower-cost alternative

Rate of return and incremental rate of return analysis are illustrated by Examples 7-5 to 7-8.

EXAMPLE 7-5

If an electromagnet is installed on the input conveyor of a coal-processing plant, it will pick up scrap metal in the coal and thereby save an estimated $1200 a year in costs resulting from damage to the machinery by the scrap metal. The electromagnetic equipment has an estimated useful life of 5 years and no salvage value. Two suppliers have been contacted: Leaseco will provide the equipment in return for three beginning-of-year annual payments of $1000 each; Saleco will provide the equipment for $2783. If the MARR is 10%, which supplier should be chosen.

SOLUTION

Since both suppliers will provide equipment with the same useful life and benefits, this is a fixed-output situation. In rate of return analysis, the method of solution is to examine the differences between the alternatives. By taking Saleco − Leaseco, we obtain an increment of investment.

Year	Leaseco	Saleco	Difference between Alternatives: Saleco − Leaseco
0	−$1000	−$2783	−$1783
1	$\begin{cases} -1000 \\ +1200 \end{cases}$	+1200	+1000
2	$\begin{cases} -1000 \\ +1200 \end{cases}$	+1200	+1000
3	+1200	+1200	0
4	+1200	+1200	0
5	+1200	+1200	0

Compute the NPW at various interest rates on the increment of investment represented by the difference between the alternatives.

Year n	Cash Flow: Saleco−Leaseco	PW* At 0%	At 8%	At 20%	At ∞%
0	−$1783	−$1783	−$1783	−$1783	−$1783
1	+1000	+1000	+926	+833	0
2	+1000	+1000	+857	+694	0
3	0	0	0	0	0
4	0	0	0	0	0
5	0	0	0	0	0
NPW =		+217	0	−256	−1783

*Each year the cash flow is multiplied by $(P/F, i, n)$.

At 0%: $(P/F, 0\%, n) = 1$ for all values of n

At ∞%: $(P/F, \infty\%, 0) = 1$

$(P/F, \infty\%, n) = 0$ for all other values of n

From the plot of these data in Figure 7-5, we see that NPW = 0 at $i = 8\%$.

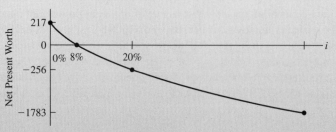

FIGURE 7-5 NPW plot for Example 7-5.

Thus, the incremental rate of return—ΔIRR—of selecting Saleco rather than Leaseco is 8%. This is less than the 10% MARR. Select Leaseco.

EXAMPLE 7-6

You must choose one of two mutually exclusive alternatives. (*Note*: Engineering economists often use the term 'mutually exclusive alternatives' to emphasize that selecting one alternative precludes selecting any other.) The alternatives are as follows:

Year	Alt. 1	Alt. 2
0	−$10	−$20
1	+15	+28

Any money not invested here may be invested elsewhere at the MARR of 6%. If you can choose only one alternative one time, which one would you choose if you were using the internal rate of return (IRR) analysis method?

SOLUTION

Using the IRR analysis method, we will select the lesser-cost alternative (Alt. 1), unless we find that the additional cost of Alt. 2 produces enough additional benefits to make it preferable instead. If we consider Alt. 2 in relation to Alt. 1, then

$$\begin{bmatrix} \text{Higher-cost} \\ \text{Alt. 2} \end{bmatrix} = \begin{bmatrix} \text{Lower-cost} \\ \text{Alt. 1} \end{bmatrix} + \begin{bmatrix} \text{Differences between} \\ \text{Alt. 1 and Alt. 2} \end{bmatrix}$$

or

$$\text{Differences between Alt. 1 and Alt. 2} = \begin{bmatrix} \text{Higher-cost} \\ \text{Alt. 2} \end{bmatrix} - \begin{bmatrix} \text{Lower-cost} \\ \text{Alt. 1} \end{bmatrix}$$

The choice between the two alternatives is reduced to an examination of the differences between them. We can compute the rate of return on the differences between the alternatives. Writing the alternatives again,

Year	Alt. 1	Alt. 2	Alt. 2 − Alt. 1
0	−$10	−$20	−$20 − (−$10) = −$10
1	+15	+28	+28 − (+15) = +13

PW of cost of differences (Alt. 2 − Alt. 1) = PW of benefit of differences (Alt. 2 − Alt. 1)

$$10 = 13(P/F, i, 1)$$

Thus,

$$(P/F, i, 1) = \frac{10}{13} = 0.7692$$

One can see that if $10 increases to $13 in one year, the interest rate must be 30%. The compound interest tables confirm this conclusion. The 30% rate of return on the difference between the alternatives is far higher than the 6% MARR. The additional $10 investment to obtain Alt. 2 is superior to investing the $10 elsewhere at 6%. To obtain this desirable increment of investment, with its 30% rate of return, Alt. 2 is selected.

To understand more about Example 7-6, compute the rate of return for each alternative.

Alternative 1

PW of cost of Alt. 1 = PW of benefit of Alt. 1

$$\$10 = \$15(P/F, i, 1)$$

Thus,

$$(P/F, i, 1) = \frac{10}{15} = 0.6667$$

From the compound interest tables: rate of return = 50%.

Alternative 2

PW of cost of Alt. 2 = PW of benefit of Alt. 2

$$\$20 = \$28(P/F, i, 1)$$

Thus,

$$(P/F, i, 1) = \frac{20}{28} = 0.7143$$

From the compound interest tables: rate of return = 40%.

One may be tempted to select Alt. 1 on the basis of these rate of return computations. We have already seen, however, that this is not the correct solution. Solve the problem again, this time using present worth analysis.

Present Worth Analysis
Alternative 1

$$NPW = -10 + 15(P/F, 6\%, 1) = -10 + 15(0.9434) = +\$4.15$$

Alternative 2

$$NPW = -20 + 28(P/F, 6\%, 1) = -20 + 28(0.9434) = +\$6.42$$

Alternative 1 has a 50% rate of return and an NPW (at the 6% MARR) of +$4.15. Alternative 2 has a 40% rate of return on a larger investment, with the result that its NPW (at the 6% MARR) is +$6.42. Our economic criterion is to maximize the return, rather than the rate of return. To maximize NPW, select Alt. 2. This agrees with the rate of return analysis on the differences between the alternatives.

EXAMPLE 7-7

If the computations for Example 7-6 do not convince you, and you still think Alternative 1 would be preferable, try this problem.

You have $20 in your wallet and two alternative ways of lending Bill some money.

(a) Lend Bill $10 with his promise of a 50% return. That is, he will pay you back $15 at the agreed time.

(b) Lend Bill $20 with his promise of a 40% return. He will pay you back $28 at the same agreed time.

You can choose whether to lend Bill $10 or $20. This is a one-time situation, and any money not lent to Bill will remain in your wallet. Which alternative do you choose?

SOLUTION

So you see that a 50% return on the smaller sum is less rewarding to you than 40% on the larger sum? Since you would prefer to have $28 than $25 ($15 from Bill plus $10 remaining in your wallet) after the loan is paid, lend Bill $20.

EXAMPLE 7-8

Solve Example 7-6 again, but this time compute the interest rate on the increment (Alt. 1 − Alt. 2). How do you interpret the results?

SOLUTION

This time the problem is being viewed as follows:

$$\text{Alt. } 1 = \text{Alt. } 2 + [\text{Alt. } 1 - \text{Alt. } 2]$$

Year	Alt. 1	Alt. 2	[Alt. 1 − Alt. 2]
0	−$10	−$20	−$10 − (−$20) = +$10
1	+15	+28	+15 − (+28) = −13

We can write one equation in one unknown:

$$\text{NPW} = \text{PW of benefit of differences} - \text{PW of cost of differences} = 0$$

$$+10 - 13(P/F, i, 1) = 0$$

Thus,

$$(P/F, i, 1) = \frac{10}{13} = 0.7692$$

Once again the interest rate is found to be 30%. The critical question is, what does the 30% represent? Looking at the increment again:

Year	Alt. 1 − Alt. 2
0	+$10
1	−13

The cash flow does *not* represent an investment; instead, it represents a loan. It is as if we borrowed $10 in Year 0 (+$10 represents a receipt of money) and repaid it in Year 1 (−$13 represents a disbursement). The 30% interest rate means this is the amount *we would pay* for the use of the $10 borrowed in Year 0 and repaid in Year 1.

Is this a desirable borrowing? Since the MARR on investments is 6%, it is reasonable to assume our maximum interest rate on borrowing would also be 6%. Here the interest rate is 30%, which means the borrowing is undesirable. Since Alt. 1 = Alt. 2 + (Alt. 1 − Alt. 2), and we do not like the (Alt. 1 − Alt. 2) increment, we should reject Alternative 1, which contains the undesirable increment. This means we should select Alternative 2—the same conclusion reached in Example 7-6.

Example 7-8 illustrated that one can analyze either **increments of investment** or **increments of borrowing.** When looking at increments of investment, we accept the increment when the incremental rate of return equals or exceeds the minimum attractive rate of return (ΔIRR $\geq$ MARR). When looking at increments of borrowing, we accept the increment when the incremental interest rate is less than or equal to the *minimum* attractive rate of return (ΔIRR $\leq$ MARR). One way to avoid much of the possible confusion is to organize the solution to any problem so that one is examining increments of investment. This is illustrated in the next example.

EXAMPLE 7-9

A firm is considering which of two devices to install to reduce costs in a particular situation. Both devices cost $1000, and both have useful lives of 5 years and no salvage value. Device A can be expected to result in $300 savings annually. Device B will provide cost savings of $400 the first year but will decline $50 annually, making the second-year savings $350, the third-year savings $300, and so forth. For a 7% MARR, which device should the firm purchase?

SOLUTION

This problem has been solved by present worth analysis (Example 5-1) and annual cost analysis (Example 6-5). This time we will use rate of return analysis. The example has fixed input ($1000) and differing outputs (savings).

In determining whether to use an (A − B) or (B − A) difference between the alternatives, we seek an increment of investment. By looking at both (A − B) and (B − A), we find that (A − B) is the one that represents an increment of investment.

Year	Device A	Device B	Difference between Alternatives: Device A − Device B
0	−$1000	−$1000	$0
1	+300	+400	−100
2	+300	+350	−50
3	+300	+300	0
4	+300	+250	+50
5	+300	+200	+100

For the difference between the alternatives, write a single equation with i as the only unknown.

$$\text{EUAC} = \text{EUAB}$$

$$[100(P/F, i, 1) + 50(P/F, i, 2)](A/P, i, 5) = [50(F/P, i, 1) + 100](A/F, i, 5)$$

The equation is cumbersome but need not be solved. Instead, we observe that the sum of the costs (−100 and −50) equals the sum of the benefits (+50 and +100). This indicates that 0% is the IRR_{A-B} on the A − B increment of *investment*. This is less than the 7% MARR; therefore, the increment is undesirable. Reject Device A and choose Device B.

As was described in Example 7-8, if the increment examined is (B − A), the interest rate would again be 0%, indicating a desirable *borrowing* situation. We would choose Device B.

Analysis Period

In discussing present worth analysis and annual cash flow analysis, an important consideration is the analysis period. This is also true in rate of return analysis. The method of solution for two alternatives is to examine the differences between the alternatives. Clearly, the examination must cover the selected analysis period. For now, we can suggest only that the assumptions made should reflect one's perception of the future as accurately as possible.

In Example 7-10 the analysis period is a common multiple of the alternative service lives and identical replacement is assumed. This problem illustrates an analysis of the differences between the alternatives over the analysis period.

EXAMPLE 7-10

Two machines are being considered for purchase. If the MARR is 10%, which machine should be bought? Use an IRR analysis comparison.

	Machine X	**Machine Y**
Initial cost	$200	$700
Uniform annual benefit	95	120
End-of-useful-life salvage value	50	150
Useful life, in years	6	12

SOLUTION

The solution is based on a 12-year analysis period and a replacement machine X that is identical to the present machine X. The cash flow for the differences between the alternatives, is as follows:

Year	Machine X	Machine Y	Machine Y − Machine X
0	−$200	−$700	−$500
1	+95	+120	+25
2	+95	+120	+25
3	+95	+120	+25
4	+95	+120	+25
5	+95	+120	+25
6	+95 +50 −200	+120	+25 +150
7	+95	+120	+25
8	+95	+120	+25
9	+95	+120	+25
10	+95	+120	+25
11	+95	+120	+25
12	+95 +50	+120 +150	+25 +100

PW of cost of differences = PW of benefits of differences

$$500 = 25(P/A, i, 12) + 150(P/F, i, 6) + 100(P/F, i, 12)$$

The sum of the benefits over the 12 years is $550, which is only a little greater than the $500 additional cost. This suggests that the rate of return is quite low. Try $i = 1\%$.

$$500 \doteq 25(11.255) + 150(0.942) + 100(0.887)$$

$$\doteq 281 + 141 + 89 = 511$$

The interest rate is too low. Try $i = 1\frac{1}{2}\%$:

$$500 \doteq 25(10.908) + 150(0.914) + 100(0.836)$$

$$\doteq 273 + 137 + 84 = 494$$

The internal rate of return on the $Y - X$ increment, IRR_{Y-X}, is about 1.3%, far below the 10% minimum attractive rate of return. The additional investment to obtain Machine Y yields an unsatisfactory rate of return; therefore X is the preferred alternative.

SPREADSHEETS AND RATE OF RETURN ANALYSIS

The spreadsheet functions covered in earlier chapters are particularly useful in calculating internal rates of returns (IRRs). If a cash flow diagram can be reduced to at most one P, one A, and/or one F, then the RATE *investment function* can be used. Otherwise the IRR *block function* is used with a cash flow in each period.

The Excel investment function is RATE(n,A,P,F,type,guess). The A, P, and F cannot all be the same sign. The F, type, and guess are optional arguments. The 'type' is end or beginning of period cash flows (for A, but not F), and the 'guess' is the starting value in the search for the IRR.

Considering Example 7-1, where $P = -8200$, $A = 2000$, and $n = 5$, the RATE function would be:

$$RATE(5, 2000, -8200)$$

which gives an answer of 7.00%, which matches that found in Example 7-1.

For Example 7-2, where $P = -700$, $A = 100$, $G = 75$, and $n = 4$, the RATE function cannot be used, since it has no provisions for the arithmetic gradient, G. Suppose the years (row 1) and the cash flows (row 2) are specified in columns B to E. The internal rate of return calculated using IRR(B2:F2) is 6.91%.

	A	B	C	D	E	F
1	Year	0	1	2	3	4
2	Cash flow	−700	100	175	250	325

Figure 7-6 illustrates the use of spreadsheet to graph the present worth of a cash flow series versus the interest rate. The interest rate with present worth equal to 0 is the IRR. The

	A	B	C	D	E	F
1	Year	Cash flow		i	PW	
2	0	−700		0%	150.0	=$B$2+NPV(D2,$B$3:$B$6)
3	1	100		2%	102.1	
4	2	175		4%	58.0	
5	3	250		6%	17.4	
6	4	325		8%	−20.0	
7				10%	−54.7	
8						
9						
10						
11						
12						
13						
14						
15						
16						
17						
18						
19						

FIGURE 7-6 Graphing present worth versus i.

y axis on this graph has been modified so that the x axis intersects at a present worth of −50 rather than at 0. To do this, click on the y axis, then right-click to bring up the 'format axis' option. Select this and then select the tab for the *scale* of the axis. This has a selection for the intersection of the x axis. This process ensures that the x-axis labels are outside the graph.

SUMMARY

Rate of return may be defined as the interest rate paid on the unpaid balance of a loan such that the loan is exactly repaid if the schedule of payments is followed. On an investment, rate of return is the interest rate earned on the unrecovered investment such that the payment schedule makes the unrecovered investment equal to zero at the end of the life of the investment. Although the two definitions of rate of return are stated differently, there is only one fundamental concept being described. It is that the rate of return is the interest rate i at which the benefits are equivalent to the costs.

There are a variety of ways of writing the cash flow equation in which the rate of return i may be the single unknown. Five of them are as follows:

$$\text{PW of benefits} - \text{PW of costs} = 0$$

$$\frac{\text{PW of benefits}}{\text{PW of costs}} = 1$$

$$\text{NPW} = 0$$

$$\text{EUAB} - \text{EUAC} = 0$$

$$\text{PW of costs} = \text{PW of benefits}$$

Rate of return analysis: Rate of return analysis is the most frequently used method in industry, as the resulting rate of return is readily understood. Also, the difficulties in selecting a suitable interest rate to use in present worth and annual cash flow analysis are avoided.

Criteria
Two Alternatives
Compute the incremental rate of return—ΔIRR—on the increment of *investment* between the alternatives. Then,

- if ΔIRR ≥ MARR, choose the higher-cost alternative, or,
- if ΔIRR < MARR, choose the lower-cost alternative

When an increment of *borrowing* is examined, where ΔIRR is the incremental interest rate,

- if ΔIRR ≤ MARR, the increment is acceptable, or
- if ΔIRR > MARR, the increment is not acceptable

Three or More Alternatives
Incremental analysis is needed, which is described in Chapter 8.

Looking Ahead

Rate of return is further described in Appendix 7A. This material concentrates on the difficulties that occur with some cash flows that yield more than one root for the rate of return equation.

PROBLEMS

7-1 For the following diagram, compute the IRR to within ½%.

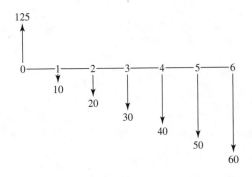

7-2 For the following diagram, compute the interest rate at which the costs are equivalent to the benefits.

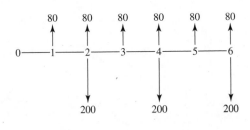

(Answer: 50%)

7-3   For the following diagram, compute the rate of return.

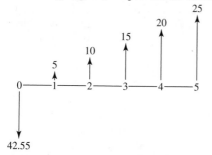

7-4   For the following diagram, compute the rate of return on the $3810 investment.

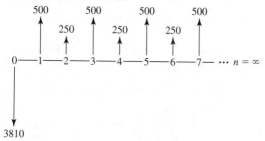

7-5   For the following diagram, compute the rate of return.

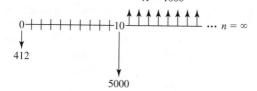

7-6   Consider the following cash flow:

Year	Cash Flow
0	−$500
1	+200
2	+150
3	+100
4	+50

Compute the rate of return represented by the cash flow.

7-7   Consider the following cash flow:

Year	Cash Flow
0	−$1000
1	0
2	+300
3	+300
4	+300
5	+300

Compute the rate of return on the $1000 investment to within 0.1%.   (*Answer:* 5.4%)

7-8   Consider the following cash flow:

Year	Cash Flow
0	−$400
1	0
2	+200
3	+150
4	+100
5	+50

Write one equation, with i as the only unknown, for the cash flow. In the equation you are not to use more than two single payment compound interest factors. (You may use as many other factors as you wish.) Then solve your equation for i.

7-9   Compute the rate of return for the following cash flow to within $\frac{1}{2}$%.

Year	Cash Flow
0	−$100
1–10	+27

(*Answer:* 23.9%)

7-10   Solve the following cash flow for the rate of return to within an $\frac{1}{2}$%.

Year	Cash Flow
0	−$500
1	−100
2	+300
3	+300
4	+400
5	+500

7-11   Compute the rate of return for the following cash flow.

Year	Cash Flow
1–5	−$233
6–10	+1000

7-12   Compute the rate of return for the following cash flow to within 0.5%.

Year	Cash Flow
0	−$640
1	0
2	100
3	200
4	300
5	300

(*Answer:* 9.3%)

7-13   Compute the rate of return for the following cash flow.

Year	Cash Flow
0	−$500
1–3	0
4	+4500

7-14   A woman went to the Beneficial Loan Company and borrowed $3000. She must pay $119.67 at the end of each month for the next 30 months.
(a) Calculate the nominal annual interest rate she is paying to within ±0.15%.
(b) What effective annual interest rate is she paying?

7-15   Helen is buying a $12,375 car with a $3000 down payment, followed by 36 monthly payments of $325 each. The down payment is paid immediately, and the monthly payments are due at the end of each month. What nominal annual interest rate is Helen paying? What effective interest rate?
(*Answers:* 15%; 16.08%)

7-16   Peter Minuit bought an island from the Manhattoes Indians in 1626 for $24 worth of glass beads and trinkets. The 1991 estimate of the value of land on this island was $12 billion. What rate of return would the Indians have received if they had retained title to the island rather than selling it for $24?

7-17   A man buys a corporate bond from a bond brokerage house for $925. The bond has a face value of $1000 and pays 4% of its face value each year. If the bond is paid off at the end of 10 years, what rate of return will the man receive?   (*Answer:* 4.97%)

7-18   A well-known industrial firm has issued $1000 bonds that carry a 4% nominal annual interest, rate paid semi-annually. The bonds mature 20 years from now, at which time the industrial firm will redeem them for $1000 plus the terminal semi-annual interest payment. From the financial pages of your newspaper you learn that the bonds may be purchased for $715 each ($710 for the bond plus a $5 sales commission). What nominal annual rate of return would you receive if you purchased the bond now and held it to maturity 20 years hence?   (*Answer:* 6.6%)

7-19   One aspect of obtaining further education is the prospect of improved future earnings in comparison to non-university graduates. Sharon Shay estimates that a university education has a $28,000 equivalent cost at graduation. She believes the benefits of her education will occur throughout 40 years of employment. She thinks that during her first 10 years out of university, her income will be higher than that of a non-university graduate by $3000 a year. During the subsequent 10 years, she expects an annual income that is $6000 a year higher. During the last 20 years of employment, she expects an annual salary that is $12,000 above the level of the non-university graduate. If her estimates are correct, what rate of return will she receive as a result of her investment in a university education?

7-20   An investor bought a one-acre lot on the outskirts of a city for $9000 cash. Each year he paid $80 of property taxes. At the end of 4 years, he sold the lot. After deducting his selling expenses, he received $15,000. What rate of return did he receive on his investment? (*Answer:* 12.92%)

7-21   A popular magazine offers a lifetime subscription for $200. Such a subscription may be given as a gift to an infant at birth (the parents can read it in those early years) or taken out by an individual for himself. Normally, the magazine costs $12.90 a year. Knowledgeable people say it will probably continue indefinitely at this $12.90 rate. What rate of return would the parents obtain if they bought a life subscription, rather than paying $12.90 per year beginning immediately? You may make any reasonable assumptions, but the compound interest factors must be used *correctly*.

7-22   On April 2, 1988, an engineer bought a $1000 bond of a Canadian airline for $875. The bond paid 6% on its principal amount of $1000, half in each of its April 1 and October 1 semi-annual payments; it repaid the $1000 principal sum on October 1, 2001. What nominal rate of return did the engineer receive from the bond if he held it to its maturity (on October 1, 2001)? (*Answer:* 7.5%)

7-23   The cash price of a machine tool is $3500. The dealer is willing to accept a $1200 down payment and 24 end-of-month monthly payments of $110 each. At what effective interest rate are these terms equivalent? (*Answer:* 14.4%)

7-24   A local bank makes automobile loans. It charges 4% a year in the following manner: if $3600 is borrowed to be repaid over a 3-year period, the bank interest charge is $3600 × 0.04 × 3 years = $432. The bank deducts the $432 of interest from the $3600 loan and gives the customer $3168 in cash. The customer must repay the loan by paying $1/36$ of $3600, or $100, at the end of each month for 36 months. What nominal annual interest rate is the bank actually charging for this loan?

7-25 Upon graduation, every engineer must decide whether to go on to graduate school. Estimate the costs of going full time to university to obtain a master of science degree. Then estimate the resulting costs and benefits. Combine the various consequences into a cash flow table and compute the rate of return. Nonfinancial benefits are probably relevant here too.

7-26 A table saw costs $175 at a local store. You may either pay cash for it or pay $35 now and $12.64 a month for 12 months beginning 30 days hence. If you choose the time payment plan, what nominal annual interest rate will you be charged?   (*Answer:* 15%)

7-27 An investment of $5000 in Biotech common stock proved to be very profitable. At the end of 3 years the stock was sold for $25,000. What was the rate of return on the investment?

7-28 The Southern Guru Copper Company operates a large mine in a South American country. A legislator said in the National Assembly that most of the capital for the mining operation was provided by loans from the World Bank; in fact, Southern Guru has only $500,000 of its own money actually invested in the property. The cash flow for the mine is as follows:

Year	Cash Flow
0	$0.5 million investment
1	3.5 million profit
2	0.9 million profit
3	3.9 million profit
4	8.6 million profit
5	4.3 million profit
6	3.1 million profit
7	6.1 million profit

The legislator divided the $30.4 million total profit by the $0.5 million investment. This produced, he said, a 6080% rate of return on the investment. Southern Guru claims the actual rate of return is much lower. They ask you to compute their rate of return.

7-29 An insurance company is offering to sell an annuity for $20,000 cash. In return the firm will guarantee to pay the purchaser 20 annual end-of-year payments, with the first payment amounting to $1100. Subsequent payments will increase at a uniform 10% rate each year (second payment is $1210; third payment is $1331, etc.). What rate of return will the purchaser receive if he buys the annuity?

7-30 A bank proudly announces that it has changed its interest computation method to continuous compound-

ing. Now $2000 left in the bank for 9 years will double to $4000.
(a) What nominal interest rate, compounded continuously, is the bank paying?
(b) What effective interest rate is it paying?

7-31 Fifteen families live in the village of Stony Hill. Although several water wells have been drilled, none has produced water. The residents take turns driving a water truck to a fire hydrant in a nearby town. They fill the truck with water and then haul it to a storage tank in Stony Hill. Last year, truck fuel and maintenance cost $3180. This year the residents are seriously considering spending $100,000 to install a pipeline from the nearby town to their storage tank. What rate of return would the Stony Hill residents receive on their new water supply pipeline if the pipeline is considered to last
(a) Forever?
(b) 100 years?
(c) 50 years?
(d) Would you recommend that the pipeline be installed? Explain.

7-32 Jan bought 100 shares of Peach Computer stock for $18 per share, plus a $45 brokerage commission. Every 6 months she received a dividend from Peach of 50 cents per share. At the end of 2 years, just after receiving the fourth dividend, she sold the stock for $23 per share and paid a $58 brokerage commission from the proceeds. What annual rate of return did she receive on her investment?

7-33 The Diagonal Stamp Company, which sells used postage stamps to collectors, advertises that its average price has increased from $1 to $5 in the last 5 years. Thus, management states, investors who had purchased stamps from Diagonal 5 years ago would have received a 100% rate of return each year.
(a) To check their calculations, compute the annual rate of return.
(b) Why is your computed rate of return less than 100%?

7-34 An investor purchased 100 shares of Omega common stock for $9000. He held the stock for 9 years. For the first 4 years he received annual end-of-year dividends of $800. For the next 4 years he received annual dividends of $400. He received no dividend for the ninth year. At the end of the ninth year he sold his stock for $6000. What rate of return did he receive on his investment?

7-35 You spend $1000 and in return receive two payments of $1094.60—one at the end of 3 years and the other at the end of 6 years. Calculate the resulting rate of return.

7-36 A mine is for sale for $240,000. It is believed the mine will produce a profit of $65,000 the first year, but the profit will decline by $5000 a year after that, eventually reaching zero, whereupon the mine will be worthless. What rate of return would this $240,000 investment produce for the purchaser of the mine?

7-37 Fred, our cat, just won the local feline lottery to the tune of 3000 cans of 9-Lives cat food (assorted flavours). A local grocer offers to take the 3000 cans and in return, supply Fred with 30 cans a month for the next 10 years. What rate of return, in terms of nominal annual rate, will Fred realize on this deal? (Compute to nearest 0.01%.)

7-38 An apartment building in your neighbourhood is for sale for $140,000. The building has four units, which are rented at $500 a month each. The tenants have long-term leases that expire in 5 years. Maintenance and other expenses for care and upkeep are $8000 annually. A new university is being built in the vicinity, and it is expected that the building could be sold for $160,000 after 5 years.
(*a*) What is the internal rate of return for this investment?
(*b*) Should this investment be accepted if your other options have a rate of return of 12%?

7-39 A new machine can be purchased today for $300,000. The annual revenue from the machine is calculated to be $67,000, and the equipment will last 10 years. Expect the maintenance and operating costs to be $3000 a year and to increase $600 per year. The salvage value of the machine will be $20,000. What is the rate of return for this machine?

7-40 Al Larson asked a bank to lend him money on January 1; the loan had the following repayment plan: the first payment would be $2 on February 28, with subsequent monthly payments increasing by $2 a month on an arithmetic gradient. (The March 31 payment would be $4; the April 30 payment would be $6, etc.) The payments were to continue for 11 years, making a total of 132 payments.
(*a*) Compute the total amount of money Al would pay the bank under the proposed repayment plan.
(*b*) If the bank charges interest at 12% nominal per year, compounded monthly, how much would it be willing to lend Al on the proposed repayment plan?

7-41 The following advertisement appeared in the *Wall Street Journal* on Thursday, February 9, 1995:

> 'There's nothing quite like the Seville SmartLease. Seville SLS $0 down, $599 a month/36 months.'

First month's lease payment of $599 plus $625 refundable security deposit and a consumer down payment of $0 for a total of $1224 due at lease signing. Monthly payment is based on a net capitalized cost of $39,264 for total monthly payments of $21,564. Payment examples based on a 1995 Seville SLS: $43,658 MSRP including destination charge. Tax, license, title fees, and insurance extra. Option to purchase at lease end for $27,854. Mileage charge of $0.15 per mile over 36,000 miles.
(*a*) Set up the cash flows.
(*b*) Determine the interest rate (nominal and effective) for the lease.

7-42 After 15 years of working for one employer, you transfer to a new job. During these years your employer contributed (that is, she diverted from your salary) $1500 each year to an account for your retirement (a fringe benefit), and you contributed a matching amount each year. The whole fund was invested at 5% during that time, and the value of the account now stands at $30,000. You are now faced with two alternatives. (1) You may leave both contributions in the fund until you retire in 35 years, during which time you will get the future value of this amount at 5% interest per year. (2) You may take out the total value of 'your' contributions, which is $15,000 (one-half of the total $30,000). You can do as you wish with the money you take out, but the other half will be lost as far as you are concerned. In other words, you can give up $15,000 today for the sake of getting now the other $15,000. Otherwise, you must wait 35 years more to get the accumulated value of the entire fund. Which alternative is more attractive? Explain your choice.

7-43 A finance company is using the following 'Money by Mail' offer. Calculate the yearly nominal IRR received by the company if a customer chooses the loan of $2000 and accepts the credit insurance (Life and Dis.).

Money by Mail	Non-negotiable INE	
	1/96	
For the		To borrow $3000,
Amount of **$3000 or $2000 or $1000** **Dollars**		$2000, or $1000
Pay to the Order of **I Feel Rich**		
Limited Time Offer		

For the Amount of $3000 Dollars	APR 18.95%	
Pay to the	Finance Charge $1034.29	
Order of _____ **I Feel Rich**		
Total of Payments **$4,280.40**	Monthly Payment $118.90	$3000 loan terms
Number of Monthly Payments	Credit Line Premium	
36 Months	$83.46*	
Amount Financed $3,246.25	Credit Disability Premium $162.65*	

For the Amount of **$2000 Dollars**	APR 19.95%	
Pay to the	Finance Charge $594.25	
Order of _____ **I Feel Rich**		
Total of Payments $2,731.50	Monthly Payment $91.05	$2000 loan terms
Number of Monthly Payments	Credit Line Premium	
30 Months	$44.38*	
Amount Financed $2,137.25	Credit Disability Premium $92.87*	

For the Amount of **$1000 Dollars**	APR 20.95%	
Pay to the	Finance Charge $245.54	
Order of _____ **I Feel Rich**		
Total of Payments $1,300.80	Monthly Payment $54.20	$1000 loan terms
Number of Monthly Payments	Credit Line Premium	
24 Months	$16.91*	
Amount Financed $1055.26	Credit Disability Premium $39.02*	

*Credit insurance. If selected, premium will be paid from amount financed. If not selected, cash advance is total amount financed.

7-44 A finance company is using the 'Money by Mail' offer shown in Problem 7-43. Calculate the yearly nominal IRR received by the company if a customer chooses the $3000 loan but declines the credit insurance.

7-45 In his will, Frank's uncle has given Frank the choice between two alternatives:

Alternative 1	$2000 cash
Alternative 2	$150 cash now plus $100 per month for 20 months beginning the first day of next month

(a) At what rate of return are the two alternatives equivalent?

(b) If Frank thinks the rate of return in (a) is too low, which alternative should he select?

7-46 The owner of a corner lot wants to find a use that will yield a desirable return on his investment. After much study and calculation, he decides that the two best alternatives are:

	Build Gas Station	Build Soft Ice Cream Stand
First cost	$80,000	$120,000
Annual property taxes	3,000	5,000
Annual income	11,000	16,000
Life of building, in years	20	20
Salvage value	0	0

If the owner wants a minimum attractive rate of return on his investment of 6%, which of the two alternatives would you recommend?

7-47 Two alternatives are as follows:

Year	A	B
0	−$2000	−$2800
1	+800	+1100
2	+800	+1100
3	+800	+1100

If 5% is considered the minimum attractive rate of return, which alternative should be selected?

7-48 Consider two mutually exclusive alternatives:

Year	X	Y
0	−$100	−$50.0
1	35	16.5
2	35	16.5
3	35	16.5
4	35	16.5

If the minimum attractive rate of return is 10%, which alternative should be selected?

7-49 Consider these two mutually exclusive alternatives:

Year	A	B
0	−$50	−$53
1	17	17
2	17	17
3	17	17
4	17	17

At a MARR of 10%, which alternative should be chosen?   (*Answer:* A)

7-50 Two mutually exclusive alternatives are being considered. Both have a 10-year useful life. If the MARR is 8%, which alternative is preferred?

	A	B
Initial cost	$100.00	$50.00
Uniform annual benefit	19.93	11.93

7-51 Consider two mutually exclusive alternatives:

Year	X	Y
0	−$5000	−$5000
1	−3000	+2000
2	+4000	+2000
3	+4000	+2000
4	+4000	+2000

If the MARR is 8%, which alternative should be selected?

7-52 Two mutually exclusive alternatives are being considered.

Year	A	B
0	−$2500	−$6000
1	+746	+1664
2	+746	+1664
3	+746	+1664
4	+746	+1664
5	+746	+1664

If the minimum attractive rate of return is 8%, which alternative should be chosen? Solve the problem by
(a) Present worth analysis
(b) Annual cash flow analysis
(c) Rate of return analysis

7-53 A contractor is considering whether to buy or lease a new machine for his layout site work. Buying a new machine will cost $12,000, and the machine will have a salvage value of $1200 at the end of its useful life of 8 years. On the other hand, leasing requires an annual lease payment of $3000. On the basis of a 15% MARR and an internal rate of return analysis, which alternative should the contractor be advised to accept? The cash flows are as follows:

Year (*n*)	Alt. A (purchase)	Alt. B (lease)
0	−$12,000	0
1		−$3000
2		−3000
3		−3000
4		−3000
5		−3000
6		−3000
7		−3000
8	+1200	−3000

7-54 Two hazardous environment facilities are being evaluated, with the projected life of each facility being 10 years. The cash flows are as follows:

	Alt. A	Alt. B
First cost	$615,000	$300,000
Maintenance and operating cost	10,000	25,000
Annual benefits	158,000	92,000
Salvage value	65,000	−5,000

The company uses a MARR of 15%. Using rate of return analysis, which alternative should be selected?

7-55 Two alternatives are being considered:

	A	B
First cost	$9200	$5000
Uniform annual benefit	$1850	$1750
Useful life, in years	8	4

If the minimum attractive rate of return is 7%, which alternative should be selected?

7-56 Jean has decided it is time to buy a new battery for her car. Her choices are:

	Zappo	Kicko
First cost	$56	$90
Guarantee period, in months	12	24

Jean believes the batteries can be expected to last only for the guarantee period. She does not want to invest extra money in a battery unless she can expect a 50% rate of return. If she plans to keep her present car another 2 years, which battery should she buy?

7-57 Two alternatives are being considered:

	A	B
Initial cost	$9200	$5000
Uniform annual benefit	1850	1750
Useful life, in years	8	4

Base your computations on a MARR of 7% and an 8-year analysis period. If identical replacement is assumed, which alternative should be chosen?

7-58 Two investment opportunities are as follows:

	A	B
First cost	$150	$100
Uniform annual benefit	25	22.25
End-of-useful-life salvage value	20	0
Useful life, in years	15	10

At the end of 10 years, Alt. B is not replaced. Thus, the comparison is 15 years of A versus 10 years of B. If the MARR is 10%, which alternative should be selected?

APPENDIX 7A Difficulties in Solving for an Interest Rate

After completing this chapter appendix, students should be able to:

- Explain why the cash flows of some projects cannot be solved for a single positive interest rate.
- Identify the patterns where multiple roots can occur.
- Evaluate how many potential roots exist for a particular project.
- Use the *modified internal rate of return (MIRR)* methodology in multiple-root cases.

Example 7A-1 illustrates the situation.

EXAMPLE 7A-1

The Going Aircraft Company has an opportunity to supply a large airplane to Interair, a foreign airline. Interair will pay $19 million when the contract is signed and $10 million one year later. Going estimates its second- and third-year costs at $50 million each. Interair will take delivery of the airplane during Year 4, and it agrees to pay $20 million at the end of that year and the $60 million balance at the end of Year 5. Compute the rate of return on this project.

SOLUTION

The PW of each cash flow can be computed at various interest rates. For example, for Year 2 and $i = 10\%$: PW $= -50(P/F, 10\%, 2) = -50(0.826) = -41.3$.

Year	Cash Flow	0%	10%	20%	40%	50%
0	+$19	+$19	+$19	+$19	+$19	+$19
1	+10	+10	+9.1	+8.3	+7.1	+6.7
2	−50	−50	−41.3	−34.7	−25.5	−22.2
3	−50	−50	−37.6	−28.9	−18.2	−14.8
4	+20	+20	+13.7	+9.6	+5.2	+3.9
5	+60	+60	+37.3	+24.1	+11.2	+7.9
	PW =	+$9	+$0.1	−$2.6	−$1.2	+$0.5

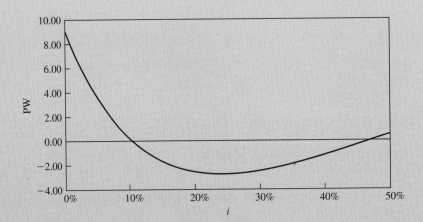

FIGURE 7A-1 PW plot.

The PW plot for this cash flow is represented in Figure 7A-1. We see that this cash flow produces *two* points at which PW = 0: one at 10.24% and the other at 47.30%.

Why Multiple Solutions Can Occur

Example 7A-1 produced unexpected results. This could happen because there were two changes in the signs of the cash flows. Years 0 and 1 were positive, Years 2 and 3 were negative, and Years 4 and 5 were positive. The cash flow series went from positive cash flows to negative cash flows to positive cash flows.

Most cash flow series have only one shift in sign. Investments start with one or more years of negative cash flows followed by many years of positive cash flows. Loans begin with a positive cash flow that is repaid with later negative cash flows. These problems have a unique rate of return because they have a single change in the sign of the cash flows.

Having more than one sign change in the cash flow series can produce multiple points or roots at which the PW equals 0. To see how many roots are possible, we link solving an economic analysis problem to solving a mathematical equation.

A project's cash flows are the values from CF_0 to CF_n.

Year	Cash Flow
0	CF_0
1	CF_1
2	CF_2
⋮	⋮
n	CF_n

The equation to find the internal rate of return, where PW = 0, is written as follows:

$$\text{PW} = 0 = CF_0 + CF_1(1+i)^{-1} + CF_2(1+i)^{-2} + \cdots + CF_n(1+i)^{-n} \qquad (7A\text{-}1)$$

If we let $x = (1+i)^{-1}$, then Equation 7A-1 may be written

$$0 = CF_0 + CF_1 x^1 + CF_2 x^2 + \cdots + CF_n x^n \qquad (7A\text{-}2)$$

Equation 7A-2 is an nth-order polynomial, and *Descartes' rule* describes the number of positive roots for x. The rule is:

> If a polynomial with real coefficients has m sign changes, then the number of positive roots will be $m - 2k$, where k is an integer between 0 and $m/2$.

A sign change exists when successive non-zero terms written according to ascending powers of x have different signs. If x is greater than zero, then the number of sign changes in the cash flows equals the number in the equation. Descartes' rule means that the number of positive roots (values of x) of the polynomial cannot exceed m, the number of sign changes. The number of positive roots for x must either be equal to m or less by an even integer.

Thus, Descartes' rule for polynomials gives the following:

Number of Sign Changes, m	Number of Positive Values of x	Number of Positive Values of i
0	0	0
1	1	1 or 0
2	2 or 0	2, 1, or 0
3	3 or 1	3, 2, 1, or 0
4	4, 2, or 0	4, 3, 2, 1, or 0

If x is greater than 1, the corresponding value of i is negative. If there is only one root and it is negative, then it is a valid IRR. One example of a valid negative IRR comes from a bad outcome, such as the IRR for a failed research and development project. Another example would be a privately sponsored student loan set up so that if a student volunteers for Crossroads Canada World Youth, only half of the principal need be repaid.

If a project has a negative and a positive root for i, for most projects only the positive root of i is used.

Projects with Multiple Sign Changes

Example 7A-1 is representative of projects for which there are initial payments (e.g., when the order for the ship, airplane, or building is signed), then the bulk of the costs occur, and then there are more payments on completion.

Projects of other types often have two or more sign changes in their cash flows. Projects with a salvage cost usually have two sign changes. This salvage cost can be large for environmental restoration at termination. Examples include pipelines, open-pit mines, and nuclear power plants. The following cash flow diagram is representative.

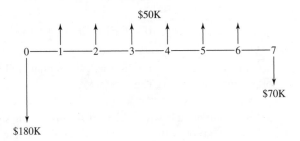

Many enhancement projects for mines and deposits have a pattern of two sign changes. Example 7A-2 describes an oil well in an existing field. The initial investment recovers more of the resource and speeds recovery of resources that would have been recovered eventually. The resources shifted for earlier recovery can lead to two sign changes.

In Example 7A-3, we consider staged construction, where three sign changes are common.

Other examples can be found in incremental comparisons of alternatives with unequal lives. In the next section we learn how to determine whether each of these examples has multiple roots.

EXAMPLE 7A-2

Adding an oil well to an existing field costs $4 million ($4M). It will increase recovered oil by $3.5M, and it shifts $4.5M worth of production from Years 5, 6, and 7 to earlier years. Thus, the cash flows for Years 1 to 4 total $8M and for Years 5 to 7 total −$4.5M. If the well is justified, one reason is that the oil is recovered sooner. How many roots for the PW equation are possible?

SOLUTION

The first step is to draw the cash flow diagram and count the number of sign changes. The following pattern is representative, although most wells have a longer life.

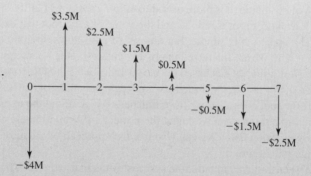

There are two sign changes; thus there may be 0, 1, or 2 positive roots for the PW = 0 equation. The additional recovery corresponds to an investment, and the shifting of recovery to earlier years corresponds to a loan (positive cash flow now and negative later). Thus, the oil wells are neither an investment nor a loan; they are a combination of both.

EXAMPLE 7A-3

A project has a first cost of $120,000. Net revenues begin at $30,000 in Year 1 and then increase by $2000 per year. In Year 5 the facility is expanded at a cost of $60,000 so that demand can continue to expand at $2000 per year. How many roots for the PW equation are possible?

SOLUTION

The first step is to draw the cash flow diagram. Then counting the three sign changes is easy.

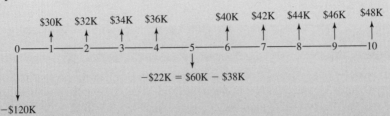

With three sign changes, there may be 0, 1, 2, or 3 positive roots for the PW = 0 equation.

Evaluating How Many Roots There Are

The number of sign changes tells us how many roots are possible—not how many roots there are. Rather than covering the many mathematical approaches that *may* tell us if the root will be unique, it is more useful to employ the power of the spreadsheet. A spreadsheet can show us if a root is unique and the value of each root that exists.

The approach is simple. For any set of cash flows, graph the PW as a function of the interest rate. We are interested in positive interest rates, so the graph usually starts at $i = 0$. Since *Descartes' rule* is based on $x > 0$ and $x = (1 + i)^{-1}$, sometimes the graph is started close to $i = -1$. This will identify any negative values of i that solve the equation.

If the root is unique, we have finished. We've found the internal rate of return. If there are multiple roots, then we know the project's PW at all interest rates—including the one our organization uses. We can also use the approach of the next section to find a *modified internal rate of return*.

The easiest way to build the spreadsheet is to use the NPV(i,values) function in Excel. Remember that this function applies to cash flows in periods 1 to N, so that the cash flow at time 0 must be added in separately. The following examples use a spreadsheet to answer the question of how many roots there are in each cash flow diagram in the last section.

EXAMPLE 7A-4

This project is representative of ones with a salvage cost. How many roots for the PW equation exist?

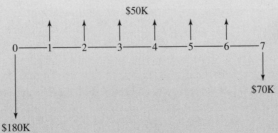

SOLUTION

Figure 7A-2 shows the spreadsheet calculations and the graph of PW versus i.

	A	B	C	D	E	F	G
1	Year	Cash Flow		i	PW *		
2	0	−180		−40%	−126.39	=$B$2+NPV(D2,$B$3:$B$9)	
3	1	50		−30%	219.99		
4	2	50		−20%	189.89		
5	3	50		−10%	114.49		
6	4	50		0%	50.00		
7	5	50		10%	1.84		
8	6	50		20%	−33.26		
9	7	−70		30%	−59.02		
10				40%	−78.24		
11	IRR	10.45%		50%	−92.88		
12	root	−38.29%					

FIGURE 7A-2 PW versus i for project with salvage cost.

In this case, there is 1 positive root of 10.45%. The value can be used as an IRR. There is also a negative root of $i = -38.29\%$. This root is not useful. With two sign changes, other similar diagrams may have 0, 1, or 2 positive roots for the PW = 0 equation. The larger the value of the final salvage cost, the more likely it is that 0 or 2 positive roots will occur.

EXAMPLE 7A-5 (7A-2 revisited)

Adding an oil well to an existing field had the following cash flow diagram. How many roots for the PW equation exist and what are they?

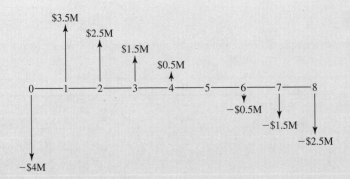

SOLUTION

Figure 7A-3 shows the spreadsheet calculations and the graph of PW versus i.

	A	B	C	D	E	F	G
1	Year	Cash Flow		i	PW		
2	0	−4.00		0%	−0.50	=$B$2+NPV(D2,$B$3:$B$9)	
3	1	3.50		5%	0.02		
4	2	2.50		10%	0.28		
5	3	1.50		15%	0.37		
6	4	0.50		20%	0.36		
7	5	−0.50		25%	0.29		
8	6	−1.50		30%	0.19		
9	7	−2.50		35%	0.06		
10				40%	−0.08		
11	root	4.73%		45%	−0.22		
12	root	37.20%		50%	−0.36		
13							
14							
15							
16							
17							
18							
19							
20							
21							
22							

FIGURE 7A-3 PW versus i for oil well.

In this case, there are 2 positive roots at 4.73 and 37.20%. These roots are not useful. This project is a combination of an investment and a loan, so we don't even know if we want a high rate or a low rate. If our interest rate is about 20%, then the project has a positive PW. However, small changes in the data can make for large changes in these results.

It is useful to apply the modified internal rate of return described in the next section.

EXAMPLE 7A-6 (7A-3 revisited)

A project has a first cost of $120,000. Net revenues begin at $30,000 in Year 1, and then increase by $2000 a year. In Year 5 the facility is expanded at a cost of $60,000 so that demand can continue to expand at $2000 a year. How many roots for the PW equation have a positive value for i?

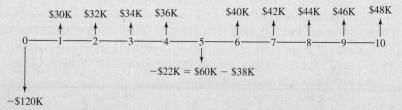

SOLUTION

Figure 7A-4 shows the spreadsheet calculations and the graph of PW versus i.

	A	B	C	D	E	F	G
1	Year	Cash Flow		i	PW		
2	0	−120.00		−80%	579646330	=$B$2+NPV(D2,$B$3:$B$12)	
3	1	30.00		−60%	735723		
4	2	32.00		−40%	17651		
5	3	34.00		−30%	1460		
6	4	36.00		0%	210		
7	5	−22.00		10%	73		
8	6	40.00		20%	7		
9	7	42.00		30%	−28		
10	8	44.00		40%	−48		
11	9	46.00		50%	−62		
12	10	48.00		60%	−71		
13				70%	−78		
14	IRR	21.69%		80%	−83		
15				90%	−87		
16							
17							
18							
19							
20							
21							
22							
23							
24							
25							
26							

FIGURE 7A-4 PW versus i for project with staged construction.

In this case, there is 1 positive root of 21.69%. The value can be used as an IRR, and the project is very attractive.

When there are two or more sign changes in the cash flow, we know that there are several possibilities concerning the number of positive values of i. Probably the greatest danger in this situation is to fail to recognize the multiple possibilities and to solve for a value of i. The approach of constructing the PW plot establishes whether there are multiple roots and what their values are. This may be tedious to do by hand but is very easy with a spreadsheet (or a graphing calculator).

If there is a single positive value of i, we have no problem. On the other hand, if there is no positive value of i, or if there are multiple positive values, the situation may be attractive, unattractive, or confusing. When there are multiple positive values of i, none of them should be considered a suitable measure of the rate of return or attractiveness of the cash flow.

Modified Internal Rate of Return (MIRR)

Two external rates of return can be used to ensure that the resulting equation is solvable for a unique rate of return—the MIRR. The MIRR is a measure of the attractiveness of the cash flows, but it is also a function of the two external rates of return.

The rates that are *external* to the project's cash flows are (1) the rate at which the organization normally invests and (2) the rate at which it normally borrows. These are external rates for investing, e_{inv}, and for financing, e_{fin}. Because profitable firms invest at higher rates than they borrow at, the rate for investing is generally higher than the rate for financing. Sometimes a single external rate is used for both, but this requires the questionable assumption that investing and financing happen at the same rate.

The approach is as follows:

1. Combine cash flows in each period into a single net receipt, R_t, or net expense, E_t.
2. Find the present worth of the expenses with the financing rate.
3. Find the future worth of the receipts with the investing rate.
4. Find the MIRR which makes the present and future worths equivalent.

The result is Equation 7A-3. This equation will have a unique root, since it has a single negative present worth and a single positive future worth. There is only one sign change in the resulting series.

$$(F/P, \text{MIRR}, n) \sum_t E_t(P/F, e_{fin}, t) = \sum_t R_t(F/P, e_{inv}, n - t) \qquad (7A\text{-}3)$$

There are other external rates of return, but the MIRR has historically been the most clearly defined. Since all of the external rates of return are affected by the assumed values for the investing and financing rates, none is a *true* rate of return on the project's cash flow. The MIRR also has an Excel function, which can now be used very easily. Example 7A-7 illustrates the calculation, which is also summarized in Figure 7A-5.

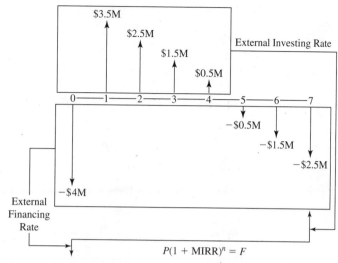

FIGURE 7A-5 MIRR for the oil well.

EXAMPLE 7A-7 (7A-2 and 7A-5 revisited)

Adding an oil well to an existing field had the cash flows summarized in Figure 7A-5. If the firm normally borrows money at 8% and invests at 15%, find the modified internal rate of return (MIRR).

SOLUTION

Figure 7A-6 shows the spreadsheet calculations.

	A	B	C	D	E	F	G	H	I
1	8%	external financing rate							
2	15%	external investing rate							
3	Year	0	1	2	3	4	5	6	7
4	Cash Flow	−4.00	3.50	2.50	1.50	0.50	−0.50	−1.50	−2.50
5	13.64%		Cell A5 contains = MIRR(A4:I4,A1,A2)						

FIGURE 7A-6 MIRR for oil well.

It is also possible to calculate the MIRR by hand. While more work, the process does clarify what the MIRR function is doing.

1. Each period's cash flow is already a single net receipt or expenditure.
2. Find the present worth of the expenses with the financing rate.

$$PW = -4M - 0.5M(P/F, 8\%, 5) - 1.5M(P/F, 8\%, 6) - 2.5M(P/F, 8\%, 7)$$

$$= -4M - 0.5M(0.6806) - 1.5M(0.6302) - 2.5M(0.5835) = -6.744M$$

3. Find the future worth of the receipts with the investing rate.

$$FW = 3.5M(F/P, 15\%, 6) + 2.5M(F/P, 15\%, 5) + 1.5M(F/P, 15\%, 4) + 0.5M(F/P, 15\%, 3)$$

$$= 3.5M(2.313) + 2.5M(2.011) + 1.5M(1.749) + 0.5M(1.521) = 16.507M$$

4. Find the MIRR that makes the present worth and future worth equivalent.

$$0 = (1 + MIRR)^n \cdot PW + FW$$

$$0 = (1 + MIRR)^7(-6.744M) + 16.507M$$

$$(1 + MIRR)^7 = 16.507M/6.744M = 2.448$$

$$(1 + MIRR) = 2.448^{1/7} = 1.1364$$

$$MIRR = 13.64\%$$

The MIRR does allow calculation of a rate of return for *any* set of cash flows. However, the result is only as realistic as the external rates that are used. The MIRR value can depend as much on the external rates that are used as it does on the cash flows that it is describing.

SUMMARY

In cash flows with more than one sign change, we find that solving the cash flow equation can result in more than one positive rate of return. Typical situations include a new oil well in an existing field, a project with a significant salvage cost, and staged construction.

In a sign change, successive non-zero values in the cash flow have different signs (that is, they change from + to −, or vice versa). Zero sign changes indicates there is no rate of return, as the cash flow is either all disbursements or all receipts.

One sign change is the usual situation, and a single positive rate of return generally results. There will be a negative rate of return whenever loan repayments are less than the loan or an investment fails to return benefits at least equal to the investment.

Multiple sign changes may result in multiple positive roots for *i*. When they occur, none of the positive multiple roots is a suitable measure of the project's economic desirability. If multiple roots are identified by a graphing of the present worth versus the interest rate, then the modified internal rate of return can be used to evaluate the project.

Graphing the present worth versus the interest rate ensures that the analyst recognizes that the cash flow has multiple sign changes. Otherwise a rate could be found and used that is not in fact a meaningful descriptor of the project.

The modified internal rate of return (MIRR) relies on rates for investing and borrowing that are external to the project. The number of sign changes are reduced to one, ensuring that the MIRR can be found.

PROBLEMS

Unless the problem asks a different question or provides different data, (1) determine how many roots are possible, and (2) graph the PW versus the interest rate to see whether multiple roots occur. If the root is a unique IRR, it is the project's rate of return. If there are multiple roots, then use an *external investing rate* of 12% and an *external borrowing rate* of 6%. Compute and use the MIRR as the project's rate of return.

7A-1 Find the rate of return for the following cash flow:

Year	Cash Flow
0	−$15,000
1	10,000
2	−8,000
3	11,000
4	13,000

(*Answer:* 21.2% IRR)

7A-2 A group of businessmen formed a partnership to buy and race an Indianapolis-type racing car. They agreed to pay an individual $50,000 for the car and associated equipment. The payment was to be in a lump sum at the end of the year. In what must have been beginner's luck, the group won a major race the first week and $80,000. The rest of the first year, however, was not as good: at the end of the first year, the group had to pay out $35,000 for expenses plus the $50,000 for the car and equipment. The second year was a poor one: the group had to pay $70,000 just to clear up the racing debts at the end of the year. During the third and fourth years, racing income just equalled costs. When the group was approached by a prospective buyer for the car, they readily accepted $80,000 cash, which was paid at the end of the fourth year. What rate of return did the businessmen obtain from their racing venture?   (*Answer:* 9.6% MIRR)

7A-3 A student organization, at the beginning of the fall quarter, purchased and operated a soft-drink vending machine as a means of helping finance its activities. The vending machine cost $75 and was installed at a gasoline station near the university. The student organization pays $75 every 3 months to the station owner for the right to keep the vending machine at the station. During the year that the student organization owned the machine, they received the following quarterly income from it, before making the $75 quarterly payment to the station owner:

Quarter	Income
Fall	$150
Winter	25
Spring	125
Summer	150

At the end of one year, the student group resold the machine for $50. Determine the quarterly cash flow.

Then determine a quarterly rate of return, a nominal annual rate, and an effective annual rate.

7A-4 Given the following cash flow, determine the rate of return on the project.   (*Answer:* 11.3% MIRR)

Year	Cash Flow
0	−$500
1	+2000
2	−1200
3	−300

7A-5 Given the following cash flow, determine the rate of return on the investment.   (*Answer:* IRR = 21.1% IRR)

Year	Cash Flow
0	−$500
1	200
2	−500
3	1200

7A-6 Given the following cash flow, determine the rate of return on the investment.

Year	Cash Flow
0	−$100
1	360
2	−570
3	360

7A-7 Compute the rate of return on the investment characterized by the following cash flow.

Year	Cash Flow
0	−$110
1	−500
2	300
3	−100
4	400
5	500

7A-8 Compute the rate of return on the investment characterized by the following cash flow.

Year	Cash Flow
0	−$50.0
1	+20.0
2	−40
3	+36.8
4	+36.8
5	+36.8

(*Answer:* IRR = 15.4%)

7A-9 A firm invested $15,000 in a project that appeared to have excellent potential. Unfortunately, a lengthy labour dispute in Year 3 resulted in costs that exceeded benefits by $8000. The cash flow for the project is as follows:

Year	Cash Flow
0	−$15,000
1	+10,000
2	+6,000
3	−8,000
4	+4,000
5	+4,000
6	+4,000

Compute the rate of return for the project.

7A-10 The following cash flow has no positive interest rate. The project, which had a projected life of 5 years, was terminated early.

Year	Cash Flow
0	−$50
1	+20
2	+20

There is, however, a negative interest rate. Compute its value.   (*Answer:* −13.7%)

7A-11 For the following cash flow, compute the rate of return.

Year	Cash Flow
0	−$20
1	0
2	−10
3	+20
4	−10
5	+100

7A-12 Given the following cash flow:

Year	Cash Flow
0	−$800
1	+500
2	+500
3	−300
4	+400
5	+275

What is the rate of return on the investment? (*Answer:* 26.55% IRR)

7A-13 Consider the following cash flow.

Year	Cash Flow
0	−$100
1	240
2	−143

If the minimum attractive rate of return is 12%, should the project be undertaken?

7A-14 Refer to the strip-mining project in Example 5-10. Compute the rate of return for the project.

7A-15 Consider the following cash flow.

Year	Cash Flow
0	−$500
1	800
2	170
3	−550

Compute the rate of return on the investment.

7A-16 Determine the rate of return on the investment for the following cash flow.

Year	Cash Flow
0	−$100
1	+360
2	−428
3	168

7A-17 Compute the rate of return on the investment on the following cash flow.

Year	Cash Flow
0	−$1200
1	+358
2	+358
3	+358
4	+358
5	+358
6	−394

7A-18 Determine the rate of return on the investment on the following cash flow.

Year	Cash Flow
0	−$3570
1–3	+1000
4	−3170
5–8	+1500

7A-19 Bill purchased a vacation lot he saw advertised on television for an $800 down and monthly payments of $55. When he visited the lot, he found it was not something he wanted to own. After 40 months he was finally able to sell the lot. The new purchaser assumed the balance of the loan on the lot and paid Bill $2500. What rate of return did Bill receive on his investment?

7A-20 Compute the rate of return on an investment having the following cash flow.

Year	Cash Flow
0	−$850
0	+600
2–9	+200
10	−1800

7A-21 Assume that the following cash flows are associated with a project.

Year	Cash Flow
0	−$16,000
1	−8,000
2	11,000
3	13,000
4	−7,000
5	8,950

Compute the rate of return for this project.

7A-22 Compute the rate of return for the following cash flow.

Year	Cash Flow
0	−$200
1	+100
2	+100
3	+100
4	−300
5	+100
6	+200
7	+200
8	−124.5

(*Answer:* 20%)

7A-23 Following are the annual cost data for a tomato press.

Year	Cash Flow
0	−$210,000
1	88,000
2	68,000
3	62,000
4	−31,000
5	30,000
6	55,000
7	65,000

What is the rate of return associated with this project?

7A-24 A project has been in operation for 5 years, yielding the following annual cash flows:

Year	Cash Flow
0	−$103,000
1	102,700
2	−87,000
3	94,500
4	−8,300
5	38,500

Calculate the rate of return and state whether it has been an acceptable rate of return.

7A-25 Consider the following situation.

Year	Cash Flow
0	−$200
1	+400
2	−100

What is the rate of return?

7A-26 An investor is considering two mutually exclusive projects. He can obtain a 6% before-tax rate of return on external investments, but he requires a minimum attractive rate of return of 7% for these projects. Use a 10-year analysis period to compute the incremental rate of return from investing in Project A rather than Project B.

	Project A: Build Drive-Up Photo Shop	Project B: Buy Land in Hawaii
Initial capital investment	$58,500	$ 48,500
Net uniform annual income	6,648	0
Salvage value 10 years hence	30,000	138,000
Computed rate of return	8%	11%

7A-27 In January, 2003, an investor bought a convertible debenture bond issued by the XLA Corporation. The bond cost $1000 and paid $60 per year interest in annual payments on December 31. Under the convertible feature of the bond, it could be converted into 20 shares of common stock by tendering the bond, together with $400 cash. The day after the investor received the December 31, 2005, interest payment, she submitted the bond together with $400 to the XLA Corporation. In return, she received the 20 shares of common stock. The common stock paid no dividends. On December 31, 2007, the investor sold the stock for $1740, terminating her 5-year investment in XLA Corporation. What rate of return did she receive?

CHAPTER 8

Incremental Analysis

Tapping into Controversy

When New York's Tappan Zee Bridge opened in 1955, it seemed big enough to handle any amount of traffic that was likely to come its way. The bridge spanned the Hudson River at a crossing in Westchester County, well north of New York City, near towns seemingly so removed from Manhattan that they were referred to as 'satellite communities'. Invoking visions of westward expansion and unpopulated vistas, state government officials proudly told voters that the $81 million bridge project would open up 'new frontiers'.

During its first full year of operation, the bridge carried only 18,000 vehicles per day, far below its design capacity of 100,000. By 2000, the satellite communities were called 'suburbs', and the Tappan Zee Bridge was being used by 135,000 vehicles every day. That number is expected to increase to 175,000 over the next two decades.

Clearly, something must be done to ease the growing congestion on the bridge. But what? The bridge's usable space has already been expanded once by converting a large median strip into a reversible traffic lane. And the bridge itself is nearing the end of its useful life, which originally was intended to be 50 years.

Upgrading the existing bridge would likely cost over $1 billion and would not adequately alleviate traffic congestion. Building a new bridge could cost as much as $4 billion dollars—quite a jump from the original price tag.

Why so costly? Inflation explains part of it, of course. The original Tappan Zee Bridge would cost almost $600 million in today's dollars. The new bridge would also be bigger. It would have eight lanes instead of six, a 33% increase.

But much of the cost would come from additional expenses that could not have been contemplated 50 years ago. Backers of the original bridge happily predicted that it would increase property values and attract new homebuyers to 'outlying areas'. And so it did. But nowadays, this effect is called suburban sprawl, and most taxpayers want less of it, not more. Many local residents fear that enlarging the bridge would simply encourage even more traffic—soon leading back to the same quandary they are now in.

To address these concerns, some government officials have suggested incorporating light-rail tracks and bus lanes into the new structure, in an attempt to encourage greater use of mass transit. Others have suggested installation of a more sophisticated toll-collection system that would give discounts to off-peak travellers. Any such additions to the bridge would considerably drive up the cost of construction.

After Completing This Chapter...

The student should be able to:

- Define *incremental analysis* and differentiate it from a standard present worth, annual worth, and internal rate of return analyses.
- Use a *graphical technique* to visualize and solve problems involving mutually exclusive choices.
- Use spreadsheets to solve incremental analysis problems.

QUESTIONS TO CONSIDER

1. In today's more environmentally aware decision-making climate, costly features such as the proposed light-rail lanes may have to be included in any proposal for new bridge construction just to get the project off the drawing board. How does this affect the "threshold" cost of public works projects?

2. Any bridge that is built would have to undergo extensive environmental impact review and a public comment period before being approved. This is likely to result in pressure for additional environmentally friendly components to be built into the project, including methods for limiting emissions from cars that are waiting to cross the bridge. How would this affect the ultimate cost of the project?

We have seen how to solve problems by each of three major methods, with one exception: for three or more alternatives, no rate of return solution was given. The reason is that under these circumstances, incremental analysis is required and it has not been discussed. This chapter will show how to solve that problem.

Incremental analysis can be defined as the examination of the differences between alternatives. By emphasizing alternatives, we are really deciding whether or not differential costs are justified by differential benefits.

In retrospect, we see that the simplest form of incremental analysis was presented in Chapter 7. We did incremental analysis by the rate of return evaluation of the differences between two alternatives. We recognized that the two alternatives could be related as follows:

$$\begin{bmatrix} \text{Higher-cost} \\ \text{alternative} \end{bmatrix} = \begin{bmatrix} \text{Lower-cost} \\ \text{alternative} \end{bmatrix} + \begin{bmatrix} \text{Differences} \\ \text{between them} \end{bmatrix}$$

We will see that incremental analysis can be examined either graphically or numerically. We will first look at graphical representations of problems, proceed with numerical solutions of rate of return problems, and see that a graphical representation may be useful in examining problems whether we are using incremental analysis or not.

GRAPHICAL SOLUTIONS

In the last chapter, we examined problems with two alternatives. Our method of solution represented a form of incremental analysis. A graphical review of that situation will help to introduce incremental analysis.

EXAMPLE 8-1

This is a review of Example 7-6, featuring two mutually exclusive alternatives:

Year	Alt. 1	Alt. 2
0	−$10	−$20
1	+$15	+$28

If 6% interest is assumed, which alternative should be chosen?

SOLUTION

To examine this problem we will use a spreadsheet to plot the two alternatives on graphs showing PW at different rates of interest. Figure 8-1 shows the PW of Alternative 1. Using the general rule from Chapter 5, *maximize the net present worth,* we can observe that Alternative 1 is acceptable for the range of interest rates all the way from 0% to when it crosses the *x* axis at its IRR of 50%. Similarly we can see from Figure 8-2 that Alternative 2 is an acceptable investment over the range

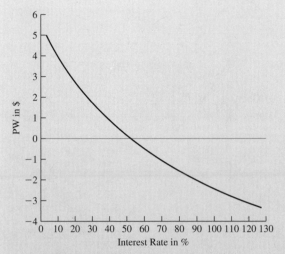

FIGURE 8-1 PW of Alternative 1.

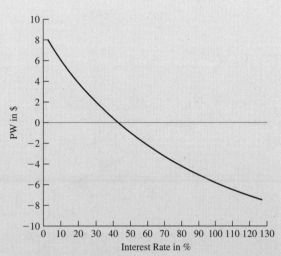

FIGURE 8-2 PW of Alternative 2.

from 0% to its IRR of 40%. The question we wish to answer, however, is not whether or not they are acceptable, but rather which of the two, at an interest rate of 6%, is the better choice.

Which of the two alternatives is best can be understood by looking at the graph that results when we plot both PW curves on the same axis.

Figure 8-3 shows that for interest rates from 0% to 30%, Alternative 2 has the higher NPW, and for rates from 30% to 50%, the NPW of Alternative 1 is higher. Above 50%, both alternatives have a negative NPW and so are not desirable when compared to the do-nothing or zero alternative.

Another way of visualizing this is by looking at the NPW graph of the incremental investment between Alternative 1 and Alternative 2.

$$\text{NPW}(\Delta) = \text{NPW (Alt. 2)} - \text{NPW (Alt. 1)}$$

$$= [-20 + 28(\text{P/F}, i, 1)] - [-10 + 15(\text{P/F}, i, 1)]$$

$$= -10 + 13(\text{P/F}, i, 1)$$

The graph of the increment shows clearly that, for any interest rate in the range 0% to 30%, the NPW is positive, and so the incremental investment meets the economic standard. Accepting the increment means that you choose the higher-initial-cost alternative, in this situation Alternative 2. Above the interest rate of 30% the NPW of the incremental investment is negative; therefore it should be rejected and the lower-initial-cost Alternative 1 is the economic choice.

Now the original question of which alternative is desirable at an interest rate of 6% is easily answered from the graphs—Alternative 2 is the correct choice. This could also be determined

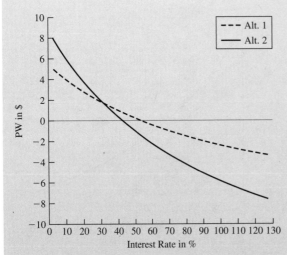

FIGURE 8-3 Graph of NPW of both alternatives.

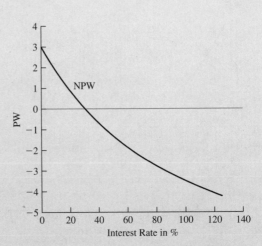

FIGURE 8-4 NPW of increment (Alt. 2− Alt. 1).

analytically by just calculating the NPW at 6%, or by calculating the IRR of the incremental investment and comparing it to 6%.

$$\text{NPW}(\Delta@6\%) = \text{NPW (Alt. 2)} - \text{NPW (Alt. 1)}$$
$$= [-20 + 28(P/F, 6\%, 1)] - [-10 + 15(P/F, 6\%, 1)]$$
$$= -10 + 13(P/F, 6\%, 1)$$
$$= -10 + 13(0.9434) = \$3.20$$

The NPW of the increment is positive at 6%; therefore Alternative 2 should be accepted.

$$\Delta\text{IRR} \quad \text{NPW (Alt. 2)} - \text{NPW (Alt. 1)} = 0$$
$$[-20 + 28(P/F, i, 1)] - [-10 + 15(P/F, i, 1)] = 0$$
$$(P/F, i, 1) = 10/14 = 0.7143 \qquad i = 30\%$$

Since the ΔIRR of the increment is greater than 6%, the incremental investment is preferable, and Alternative 2 should be accepted.

The calculations, however, answer the question only for a single interest rate. The graphical method provides an entire spectrum of information. This is especially useful if the decision maker is not sure of the accuracy of the chosen interest rate. Suppose, for example, that MARR was thought to be somewhere in the range of 5% to 10%. Where the rate falls in that range is not relevant because Alternative 2 is the best choice for any rate from 0% to 30%.

EXAMPLE 8-2

Solve Example 7-10 by means of an NPW graph. Two machines are being considered for purchase. If the minimum attractive rate of return (MARR) is 10%, which machine should be bought?

	Machine X	**Machine Y**
Initial cost	$200	$700
Uniform annual benefit	$ 95	$120
End-of-useful-life-salvage value	$ 50	$150
Useful life, in years	6	12

SOLUTION

Since the useful lives of the two alternatives are different, for an NPW analysis we must adjust them to the same analysis period. If the need seems continious, then the like-for-like assumption is reasonable and a 12-year analysis period can be used. The annual cash flows and the incremental cash flows are as follows:

End of Year	Cash Flows Machine X	Machine Y	Y − X
0	$(200)	$(700)	$(500)
1	95	120	25
2	95	120	25
3	95	120	25
4	95	120	25
5	95	120	25
6	(55)	120	175
7	95	120	25
8	95	120	25
9	95	120	25
10	95	120	25
11	95	120	25
12	145	270	125

By means of a spreadsheet program, the NPWs are calculated for a range of interest rates and then plotted on an NPW graph.

For a MARR of 10%, Machine X is clearly the superior choice. In fact, as the graph clearly illustrates, Machine X is the correct choice for most values of MARR. The intersection point of the two graphs can be found by calculating ΔIRR, the rate of return on the incremental investment. From the spreadsheet IRR function:

$$\Delta IRR = 1.3\%$$

FIGURE 8-5 NPW graph.

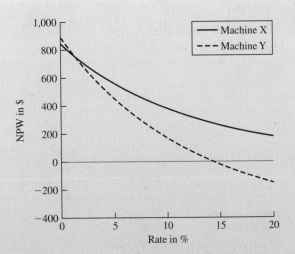

Then we can see from the graph that for MARR greater than 1.3%, Machine X the right choice, and for MARR values less than 1.3%, Machine Y is the right choice.

	NPW	
Rate	Machine X	Machine Y
0%	$840.00	$890.00
2	710.89	687.31
4	604.26	519.90
6	515.57	380.61
8	441.26	263.90
10	378.56	165.44
12	325.30	81.83
14	279.77	10.37
16	240.58	(51.08)
18	206.65	(104.23)
20	177.10	(150.47)

The two examples of problems show us some aspects of graphical analysis and incremental analysis. We will now examine some multiple-alternative problems. These problems can be solved by means of present worth and annual cash flow analysis without any difficulties. With rate of return analysis it is more complex. We can still use the incremental analysis, but the problem is which increments to analyze. When we had two alternatives, A and B, with B requiring the large investment, it was necessary only to examine the ΔIRR of the increment (B − A). But when we have three alternatives, A, B, and C, in order of increasing investment, there are the increments (B − A) and (C − B), but also the increment (C − A). When there are four alternatives, there are six incremental IRRs to calculate; and in the case of five alternatives, the ΔIRRs increase to 10. Fortunately the NPW graph can be used to show which ΔIRRs need to be calculated, and the problem becomes both more straightforward and more easily understood.

EXAMPLE 8-3

Consider the three mutually exclusive alternatives:

	A	B	C
Initial cost	$2,000	$4,000	$5,000
Uniform annual benefit	(410)	(639)	(700)

Each alternative has a 20-year life and no salvage value. If the MARR is 6%, which alternative should be selected?

SOLUTION

Using a spreadsheet to plot the NPW of the three alternatives produces Figure 8-6.

FIGURE 8-6 NPW graph of Alternatives A, B, and C.

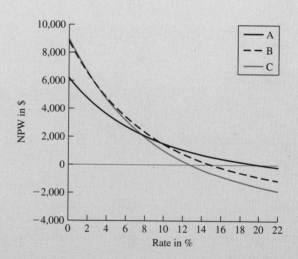

The highest line at any interest rate shows which alternative will provide the maximum NPW at that rate. From the graph we see that Alternative C has the maximum, then B, and finally A. The precise breakpoints can be found by calculating the ΔIRR of the corresponding increment.

$$NPW(C - B) = -\$5000 + \$4000 + (\$700 - \$639)(P/A, i, 20) = 0$$

$$\Delta IRR(C - B) = 2\%$$

$$NPW(B - A) = -\$4000 + \$2000 + (\$639 - \$410)(P/A, i, 20) = 0$$

$$\Delta IRR(B - A) = 9.6\%$$

The information from the graph can be presented in a Choice Table as follows:

If		MARR >	9.6%	Choose Alt. A
If	10%	> MARR >	2%	Choose Alt. B
If	2%	> MARR >	0%	Choose Alt. C

and the answer to the original question is select Alternative B if MARR is 6%.

EXAMPLE 8-4

Further study of the three alternatives of Example 8-3 reveals that the uniform annual benefit of Alternative A was overstated. It is now projected to be $122 rather than $410. Replot the NPW graph for this changed situation.

	A	B	C
Initial cost	$2,000	$4,000	$5,000
Uniform annual benefit	(122)	(639)	(700)

Each alternative has a 20-year life and no salvage value. If the MARR is 6%, which alternative should be chosen?

SOLUTION

Using a spreadsheet to re-plot the NPW of the three alternatives produces Figure 8-7.

FIGURE 8-7 NPW graph of Alternatives A, B, and C.

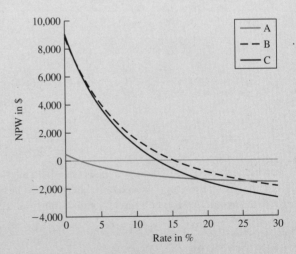

What is immediately obvious from this graph is that at MARR values greater than 15%, none of the alternatives will be chosen if the zero alternative is available. And for the range calculated, Alternative A will never be chosen because its positive NPW range is dominated by both B and C.

The increments of interest can easily be found since, first, C has the highest NPW, then B, and finally A. Thus the rates to be calculated are $\Delta IRR(C - B)$ and $\Delta IRR(B - A)$.

$$NPW(C - B) = -\$5000 + \$4000 + (\$700 - \$639)(P/A, i, 20) = 0$$

$$\Delta IRR(C - B) = 2\%$$

$$NPW(B - A) = -\$4000 + \$2000 + (\$639 - \$122)(P/A, i, 20) = 0$$

$$\Delta IRR(B - A) = 26\%$$

A complete description of the solution can be presented in the form of a Choice Table. If the zero (do nothing) alternative is available:

If		MARR >	15%	do nothing
If	15%	> MARR >	2%	choose Alt. B
If	2%	> MARR >	0%	choose Alt. C

If the zero alternative is not available (that is, you must choose one):

If		MARR >	26%	choose Alt. A
If	26%	> MARR >	2%	choose Alt. B
If	2%	> MARR >	0%	choose Alt. C

and the answer to the original problem—choose B if MARR = 6%.

EXAMPLE 8-5

The following information refers to three mutually exclusive alternatives. The decision maker wishes to choose the right machine but is uncertain what MARR to use. Create a Choice Table that will help the decision maker to make the correct economic decision.

	Machine X	Machine Y	Machine Z
Initial cost	$200	$700	$425
Uniform annual benefit	65	110	100
Useful life, in years	6	12	8

SOLUTION

In this question, the lives of the three alternatives are different. Thus we cannot directly use the present worth criterion but must first make some assumptions about the period of use and the analysis period. As we saw in Chapter 6, when the service period is expected to be continuous and the assumption of like-for-like replacement reasonable, we can assume a series of replacements and compare annual worth values just as we did with present worth values. There is a direct

FIGURE 8-8 EUAW graph

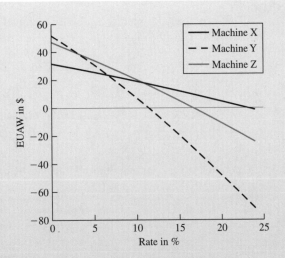

relationship between the present worth of cost and the equivalent uniform annual worth. It is

$$EUAW = (NPW)(A/P, i, n).$$

We will make these assumptions in this instance, and instead of plotting NPW, we shall plot EUAW. This was done on a spreadsheet, and the result is Figure 8-8.

The EUAW graph shows that for low values of MARR, Y is the correct choice. Then, as MARR increases, Z and finally X become the preferred machines. This can be expressed in a Choice Table read directly from the graph. If the zero or 'do nothing' alternative is available, the table is the following:

Choice Table with Zero Alternative

If		MARR >	23%	do nothing
If	23%	> MARR >	10%	choose X
If	10%	> MARR >	3.5%	choose Z
If	3.5%	> MARR		choose Y

If the 'do nothing' alternative is available, then the table reads

Choice Table (must select one)

If		MARR >	10%	choose X
If	10%	> MARR >	3.5%	choose Z
If	3.5%	> MARR		choose Y

The choice now is back in the decision maker's hands. There is still the need to determine MARR, but if the uncertainty was, for example, that MARR was some value in the range 12% to 18%, it can be seen that for this problem it doesn't matter. The answer is Machine X in any event. If, however, the uncertainty of MARR were in the 7% to 13% range, it is clear the decision maker would have to determine MARR with greater accuracy before solving this problem.

EXAMPLE 8-6

The following information is for five mutually exclusive alternatives that have 20-year useful lives. The decision maker may choose any one of the options or reject them all. Prepare a Choice Table.

	Alternatives				
	A	B	C	D	E
Cost	$4,000	$2,000	$6,000	$1,000	$9,000
Uniform annual benefit	(639)	(410)	(761)	(117)	(785)

SOLUTION

Figure 8-9 is an NPW graph of the alternatives constructed by means of a spreadsheet.

FIGURE 8-9 NPW graph.

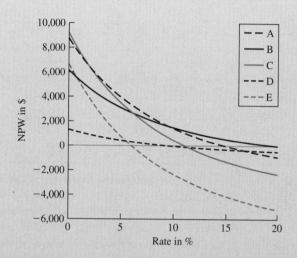

The graph clearly shows that Alternatives D and E are never part of the solution. They are dominated by the other three. The crossover points can either be read from the graph (if you have plotted it at a large enough scale) or found by calculating the ΔIRR of the intersecting graphs.

Calculating the incremental interest rates:

$$\Delta \text{IRR (C} - \text{A)}$$

$$\$6000 - \$4000 = (\$761 - \$639)\,(P/A, i, 20) \qquad i = 2\%$$

$$\Delta \text{IRR (A} - \text{B)}$$

$$\$4000 - \$2000 = (\$639 - \$410)\,(P/A, i, 20) \qquad i = 11\%$$

and to find where the NPW of A crosses the 0 axis

$$IRR\ (A)$$

$$\$4000 = \$639\ (P/A, i, 20) \qquad i = 20\%$$

Placing these numbers into a choice table:

If		MARR >	20%	do nothing
If	20%	> MARR >	11%	select B
If	11%	> MARR >	2%	select A
If	2%	> MARR		select C

A final point to note on this example is that if we view the IRRs of the five alternatives, the only information we can gleam from them is that if the

	A	B	C	D	E
IRR	15%	20%	11%	10%	6%

'do nothing' alternative is available, it will be chosen if MARR is greater than that of the largest IRR. There is nothing in the numbers to tell us which alternatives will be in the final solution set; only the NPW graph will show us where the change points will be.[1]

ELEMENTS IN INCREMENTAL RATE OF RETURN ANALYSIS

1. **Be sure all the alternatives are identified.** In textbook problems the alternatives will be well defined, but industrial problems may be less clear. Before proceeding, one must have all the mutually exclusive alternatives tabulated, including the do-nothing or 'keep doing the same thing' alternative, if appropriate.

2. **Construct an NPW graph showing all alternatives plotted on the same axes.** This would be a difficult task were it not for spreadsheets with embedded functions and wizards.

3. **Examine the line of maximum NPW and determine which alternatives create it, and over what range.**

4. **Determine the changeover points** where the line of maximum NPW changes from one alternative to another. These can either be read directly off the graph or calculated since they are the intersection points of the two curves and, what is more important and meaningful for engineering economy, they are **the ΔIRR of the incremental investment** between the two alternatives.

---

[1] There are analytical techniques for determining which incremental investments enter the solution set, but they are beyond the scope of this book and, in any case, are redundant in this era of spreadsheets. Interested readers can consult *Economic Analysis for Engineers and Managers* by Sprague and Whittaker, Prentice-Hall 1987.

5. Create a Choice Table to present the information in compact and easily understandable form.

CHOOSING AN ANALYSIS METHOD

At this point, we have examined in detail the three major economic analysis techniques: present worth analysis, annual cash flow analysis, and rate of return analysis. A practical question is, which method should be used for a particular problem?

While the obvious answer is to use the method requiring the least computations, a number of factors may affect the decision.

1. Unless the MARR—minimum attractive rate of return (or minimum required interest rate for invested money)—is known, neither present worth analysis nor annual cash flow analysis is possible.

2. Present worth analysis and annual cash flow analysis often require far less computation than rate of return analysis.

3. In some situations, a rate of return analysis is easier to explain to people unfamiliar with economic analysis. At other times, an annual cash flow analysis may be easier to explain.

4. Business enterprises generally adopt one, or at most two, analysis techniques for broad categories of problems. If you work for a corporation and the policy manual specifies rate of return analysis, you would appear to have no choice in the matter.

Since one may not always be able to choose the analysis technique computationally best suited to the problem, this book illustrates how to use each of the three methods in all feasible situations. Ironically, the most difficult method—rate of return analysis—is the one used most frequently by engineers in industry!

SPREADSHEETS AND GOAL SEEK

An incremental analysis of two alternatives is easily done with the RATE or IRR functions when the lives of the alternatives are the same. The problem is more difficult, however, when the lives are different. As was discussed in Chapters 5 and 6, when comparing alternatives having lives of different lengths, the usual approach is to assume that the alternatives are repeated until the least common multiple of their lives. This can be done with a spreadsheet, but Excel supports an easier approach.

Excel has a tool called GOAL SEEK that identifies a formula cell, a target value, and a variable cell. This tool causes the variable cell to be changed automatically until the formula cell equals the target value. To find an IRR for an incremental analysis, the formula cell can be the difference between two equivalent annual worths with a target value of 0. Then if the variable cell is the interest rate, GOAL SEEK will find the IRR.

In Excel this tool is accessed by selecting T(ools) on the main toolbar or menu and G(oal seek) on the sub-menu. As shown in Example 8-10, the variable cell (with the interest rate) must somehow affect the formula cell (difference in equivalent uniform annual worths or costs (EUAWs or EUACs)), although it need not appear directly in the formula cell. In

Figures 8-10a and 8-10b the interest rate (cell A1) appears in the EUAC formulas (cells D3 and D4) but not in the formula cell (cell D5).

EXAMPLE 8-7

Two different asphalt mixes can be used on a highway. The good mix will last 6 years and will cost $600,000 to buy and lay down. The better mix will last 10 years and will cost $800,000 to buy and lay down. Find the incremental IRR for using the more expensive mix.

SOLUTION

This example would be difficult to solve without the GOAL SEEK tool. The least common multiple of 6 and 10 is 30 years, which is the comparison period. With the GOAL SEEK tool a very simple spreadsheet does the job. In Figure 8-10a the spreadsheet is shown before GOAL SEEK. Figure 8-10b shows the result after the goal (D5) has been set $= 0$ and A1 has been chosen as the variable cell to change.

	A	B	C	D	E	F
1	8.00%	Interest rate				
2	Alternative	Cost	Life	EUAC		
3	Good	600000	6	129,789	= −PMT($A$1,C3,B3)	
4	Better	800000	10	119,224	= −PMT($A$1,C4,B4)	
5			difference =	10,566	= D3−D4	

FIGURE 8-10a Spreadsheet before GOAL SEEK.

	A	B	C	D	E	F
1	14.52%	Interest rate				
2	Alternative	Cost	Life	EUAC		
3	Good	600000	6	156,503	= −PMT($A$1,C3,B3)	
4	Better	800000	10	156,503	= −PMT($A$1,C4,B4)	
5			difference =	0	= D3−D4	

FIGURE 8-10b Spreadsheet after GOAL SEEK.

The incremental IRR found by GOAL SEEK is 14.52%.

SUMMARY

For choosing from a set of mutually exclusive alternatives, the rate of return technique is more complex than the present worth or annual cash flow techniques because in the last two techniques the numbers can be compared directly, whereas with the rate of return it is necessary to consider the *increment of investment*. This is fairly straightforward if there are only two alternatives, but it becomes more and more complex as the number of alternatives increases.

A visual display of the problem can be created by using a spreadsheet to draw the NPW graphs of the alternatives. The steps are as follows:

1. Be sure all the alternatives are identified.
2. Construct an NPW graph showing all alternatives plotted on the same axes.
3. Examine the line of maximum NPW and determine which alternatives create it, and over what range.
4. Determine the changeover points (ΔIRRs).
5. Create a choice table.

This *incremental analysis* approach, although a more arduous procedure, is a more powerful one because, by allowing the decision maker to see the range over which the choices are valid, it provides a form of sensitivity analysis.

PROBLEMS

Unless otherwise noted, all problems for Chapter 8 should be solved by rate of return analysis.

8-1 A firm is considering moving its manufacturing plant from Red Deer to a new location. The industrial engineering department was asked to identify the various alternatives together with the costs to relocate the plant, and the benefits. The engineers examined six likely sites, together with the do-nothing alternative of keeping the plant at its present location. Their findings are summarized as follows:

Plant Location	First Cost ($000s)	Uniform Annual Benefit ($000s)
Halifax	$300	$ 52
Edmonton	550	137
Toronto	450	117
Vancouver	750	167
Calgary	150	18
Regina	200	49
Red Deer	0	0

The annual benefits are expected to be constant over the 8-year analysis period. If the firm uses 10% annual interest in its economic analysis, where should the manufacturing plant be located? (*Answer:* Edmonton)

8-2 In a particular situation, four mutually exclusive alternatives are being considered. Each of the alternatives costs $1300 and has no end-of-useful-life salvage value.

Alternative	Annual Benefit	Useful Life (years)	Calculated Rate of Return
A	$100 at end of first year; increasing by $30 per year thereafter	10	10.0%
B	$10 at end of first year; increasing by $50 per year thereafter	10	8.8%
C	Annual end-of-year benefit = $260	10	15.0%
D	$450 at end of first year; declining $50 per year thereafter	10	18.1%

For what range of MARR would D be the best choice?

8-3 A more detailed examination of the situation in Problem 8-2 reveals that there are two additional mutually exclusive alternatives to be considered. Both cost more than the $1300 for the four original alternatives.

Alternative	Cost	Annual End-of-Years Benefit	Useful Life (years)	Calculated Rate of Return
E	$3000	$ 488	10	10.0%
F	5850	1000	10	11.2%

If the MARR remains at 8%, which one of the six alternatives should be selected? Neither Alt. E nor F has any end-of-useful-life salvage value. (*Answer:* Alt. F)

8-4 The owner of a downtown parking lot has employed a civil engineering consulting firm to advise him on the economic feasibility of constructing an office building on the site. Luc Tremblay, a newly hired civil engineer, has been assigned to make the analysis. He has assembled the following data:

Alternative	Total Investment*	Total Net Annual Revenue from Property
Sell parking lot	$ 0	$ 0
Keep parking lot	200,000	22,000
Build 1-storey building	400,000	60,000
Build 2-storey building	555,000	72,000
Build 3-storey building	750,000	100,000
Build 4-storey building	875,000	105,000
Build 5-storey building	1,000,000	120,000

*Includes the value of the land.

The analysis period is to be 15 years. For all alternatives, the property has an estimated resale (salvage) value at the end of 15 years equal to the present total investment. If the MARR is 10%, what recommendation should Luc make?

8-5 An oil company plans to buy a piece of vacant land on the corner of two busy streets for $70,000. On properties of this type, the company installs businesses of four different types.

Plan	Cost of Improvements*	Type of Business
A	$ 75,000	Conventional gas station with service facilities for lubrication, oil changes, etc.
B	230,000	Automatic carwash facility with gasoline pump island in front
C	30,000	Discount gas station (no service bays)
D	130,000	Gas station with low-cost, quick-carwash facility

*Cost of improvements does not include the $70,000 cost of land.

In each case, the estimated useful life of the improvements is 15 years. The salvage value for each is estimated to be the $70,000 cost of the land. The net annual income, after all operating expenses are paid for, is projected as follows:

Plan	Net Annual Income
A	$23,300
B	44,300
C	10,000
D	27,500

If the oil company expects a 10% rate of return on its investments, which plan (if any) should it choose?

8-6 A firm is considering three mutually exclusive alternatives as part of a production improvement program. The alternatives are as follows:

	A	B	C
Installed cost	$10,000	$15,000	$20,000
Uniform annual benefit	1,625	1,625	1,890
Useful life, in years	10	20	20

For each alternative, the salvage value at the end of useful life is zero. At the end of 10 years, Alt. A could be replaced by another A with identical cost and benefits. The MARR is 6%. If the analysis period is 20 years, which alternative should be selected?

8-7 Given the following four mutually exclusive alternatives and using 8% for the MARR, which alternative should be selected?

	A	B	C	D
First cost	$75	$50	$50	$85
Uniform annual benefit	16	12	10	17
Useful life, in years	10	10	10	10
End-of-useful-life salvage value	0	0	0	0
Computed rate of return	16.8%	20.2%	15.1%	15.1%

(*Answer:* A)

8-8 Consider the following three mutually exclusive alternatives:

	A	B	C
First cost	$200	$300	$600
Uniform annual benefit	59.7	77.1	165.2
Useful life, in years	5	5	5
End-of-useful-life salvage value	0	0	0
Computed rate of return	15%	9%	11.7%

For what range of values of MARR is Alt. C the preferable alternative? Put your answer in the following form: 'Alternative C is preferred when ____% ≤ MARR ≤ ____%.'

8-9 Consider four mutually exclusive alternatives, each having an 8-year useful life:

	A	B	C	D
First cost	$1000	$800	$600	$500
Uniform annual benefit	122	120	97	122
Salvage value	750	500	500	0

If the minimum attractive rate of return is 8%, which alternative should be selected?

8-10 Three mutually exclusive projects are being considered:

	A	B	C
First cost	$1000	$2000	$3000
Uniform annual benefit	150	150	0
Salvage value	1000	2700	5600
Useful life, in years	5	6	7

When each project reaches the end of its useful life, it would be sold for its salvage value and there would be no replacement. If 8% is the desired rate of return, which project should be selected?

8-11 Consider three mutually exclusive alternatives:

Year	Buy X	Buy Y	Do Nothing
0	−$100.0	−$50.0	0
1	+31.5	+16.5	0
2	+31.5	+16.5	0
3	+31.5	+16.5	0
4	+31.5	+16.5	0

Which alternative should be selected:
(a) If the minimum attractive rate of return equals 6%?
(b) If MARR = 9%?
(c) If MARR = 10%?
(d) If MARR = 14%?
(*Answers:* (a) X; (b) Y; (c) Y; (d) Do nothing)

8-12 Consider the three alternatives:

Year	A	B	Do Nothing
0	−$100	−$150	0
1	+30	+43	0
2	+30	+43	0
3	+30	+43	0
4	+30	+43	0
5	+30	+43	0

Which alternative should be selected:
(a) If MARR = 6%?
(b) If MARR = 8%?
(c) If MARR = 10%?

8-13 A firm is considering two alternatives:

	A	B
Initial cost	$10,700	$5500
Uniform annual benefits	2,100	1800
Salvage value at end of useful life	0	0
Useful life, in years	8	4

At the end of 4 years, another B may be purchased with the same cost, benefits, and so forth. If the MARR is 10%, which alternative should be selected?

8-14 Consider the following alternatives:

	A	B	C
Initial cost	$300	$600	$200
Uniform annual benefits	41	98	35

Each alternative has a 10-year useful life and no salvage value. If the MARR is 8%, which alternative should be selected?

8-15 Given the following:

Year	X	Y
0	−$10	−$20
1	+15	+28

Over what range of values of MARR is Y the preferable alternative?

8-16 Consider four mutually exclusive alternatives:

	A	B	C	D
Initial cost	$770.00	$1406.30	$2563.30	0
Uniform annual benefit	420.00	420.00	420.00	0
Useful life, in years	2	4	8	0
Computed rate of return	6.0%	7.5%	6.4%	0

The analysis period is 8 years. At the end of 2, 4, and 6 years, Alt. A will have an identical replacement. Alternative B will have a single identical replacement at the end of 4 years. Over what range of values of MARR is Alt. B the preferred alternative?

8-17 Consider the three alternatives:

	A	B	C
Initial cost	$1500	$1000	$2035
Annual benefit in each of first 5 years	250	250	650
Annual benefit in each of subsequent 5 years	450	250	145

Each alternative has a 10-year useful life and no salvage value. At a MARR of 15%, which alternative should be selected? Where appropriate, use an external interest rate of 10% to transform a cash flow to one sign change before proceeding with rate of return analysis.

8-18 A new 10,000-square-metre warehouse next door to the Tyre Corporation is for sale for $450,000. The terms offered are $100,000 down with the balance being paid in 60 equal monthly payments based on 15% interest. It is estimated that the warehouse would have a resale value of $600,000 at the end of 5 years.

Tyre has the necessary cash available and could buy the warehouse but does not need all the warehouse space at this time. The Johnson Company has offered to lease half the new warehouse for $2500 a month.

Tyre presently rents and uses 7000 square metres of warehouse space for $2700 a month. It has the option of reducing the rented space to 2000 square metres, in which case the monthly rent would be $1000 a month. Or Tyre could cease renting warehouse space entirely. Tom Clay, the Tyre Corp. plant engineer, is considering three alternatives:

1. Buy the new warehouse and rent half the space to the Johnson Company. In turn, the space that Tyre rents would be reduced to 2000 square metres.
2. Buy the new warehouse and cease renting any warehouse space.
3. Continue as is, with 7000 square metres of rented warehouse space.

On the basis of a 20% minimum attractive rate of return, which alternative should be selected?

8-19 Consider the following alternatives:

	A	B	C
Initial cost	$100.00	$150.00	$200.00
Uniform annual benefit	10.00	17.62	55.48
Useful life, in years	Infinite	20	5

Use present worth analysis, an 8% interest rate, and an infinite analysis period. Which alternative should be chosen in each of the two following situations?

1. Alternatives B and C are replaced at the end of their useful lives with identical replacements.
2. Alternatives B and C are replaced at the end of their useful lives with alternatives that provide an 8% rate of return.

8-20 A problem often discussed in the engineering economy literature is the "oil-well pump problem."[1] Pump 1 is a small pump; Pump 2 is a larger pump that costs more, will produce slightly more oil, and will produce it more rapidly. If the MARR is 20%, which pump should be selected? Assume that any temporary external investment of money earns 10% per year and that any temporary financing is done at 6%.

Year	Pump 1 ($000s)	Pump 2 ($000s)
0	–$100	–$110
1	+70	+115
2	+70	+30

---

[1] One of the more interesting exchanges of opinion about this problem is in Prof. Martin Wohl's "Common Misunderstandings About the Internal Rate of Return and Net Present Value Economic Analysis Methods;" and the associated discussion by Professors Winfrey, Leavenworth, Steiner, and Bergmann, published in *Evaluating Transportation Proposals,* Transportation Research Record 731, Transportation Research Board, Washington, D.C. See also Appendix 7A in Chapter 7.

8-21 Three mutually exclusive alternatives are being studied. If the MARR is 12%, which alternative should be selected?

Year	A	B	C
0	−$20,000	−$20,000	−$20,000
1	+10,000	+10,000	+5,000
2	+5,000	+10,000	+5,000
3	+10,000	+10,000	+5,000
4	+6,000	0	+15,000

8-22 The South End bookstore has an annual profit of $170,000. The owner is considering opening a second bookstore on the north side of the campus. He can rent an existing building for 5 years with an option to continue the lease for a second 5-year period. If he opens the second bookstore, he expects the existing store will lose some business, which will be gained by The North End, the new bookstore. It will take $500,000 of store fixtures and inventory to open The North End. He believes that the two stores will have a combined profit of $260,000 a year after all the expenses of both stores have been paid.

The owner's economic analysis is based on a 5-year period. He will be able to recover this $500,000 investment at the end of 5 years by selling the store fixtures and inventory. The owner will not open The North End unless he can expect a 15% rate of return. What should he do? Show computations to justify your decision.

8-23 A paper mill is considering two types of pollution control equipment.

	Neutralization	Precipitation
Initial cost	$700,000	$500,000
Annual chemical cost	40,000	110,000
Salvage value	175,000	125,000
Useful life, in years	5	5

The firm wants a 12% rate of return on any avoidable increments of investment. Which equipment should be purchased?

8-24 A stockbroker has proposed two investments in low-rated corporate bonds paying high interest rates and selling below their stated value (in other words, junk bonds). The two bonds are rated as equally risky. Which, if any, of the bonds should you buy if your MARR is 25%?

Bond	Stated Value	Annual Interest Payment	Current Market Price, Including Buying Commission	Bond Maturity*
Gen Dev	$1000	$ 94	$480	15 years
RJR	1000	140	630	15

*At maturity the bondholder receives the last interest payment plus the bond stated value.

8-25 Three mutually exclusive alternatives are being considered.

	A	B	C
Initial investment	$50,000	$22,000	$15,000
Annual net income	5,093	2,077	1,643
Computed rate of return	8%	7%	9%

Each alternative has a 20-year useful life with no salvage value. If the minimum attractive rate of return is 7%, which alternative should be selected?

8-26 A firm is considering the following alternatives, as well as a fifth choice: do nothing.

	1	2	3	4
Initial cost	$100,000	$130,000	$200,000	$330,000
Uniform annual net income ($1000s)	26.38	38.78	47.48	91.55
Computed rate of return	10%	15%	6%	12%

Each alternative has a 5-year useful life. The firm's minimum attractive rate of return is 8%. Which alternative should be selected?

8-27

	Alternatives			
	A	B	C	D
Initial cost	$2000	5000	4000	3000
Annual benefit	800	500	400	1300
Salvage value	2000	1500	1400	3000
Life, in years	5	6	7	4
MARR required	6%	6%	6%	6%

Using incremental IRR analysis, find the best alternative.

8-28 Our cat Fred's summer kitty-cottage needs a new roof. He's considering the following two proposals and feels a 15-year analysis period is in line with his remaining lives. Which roof should he choose if his MARR is 12%? What is the actual value of the IRR on the incremental cost? (There is no salvage value for old roofs.)

	Thatch	Slate
First cost	$20	$40
Annual upkeep	5	2
Service life, in years	3	5

8-29 Don Garlits is a landscaper. He is considering the purchase of a new commercial lawn mower, either the Atlas or the Zippy. The minimum attractive rate of return is 8%. The table provides all the necessary information for the two machines.

	Atlas	Zippy
Initial cost	$6,700	$16,900
Annual operation and maintenance cost	1,500	1,200
Annual benefit	4,000	4,500
Salvage value	1,000	3,500
Useful life, in years	3	6

(a) Determine the rate of return on the Atlas mower (to the nearest 1%).

(b) Does the rate of return on the Zippy mower exceed the MARR?

(c) Use incremental rate of return analysis to decide which machine should be bought.

8-30 QZY, Inc. is evaluating new widget machines offered by three companies. The machines have the following characteristics:

	Company A	Company B	Company C
First cost	$15,000	$25,000	$20,000
Maintenance and operating	1,600	400	900
Annual benefit	8,000	13,000	9,000
Salvage value	3,000	6,000	4,500
Useful life, in years	4	4	4

MARR = 15%. Use rate of return analysis to decide from which company, if any, you should purchase the widget machine.

8-31 The Croc Co. is considering a new milling machine from among three alternatives:

	Alternative		
	Deluxe	Regular	Economy
First cost	$220,000	$125,000	$75,000
Annual benefit	79,000	43,000	28,000
Maintenance and operating costs	38,000	13,000	8,000
Salvage value	16,000	6,900	3,000

All machines have a life of 10 years, and MARR = 15%. Using incremental rate of return analysis, which alternative, if any, should the company choose?

8-32 Wayward Airfreight, Inc. has asked you to recommend a new automatic parcel sorter. You have obtained the following bids:

	Ship-R	Sort-Of	U-Sort-M
First cost	$184,000	$235,000	$180,000
Salvage value	38,300	44,000	14,400
Annual benefit	75,300	89,000	68,000
Yearly maintenance and operating cost	21,000	21,000	12,000
Useful life, in years	7	7	7

Using an MARR of 15% and a rate of return analysis, decide which alternative, if any, should be selected.

8-33 A firm must decide which of three alternatives to adopt to expand its capacity. The firm wishes a minimum annual profit of 20% of the initial cost of each separable increment of investment. Any money not invested in capacity expansion can be invested elsewhere for an annual yield of 20% of initial cost.

Alt.	Initial Cost	Annual Profit	Profit Rate
A	$100,000	$30,000	30%
B	300,000	66,000	22%
C	500,000	80,000	16%

Which alternative should be selected? Use a rate of return analysis.

8-34 The Pure White Soap Company is considering adding some processing equipment to the plant to aid in the removal of impurities from some raw materials. By adding the processing equipment, the firm can purchase lower-grade raw material at reduced cost and upgrade it for use in its products.

Four different pieces of processing equipment are being considered:

	A	B	C	D
Initial investment	$10,000	$18,000	$25,000	$30,000
Annual saving in materials costs	4,000	6,000	7,500	9,000
Annual operating cost	2,000	3,000	3,000	4,000

The company can obtain a 15% annual return on its investment in other projects and is willing to invest money on the processing equipment only as long as it can obtain 15% annual return on each increment of money invested. Which one, if any, of the alternatives should be selected? Use a rate of return analysis.

8-35 Build a spreadsheet to find the EAC of each roof in Problem 8-28. Use the GOAL SEEK tool of Excel to find the IRR of the incremental investment.

8-36 Build a spreadsheet to find the EAW of each lawn-mower in Problem 8-29. Use the GOAL SEEK tool of Excel to find the IRR of the incremental investment.

Other Analysis Techniques

Clean, Green, and Far Between

Designers and engineers have made great progress in developing environmentally friendly construction materials and building techniques. And 'green' office buildings offer numerous advantages. They use energy more efficiently, provide a more natural environment for workers, and can even reduce indoor air pollution.

The only problem is, no one wants to build them. In the commercial sector, barely 1% of office construction projects have bothered to apply for certification as green buildings.

Commercial developers have found that environmentally friendly features add to the expense of construction. Even though additional costs may increase overall construction expenses by only 1% or 2%, they can make it harder for the builder to recoup its investment and break even on the project.

Moreover, the extra expenditures apparently do little to attract tenants. Renters may like the idea of being cleaner and greener—but they don't want to pay extra for it.

After Completing This Chapter...

The student should be able to:

- Use future worth, benefit-cost ratio, payback period, and sensitivity analysis methods to solve engineering economy problems.
- Link the use of the *future worth* analysis to the present worth and annual worth methods developed earlier.
- Mathematically develop the *benefit-cost ratio,* and use this model to select alternatives and make economic choices.
- Understand the concept of the *payback period* of an investment, and be able to calculate this quantity for prospective projects.
- Demonstrate a basic understanding of *sensitivity* and *break-even analyses* and the use of these tools in an engineering economic analysis.
- Use a spreadsheet to perform *sensitivity* and *break-even analyses.*

QUESTIONS TO CONSIDER

1. Office leases frequently require the building owners, rather than the tenants, to pay heating and cooling costs. What effect might this have on the decision making of potential tenants for green buildings?
2. The green building movement has had more success among developers who hold on to their buildings for years and rent them out, rather than selling them as soon as they are constructed. What factors might influence their views?
3. Many environmentally friendly buildings are architecturally distinctive and feature better-quality materials and workmanship than traditional commercial structures. Environmental advocates hope these characteristics will help green buildings attract a rent 'premium'. How might these features make the buildings more attractive to tenants?

Chapter 9 examines four topics:

- Future worth analysis
- Benefit-cost ratio analysis
- Payback period
- Sensitivity and break-even analysis

Future worth analysis is very much like present worth analysis, dealing with *then* (future worth) rather than with *now* (present worth) situations.

Previously, we have written economic analysis relationships based on either:

$$\text{PW of cost} = \text{PW of benefit} \quad \text{or} \quad \text{EUAC} = \text{EUAB}$$

Instead of writing it in this form, we could define these relationships as

$$\frac{\text{PW of benefit}}{\text{PW of cost}} = 1 \quad \text{or} \quad \frac{\text{EUAB}}{\text{EUAC}} = 1$$

When economic analysis is based on these ratios, the calculations are called benefit-cost ratio analysis.

Payback period is an approximate analysis technique, generally defined as the time required for cumulative benefits to equal cumulative costs. Sensitivity describes the relative magnitude of a particular variation in one or more elements of a problem that is sufficient to change a particular decision. Closely related is break-even analysis, which determines the conditions where two alternatives are equivalent. Thus, break-even analysis is a form of sensitivity analysis.

FUTURE WORTH ANALYSIS

We have seen how economic analysis techniques resolved alternatives into comparable units. In present worth analysis, the comparison was made in terms of the present consequences of taking the feasible courses of action. In annual cash flow analysis, the comparison was in terms of equivalent uniform annual costs (or benefits). We saw that we could easily convert from present worth to annual cash flow, and vice versa. But the concept of resolving alternatives into comparable units is not restricted to a present or annual comparison. The comparison may be made at any point in time. In many situations we would like to know what the *future* situation will be, if we take some particular course of action *now*. This is called **future worth analysis.**

EXAMPLE 9-1

Ron Chan, a 20-year-old university student, smokes about a carton of cigarettes a week. He wonders how much money he could accumulate by age 65 if he quit smoking now and put his cigarette money into a savings account. Cigarettes cost $85 a carton. Ron expects that a savings account would earn 5% interest, compounded semi-annually. Compute Ron's future worth at age 65.

SOLUTION

$$\text{Semi-annual saving } \$85/\text{carton} \times 26 \text{ weeks} = \$2210$$

$$\text{Future worth (FW)} = A(F/A, 2^{1}/_{2}\%, 90) = 2210(329.2) = \$727,532$$

EXAMPLE 9-2

An east coast firm has decided to establish a second plant in the west. There is a factory for sale for $850,000 that, with extensive remodelling, could be used. As an alternative, the company could buy vacant land for $85,000 and have a new plant constructed there. Either way, it will be 3 years before the company will be able to get a plant into production. The timing and cost of the various components for the factory are given in the following cash flow table.

Year	Construct New Plant		Remodel Available Factory	
0	Buy land	$ 85,000	Purchase factory	$850,000
1	Design and initial construction costs	200,000	Design and remodelling costs	250,000
2	Balance of construction costs	1,200,000	Additional remodelling costs	250,000
3	Set-up of production equipment	200,000	Set-up of production equipment	250,000

If interest is 8%, which alternative results in the lower equivalent cost when the firm begins production at the end of the third year?

SOLUTION

New Plant

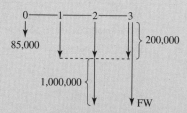

$$\text{Future worth of cost (FW)} = 85,000(F/P, 8\%, 3) + 200,000(F/A, 8\%, 3)$$

$$+ 1,000,000(F/P, 8\%, 1) = \$1,836,000$$

Remodel Available Factory

$$\text{Future worth of cost (FW)} = 850,000(F/P, 8\%, 3) + 250,000(F/A, 8\%, 3)$$

$$= \$1,882,000$$

The total cost of remodelling the available factory ($1,600,000) is smaller than the total cost of a new plant ($1,685,000). The timing of the expenditures, however, is less favourable than building the new plant. The new plant is projected to have the smaller future worth of cost and thus is the preferred alternative.

BENEFIT-COST RATIO ANALYSIS

At a given minimum attractive rate of return (MARR), we would consider an alternative acceptable, provided

$$\text{PW of benefits} - \text{PW of costs} \geq 0 \quad \text{or} \quad \text{EUAB} - \text{EUAC} \geq 0$$

These could also be stated as a ratio of benefits to costs, or

$$\text{Benefit-cost ratio } \frac{B}{C} = \frac{\text{PW of benefit}}{\text{PW of costs}} = \frac{\text{EUAB}}{\text{EUAC}} \geq 1$$

Rather than using present worth or annual cash flow analysis to solve problems, we can base the calculations on the benefit-cost ratio, B/C. The criteria are presented in Table 9-1. We will illustrate 'B/C analysis' by solving the same example problems worked by other economic analysis methods.

TABLE 9-1 Benefit-Cost Ratio Analysis

	Situation	Criterion
Fixed input	Amount of money or other input resources are fixed	Maximize B/C
Fixed output	Fixed task, benefit, or other output to be accomplished	Maximize B/C
Neither input nor output fixed	Neither amount of money or other inputs nor amount of benefits or other outputs are fixed	*Two alternatives:* Compute incremental benefit-cost ratio ($\Delta B/\Delta C$) on the increment of investment between the alternatives. If $\Delta B/\Delta C \geq 1$, choose higher-cost alternative; otherwise, choose lower-cost alternative. *Three or more alternatives:* Solve by benefit-cost ratio incremental analysis

EXAMPLE 9-3

A firm is trying to decide which of two devices to install to reduce costs in a particular situation. Both devices cost $1000 and have useful lives of 5 years and no salvage value. Device A can be expected to result in $300 savings annually. Device B will provide cost savings of $400 the first year, but savings will decline by $50 annually, making the second-year savings $350, the third-year savings $300, and so forth. With interest at 7%, which device should the firm purchase?

SOLUTION

We have used three types of analysis thus far to solve this problem: present worth in Example 5-1, annual cash flow in Example 6-5, and rate of return in Example 7-9.

Device A

$$PW \text{ of cost} = \$1000$$

$$PW \text{ of benefits} = 300(P/A, 7\%, 5)$$

$$= 300(4.100) = \$1230$$

$$\frac{B}{C} = \frac{PW \text{ of benefit}}{PW \text{ of costs}} = \frac{1230}{1000} = 1.23$$

Device B

$$PW \text{ of cost} = \$1000$$

$$PW \text{ of benefit} = 400(P/A, 7\%, 5) - 50(P/G, 7\%, 5)$$

$$= 400(4.100) - 50(7.647) = 1640 - 382 = 1258$$

$$\frac{B}{C} = \frac{PW \text{ of benefit}}{PW \text{ of costs}} = \frac{1258}{1000} = 1.26$$

To maximize the benefit-cost ratio, select Device B.

EXAMPLE 9-4

Two machines are being considered for purchase. Assuming 10% interest, which machine should be bought?

	Machine X	Machine Y
Initial cost	$200	$700
Uniform, annual benefit	95	120
End-of-useful-life salvage value	50	150
Useful life, in years	6	12

SOLUTION

Assuming a 12-year analysis period, the cash flow table is as follows:

Year	Machine X	Machine Y
0	−$200	−$700
1–5	+95	+120
6	+95 −200 +50	+120
7–11	+95	+120
12	+95 +50	+120 +150

We will solve the problem using

$$\frac{B}{C} = \frac{EUAB}{EUAC}$$

and considering the salvage value of the machines to be reductions in cost, rather than increases in benefits. This choice affects the ratio value but not the decision.

Machine X

$$EUAC = 200(A/P, 10\%, 6) - 50(A/F, 10\%, 6)$$

$$= 200(0.2296) - 50(0.1296) = 46 - 6 = \$40$$

$$EUAB = \$95$$

Note that this assumes the replacement for the last 6 years has identical costs. Under these circumstances, the EUAC for the first 6 years equals the EUAC for all 12 years.

Machine Y

$$EUAC = 700(A/P, 10\%, 12) - 150(A/F, 10\%, 12)$$

$$= 700(0.1468) - 150(0.0468) = 103 - 7 = \$96$$

$$EUAB = \$120$$

Machine Y − Machine X

$$\frac{\Delta B}{\Delta C} = \frac{120 - 95}{96 - 40} = \frac{25}{56} = 0.45$$

Since the incremental benefit-cost ratio is less than 1, it represents an undesirable increment of investment. We therefore choose the lower-cost alternative—Machine X. If we had computed benefit-cost ratios for each machine, they would have been:

Machine X
$$\frac{B}{C} = \frac{95}{40} = 2.38$$

Machine Y
$$\frac{B}{C} = \frac{120}{96} = 1.25$$

Although $B/C = 1.25$ for Machine Y (the higher-cost alternative), we must not use this fact as the basis for selecting the more expensive alternative. The incremental benefit-cost ratio, $\Delta B/\Delta C$, clearly shows that Y is a less desirable alternative than X. Moreover, we must not jump to the conclusion that the best alternative is always the one with the largest B/C ratio. This, too, may lead to incorrect decisions—as we shall see when we examine problems with three or more alternatives.

EXAMPLE 9-5

Consider the following six mutually exclusive alternatives. They have 20-year useful lives and no salvage value. If the minimum attractive rate of return is 6%, which alternative should be selected?

	A	B	C	D	E	F
Cost	$4,000	$2,000	$6,000	$1,000	$9,000	$10,000
PW of benefit	7,330	4,700	8,730	1,340	9,000	9,500
$\dfrac{B}{C} = \dfrac{\text{PW of benefits}}{\text{PW of cost}}$	1.83	2.35	1.46	1.34	1.00	0.95

SOLUTION

Incremental analysis is needed to solve the problem. The steps in the solution involve paired comparisons of incremental amounts of investment with the criterion of $\Delta B/\Delta C$, and a cut-off of 1.

The steps are:

1. Be sure all the alternatives are identified.
2. Compute the B/C ratio for each alternative. Since there are alternatives for which $B/C \geq 1$, we will discard any with $B/C < 1$. Discard Alt. F.
3. Arrange the remaining alternatives in ascending order of investment.

	D	B	A	C	E
Cost (= PW of cost)	$1000	$2000	$4000	$6000	$9000
PW of benefits	1340	4700	7330	8730	9000
B/C	1.34	2.35	1.83	1.46	1.00

	Increment B–D	Increment A–B	Increment C–A
ΔCost	$1000	$2000	$2000
ΔBenefit	3360	2630	1400
$\Delta B/\Delta C$	3.36	1.32	0.70

4. Examine each separable increment of investment. If $\Delta B/\Delta C < 1$, the increment is not attractive. If $\Delta B/\Delta C \geq 1$, the increment of investment is desirable. The increments B–D and A–B are desirable. Thus, of the first three alternatives (D, B, and A), Alt. A is the preferred alternative. Increment C–A is not attractive as $\Delta B/\Delta C = 0.70$, which indicates that of the first four alternatives (D, B, A, and C), A continues as the best of the four. Now we want to decide between A and E, which we'll do by examining the increment of investment that represents the difference between these alternatives.

Increment E–A

ΔCost	$5000
ΔBenefit	1670
$\Delta B/\Delta C$	0.33

The increment is undesirable. We choose Alt. A as the best of the six alternatives. One should note that the best alternative in this example does not have the highest B/C ratio.

Benefit-cost ratio analysis may be graphically represented. Figure 9.1 is a graph of Example 9-5. We see that F has a B/C < 1 and can be discarded. Alternative D is the starting point for examining the separable increments of investment. The slope of line $B–D$ indicates a $\Delta B/\Delta C$ ratio of >1. This is also true for line $A–B$. Increment $C–A$ has a slope much flatter than B/C = 1, indicating an undesirable increment of investment. Alternative C is therefore discarded and A retained. Increment $E–A$ is similarly unattractive. Alternative A is therefore the best of the six alternatives.

Note particularly two additional things about Figure 9-1: first, even if alternatives with B/C ratio < 1 had not been initially excluded, they would have been systematically eliminated in the incremental analysis. Since this is the case, it is not essential to compute the B/C ratio for each alternative as a first step in incremental analysis. Nevertheless, it seems like an orderly and logical way to approach a multiple-alternative problem. Second, Alt. B had the highest B/C ratio (B/C = 2.35), but it is not the best of the six alternatives. We saw the same situation in rate of return analysis of three or more alternatives. The reason is the same in both analysis situations. We seek to maximize the *total* profit, not the profit rate.

FIGURE 9-1 Benefit-cost ratio graph of Example 9-5.

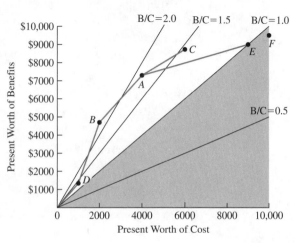

Continuous Alternatives

At times the feasible alternatives are a continuous function. For example, when we consider the height of a dam, it is possible to build the dam anywhere from 200 to 500 feet high.

In many situations, the projected capacity of an industrial plant can be varied continuously over some feasible range. In these cases, we seek to add increments of investment where $\Delta B/\Delta C \geq 1$ and avoid increments where $\Delta B/\Delta C < 1$. The optimal size of such a project is where $\Delta B/\Delta C = 1$. Figure 9-2a shows the line of feasible alternatives with

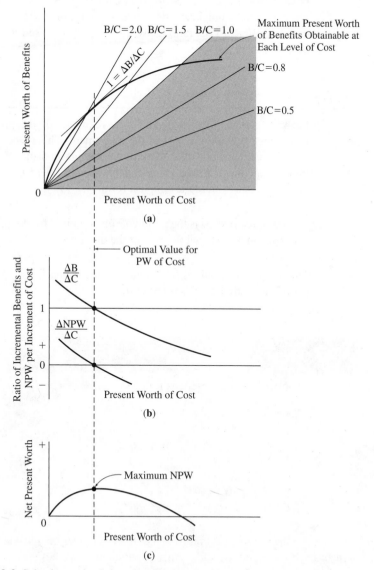

FIGURE 9-2 Selecting optimal size of project: (a) feasible alternatives, (b) changes in $\Delta B/\Delta C$, and (c) total NPW plotted versus size of the project.

their costs and benefits. This may represent a large number of calculations to locate points through which the line passes.

Figure 9-2b shows how the incremental benefit-cost ratio ($\Delta B / \Delta C$) changes as one moves along the line of feasible alternatives. Figure 9-2b also plots the ratio of incremental net present worth to incremental cost ($\Delta NPW / \Delta C$). As expected, we are adding increments of NPW as long as $\Delta B / \Delta C > 1$. Finally, in Figure 9-2c, we see the plot of (total) NPW versus the size of the project.

This three-part figure demonstrates that present worth analysis and benefit-cost ratio analysis lead to the same optimal decision. We saw in Chapter 8 that rate of return and present worth analysis led to identical decisions. Any of the exact analysis methods—present worth, annual cash flow, rate of return, or benefit-cost ratio—will lead to the same decision. Benefit-cost ratio analysis is extensively used in economic analysis at all levels of government.

PAYBACK PERIOD

Payback period is the period of time required for the profit or other benefits from an investment to equal the cost of the investment. This is the general definition for payback period, but there are other definitions. Others consider depreciation of the investment, interest, and income taxes; they, too, are simply called 'payback period'. For now, we will limit our discussion to the simplest form.

> **Payback period** is the period of time required for the profit or other benefits of an investment to equal the cost of the investment.

The rule in all situations is to minimize the payback period. The computation of payback period is illustrated in Examples 9-6 and 9-7.

EXAMPLE 9-6

The cash flows for two alternatives are as follows:

Year	A	B
0	−$1000	−$2783
1	+200	+1200
2	+200	+1200
3	+1200	+1200
4	+1200	+1200
5	+1200	+1200

You may assume the benefits occur throughout the year rather than just at the end of the year. On the basis of payback period, which alternative should be selected?

SOLUTION

Alternative A

Payback period is the period of time required for the profit or other benefits of an investment to equal the cost of the investment. In the first 2 years, only $400 of the $1000 cost is recovered. The remaining $600 cost is recovered in the first half of Year 3. Thus the payback period for Alt. A is 2.5 years.

Alternative B

Since the annual benefits are uniform, the payback period is simply

$$\$2783/\$1200 \text{ per year} = 2.3 \text{ years}$$

To minimize the payback period, choose Alt. B.

EXAMPLE 9-7

A firm is trying to decide which of two weighing scales it should install to check a package-filling operation in the plant. If both scales have a 6-year life, which one should be selected? Assume an 8% interest rate.

Alternative	Cost	Uniform Annual Benefit	End-of-Useful-Life Salvage Value
Atlas scale	$2000	$450	$100
Tom Thumb scale	3000	600	700

SOLUTION

Atlas Scale

$$\text{Payback period} = \frac{\text{Cost}}{\text{Uniform annual benefit}}$$

$$= \frac{2000}{450} = 4.4 \text{ years}$$

Tom Thumb Scale

$$\text{Payback period} = \frac{\text{Cost}}{\text{Uniform annual benefit}}$$

$$= \frac{3000}{600} = 5 \text{ years}$$

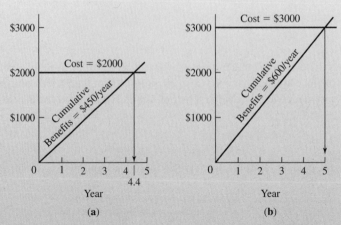

FIGURE 9-3 Payback period plots for Example 9-7: (a) Atlas scale and (b) Tom Thumb scale.

Figure 9-3 illustrates the situation. To minimize payback period, select the Atlas scale.

There are four important points to be understood about payback period calculations:

1. This is an approximate, rather than an exact, economic analysis calculation.
2. All costs and all profits, or savings of the investment, before payback are included *without* considering differences in their timing.
3. All the economic consequences beyond the payback period are completely ignored.
4. Being an approximate calculation, payback period may or may not select the correct alternative. This is, the payback period calculations may select an alternative different from that found by exact economic analysis techniques.

This last point—that payback period may result in the selection of the *wrong* alternative—was illustrated by Example 9-7. When payback period is used, the Atlas scale appears to be the more attractive alternative. Yet, when the same problem was solved earlier by the present worth method (Example 5-4), the Tom Thumb scale was the alternative chosen. A review of the problem reveals the reason for the different conclusions. The $700 salvage value at the end of 6 years for the Tom Thumb scale is a significant benefit. The salvage value occurs after the payback period; so it was ignored in the payback calculation. It *was* considered in the present worth analysis, with the result that the Tom Thumb scale was in fact more desirable.

But if payback period calculations are approximate and may lead to the selection of the wrong alternative, why is the method used at all? There are two primary answers: first, the calculations can be made readily by people unfamiliar with economic analysis. One does not need to know how to use gradient factors, or even to have a set of compound interest tables. Second, payback period is an easily understood concept. Earlier we pointed out that this was also an advantage to rate of return.

Moreover, payback period *does* give us a useful measure, telling us how long it will take for the cost of the investment to be recovered from the benefits of the investment. Businesses and industrial firms are often very interested in this time period: a rapid return of invested capital means that it can be reused sooner for other purposes. But one must not confuse the *speed* of the return of the investment, as measured by the payback period, with economic *efficiency*. They are two distinctly separate concepts. The former emphasizes the quickness with which invested funds return to a firm; the latter considers the overall profitability of the investment.

We can create another situation to illustrate how it may be unwise to choose between alternatives by the payback period rule.

EXAMPLE 9-8

A firm is buying production equipment for a new plant. Two different machines are being considered for a particular operation.

	Tempo Machine	**Dura Machine**
Installed cost	$30,000	$35,000
Net annual benefit after all annual expenses have been deducted	$12,000 the first year, *declining* $3000 per year thereafter	$1000 the first year, increasing $3000 per year thereafter
Useful life, in years	4	8

Neither machine has any salvage value. Compute the payback period for each of the machines.

SOLUTION BASED ON PAYBACK PERIOD

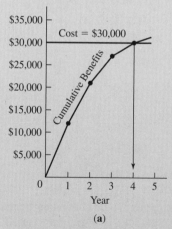
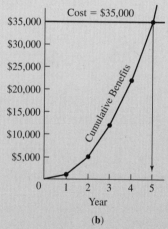

FIGURE 9-4 Payback period plots for Example 9-7: (a) Tempo machine and (b) Dura machine.

The Tempo machine has a declining annual benefit, while the Dura has an increasing annual benefit. Figure 9-4 shows the Tempo has a 4-year payback period and the Dura has a 5-year payback period. To minimize the payback period, the Tempo is selected.

Now, as a check on the payback period analysis, compute the rate of return for each alternative. Assume the minimum attractive rate of return is 10%.

SOLUTION BASED ON RATE OF RETURN

The cash flows for the two alternatives are as follows:

Year	Tempo Machine	Dura Machine
0	−$30,000	−$35,000
1	+12,000	+1,000
2	+9,000	+4,000
3	+6,000	+7,000
4	+3,000	+10,000
5	0	+13,000
6	0	+16,000
7	0	+19,000
8	0	+22,000
	0	+57,000

Tempo Machine

Since the sum of the cash flows for the Tempo machine is zero, we see immediately that the $30,000 investment just equals the subsequent benefits. The resulting rate of return is 0%.

Dura Machine

$$35,000 = 1000(P/A, i, 8) + 3000(P/G, i, 8)$$

Try $i = 20\%$:

$$35,000 \doteq 1000(3.837) + 3000(9.883)$$
$$\doteq 3837 + 29,649 = 33,486$$

The 20% interest rate is too high. Try $i = 15\%$:

$$35,000 \doteq 1000(4.487) + 3000(12.481)$$
$$\doteq 4487 + 37,443 = 41,930$$

This time, the interest rate is too low. Linear interpolation would show that the rate of return is approximately 19%.

If we use an exact calculation—rate of return—it is clear that the Tempo is not very attractive economically. Yet it was this alternative, and not the Dura machine, that was preferable according to the payback period calculations. On the other hand, the shorter payback period for the Tempo does give a measure of the speed of the return of the investment not found in the Dura. The conclusion to be drawn is that **liquidity** and **profitability** may be two quite different criteria.

From the discussion and the examples, we see that payback period can be helpful in providing a measure of the speed of the return of the investment. This might be quite important, for example, for a company that is short of working capital or for a firm in an industry experiencing rapid changes in technology. Calculation of payback period alone, however, must not be confused with a careful economic analysis. We have shown that a short payback period does not always mean that the associated investment is desirable. Thus, payback period should not be considered a suitable replacement for accurate economic analysis calculations.

SENSITIVITY AND BREAK-EVEN ANALYSIS

Since many data gathered in solving a problem represent *projections* of future consequences, there may be considerable uncertainty regarding the accuracy of the data. Since the desired result of the analysis is decision making, a reasonable question is: to what extent do variations in the data affect my decision? When small variations in a particular estimate would change the choice of an alternative, the decision is said to be **sensitive to the estimate.** To better evaluate the impact of any particular estimate, we compute "the variation to a particular estimate that would be necessary to change a particular decision." This is called **sensitivity analysis.**

An analysis of the sensitivity of the solution to a problem to its various parameters highlights the important and significant aspects of that problem. For example, one might be concerned that the estimates for annual maintenance and future salvage value in a particular problem may vary substantially. Sensitivity analysis might show that the decision is insensitive to the salvage-value estimate over the full range of possible values. But, at the same time, we might find that the decision is sensitive to changes in the annual maintenance estimate. Under these circumstances, one should place greater emphasis on improving the annual maintenance estimate and less on the salvage-value estimate.

As indicated at the beginning of this chapter, break-even analysis is a form of sensitivity analysis. To illustrate the sensitivity of a decision between alternatives to particular estimates, break-even analysis is often presented as a **break-even chart.**

Sensitivity and break-even analysis are frequently useful in engineering problems called **stage construction.** Should a facility be constructed now to meet its future full-scale requirement, or should it be constructed in stages as the need for the increased capacity arises? Here are three examples of this situation:

- Should we install a cable with 400 circuits now or a 200-circuit cable now and another 200-circuit cable later?
- A 10 cm water main is needed to serve a new group of homes. Should it be installed now, or should a 15 cm main be installed to ensure an adequate water supply to adjoining areas later, when other homes have been built?
- An industrial firm needs a new warehouse now and estimates that it will need to double its size in 4 years. The firm could have a warehouse built now and enlarged later, or it could have the warehouse with capacity for expanded operations built right away.

Examples 9-9 and 9-10 illustrate sensitivity and break-even analysis.

EXAMPLE 9-9

Consider a project that may be constructed to full capacity now or may be constructed in two stages.

Construction Costs

Two-stage construction
Construct first stage now $100,000
Construct second stage   120,000
    n years from now
Full-capacity construction
Construct full capacity now   140,000

Other Factors

1. All facilities will last for 40 years regardless of when they are installed; after 40 years, they will have zero salvage value.
2. The annual cost of operation and maintenance is the same for both two-stage construction and full-capacity construction.
3. Assume an 8% interest rate.

Plot 'age when second stage is constructed' versus 'costs for both alternatives'. Mark the break-even point on your graph. What is the sensitivity of the decision to second-stage construction 16 or more years in the future?

SOLUTION

Since we are dealing with a common analysis period, the calculations may be either annual cost or present worth. Present worth calculations appear simpler and are used here.

Construct Full Capacity Now

$$\text{PW of cost} = \$140,000$$

Two-Stage Construction

First stage constructed now and second stage to be constructed n years hence. Compute the PW of cost for several values of n (years).

$$\text{PW of cost} = 100,000 + 120,000(P/F, 8\%, n)$$

$n = 5$ $\text{PW} = 100,000 + 120,000(0.6806) = \$181,700$
$n = 10$ $\text{PW} = 100,000 + 120,000(0.4632) = \ \ 155,600$
$n = 20$ $\text{PW} = 100,000 + 120,000(0.2145) = \ \ 125,700$
$n = 30$ $\text{PW} = 100,000 + 120,000(0.0994) = \ \ 111,900$

These data are plotted in the form of a break-even chart in Figure 9-5.

FIGURE 9-5 Break-even chart for Example 9-9.

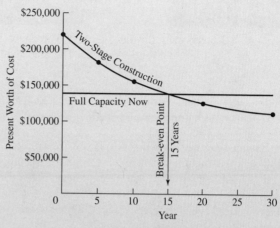

Figure 9-5 portrays the PW of cost for the two alternatives. The *x*-axis variable is the *time* when the second stage is constructed. We see that the PW of cost for two-stage construction naturally decreases as the time for the second stage is deferred. The one-stage construction (full capacity now) is unaffected by the *x*-axis variable and, hence, is a horizontal line on the graph.

The break-even point on the graph is the point at which both alternatives have equivalent costs. We see that if, in two-stage construction, the second stage is deferred for 15 years, then the PW of cost of two-stage construction is equal to one-stage construction; year 15 is the break-even point. The graph also shows that if the second stage were to be needed before Year 15, then one-stage construction, with its smaller PW of cost, would be preferred. On the other hand, if the second stage were not to be needed until after 15 years, two-stage construction would be preferred.

The decision on how to construct the project is sensitive to the age at which the second stage is needed *only* if the range of estimates includes 15 years. For example, if one estimated that the second-stage capacity would be needed between 5 and 10 years hence, the decision is insensitive to that estimate. For any value within that range, the decision does not change. The more economical thing to do is to build the full capacity now. But if the second-stage capacity were to be needed sometime between, say, 12 and 18 years, the decision would be sensitive to the estimate of when the full capacity would be needed.

One question posed by Example 9-9 is *how* sensitive the decision is to the need for the second stage at 16 years or beyond. The graph shows that the decision is insensitive. In all cases for construction on or after 16 years, two-stage construction has a lower PW of cost.

EXAMPLE 9-10

Example 8-3 posed the following situation. Three mutually exclusive alternatives are given, each with a 20-year life and no salvage value. The minimum attractive rate of return is 6%.

	A	**B**	**C**
Initial cost	$2000	$4000	$5000
Uniform annual benefit	410	639	700

In Example 8-3 we found that Alt. B was the preferred alternative. Here we would like to know how sensitive the decision is to our estimate of the initial cost of B. If B is preferred at an initial cost of $4000, it will continue to be preferred at any smaller initial cost. But *how much* higher than $4000 can the initial cost be and still have B the preferred alternative? The computations may be done several different ways. With neither input nor output fixed, maximizing net present worth is a suitable criterion.

Alternative A

$$\text{NPW} = \text{PW of benefit} - \text{PW of cost}$$

$$= 410(P/A, 6\%, 20) - 2000$$

$$= 410(11.470) - 2000 = \$2703$$

Alternative B

Let $x =$ initial cost of B.

$$\text{NPW} = 639(P/A, 6\%, 20) - x$$

$$= 639(11.470) - x$$

$$= 7329 - x$$

Alternative C

$$\text{NPW} = 700(P/A, 6\%, 20) - 5000$$

$$= 700(11.470) - 5000 = \$3029$$

For the three alternatives, we see that B will maximize NPW only as long as its NPW is greater than 3029.

$$3029 = 7329 - x$$

$$x = 7329 - 3029 = \$4300$$

Therefore, B is the preferred alternative if its initial cost does not exceed $4300.

Figure 9-6 is a break-even chart for the three alternatives. Here the rule is to maximize NPW; as a result, the graph shows that B is preferred if its initial cost is less than $4300. At an

FIGURE 9-6 Break-even chart for Example 9-10.

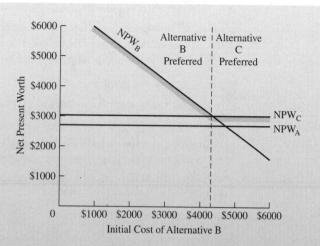

initial cost above $4300, C is preferred. We have a break-even point at $4300. When B has an initial cost of $4300, B and C are equally desirable.

Sensitivity analysis and break-even point calculations can be very useful in identifying how different estimates affect the calculations. It must be recognized that these calculations assume that all parameters except one are held constant, and the sensitivity of the decision to that one variable is evaluated. Later we will look further at the impact of parameter estimates on decision making.

GRAPHING WITH SPREADSHEETS FOR SENSITIVITY AND BREAK-EVEN ANALYSIS

Chapter 4 introduced the drawing of xy plots with spreadsheets, and Chapter 7 reviewed this procedure for plotting present worth versus i. The Chapter 7 plot is an example of break-even analysis, because it is used to determine at what interest rate the project breaks even or has a present worth of 0. This section will present some of the spreadsheet tools and options that can make the xy plots more effective and attractive.

The spreadsheet tools and options can be used to:

- Modify the x or y axis

    Specify the minimum or maximum value
    Specify at what value the other axis intersects (default is 0)

- Match line types to data

    Use line types to distinguish one curve from another
    Use markers to show real data
    Use lines without markers to plot curves (straight segments or smooth curves)

- Match chart colours to how they are displayed

    Colour defaults are fine for colour computer screen
    Colour defaults are OK for colour printers
    Black and white printing is better with editing (use line types, not colours)

- Annotate the graph

    Add text, arrows, and lines to graphs
    Add data labels

In most cases the menus of Excel are self-explanatory, and so the main step is to decide what you want to achieve. Then you just look for the way to do it. Left clicks are used to select the item to modify, and right clicks are used to bring up the options for that item. Example 9-11 illustrates this process.

EXAMPLE 9-11

The staged construction choice described in Example 9-9 used a broad range of x values for the x axis. Create a graph that focuses on the 10- to 20-year period and is designed for printing in a report. The costs are:

Year	Full Capacity	Two Stage
0	$140,000	$100,000
n	0	120,000

SOLUTION

The first step is to create a table of values that shows the present worth of the costs for different values of n—the length of time until the second stage or full capacity is needed. Notice that the full capacity is calculated at $n = 0$. The only reason to calculate the corresponding value for staged construction is to see if the formula is entered properly, since building both stages at the same time will not really cost $220,000. The values for staged construction at 5, 10, 20, and 30 years correspond to the values in Example 9-9.

The next step is to select cells A8:C13, which includes the x values and two series of y values. Then the ChartWizard tool is selected. In the first step, the xy (scatter) plot is selected with the option of smoothed lines without markers. In Step 2 no action is required since the cells A8:C13 were selected first. In Step 3 labels are added for the x and y axes. In Step 4 the chart is moved around on the worksheet page, so that it does not overlap with the data. The result is shown in Figure 9-7 (except that this is the colour screen version printed to a black and white printer).

Our first step in cleaning up the graph is to delete the formula in cell C9, since two-stage construction will not be done at Time 0. We also delete the label in the adjacent cell, which explains the formula. Then we create a new label for cell C10. As shown in the appendix, the easy way to create that label is to insert an apostrophe as the first entry in cell C10. This converts the formula to a label which we can copy to D10. Then we delete the apostrophe in cell C10.

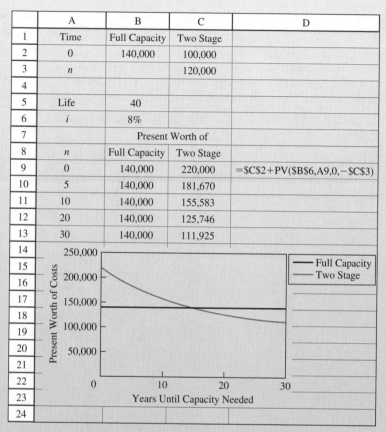

	A	B	C	D
1	Time	Full Capacity	Two Stage	
2	0	140,000	100,000	
3	n		120,000	
4				
5	Life	40		
6	i	8%		
7		Present Worth of		
8	n	Full Capacity	Two Stage	
9	0	140,000	220,000	=$C$2+PV($B$6,A9,0,−$C$3)
10	5	140,000	181,670	
11	10	140,000	155,583	
12	20	140,000	125,746	
13	30	140,000	111,925	

FIGURE 9-7 Automatic graph from spreadsheet.

The axis scales must be modified to focus on the area of concern. Select the x axis and change the minimum from automatic to 10 and the maximum to 20. Select the y axis and change the minimum to 125,000 and the maximum to 160,000.

Left-click on the plot area to select it. Then right click to bring up the options. Select Format Plot Area and change the area pattern to 'none'. This will eliminate the grey fill that made Figure 9-7 difficult to read.

Left-click on the two-stage curve to select it. Then right-click for the options. Format the data series using the 'patterns' tab. Change the line style from solid to dashed, the line colour from automatic to black, and increase the line weight. Similarly, increase the line weight for the full-capacity line. Finally, select a grid line and change the line style to dotted. The result is far easier to read in black and white.

To further improve the graph, we can replace the legend with annotations on the graph. Left-click somewhere in the white area around the graph to select 'chart area'. Right-click and then choose the chart options on the menu. The legends tab will let us delete the legend by turning 'show legend' off. Similarly, we can turn the x-axis gridlines on. The line style for these gridlines should be changed to match the y-axis gridlines. This allows us to see that the break-even time is between 14 and 15 years.

To make the graph less busy, change the scale on the x axis so that the interval is 5 years rather than automatic. Also eliminate the gridlines for the y axis (by selecting the chart area, chart options, and gridlines tabs). The graph size can be increased for easier reading, as well. This may require specifying an interval of 10,000 for the scale of the y axis.

Finally to add the labels for the full-capacity curve and the two-stage curve, find the toolbar for graphics, which is open when the chart is selected (probably along the bottom of the spreadsheet). Select the text box icon, and click on a location close to the two-stage chart. Type in the label for two-stage construction. Notice how including a return and a few spaces can shape the label to fit the slanted line. Add the label for full construction. Figure 9-8 is the result.

	A	B	C	D
1	Time	Full Capacity	Two Stage	
2	0	140,000	100,000	
3	n		120,000	
4				
5	life	40		
6	i	8%		
7		Present worth of		
8	n	Full Capacity	Two Stage	
9	0	140,000		
10	5	140,000	181,670	=$C$2+PV($B$6,A10,0,−$C$3)
11	10	140,000	155,583	
12	20	140,000	125,746	
13	30	140,000	111,925	
14				

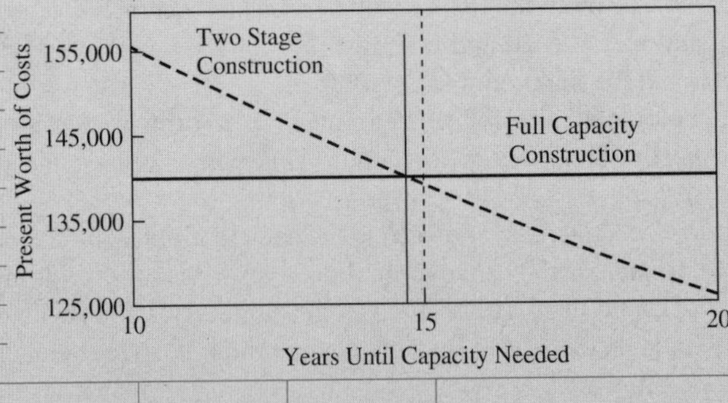

FIGURE 9-8 Spreadsheet of Figure 9-7 with improved graph.

SUMMARY

In this chapter, we have looked at four new analysis techniques.

Future worth: When the comparison between alternatives will be made in the future, the calculation is called future worth. This is very similar to present worth, which is based on the present, rather than a future point in time.

Benefit-cost ratio analysis: This technique is based on the ratio of benefits to costs by means of either present worth or annual cash flow calculations. The method is graphically similar to present worth analysis. When neither input nor output is fixed, incremental benefit-cost ratios ($\Delta B / \Delta C$) are required. The method is similar in this respect to rate of return analysis. Benefit-cost ratio analysis is often used at the various levels of government.

Payback period: Here we define payback as the period of time required for the profit or other benefits of an investment to equal the cost of the investment. Although simple to use and to understand, payback is a poor analysis technique for ranking alternatives. Although it provides a measure of the speed of the return of the investment, it is not an accurate measure of the profitability of an investment.

Sensitivity and break-even analysis: These techniques are used to see how sensitive a decision is to estimates for the various parameters. Break-even analysis is done to locate conditions under which the alternatives are equivalent. This is often presented in the form of break-even charts. Sensitivity analysis is an examination of a range of values for some parameters to determine their effect on a particular decision.

PROBLEMS

9-1 For a 12% interest rate, compute the value of F in the following diagram.

9-2 Compute F for the following diagram.

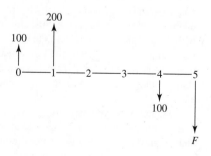

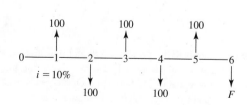

9-3   Compute F for the following diagram.

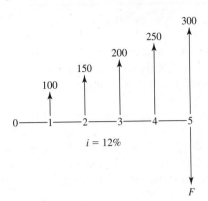

(*Answer:* $F = \$1199$)

9-4   For the following diagram, compute F.

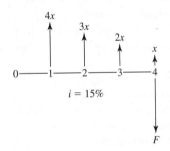

9-5   For the following diagram, compute F.

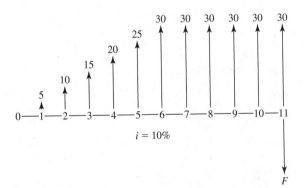

9-6   Calculate the present worth and the future worth of a series of 10 annual cash flows with the first cash flow equal to $15,000 and each successive cash flow increasing by $1200. The interest rate is 12%. The total cash flow series is a combination of Systems 1 and 2.

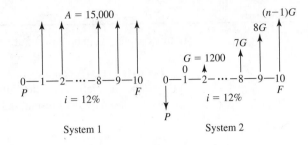

System 1   System 2

9-7   The interest rate is 16% per year and there are 48 compounding periods per year. The principal is $50,000. What is the future worth in 5 years?

9-8   A 20-year-old student decided to set aside $100 on his 21st birthday for investment. Each subsequent year through his 55th birthday, he plans to increase the sum for investment on a $100 arithmetic gradient. He will not set aside additional money after his 55th birthday. If the student can achieve a 12% rate of return, what is the future worth of the investments on his 65th birthday?   (*Answer:* $1,160,700)

9-9   You have an opportunity to purchase a piece of vacant land for $30,000 cash. If you buy the property, you plan to hold it for 15 years and then sell it at a profit. During this period, you would have to pay annual property taxes of $600. You would have no income from the property. Assuming that you would want a 10% rate of return from the investment, at what net price would you have to sell it 15 years hence? (*Answer:* $144,373)

9-10   A man's salary is now $32,000 per year and he expects to retire in 30 years. If his salary is increased by $600 each year and he deposits 10% of his yearly salary into a fund that earns 7% interest compounded annually, what is the future worth of the amount accumulated at the time of his retirement?

9-11   Stamp collecting has become an increasingly popular—and expensive—hobby. One favourite method is to save plate blocks (usually four stamps with the printing plate number in the margin) of each new stamp as it is issued by the post office. But with the rising postage rates and increased numbers of new stamps being issued, this collecting plan costs more each year.

Stamps, however, may have been a good place to invest money over the last 10 years, as the demand for stamps previously issued has caused resale prices to increase 18% each year. Suppose a collector bought

$100 worth of stamps 10 years ago, and increased his purchases by $50 a year in each subsequent year. After 10 years of stamp collecting, what is the future worth of the stamp collection?

9-12 Sally deposited $100 a month in her savings account for 24 months. For the next 5 years she made no deposits. What is the future worth in Sally's savings account at the end of the 7 years if the account earned 6% annual interest, compounded monthly? (*Answer:* $3430.78)

9-13 In the early 1980s, planners were examining alternative sites for a new airport to serve London. In their economic analysis, they computed the value of the structures that would need to be removed from various airport sites. At one airport site, the twelfth-century Norman church of St Michael's, in the village of Stewkley, would have to be demolished. The planners used the value of the fire insurance policy on the church—a few thousand pounds sterling—as the value of the structure.

An outraged antiquarian wrote to the London *Times* that an equally plausible computation would be to assume that the original cost of the church (estimated at 100 pounds sterling) should be increased at the rate of 10% a year for 800 years. According to that calculation, what would the future worth of St Michael's be? (*Note:* There was great public objection to tearing down the church, and it was spared.)

9-14 Bill made a budget and planned to deposit $150 a month in a savings account, beginning September 1. He did this, but on the following January 1, he reduced the monthly deposits to $100 a month. In all he made 18 deposits, four at $150 and 14 at $100. If the savings account paid 6% interest, compounded monthly, what was the future worth of his savings account immediately after he made the last deposit? (*Answer:* $2094.42)

9-15 A company deposits $1000 in a bank at the beginning of each year for 6 years. The account earns 8% interest, compounded every 6 months. What is the future worth of the account at the end of 6 years? Make a careful, accurate computation.

9-16 Don Ball is a 55-year-old engineer. According to mortality tables, a man at age 55 has an average life expectancy of 21 more years. In prior years, Don has accumulated $48,500, including interest, toward his retirement. He is now adding $5000 per year to his retirement fund. The fund earns 12% interest. Don's

goal is to retire when he can obtain an annual income from his retirement fund of $20,000 a year, assuming he lives to age 76. He will make no provision for a retirement income after age 76. What is the youngest age at which Don can retire, based on his criteria?

9-17 Jean invests $100 in Year 1 and doubles the amount each year after that (so the investment is $100, 200, 400, 800, . . .). If she continues to do this for 10 years, and the investment pays 10% annual interest, what is the future worth of her investment at the end of 10 years?

9-18 If you invested $2500 in a bank 24-month certificate of deposit paying 8.65%, compounded monthly, what is the future worth of the certificate of deposit when it matures in 2 years?

9-19 After receiving an inheritance of $25,000 on her 21st birthday, Ayn Rand deposited the inheritance in a savings account with an effective annual interest rate of 6%. She decided that she would make regular deposits on each future birthday, beginning with $1000 on her 22nd birthday and then increasing the amount by $200 in each following year (i.e., $1200 on her 23rd birthday, $1400 on her 24th birthday, etc.). What was the future worth of Ayn's deposits on her 56th birthday?

9-20 The Association of General Contractors (AGC) wished to establish an endowment fund of $1 million in 10 years for the Construction Engineering Technology Program at Grambling State University in Grambling, Louisiana. In doing so, the AGC established an escrow account in which 10 equal end-of-year deposits that earn 7% compound interest were made. After seven deposits, the Louisiana legislature revised laws relating to the licensing fees that the AGC can charge its members, with the result that there was no deposit at the end of Year 8. What must the amount of the remaining equal end-of-year deposits be to ensure that the $1 million is available to Grambling State for its Construction Engineering Technology Program?

9-21 On her birthday, a 25-year-old engineer is considering investing in an individual retirement account (IRA). After some research, she finds a mutual fund with an average return of 10% a year. What is the future worth of her IRA at age 65 if she makes annual investments of $2000 into the fund beginning on her 25th birthday? Assume that the fund continues to earn an annual return of 10%.

9-22 IPS Corp. will upgrade its package-labelling machinery. It costs $150,000 to buy the machinery and have it installed. Operation and maintenance costs are $1500 per year for the first 3 years and increase by $500 per year for the remaining years of the machine's 10-year life. The machinery has a salvage value of 5% of its initial cost. Interest is 10%. What is the future worth of cost of the machinery?

9-23 A company is considering buying a new bottle-capping machine. The initial cost of the machine is $325,000, and it has a 10-year life. Monthly maintenance costs are expected to be $1200 a month for the first 7 years and $2000 a month for the remaining years. The machine requires a major overhaul costing $55,000 at the end of the fifth year of service. Assume that all these costs occur at the end of the relevant period. What is the future value of all the costs of owning and operating this machine if the nominal interest rate is 7.2%?

9-24 A family starts an education fund for their son Patrick when he is 8 years old, investing $150 on his eighth birthday and increasing the yearly investment by $150 per year until Patrick is 18 years old. The fund pays 9% annual interest. What is the future worth of the fund when Patrick is 18?

9-25 A bank account pays 19.2% interest with monthly compounding. A series of deposits started with a deposit of $5000 on January 1, 1997. Deposits in the series were to occur each 6 months. Each deposit in the series is for $150 less than the one before it. The last deposit in the series will be due on January 1, 2012. What is the future worth of the account on July 1, 2014, if the balance was zero before the first deposit and no withdrawals are made?

9-26 Let's assume that a late-twentieth-century university graduate got a good job and began a savings account. He is paid monthly and authorized the bank to automatically withdraw $75 each month. The bank made the first withdrawal on July 1, 1997, and is instructed to make the last withdrawal on January 1, 2015. The bank pays a nominal interest rate of 4.5% and compounds twice a month. What is the future worth of the account on January 1, 2015?

9-27 Bob, an engineer, decided to start a university fund for his son. Bob will deposit a series of equal, semi-annual cash flows with each deposit equal to $1500. Bob made the first deposit on July 1, 1998, and will make the last deposit on July 1, 2018. Joe, a friend of Bob's, received an inheritance on April 1, 2003, and has decided to begin a university fund for his daughter. Joe wants to send his daughter to the same university as Bob's son. Therefore, Joe needs to accumulate the same amount of money on July 1, 2018, as Bob will have accumulated from his semi-annual deposits. Joe never took engineering economics and had no idea how to determine the amount that should be deposited. He decided to deposit $40,000 on July 1, 2003. Will Joe's deposit be sufficient? If not, how much should he have put in? Use a nominal interest of 7% with semi-annual compounding on all accounts.

9-28 A business executive is offered a management job at Generous Electric Company, which offers him a 5-year contract that calls for a salary of $62,000 a year, plus 600 shares of GE stock at the end of the 5 years. This executive is currently employed by Fearless Bus Company, which also has offered him a 5-year contract. It calls for a salary of $65,000, plus 100 shares of Fearless stock each year. The Fearless stock is currently worth $60 per share and pays an annual dividend of $2 per share. Assume end-of-year payments of salary and stock. Stock dividends begin one year after the stock is received. The executive believes that the value of the stock and the dividend will remain constant. If the executive considers 9% a suitable rate of return in this situation, what must the Generous Electric stock be worth per share to make the two offers equally attractive? Use the future worth analysis method in your comparison.
(*Answer:* $83.76)

9-29 A project will cost $50,000. The benefits at the end of the first year are estimated to be $10,000, increasing at a 10% uniform rate in subsequent years. Using an 8-year analysis period and a 10% interest rate, compute the benefit-cost ratio.

9-30 Each of the three alternatives shown has a 5-year useful life. If the MARR is 10%, which alternative should be selected? Solve the problem by benefit-cost ratio analysis.

	A	B	C
Cost	$600.0	$500.0	$200.0
Uniform annual benefit	158.3	138.7	58.3

(*Answer:* B)

9-31 Consider three alternatives, each with a 10-year useful life. If the MARR is 10%, which alternative should be selected? Solve the problem by benefit-cost ratio analysis.

	A	B	C
Cost	$800	$300	$150
Uniform annual benefit	142	60	33.5

9-32 An investor is considering buying some land for $100,000 and constructing an office building on it. Three different buildings are being analyzed.

	Building Height		
	2 Storeys	**5 Storeys**	**10 Storeys**
Cost of building (excluding cost of land)	$400,000	$800,000	$2,100,000
Resale value* of land and building at end of 20-year analysis period	200,000	300,000	400,000
Annual rental income after all operating expenses have been deducted	70,000	105,000	256,000

*Resale value to be considered a reduction in cost, rather than a benefit.

Using benefit-cost ratio analysis and an 8% MARR, determine which alternative, if any, should be selected.

9-33 Using benefit-cost ratio analysis, determine which one of the three mutually exclusive alternatives should be selected.

	A	B	C
First cost	$560	$340	$120
Uniform annual benefit	140	100	40
Salvage value	40	0	0

Each alternative has a 6-year useful life. Assume a 10% MARR.

9-34 Consider four alternatives, each of which has an 8-year useful life:

	A	B	C	D
Cost	$100.0	$80.0	$60.0	$50.0
Uniform annual benefit	12.2	12.0	9.7	12.2
Salvage value	75.0	50.0	50.0	0

If the MARR is 8%, which alternative should be selected? Solve the problem by benefit-cost ratio analysis.

9-35 Using benefit-cost ratio analysis, a 5-year useful life, and a 15% MARR, determine which of the following alternatives should be selected.

	A	B	C	D	E
Cost	$100	$200	$300	$400	$500
Uniform annual benefit	37	69	83	126	150

9-36 Five mutually exclusive investment alternatives have been proposed. Based on benefit-cost ratio analysis, and a MARR of 15%, which alternative should be selected?

Year	A	B	C	D	E	F
0	−$200	−$125	−$100	−$125	−$150	−$225
1–5	+68	+40	+25	+42	+52	+68

9-37 Able Plastics, an injection-moulding firm, has negotiated a contract with a national chain of department stores. Plastic pencil boxes are to be produced for a 2-year period. Able Plastics has never produced the item before and, therefore, requires all new dies. If the firm invests $67,000 for special removal equipment to unload the completed pencil boxes from the moulding machine, one machine operator can be eliminated. This would save $26,000 a year. The removal equipment has no salvage value and is not expected to be used after the 2-year production contract is completed. The equipment, although useless, would be serviceable for about 15 years. You have been asked to do a payback period analysis on whether to purchase the special removal equipment. What is the payback period? Should Able Plastics buy the removal equipment?

9-38 A cannery is considering installing an automatic case-sealing machine to replace current hand methods. If they purchase the machine for $3800 in June, at the beginning of the canning season, they will save $400 a month for the 4 months each year that the plant is in operation. Maintenance costs of the case-sealing machine are expected to be negligible. The case-sealing machine is expected to be useful for five annual canning seasons and will have no salvage value at the end of that time. What is the payback period? Calculate the nominal annual rate of return based on the estimates.

9-39 A project has the following costs and benefits. What is the payback period?

Year	Costs	Benefits
0	$1400	
1	500	
2	300	$400
3–10		300 in each year

9-40 A car dealer is leasing a small computer with software for $5000 a year. As an alternative he could buy the computer for $7000 and lease the software for $3500 a year. Any time he decided to switch to some other computer system he could cancel the software lease and sell the computer for $500. If he buys the computer and leases the software,
(a) What is the payback period?
(b) If he kept the computer and software for 6 years, what would the benefit-cost ratio be if the interest rate is 10%?

9-41 A large project requires an investment of $200 million. The construction will take 3 years: $30 million will be spent during the first year, $100 million during the second year, and $70 million during the third year of construction. Two project operation periods are being considered: 10 years with the expected net profit of $40 million a year and 20 years with the expected net profit of $32.5 million a year. For simplicity of calculations it is assumed that all cash flows occur at end of year. The company minimum required return on investment is 10%.

Calculate for each alternative:
(a) The payback periods
(b) The total equivalent investment cost at the end of the construction period
(c) The equivalent uniform annual worth of the project (use the operation period of each alternative)

Make recommendations based on the foregoing economic parameters.

9-42 Two alternatives are being considered:

	A	B
Initial cost	$500	$800
Uniform annual cost	200	150
Useful life, in years	8	8

Both alternatives provide an identical benefit.
(a) Compute the payback period if Alt. B is purchased rather than Alt. A.

(b) Use a MARR of 12% and benefit-cost ratio analysis to identify the alternative that should be selected.

9-43 Tom Sewell has gathered data on the relative costs of a solar water heater system and a conventional electric water heater. The data are based on statistics for a western city and assume that during cloudy days an electric heating element in the solar heating system will provide the necessary heat.

The installed cost of a conventional electric water tank and heater is $200. A family of four uses an average of 300 litres of hot water a day, which takes $230 of electricity a year. The glass-lined tank has a 20-year guarantee. This is probably a reasonable estimate of its actual useful life.

The installed cost of two solar panels, a small electric pump, and a storage tank with auxiliary electric heating element is $1400. It will cost $60 a year for electricity to run the pump and heat water on cloudy days. The solar system will require $180 of maintenance work every 4 years. Neither the conventional electric water heater nor the solar water heater will have any salvage value at the end of its useful life.
(a) Using Tom's data, find the payback period if the solar water heater system is installed, rather than the conventional electric water heater.
(b) Chris Cook studied the same situation and decided that the solar system will *not* require the $180 of maintenance every 4 years. Chris believes future replacements of either the conventional electric water heater or the solar water heater system can be made at the same cost and useful lives as the initial installation. Using a 10% interest rate, calculate what the useful life of the solar system must be to make it no more expensive than the electric water heater system.

9-44 Consider four mutually exclusive alternatives:

	A	B	C	D
Cost	$75.0	$50.0	$15.0	$90.0
Uniform annual benefit	18.8	13.9	4.5	23.8

Each alternative has a 5-year useful life and no salvage value. The MARR is 10%. Which alternative should be selected if one uses
(a) Future worth analysis
(b) Benefit-cost ratio analysis
(c) The payback period

9-45 Consider three alternatives:

	A	B	C
First cost	$50	$150	$110
Uniform annual benefit	28.8	39.6	39.6
Useful life, in years*	2	6	4
Computed rate of return	10%	15%	16.4%

*At the end of its useful life, an identical alternative (with the same cost, benefits, and useful life) may be installed.

All the alternatives have no salvage value. If the MARR is 12%, which alternative should be selected?
(a) Solve the problem by future worth analysis.
(b) Solve the problem by benefit-cost ratio analysis.
(c) Solve the problem by payback period.
(d) If the answers in parts (a), (b), and (c) differ, explain why this is the case.

9-46 Consider three mutually exclusive alternatives. The MARR is 10%.

Year	X	Y	Z
0	−$100	−$50	−$50
1	25	16	21
2	25	16	21
3	25	16	21
4	25	16	21

(a) For Alt. X, compute the benefit-cost ratio.
(b) Based on the payback period, which alternative should be selected?
(c) Using an exact economic analysis method, determine the preferred alternative.

9-47 The cash flows for three alternatives are as follows:

Year	A	B	C
0	−$500	−$600	−$900
1	−400	−300	0
2	200	350	200
3	250	300	200
4	300	250	200
5	350	200	200
6	400	150	200

(a) On the basis of payback period, which alternative should be selected?
(b) Using future worth analysis and a 12% interest rate, determine which alternative should be selected.

9-48 Three mutually exclusive alternatives are being considered:

	A	B	C
Initial cost	$500	$400	$300
Benefit at end of the first year	200	200	200
Uniform benefit at end of subsequent years	100	125	100
Useful life, in years	6	5	4

At the end of its useful life, an alternative is *not* replaced. If the MARR is 10%, which alternative should be selected:
(a) On the basis of the payback period?
(b) On the basis of benefit-cost ratio analysis?

9-49

Year	E	F	G	H
0	−$90	−$110	−$100	−$120
1	20	35	0	0
2	20	35	10	0
3	20	35	20	0
4	20	35	30	0
5	20	0	40	0
6	20	0	50	180

(a) Using future worth analysis, decide which of the four alternatives is preferred at 6% interest.
(b) Using future worth analysis, decide which alternative is preferable at 15% interest.
(c) Using the payback period, decide which alternative is preferred.
(d) At 7% interest, what is the benefit-cost ratio for Alt. G?

9-50 Tom Jackson is preparing to buy a new car. He knows it represents a large expenditure of money, so he wants to do an analysis to see which of two cars is more economical. Alternative A is a domestic-built compact car. It has an initial cost of $8900 and operating costs of 9¢/km, excluding depreciation. Tom checked automobile resale statistics. From them he estimates the domestic car can be resold at the end of 3 years for $1700. Alternative B is a foreign-built Fiasco. Its initial cost is $8000, the operating cost, also excluding depreciation, is 8¢/km. How low could the resale value of the Fiasco be to provide equally economical transportation? Assume Tom will drive 12,000 km/year and considers 8% as a reasonable interest rate.   (*Answer:* $175)

9-51 A newspaper is considering purchasing locked vending machines to replace open newspaper racks for the sale of its newspapers in the downtown area.

The vending machines cost $45 each. It is expected that the annual revenue from selling the same number of newspapers will increase $12 per vending machine. The useful life of the vending machine is unknown.

(a) To determine the sensitivity of rate of return to useful life, prepare a graph for rate of return versus useful life for lives up to 8 years.

(b) If the newspaper requires a 12% rate of return, what minimum useful life must it obtain from the vending machines?

(c) What would be the rate of return if the vending machines were to last indefinitely?

9-52 Data for two alternatives are as follows:

	A	B
Cost	$800	$1000
Uniform annual benefit	230	230
Useful life, in years	5	X

If the MARR is 12%, compute the value of X that makes the two alternatives equally desirable.

9-53 What is the cost of Alt. B that will make it at the break-even point with Alt. A, assuming a 12% interest rate?

	A	B
Cost	$150	$ X
Uniform annual benefit	40	65
Salvage value	100	200
Useful life, in years	6	6

9-54 Consider two alternatives:

	A	B
Cost	$500	$300
Uniform annual benefit	75	75
Useful life, in years	Infinity	X

Assume that Alt. B is not replaced at the end of its useful life. If the MARR is 10%, what must the useful life of B be to make Alternatives A and B equally desirable?

9-55 Jane Chang is making plans for a summer vacation. She will take $1000 with her in the form of traveller's cheques. From the newspaper, she finds that if she buys the cheques by May 31, she will not have to pay a service charge. That is, she will obtain $1000 worth of traveller's cheques for $1000. But if she waits to buy the cheques until just before starting her summer trip, she must pay a 1% service charge. (It will cost her $1010 for $1000 of traveller's cheques.)

Jane can obtain a 13% interest rate, compounded weekly, on her money. To help with her planning, Jane needs to know how many weeks after May 31 she can begin her trip and still justify buying the traveller's cheques on May 31. She asks you to make the computations for her. What is the answer?

9-56 Fence posts for a particular job cost $10.50 each to install, including the labour cost. They will last 10 years. If the posts are treated with a wood preservative they can be expected to have a 15-year life. Assuming a 10% interest rate, how much could one afford to pay for the wood preservative treatment?

9-57 A piece of property is bought for $10,000 and yields a $1000 yearly net profit. The property is sold after 5 years. What is its minimum price to break even with interest at 10%?

9-58 Rental equipment is for sale for $110,000. A prospective buyer estimates he would keep the equipment for 12 years and spend $6000 a year on maintaining it. Estimated annual net receipts from equipment rentals would be $14,400. It is estimated rental equipment could be sold for $80,000 at the end of 12 years. If the buyer wants a 7% rate of return on his investment, what is the maximum price he should pay for the equipment?

9-59 A motor with a 200-horsepower output is needed in the factory for intermittent use. A Graybar motor costs $7000 and has an electrical efficiency of 89%. A Blueball motor costs $6000 and has an 85% efficiency. Neither motor would have any salvage value, since the cost to remove it would equal its scrap value. The annual maintenance cost for either motor is estimated at $300 a year. Electric power costs $0.072/kilowatt hour (1 hp = 0.746 kW). If a 10% interest rate is used in the calculations, what is the minimum number of hours the higher initial cost Graybar motor must be used each year to justify its purchase?

9-60 Plan A requires a $100,000 investment now. Plan B requires an $80,000 investment now and an additional $40,000 investment at a later time. At 8% interest, compute the break-even point for the timing of the $40,000 investment.

9-61 A low-carbon-steel machine part, operating in a corrosive atmosphere, lasts 6 years and costs $350 installed. If the part is treated for corrosion resistance, it will cost $500 installed. How long must the treated part last to be the preferred alternative, assuming 10% interest?

9-62 Neither of the following machines has any net salvage value.

	A	**B**
Original cost	$55,000	$75,000
Annual expenses		
Operation	9,500	7,200
Maintenance	5,000	3,000
Taxes and insurance	1,700	2,250

At what useful life are the machines equivalent if
(a) 10% interest is used in the computations?
(b) 0% interest is used in the computations?

9-63 A machine costs $5240 and produces benefits of $1000 at the end of each year for eight years. Assume continuous compounding and a nominal annual interest rate of 10%.
(a) What is the payback period (in years)?
(b) What is the break-even point (in years)?
(c) Since the answers in (a) and (b) are different, which one is 'correct'?

CHAPTER 10

Uncertainty in Future Events

Technological Uncertainty: The Swan Hills Waste Disposal Centre

In the 1970s, the Alberta government, recognizing that increased industrial activity brought an increase in the volume of hazardous waste, started to review all available technologies for disposing of such waste. After an exhaustive study, a detailed regional assessment, an information campaign, and local referendums, the Province commissioned the construction of the largest and most advanced incinerator in Canada.

The Swan Hills Special Waste Treatment Centre opened in 1987 with the capacity to process in excess of 20,000 metric tons of waste per year. It could incinerate organic liquids and solids, treat inorganic liquids and solids, stabilize and decontaminate hazardous wastes, and landfill contaminated bulk solids. It was the most comprehensive and integrated treatment facility in North America.

But the technology was not as clean as the government had hoped. Although initially heralded as a model facility, the Swan Hills treatment centre soon earned a reputation as a serious polluter. Local Native groups sued because they were unable to eat wild fish and game due to dioxin contamination. Criminal charges against the operators of the facility resulted in convictions and fines for failing to report emissions. The facility has been plagued by explosions and leaks. Compounding these operating problems were economic ones that have wound up costing Alberta taxpayers hundreds of millions of dollars.

In 1995, the plant was taken over by Bovar Company, which has continued to operate the facility and has increased the capacity of the incinerator to 35,000 metric tons.

Growth and change bring many new technologies, ranging from waste disposal and recycling to clean coal technologies. Developing these kinds of technology in a way that makes them effective and financially viable, especially in light of the need for sustainability and the issue of climate change, is one of the toughest challenges facing the world today. An understanding of how new technologies evolve and penetrate the marketplace is central to commercializing these technologies successfully. Even with a successful innovation pathway, significant resources can be expended in moving a new technology from basic research through the research innovation chain to commercial acceptance.

—Ted Heidrick, Joanne Phillips, and Ian Potter

After Completing This Chapter...

The student should be able to:

- Use a range of estimated variables to evaluate a project.
- Describe possible outcomes with probability distributions.
- Combine probability distributions for individual variables into joint probability distributions.
- Use expected values for economic decision making.
- Use economic decision trees to describe and solve more complex problems.
- Measure and consider risk when making economic decisions.
- Understand how simulation can be used to evaluate economic decisions.

QUESTIONS TO CONSIDER

1. How can a company quantify the risk of an event such as the discovery of downwind pollution or accidental spills?
2. What are some ways in which a company might estimate the cost of such an event, should it occur?
3. Before the 1970s it was common (and legal) for manufacturers to landfill chemicals and other potentially hazardous materials. How might companies anticipate and prepare for future laws that might penalize activities that are legal today?

An assembly line is built after the engineering economic analysis has shown that the anticipated product demand will generate profits. A new motor, heat exchanger, or filtration unit is installed after analysis has shown that future cost savings will economically justify current costs. A new road, school, or other public facility is built after analysis has shown that the future demand and benefits justify the present cost of building. However, future performance of the assembly line, motor, and so on is uncertain, and demand for the product or public facility is more uncertain.

Engineering economic analysis is used to evaluate projects with long-term consequences when the time value of money matters. Thus, it must concern itself with future consequences; but describing the future accurately is not easy. In this chapter we consider the problem of evaluating the future. The easiest way to begin is to make a careful estimate. Then we examine the possibility of predicting a range of possible outcomes. Finally, we consider what happens when the probabilities of the various outcomes are known or may be estimated. We will show that the tools of probability are quite useful for economic decision making.

ESTIMATES AND THEIR USE IN ECONOMIC ANALYSIS

Economic analysis requires evaluating the future consequences of an alternative. In practically every chapter of this book, there are cash flow tables and diagrams that describe precisely the costs and benefits for future years. We don't really believe that we can exactly foretell a future cost or benefit. Instead, our goal is to choose a single value representing the *best* estimate that can be made.

We recognize that estimated future consequences are not precise and that the actual values will be somewhat different from our estimates. Even so, it is likely we have made the tacit assumption that these estimates *are* correct. We know that estimates will not always turn out to be correct; yet we treat them like facts once they are *in* the economic analysis. We do the analysis as though the values were exact. This can lead to trouble. If actual costs and benefits are different from the estimates, an undesirable alternative may be selected. This is because the variability of future consequences is concealed by assuming that the best estimates will actually occur. The problem is illustrated by Example 10-1.

EXAMPLE 10-1

Two alternatives are being considered. The best estimates for the various consequences are as follows:

	A	B
Cost	$1000	$2000
Net annual benefit	$150	$250
Useful life, in years	10	10
End-of-useful-life salvage value	$100	$400

If interest is $3\frac{1}{2}\%$, which alternative has the better net present worth (NPW)?

SOLUTION

Alternative A

$$NPW = -1000 + 150(P/A, 3\tfrac{1}{2}\%, 10) + 100(P/F, 3\tfrac{1}{2}\%, 10)$$

$$= -1000 + 150(8.317) + 100(0.7089)$$

$$= -1000 + 1248 + 71$$

$$= +\$319$$

Alternative B

$$NPW = -2000 + 250(P/A, 3\tfrac{1}{2}\%, 10) + 400(P/F, 3\tfrac{1}{2}\%, 10)$$

$$= -2000 + 250(8.317) + 400(0.7089)$$

$$= -2000 + 2079 + 284$$

$$= +\$363$$

Alternative B, with its larger NPW, would be selected.

Alternative Formation of Example 10-1

Suppose that at the end of 10 years, the actual salvage value for B were $300 instead of the $400 best estimate. If all the other estimates were correct, is B still the preferred alternative?

SOLUTION

Revised B

$$NPW = -2000 + 250(P/A, 3\tfrac{1}{2}\%, 10) + 300(P/F, 3\tfrac{1}{2}\%, 10)$$

$$= -2000 + 250(8.317) + 300(0.7089)$$

$$= -2000 + 2079 + 213$$

$$= +\$292 \rightarrow \text{A is now the preferred alternative.}$$

Example 10-1 shows that the change in the salvage value of Alternative B actually results in a change of preferred alternative. Thus, a more thorough analysis of Example 10-1 would consider (1) which values are uncertain, (2) whether the uncertainty is ±5% or −50 to +80%, and (3) which uncertain values lead to different decisions. A more thorough analysis, which is done with the tools of this chapter, determines which decision is better over the range of possibilities. Explicitly considering uncertainty lets us make better decisions. The tool of break-even analysis is illustrated in Example 10-2.

EXAMPLE 10-2

Use the data of Example 10-1 to compute the sensitivity of the decision to the Alt. B salvage value by computing the break-even value. For Alt. A, NPW = +319. For break-even between the alternatives,

$$NPW_A = NPW_B$$

$$+319 = -2000 + 250(P/A, 3\tfrac{1}{2}\%, 10) + \text{Salvage value}_B (P/F, 3\tfrac{1}{2}\%, 10)$$

$$= -2000 + 250(8.317) + \text{Salvage value}_B (0.7089)$$

At the break-even point

$$\text{Salvage value}_B = \frac{319 + 2000 - 2079}{0.7089} = \frac{240}{0.7089} = \$339$$

When Alt. B salvage value >\$339, B is preferred; when <\$339, A is preferred.

Break-even analysis, as shown in Example 10-2, is one means of examining the impact of the variability of some estimate on the outcome. It helps by answering the question, How much variability can a parameter have before the decision will be affected? While the preferred decision depends on whether the salvage value is above or below the break-even value, the economic difference between the alternatives is small when the salvage value is close to break-even. Break-even analysis does not solve the basic problem of how to take the inherent variability of parameters into account in an economic analysis. This will be considered next.

A RANGE OF ESTIMATES

It is usually more realistic to describe parameters with a range of possible values, rather than a single value. A range could include an **optimistic** estimate, the **most likely** estimate, and a **pessimistic** estimate. Then, the economic analysis can determine whether the decision is sensitive to the range of projected values.

EXAMPLE 10-3

A firm is considering an investment. The most likely data values were found during the feasibility study. Analyzing past data of similar projects shows that optimistic values for the first cost and the annual benefit are 5% better than most likely values. Pessimistic values are 15% worse. The firm's most experienced project analyst has estimated the values for the useful life and salvage value.

	Optimistic	Most Likely	Pessimistic
Cost	$950	$1000	$1150
Net annual benefit	$210	$200	$175
Useful life, in years	12	10	8
Salvage value	$100	$0	$0

Compute the rate of return for each estimate. If a 10% before-tax minimum attractive rate of return is required, is the investment justified under all three estimates? If it is only justified under some estimates, how can these results be used?

SOLUTION

Optimistic Estimate

$$\text{PW of cost} = \text{PW of benefit}$$

$$\$950 = 210(P/A, \text{IRR}_{\text{opt}}, 12) + 100(P/F, \text{IRR}_{\text{opt}}, 12)$$

$$\text{IRR}_{\text{opt}} = 19.8\%$$

Most Likely Estimate

$$\$1000 = 200(P/A, \text{IRR}_{\text{most likely}}, 10)$$

$$(P/A, \text{IRR}_{\text{most likely}}, 10) = 1000/200 = 5 \rightarrow \text{IRR}_{\text{most likely}} = 15.1\%$$

Pessimistic Estimate

$$\$1150 = 170(P/A, \text{IRR}_{\text{pess}}, 8)$$

$$(P/A, \text{IRR}_{\text{pess}}, 8) = 1150/170 = 6.76 \rightarrow \text{IRR}_{\text{pess}} = 3.9\%$$

From the calculations we conclude that the rate of return for this investment is most likely to be 15.1% but might range from 3.9% to 19.8%. The investment meets the 10% MARR criterion for two of the estimates. These estimates can be considered to be scenarios of what may happen with this project. Since one scenario suggests that the project is not attractive, we need to have a method of weighting the scenarios or considering how likely each is.

Example 10-3 made separate calculations for the sets of optimistic, most likely, and pessimistic values. The range of scenarios is useful. However, if there are more than a few uncertain variables, it is unlikely that all will prove to be optimistic (best case) or most likely or pessimistic (worst case). It is more likely that many parameters are the most likely values, while some are optimistic and some are pessimistic.

This can be addressed by using Equation 10.1 to calculate average or mean values for each parameter. Equation 10-1 puts four times the weight on the most likely value than on the other two. This equation was developed as an approximation with the beta distribution.

$$\text{Mean value} = \frac{\text{Optimistic value} + 4(\text{Most likely value}) + \text{Pessimistic value}}{6} \qquad (10\text{-}1)$$

This approach is illustrated in Example 10-4.

EXAMPLE 10-4

Solve Example 10-3 by using Equation 10-1. Compute the resulting mean rate of return.

SOLUTION

Compute the mean for each parameter:

$$\text{Mean cost} = [950 + 4 \times 1000 + 1150]/6 = 1016.7$$

$$\text{Mean net annual benefit} = [210 + 4 \times 200 + 170]/6 = 196.7$$

$$\text{Mean useful life} = [12 + 4 \times 10 + 8]/6 = 10.0$$

$$\text{Mean salvage life} = 100/6 = 16.7$$

Compute the mean rate of return:

$$\text{PW of cost} = \text{PW of benefit}$$

$$\$1016.7 = 196.7(P/A, \text{IRR}_{\text{beta}}, 10) + 16.7(P/F, \text{IRR}_{\text{beta}}, 10)$$

$$\text{IRR}_{\text{beta}} = 14.2\%$$

Example 10-3 gave a most likely rate of return (15.1%), which differed from the mean rate of return (14.2%) computed in Example 10-4. These values are different because the former is based exclusively on the most likely values and the latter takes into account the variability of the parameters.

In examining the data, we see that the pessimistic values are further away from the most likely values than are the optimistic values. This is a common occurrence. For example, a savings of 10% to 20% may be the maximum possible, but a cost overrun can be 50%, 100%, or even more. This causes the resulting weighted mean values to be less favourable than the most likely values. As a result, the mean rate of return, in this example, is less than the rate of return based on the most likely values.

PROBABILITY

We all have used probabilities. For example, what is the probability of getting a 'head' when flipping a coin? Using a model that assumes that the coin is fair, both the heads and tails come up with a probability of 50%, or $\frac{1}{2}$. This probability is the likelihood of an event in a single trial. It also describes the long-run relative frequency of getting heads in many trials (out of 50 coin flips, we expect to get 25 heads).

Probabilities can also be based on data, expert judgment, or a combination of both. Past data on weather and climate, on project completion times and costs, and on highway traffic are combined with expert judgment to forecast future events. These examples can be important in engineering economy.

Another example based on long-run relative frequency is the PW of a flood protection dam that depends on the probabilities of different-sized floods over many years. This would be based on data about past floods and would include many years of potential flooding. An example of a single event that may be estimated by expert judgment is the probability of a successful outcome for a research and development project, which will determine its PW.

All the data in an engineering economy problem may have some level of uncertainty. However, small uncertainties may be ignored, so that more analysis can be done with the large uncertainties. For example, the price of an off-the-shelf piece of equipment may vary by only $\pm 5\%$. The price could be treated as a known or deterministic value. On the other hand, demand over the next 20 years will have more uncertainty. Demand should be analyzed as a random or stochastic variable. We should establish probabilities for different values of demand.

There are also logical or mathematical rules for probabilities. If an outcome can never happen, then the probability is 0. If an outcome will certainly happen, then the probability is 1, or 100%. This means that probabilities cannot be negative or greater than 1; in other words, they must be within the interval [0, 1], as indicated below in Equation 10-2.

Probabilities are defined so that the sum of probabilities for all possible outcomes is 1 or 100% (Equation 10-3). Summing the probability of 0.5 for a head and 0.5 for a tail leads to a total of 1 for the possible outcomes from the coin flip. An exploration well drilled in a potential oil field will have three outcomes (dry hole, non-commercial quantities, or commercial quantities) whose probabilities will sum to one.

Equations 10-2 and 10-3 can be used to check that probabilities are valid. If the probabilities for all but one outcome are known, the equations can be used to find the unknown probability for that outcome (see Example 10-5).

$$0 \leq \text{Probability} \leq 1 \tag{10-2}$$

$$\sum_{j=1 \text{ to } K} P(\text{outcome}_j) = 1, \quad \text{where there are } K \text{ outcomes} \tag{10-3}$$

In a probability course many probability distributions, such as the normal, uniform, and beta are presented. These continuous distributions describe a large population of data. However, for engineering economy it is more common to use 2 to 5 outcomes with discrete probabilities—even though the 2 to 5 outcomes only represent or approximate the range of possibilities.

This is done for two reasons. First, the data are often estimated by expert judgment, so that using 7 to 10 outcomes would be false accuracy. Second, each outcome requires more analysis. In most cases the 2 to 5 outcomes represent the best trade-off between representing the range of possibilities and the amount of calculation required. Example 10-5 illustrates these calculations.

EXAMPLE 10-5

What are the probability distributions for the annual benefit and life for the following project?

The most likely value of the annual benefit is $8000 with a probability of 60%. There is a 30% probability that it will be $5000, and the highest value that is likely is $10,000. A life of 6 years is twice as likely as a life of 9 years.

SOLUTION

Probabilities are given for only two of the possible outcomes for the annual benefit. The third value is found from the fact that the probabilities for the three outcomes must sum to 1 (Equation 10-3).

$$1 = P(\text{Benefit is } \$5000) + P(\text{Benefit is } \$8000) + P(\text{Benefit is } \$10,000)$$

$$P(\text{Benefit is } \$10,000) = 1 - 0.6 - 0.3 = 0.1$$

The probability distribution can then be summarized in a table. Figure 10-1 shows the histogram, or relative frequency diagram.

Annual benefit	$5000	$8000	$10,000
Probability	0.3	0.6	0.1

FIGURE 10-1 Probability distribution for annual benefit.

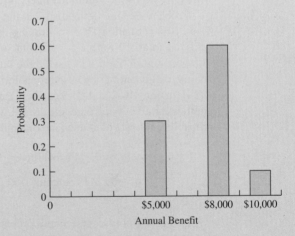

The problem statement tells us:

$$P(\text{life is 6 years}) = 2P(\text{life is 9 years})$$

Equation 10-3 can be applied to write a second equation for the two unknown probabilities:

$$P(6) + P(9) = 1$$

Combining these, we write

$$2P(9) + P(9) = 1$$
$$P(9) = 1/3$$
$$P(6) = 2/3$$

The probability distribution for the life is $P(6) = 66.7\%$ and $P(9) = 33.3\%$.

JOINT PROBABILITY DISTRIBUTIONS

Example 10-5 constructed the probability distributions for the annual benefit and life of a project. These examples show how likely each value is for the input data of the problem. We would like to construct a similar probability distribution for the project's present worth. This is the distribution that we can use to evaluate the project. That present worth depends on both input probability distributions, and so we need to construct the *joint* probability distribution for the different combinations of their values.

For this introductory text, we assume that two random variables, such as the annual benefit and life, are unrelated or statistically independent. This means that the *joint* probability of a combined event (event A defined on the first variable and event B on the second variable) is the product of the probabilities for the two events. This is Equation 10-4:

$$\text{If A and B are independent, then } P(\text{A and B}) = P(\text{A}) \times P(\text{B}) \qquad (10\text{-}4)$$

For example, flipping a coin and rolling a die are statistically independent. Thus, the probability of {flipping a head and rolling a 4} equals the probability of a {heads} $= \frac{1}{2}$ times the probability of a {4} $= \frac{1}{6}$, for a joint probability $= \frac{1}{12}$.

The number of outcomes in the joint distribution is the product of the number of outcomes in each variable's distribution. Thus, for the coin and the die, there are 2 times 6 or 12 combinations. Each of the 2 outcomes for the coin is combined with each of the 6 outcomes for the die.

Some variables are not statistically independent, and the calculation of their joint probability distribution is more complex. For example, a project with low revenues may be terminated early and one with high revenues may be kept operating as long as possible. In these cases annual cash flow and project life are not independent. While this type of relationship can sometimes be modelled with economic decision trees (covered later in this chapter), we will limit our coverage in this text to the simpler case of independent variables.

Example 10-6 uses the three values and probabilities for the annual benefit and the two values and probabilities for the life to construct the six possible combinations. Then the values and probabilities are constructed for the PW of the project.

EXAMPLE 10-6

The project described in Example 10-5 has a first cost of $25,000. The firm uses an interest rate of 10%. Assume that the probability distributions for annual benefit and life are unrelated or statistically independent. Calculate the probability distribution for the PW.

SOLUTION

Since there are three outcomes for the annual benefit and two outcomes for the life, there are six combinations. The first four columns of the following table show the six combinations of life and annual benefit. The probabilities in columns 2 and 4 are multiplied to calculate the joint probabilities in column 5. For example, the probability of a low annual benefit and a short life is $0.3 \times 2/3$, which equals 0.2 or 20%.

The PW values include the $25,000 first cost and the results of each pair of annual benefit and life. For example, the PW for the combination of high benefit and long life is:

$$PW_{\$10,000,9} = -25,000 + 10,000(P/A, 10\%, 9) = -25,000 + 10,000(5.759) = \$32,590$$

Annual Benefit	Probability	Life	Probability	Joint Probability	PW
$ 5,000	30%	6	66.7%	20.0%	−$ 3,224
8,000	60	6	66.7	40.0	9,842
10,000	10	6	66.7	6.7	18,553
5,000	30	9	33.3	10.0	3,795
8,000	60	9	33.3	20.0	21,072
10,000	10	9	33.3	3.3	32,590
				100.0%	

Figure 10-2 shows the probabilities for the PW. This is called the histogram, relative frequency distribution, or probability distribution function.

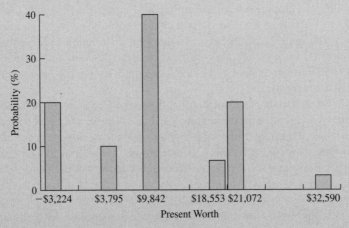

FIGURE 10-2 Probability distribution function for PW.

This probability distribution function shows that there is a 20% chance of having a negative PW. It also shows that there is a small 3.3% chance of the PW being $32,590. The three values used to describe possible annual benefits for the project and the two values for life have been combined to describe the uncertainty in the project's PW.

Creating a distribution, as in Example 10-6, gives us a much better understanding of the possible PW values along with their probabilities. The three possibilities for the annual benefit and the two for the life are representative of the much broader set of possibilities that really exist. Optimistic, most likely, and pessimistic values are a good way to represent the uncertainty about a variable.

Similarly the six values for the PW represent the much broader set of possibilities. The 20% probability of a negative PW is one measure of risk that we will talk about later in the chapter.

Some problems, such as Examples 10-3 and 10-4, have so many variables or different outcomes that constructing the joint probability distribution is arithmetically burdensome. If the values in Equation 10-1 are treated as a discrete probability distribution function, the probabilities are $1/6, 2/3, 1/6$. With an optimistic, most likely, and pessimistic outcome for each of 4 variables, there are $3^4 = 81$ combinations. In Examples 10-3 and 10-4, the salvage value has only two distinct values, and so there are still $3 \times 3 \times 3 \times 2 \doteq 54$ combinations.

When the problem is important enough, the effort to construct the joint probability distribution is worthwhile. It gives the analyst and the decision maker a better understanding of what may happen. It is also needed to calculate measures of the risk of a project. While spreadsheets can automate the arithmetic, simulation (described at the end of the chapter) can be a better choice when there are a large number of variables and combinations.

EXPECTED VALUE

For any probability distribution we can compute the **expected value (EV)** or arithmetic average (mean). To calculate the EV, each outcome is weighted by its probability, and the results are summed. This is *not* the simple average or unweighted mean. When the class average on a test is computed, this is an unweighted mean. Each student's test has the same weight. This simple 'average' is the one that is shown by the button $\bar{x}$ on many calculators.

The expected value is a weighted average, like a student's grade point average (GPA). For a GPA the grade in each class is weighted by the number of credits. For the expected value of a probability distribution, the weights are the probabilities.

This is described in Equation 10-5. We saw in Example 10-4 that these expected values can be used to compute a rate of return. They can also be used to calculate a present worth as in Example 10-7.

$$\text{Expected value} = \text{Outcome}_A \times P(A) + \text{Outcome}_B \times P(B) + \cdots \qquad (10\text{-}5)$$

EXAMPLE 10-7

The first cost of the project in Example 10-5 is $25,000. Use the expected values for annual benefits and life to estimate the present worth. Use an interest rate of 10%.

SOLUTION

$$EV_{benefit} = 5000(0.3) + 8000(0.6) + 10,000(0.1) = \$7300$$

$$EV_{life} = 6(2/3) + 9(1/3) = 7 \text{ years}$$

The PW using these values is

$$PW(EV) = -25,000 + 7300(P/A, 10\%, 7) = -25,000 + 6500(4.868) = \$10,536$$

(*Note:* This is the present worth of the expected values, PW(EV), not the expected value of the present worth, EV(PW). It is an easy value to calculate that approximates the EV(PW), which will be computed from the joint probability distribution found in Example 10.6.)

Example 10-7 is a simple way to approximate the project's expected PW. But the true expected value of the PW is somewhat different. To find it, we must use the joint probability distribution for benefit and life and the resulting probability distribution function for PW that was derived in Example 10-6. Example 10-8 shows the expected value of the PW or the EV(PW).

EXAMPLE 10-8

Use the probability distribution function of the PW that was derived in Example 10-6 to calculate the EV(PW). Does this indicate an attractive project?

SOLUTION

The table from Example 10-6 can be reused with one additional column for the weighted values of the PW (= PW × probability). Then, the expected value of the PW is calculated by summing the column of present worth values that have been weighted by their probabilities.

Annual Benefit	Probability	Life (years)	Probability	Joint Probability	PW	PW × Joint Probability
$ 5,000	30%	6	66.7%	20.0%	−$ 3,224	−$ 645
8,000	60	6	66.7	40.0	9,842	3,937
10,000	10	6	66.7	6.7	18,553	1,237
5,000	30	9	33.3	10.0	3,795	380
8,000	60	9	33.3	20.0	21,072	4,214
10,000	10	9	33.3	3.3	32,590	1,086
				100.0%		EV(PW) = $10,209

With an expected PW of $10,209, this is an attractive project. While there is a 20% chance of a negative PW, the possible positive outcomes are larger and more likely. Having analyzed the project under uncertainty, we are much more knowledgeable about the potential result of the decision to proceed.

The $10,209 value is more accurate than the approximate value calculated in Example 10-7. The values differ because PW is a non-linear function of the life. The more accurate value of $10,209 is lower because the annual benefit values for the longer life are discounted by $1/(1+i)$ for more years.

In Examples 10-7 and 10-8, the question was whether the project had a positive PW. With two or more alternatives, the rule would have been to maximize the PW. With equivalent uniform annual costs (EUACs) the goal is to minimize the EUAC. Example 10-9 uses the rule of minimizing the EV of the EUAC to choose the best height for a dam.

EXAMPLE 10-9

A dam is being considered to reduce river flooding. But if a dam is built, what height should it be? Increasing the dam's height will (1) reduce the probability of a flood, (2) reduce the damage when floods occur, and (3) cost more. Which dam height minimizes the expected total annual cost? The province uses an interest rate of 5% for flood protection projects, and all the dams should last 50 years.

Dam Height (ft)	First Cost	Annual P (flood) > Height	Damages If Flood Occurs
No dam	$ 0	0.25	$800,000
20	700,000	0.05	500,000
30	800,000	0.01	300,000
40	900,000	0.002	200,000

SOLUTION

The easiest way to solve this problem is to choose the dam height with the lowest equivalent uniform annual cost (EUAC). Calculating the EUAC of the first cost requires us to multiply the first cost by $(A/P, 5\%, 50)$. For example, for the dam 20 ft high, this is $700,000(A/P, 5\%, 50) = \$38,344$.

Calculating the annual expected flood damage cost for each alternative is simplified because the term for the P(no flood) drops out, because the damages for no flood are $0. Thus we need to calculate only the term for flooding. This is done by multiplying the P(flood) times the damages if a flood happens. For example, the expected annual flood damage cost with no levee is $0.25 \times \$800,000$, or $200,000.

Then the EUAC of the first cost and the expected annual flood damage are added together to find the total EUAC for each height. The 30 ft dam is somewhat cheaper than the 40 ft dam.

Height of Dam (ft)	EUAC of First Cost	Expected Annual Flood Damage	Total Expected EUAC
No dam	$ 0	$200,000	$200,000
20	38,344	25,000	63,344
30	43,821	3000	46,821
40	49,299	400	49,699

ECONOMIC DECISION TREES

Some engineering projects are more complex, and evaluating them properly is correspondingly more complex. For example, consider a new product with potential sales volumes ranging from low to high. If the sales volume is low, then the product may be discontinued early in its potential life. On the other hand, if sales volume is high, additional capacity may be added to the assembly line and new product variations may be added. This can be modelled with a decision tree.

The following symbols are used to model decisions with decision trees:

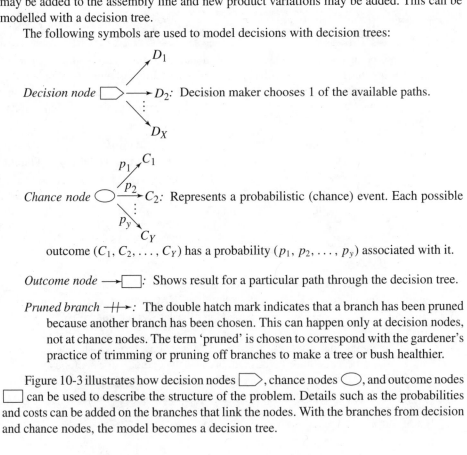

Decision node: Decision maker chooses 1 of the available paths.

Chance node: Represents a probabilistic (chance) event. Each possible outcome $(C_1, C_2, \ldots, C_Y)$ has a probability $(p_1, p_2, \ldots, p_y)$ associated with it.

Outcome node ⟶☐: Shows result for a particular path through the decision tree.

Pruned branch ⊣⊢▶: The double hatch mark indicates that a branch has been pruned because another branch has been chosen. This can happen only at decision nodes, not at chance nodes. The term 'pruned' is chosen to correspond with the gardener's practice of trimming or pruning off branches to make a tree or bush healthier.

Figure 10-3 illustrates how decision nodes ▷, chance nodes ◯, and outcome nodes ☐ can be used to describe the structure of the problem. Details such as the probabilities and costs can be added on the branches that link the nodes. With the branches from decision and chance nodes, the model becomes a decision tree.

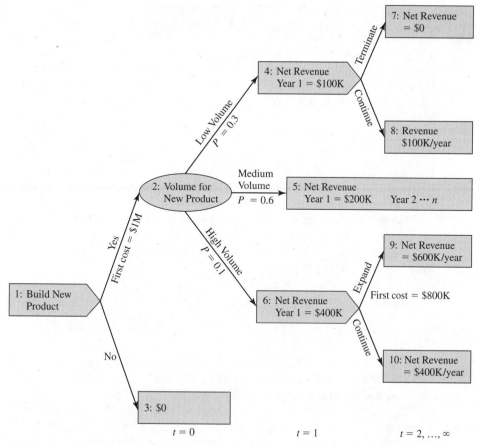

FIGURE 10-3 Economic decision tree for new product.

Figure 10-3 illustrates that decision trees describe the problem by starting at the decision that must be made and then adding chance and decision nodes in the proper logical sequence. Thus describing the problem starts at the first step and goes forward in time with sequences of decision and chance nodes.

To make the decision, calculations begin with the final nodes in the tree. Since they are the final nodes, enough information is available to evaluate them. At decision nodes the rule is either to maximize PW or to minimize EUAC. At chance nodes an expected value for PW or EUAC is calculated.

Once all nodes that branch from a node have been evaluated, the originating node can be evaluated. If the originating node is a decision node, choose the branch with the best PW or EUAC and place that value in the node. If the originating node is a chance node, calculate the expected value and place that value in the node. This process 'rolls back' values from the terminal nodes in the tree to the initial decision. Example 10-10 illustrates this.

EXAMPLE 10-10

What decision should be made on the new product summarized in Figure 10-3? What is the expected value of the product's PW? The firm uses an interest rate of 10% to evaluate projects. If the product is terminated after one year, the capital equipment has a salvage value of $550,000 for use with other new products. If the equipment is used for 8 years, the salvage value is $0.

SOLUTION

Evaluating decision trees is done by starting with the end outcome nodes and the decisions that lead to them. In this case the decisions are whether to terminate after 1 year if sales volume is low and whether to expand after 1 year if sales volume is high.

The decision to terminate the product depends on which is more valuable, the $550,000 salvage value of the equipment or the revenue of $100,000 per year for 7 more years. The worth (PW_1) of the salvage value is $550,000. The worth ($PW_1$) of the revenue stream at the end of Year 1 shown in node 8 is:

$$PW_1 \text{ for node } 8 = 100,000(P/A, 10\%, 7)$$

$$= 100,000(4.868) = \$486,800$$

Thus, terminating the product and using the equipment for other products is better. We enter the two 'present worth' values at the end of Year 1 in nodes 7 and 8. We make the arc to node 7 bold, and use a double hatch mark to show that we're pruning the arc to node 8.

The decision to expand at node 6 could be based on whether the $800,000 first cost for expansion can be justified by an annual increase in revenues of $200,000 for 7 years. However, this is difficult to show on the tree. It is easier to calculate the 'present worth' values at the end of Year 1 for each of the two choices. The worth (PW_1) of node 9 (expand) is:

$$PW_1 \text{ for node } 9 = -800,000 + 600,000(P/A, 10\%, 7)$$

$$= -800,000 + 600,000(4.868)$$

$$= \$2,120,800$$

The value of node 10 (continue without expanding) is:

$$PW_1 \text{ for node } 10 = 400,000(P/A, 10\%, 7)$$

$$= 400,000(4.868)$$

$$= \$1,947,200$$

This is $173,600 less than the expansion node, and so the expansion should happen if volume is high. Figure 10-4 summarizes what we know at this stage of the process.

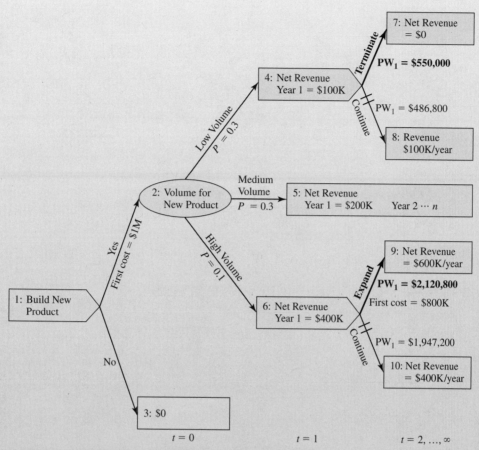

FIGURE 10-4 Partially solved decision tree for new product.

The next step is to calculate the PW at nodes 4, 5, and 6.

PW at node 4 = $(100,000 + 550,000)(P/F, 10\%, 1) = 650,000(0.9091) = \$590,915$

PW at node 5 = $(200,000)(P/A, 10\%, 8) = 200,000(5.335) = \$1,067,000$

PW at node 6 = $[400,000 - 800,000 + 600,000(P/A, 10\%, 7)](P/F, 10\%, 1)$

$= [-400,000 + 600,000(4.868)](0.9091) = \$2,291,660$

Now the expected value at node 2 can be calculated

EV at node 2 = $0.3(590,915) + 0.6(1,067,000) + 0.1(2,291,660) = \$1,046,640$

Since the cost of selecting node 2 is $1,000,000, the expected PW of proceeding with the product is $46,640. This is greater than the $0 for not building the project. So the decision is to build. Figure 10-5 is the decision tree at the final stage.

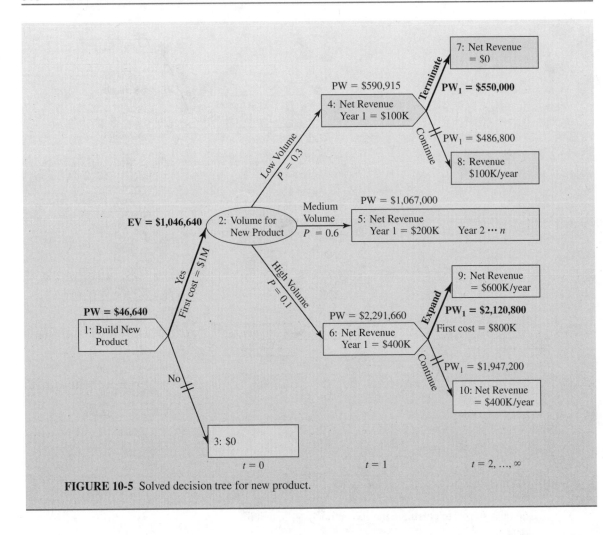

FIGURE 10-5 Solved decision tree for new product.

Example 10-10 is representative of many problems in engineering economy. The main criteria is maximizing PW or minimizing EUAC. However, as shown in Example 10-11, other criteria, such as risk, are used in addition to expected value.

EXAMPLE 10-11

Consider the economic evaluation of collision and comprehensive (fire, theft, etc.) insurance for a car. This insurance is usually required by lenders, but once the car has been paid for, this insurance is not required. (Liability insurance is a legal requirement.)

Figure 10-6 begins with a decision node with two alternatives for the next year. Insurance will cost $800 per year with a $500 deductible if a loss occurs. The other option is to self-insure, which means to go without buying collision and comprehensive insurance. Then if there is a loss,

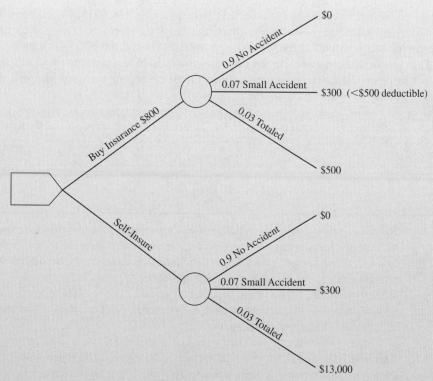

$0

0.9 No Accident

0.07 Small Accident $300 (<$500 deductible)

0.03 Totaled

Buy Insurance $800

$500

$0

Self-Insure

0.9 No Accident

0.07 Small Accident $300

0.03 Totaled

$13,000

FIGURE 10-6 Decision tree for buying auto collision insurance.

the owner must replace the vehicle with money from savings or a loan, or do without a vehicle until he or she can afford to replace it.

Three accident severities are used to represent the range of possibilities: a 90% chance of no accident, a 7% chance of a small accident (at a cost of $300, which is less than the deductible), and a 3% chance of totalling the $13,000 vehicle. Since our driving habits are likely to be the same with and without insurance, the accident probabilities are the same for both chance nodes.

Even though this is a text on engineering economy, we have simplified the problem and ignored the difference in timing of the cash flows. Insurance payments are made at the beginning of the covered period, and accident costs occur during the covered period. Since car insurance is usually paid semi-annually, the results of the economic analysis are not changed significantly by the simplification. We focus on the new concepts of expected value, economic decision trees, and risk.

What are the expected values for each alternative, and what decision is recommended?

SOLUTION

The expected values are computed by using Equation 10-5. If insured, the maximum cost equals the deductible of $500. If self-insured, the cost is the cost of the accident.

$$\text{EV}_{\text{accident w/ins.}} = (0.9 \times 0) + (0.07 \times 300) + (0.03 \times 500) = \$36$$

$$\text{EV}_{\text{accident w/o ins.}} = (0.9 \times 0) + (0.07 \times 300) + (0.03 \times 13{,}000) = \$411$$

Thus, buying insurance lowers the expected cost of an accident by $375. To evaluate whether we should buy insurance, we must also account for the cost of the insurance. Thus, these expected costs are combined with the $0 for self-insuring (total $411) and the $800 for insuring (total $836). Thus self-insuring has an expected value cost that is $425 less per year (= $836 − $411). This is not surprising, since the premiums collected must cover both the costs of operating the insurance company and the expected value of the payouts.

This is also an example of *expected values alone not determining the decision*. Buying insurance has an expected cost that is $425 per year higher, but that insurance limits the maximum loss to $500 rather than $13,000. The $425 may be worth spending to avoid that risk.

RISK

Risk can be thought of as the chance of getting an outcome other than the expected value—with an emphasis on something negative. One common measure of risk is the probability of a loss (see Example 10-6). The other common measure is the **standard deviation** (σ), which measures the dispersion of outcomes about the expected value. For example, many students have used the normal distribution in other classes. The normal distribution has 68% of its probable outcomes within ±1 standard deviation of the mean and 95% within ±2 standard deviations of the mean.

Mathematically, the standard deviation is defined as the square root of the variance. This term is defined as the weighted average of the squared difference between the outcomes of the random variable X and its mean. Thus the larger the difference between the mean and the values, the larger are the standard deviation and the variance. This is Equation 10-6:

$$\text{Standard deviation } (\sigma) = \sqrt{[\text{EV}(X - \text{mean})^2]} \qquad (10\text{-}6)$$

Squaring the differences between individual outcomes and the EV ensures that positive and negative deviations receive positive weights. Consequently, negative values for the standard deviation are impossible, and they instantly indicate arithmetic mistakes. The standard deviation equals 0 if only one outcome is possible. Otherwise, the standard deviation is positive.

This is not the standard deviation formula built into most calculators, just as the weighted average is not the simple average built into most calculators. The calculator formulas are for N equally likely data points from a randomly drawn sample, so that each probability is $1/N$. In economic analysis we will use a weighted average for the squared deviations since the outcomes are not equally likely.

The second difference is that for calculations (by hand or the calculator), it is easier to use Equation 10-7, which is shown to be equivalent to Equation 10-6 in introductory probability and statistics texts.

$$\text{Standard deviation } (\sigma) = \sqrt{\{\text{EV}(X^2) - [\text{EV}(X)]^2\}} \qquad (10\text{-}7)$$

$$= \sqrt{\{\text{Outcome}_A^2 \times P(\text{A}) + \text{Outcome}_B^2 \times P(\text{B}) + \cdots - \text{expected value}^2\}} \qquad (10\text{-}7')$$

This equation is the square root of the difference between the average of the squares and the square of the average. The standard deviation is used instead of the *variance* because the standard deviation is measured in the same units as the expected value. The variance is measured in 'squared dollars'—whatever they are.

The calculation of a standard deviation by itself is only a descriptive statistic of limited value. However, as shown in the next section on risk-versus-return trade-offs, it is useful when the standard deviation of each alternative is calculated and these are compared. But first, here are some examples of calculating the standard deviation.

EXAMPLE 10-12 (Example 10-11 continued)

Consider the economic evaluation of collision and comprehensive (fire, theft, etc.) insurance for a car. One example was described in Figure 10-6. The probabilities and outcomes are summarized in the calculation of the expected values, which was done with Equation 10-5.

$$EV_{accident\ w/ins.} = (0.9 \times 0) + (0.07 \times 300) + (0.03 \times 500) = \$36$$

$$EV_{accident\ w/o\ ins.} = (0.9 \times 0) + (0.07 \times 300) + (0.03 \times 13,000) = \$411$$

Calculate the standard deviations for insuring and not insuring.

SOLUTION

The first step is to calculate the EV(outcome^2) for each.

$$EV^2_{accident\ w/ins.} = (0.9 \times 0^2) + (0.07 \times 300^2) + (0.03 \times 500^2) = \$13,800$$

$$EV^2_{accident\ w/o\ ins.} = (0.9 \times 0^2) + (0.07 \times 300^2) + (0.03 \times 13,000^2) = \$5,076,300$$

Then the standard deviations can be calculated.

$$\sigma_{w/ins.} = \sqrt{(13,800 - 36^2)} = \sqrt{12,504} = \$112$$

$$\sigma_{w/o\ ins.} = \sqrt{(5,076,300 - 411^2)} = \sqrt{4,907,379} = \$2215$$

As described in Example 10-11, the expected value cost of insuring is \$836 (= \$36 + \$800) and the expected value cost of self-insuring is \$411. Thus the expected cost of not insuring is about half the cost of insuring. But the standard deviation of self-insuring is 20 times larger. It is clearly riskier.

Which choice is preferred depends on how much risk one is comfortable with.

As stated before, this is an example of *expected values alone not determining the decision.* Buying insurance has an expected cost that is \$425 per year higher, but that insurance limits the maximum loss to \$500 rather than \$13,000. The \$425 may be worth spending to avoid that risk.

EXAMPLE 10-13 (Example 10-6 continued)

Using the probability distribution for the PW from Example 10-6, calculate the standard deviation of the PW.

SOLUTION

The following table adds a column for (PW^2) (probability) to calculate the $EV(PW^2)$.

Annual Benefit	Probability	Life (years)	Probability	Joint Probability	PW	PW × Probability	PW² × Probability
$ 5,000	30%	6	66.7%	20.0%	−$ 3,224	−$ 645	$ 2,079,480
8,000	60	6	66.7	40.0	9,842	3,937	38,747,954
10,000	10	6	66.7	6.7	18,553	1,237	22,950,061
5,000	30	9	33.3	10.0	3,795	380	1,442,100
8,000	60	9	33.3	20.0	21,072	4,214	88,797,408
10,000	10	9	33.3	3.3	32,590	1,086	35,392,740
					EV	$10,209	$189,409,745

$$\text{Standard deviation} = \sqrt{\{EV(X^2) - [EV(X)]^2\}}$$

$$\sigma = \sqrt{\{189,405,745 - [10,209]^2\}} = \sqrt{85,182,064} = \$9229$$

For those with stronger backgrounds in probability than this chapter assumes, let us consider how the standard deviation in Example 10-13 depends on the assumption of independence between the variables. While exceptions exist, a positive statistical dependence between variables often increases the PW's standard deviation. Similarly, a negative statistical dependence between variables often decreases the standard deviation of the PW.

RISK VERSUS RETURN

A graph of risk versus return is one way to consider these items together. Figure 10-7 in Example 10-14 illustrates the most common format. Risk measured by standard deviation is placed on the x axis, and return measured by expected value is placed on the y axis. This is usually done with the internal rates of return of the alternatives or projects.

EXAMPLE 10-14

A large firm is discontinuing an older product, and so some facilities are becoming available for other uses. The following table summarizes 8 new projects that would use the facilities. Considering expected return and risk, which projects are good candidates? The firm believes it can earn 4% on a risk-free investment in government securities (labelled as Project F).

Project	IRR	Standard Deviation
1	13.1%	6.5%
2	12.0	3.9
3	7.5	1.5
4	6.5	3.5
5	9.4	8.0
6	16.3	10.0
7	15.1	7.0
8	15.3	9.4
F	4.0	0.0

SOLUTION

Answering the question is far easier if we use Figure 10-7. Since a larger expected return is better, we want to select projects that are as high up as possible. Since a lower risk is better, we want to select projects that are as far left as possible. The graph lets us examine the trade-off of accepting more risk for a higher return.

FIGURE 10-7 Risk-versus-return graph.

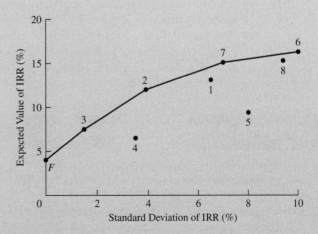

First, we can eliminate Projects 4 and 5. They are **dominated projects.** Dominated alternatives are no better than another alternative on all measures and inferior on at least one measure. Project 4 is dominated by Project 3, which has a higher expected return and a lower risk. Project 5 is dominated by Projects 1, 2, and 7. All three have a higher expected return and a lower risk.

Second, we look at the **efficient frontier.** This is the line in Figure 10-7 that connects Projects F, 3, 2, 7, and 6. Depending on the trade-off that we want to make between risk and return, any of these could be the best choice.

Project 1 appears to be inferior to Projects 2 and 7. Project 8 appears to be inferior to Projects 7 and 6. Projects 1 and 8 are inside and not on the efficient frontier.

There are models of risk and return that can allow us to choose between Projects F, 3, 2, 7, and 6; but those models are beyond what is covered here.

SIMULATION

Simulation is a more advanced approach to considering risk in engineering economy problems. As such, the following discussion focuses on what it is. As the examples show, spreadsheet functions and add-in packages make simulation easier to use for economic analysis.

Economic **simulation** uses random sampling from the probability distributions of one or more variables to analyze an economic model for many iterations. For each iteration, all variables with a probability distribution are randomly sampled. These values are used to calculate the PW, IRR, or EUAC. Then the results of all iterations are combined to create a probability distribution for the PW, IRR, or EUAC.

Simulation can be done by hand, by means of a table of random numbers—if there are only a few random variables and iterations. However, results are more reliable as the number of iterations increases, so in practice this is usually computerized. This can be done in Excel using the RAND() function to generate random numbers, as shown in Example 10-15.

Because we were analyzing each possible outcome, the probability distributions earlier in this chapter (and in the end-of-chapter problems) used two or three discrete outcomes. This limited the number of combinations that we needed to consider. Simulation makes it easy to use continuous probability distributions like the uniform, normal, exponential, log normal, binomial, and triangular. Examples 10-15 and 10-16 use the normal and the discrete uniform distributions.

EXAMPLE 10-15

ShipM4U is considering installing a new, more accurate scale, which will reduce the error in computing postage charges and save $250 a year. The useful life of the scale is believed to be uniformly distributed over 12, 13, 14, 15, and 16 years. The initial cost of the scale is estimated to be normally distributed with a mean of $1500 and a standard deviation of $150.

Use Excel to simulate 25 random samples of the problem, and compute the rate of return for each sample. Construct a graph of rate of return versus frequency of occurrence.

SOLUTION

This problem is simple enough that a table with each iteration's values of the life and the first cost can be constructed. From these values and the annual savings of $250, the IRR for each iteration can be calculated with the RATE function. These are shown in Figure 10-8. The IRR values are summarized in a relative frequency diagram in Figure 10-9.

(*Note:* Each time Excel recalculates the spreadsheet, different values for all the random numbers are generated. Thus the results depend on the set of random numbers, and your results will be different if you create this spreadsheet.)

FIGURE 10-8 Excel spreadsheet for simulation ($N = 25$).

	A	B	C	D
1	250	Annual Savings		
2		Life	First Cost	
3	Min	12	1500	Mean
4	Max	16	150	Std dev
5				
6	Iteration			IRR
7	1	12	1277	16.4%
8	2	15	1546	13.9%
9	3	12	1523	12.4%
10	4	16	1628	13.3%
11	5	14	1401	15.5%
12	6	12	1341	15.2%
13	7	12	1683	10.2%
14	8	14	1193	19.2%
15	9	15	1728	11.7%
16	10	12	1500	12.7%
17	11	16	1415	16.0%
18	12	12	1610	11.2%
19	13	15	1434	15.4%
20	14	12	1335	15.4%
21	15	14	1468	14.5%
22	16	13	1469	13.9%
23	17	14	1409	15.3%
24	18	15	1484	14.7%
25	19	14	1594	12.8%
26	20	15	1342	16.8%
27	21	14	1309	17.0%
28	22	12	1541	12.1%
29	23	16	1564	14.0%
30	24	13	1590	12.2%
31	25	16	1311	17.7%
32				
33	Mean	14	1468	14.4%
34	Std dev	2	135	2.2%

Note for students who have had a course in probability and statistics: Creating the random values for life and first cost is done as follows. Select a random number in (0, 1) using Excel's RAND function. This is the value of the cumulative distribution function for the variable. Convert this to the variable's value by using an inverse function from Excel, or build the inverse function. For the discrete uniform life the function is = min life + INT(range∗ RAND()). For the normally distributed first cost the functions is = NORMINV(RAND(), mean, standard deviation).

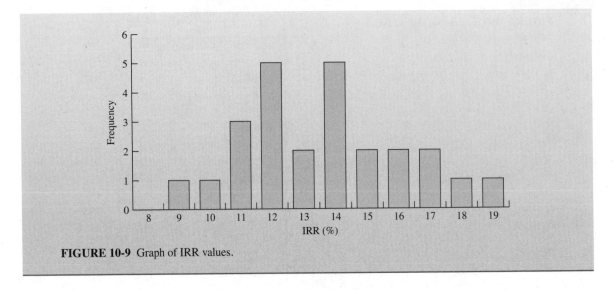

FIGURE 10-9 Graph of IRR values.

Stand-alone simulation programs and commercial spreadsheet add-in packages, such as @Risk and Crystal Ball, have probability distribution functions to use for each input variable. In Example 10-16 the functions RiskUniform and RiskNormal are used. The packages also collect values for the output variables, such as the IRR for Example 10-16. In other problems the PW or EUAC could be collected. These values form a probability distribution for the PW, IRR, or EUAC. From this distribution the simulation package can calculate the expected return, P(loss), and the standard deviation of the return.

Example 10-16 uses @Risk to simulate 1000 iterations of PW for the data in Example 10-15. A simulation package makes it easy to do more iterations. More important still, since it is much easier to use different probability distributions and parameters, more accurate models can be built. Because the models are easier to build, they are less likely to contain errors.

EXAMPLE 10-16

Consider the scale described in Example 10-15. Generate 1000 iterations and construct a frequency distribution for the scale's rate of return.

SOLUTION

The first IRR (cell A8) of 14.01% that is computed in Figure 10-10 is based on the average life and the average first cost. The second IRR (cell A11) of 14.01% is computed by @Risk using the average of each distribution. The cell *content* is the RATE formula with its RiskUniform and

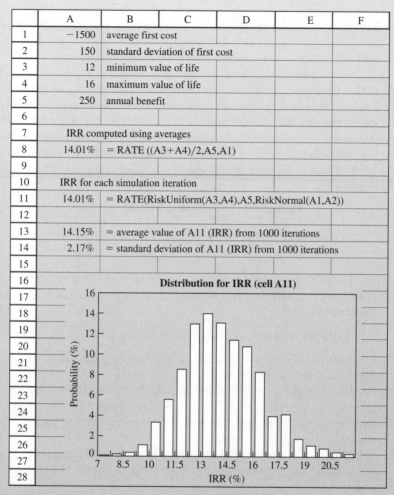

	A	B	C	D	E	F
1	−1500	average first cost				
2	150	standard deviation of first cost				
3	12	minimum value of life				
4	16	maximum value of life				
5	250	annual benefit				
6						
7		IRR computed using averages				
8	14.01%	= RATE ((A3+A4)/2,A5,A1)				
9						
10		IRR for each simulation iteration				
11	14.01%	= RATE(RiskUniform(A3,A4),A5,RiskNormal(A1,A2))				
12						
13	14.15%	= average value of A11 (IRR) from 1000 iterations				
14	2.17%	= standard deviation of A11 (IRR) from 1000 iterations				
15						
16		**Distribution for IRR (cell A11)**				

FIGURE 10-10 Simulation spreadsheet for Example 10-15 and 10-16.

RiskNormal function, however spreadsheets with @Risk functions *display* by default the results of using average values.

The RATE function contains two @Risk functions: RiskUniform and RiskNormal. The uniform distribution has the minimum and maximum values as parameters. The normal distribution has the average and standard deviation as parameters.

The third IRR (cell A13) is the average for 1000 iterations. It will change each time the simulation is done. The graph in Figure 10-10 with 1000 iterations is much smoother than the graph from Example 10-15, where 25 iterations were done.

SUMMARY

Estimating the future is required for economic analysis, and there are several ways to do this. Precise estimates will not ordinarily be exactly correct, but they are considered to be the best single values to represent what we think will happen.

A simple way to represent uncertainty is through a range of estimates for each variable, such as optimistic, most likely, and pessimistic. The full range of prospective results may be examined by using the optimistic values to solve the problem and then using the pessimistic values. Solving the problem with the most likely values is a good single value estimate. However, the extremes with all optimistic values or all pessimistic values are less likely—it is more likely that a mix of optimistic, most likely, and pessimistic values will occur.

One approach uses weighted values instead of a range of estimates. One set of weights suggested is:

Estimate	Relative Weight
Optimistic	1
Most likely	4
Pessimistic	1

The most commonly used approach for decision making relies on **expected values.** Here, known or estimated probabilities for future events are used as weights for the corresponding outcomes.

$$\text{Expected value} = \text{Outcome}_A \times \text{Probability}_A + \text{Outcome}_B \times \text{Probability}_B + \cdots$$

Expected value is the most useful and the most frequently used technique for estimating the attractiveness of a project.

However, risk as measured by standard deviation and the probability of a loss is also important in evaluating projects. Since projects with higher expected returns also frequently have higher risk, evaluating the trade-offs between risk and return is useful in decision making.

More complicated problems can be summarized and analyzed by using decision trees, which allow logical evaluation of problems with sequential chance, decision, and outcome nodes.

Where the elements of an economic analysis are stated in terms of probability distributions, a repetitive analysis of a random sample is often done. This simulation-based approach relies on the premise that a random sampling of increasing size becomes a better and better estimate of the possible outcomes. The large number of computations means that simulation is usually computerized.

PROBLEMS

10-1 Telephone poles exemplify items that have varying useful lives. Telephone poles, once installed in a location, remain in useful service until one of a variety of events occur.

(a) Name three reasons why a telephone pole might be removed from useful service at a particular location.

(b) You are to estimate the total useful life of telephone poles. If the pole is removed from an original location while it is still serviceable, it will be installed elsewhere. Estimate the optimistic life, most likely life, and pessimistic life for telephone poles. What percentage of all telephone poles would you expect to have a total useful life greater than your estimated optimistic life?

10-2 The purchase of a used pick-up for $9000 is being considered. Records for other vehicles show that costs for oil, tires, and repairs about equal the cost for fuel. Fuel costs are $990 per year if the truck is driven 16,000 km. The salvage value after 5 years of use drops by about $0.05 per km. Find the equivalent uniform annual cost if the interest rate is 8%. How much does this change if the annual mileage is 24,000? 8000?

10-3 A heat exchanger is being installed as part of a plant modernization program. It costs $80,000, including installation, and is expected to reduce the overall plant fuel cost by $20,000 per year. Estimates of the useful life of the heat exchanger range from an optimistic 12 years to a pessimistic 4 years. The most likely value is 5 years. Using the range of estimates to compute the mean life, determine the estimated before-tax rate of return. Assume the heat exchanger has no salvage value at the end of its useful life.

10-4 For the data in Problem 10-2 assume that the 8000, 16,000, and 24,000 kilometre values are, respectively, pessimistic, most likely, and optimistic estimates. Use a weighted estimate to calculate the equivalent annual cost.

10-5 When a pair of dice are tossed, the results may be any whole number from 2 to 12. In the game of craps one can win by tossing either a 7 or an 11 on the first roll. What is the probability of doing this? (*Hint:* There are 36 ways that a pair of six-sided dice can be tossed. What portion of them result in either a 7 or an 11?) (*Answer:* $8/36$)

10-6 Annual savings due to an energy efficiency project have a most likely value of $30,000. The high estimate of $40,000 has a probability of 0.2, and the low estimate of $20,000 has a probability of 0.30. What is the expected value for the annual savings? (*Answer:* $29,000)

10-7 Over the last 10 years, the hurdle or discount rate for projects from the firm's research and development division has been 10% twice, 15% three times, and 20% the rest of the time. There is no recognizable pattern. Calculate the probability distribution and the expected value for next year's discount rate.

10-8 The construction time for a bridge depends on the weather. The project is expected to take 250 days if the weather is dry and hot. If the weather is damp and cool, the project is expected to take 350 days. Otherwise, it is expected to take 300 days. Historical data suggest that the probability of cool, damp weather is 30% and that of dry, hot weather is 20%. Find the probability distribution and expected completion time for the project.

10-9 You recently had a car accident that was your fault. If you have another accident or receive a another moving violation within the next 3 years, you will become part of the 'assigned risk' pool, and you will pay an extra $600 per year for insurance. If the probability of an accident or moving violation is 20% a year, what is the probability distribution of your 'extra' insurance payments over the next 4 years? Assume that insurance is purchased annually and that violations register at the end of the year—just in time to affect next year's insurance premium.

10-10 Two instructors announced that they 'grade on the curve', that is, give a fixed percentage of each of the various letter grades to each of their classes. Their curves are as follows:

Grade	Instructor A	Instructor B
A	10%	15%
B	15	15
C	45	30
D	15	20
F	15	20

If a random student came to you and said that his object was to enroll in the class in which he could expect the higher grade point average, which instructor would you recommend? (*Answer:* GPA$_B$ = 1.95, Instructor A)

10-11 A man wants to decide whether or not to invest $1000 in a friend's speculative venture. He will do so if he thinks he can get his money back in one year. He believes the probabilities of the various outcomes at the end of one year are as follows:

Result	Probability
$2000 (double his money)	0.3
1500	0.1
1000	0.2
500	0.3
0 (lose everything)	0.1

What would be his expected outcome if he invests the $1000?

10-12 A university football team has 10 games scheduled for next season. The business manager wishes to estimate how much money the team can be expected to have left over after paying the season's expenses, including any post-season exhibition game expenses. From records for the past season and estimates by informed people, the business manager has assembled the following data:

Situation	Prob-ability	Situation	Net Income
Regular season		Regular season	
Win 3 games	0.10	Win 5 or	$250,000
Win 4 games	0.15	fewer	
Win 5 games	0.20	games	
Win 6 games	0.15	Win 6 to 8	400,000
Win 7 games	0.15	games	
Win 8 games	0.10		
Win 9 games	0.07	Win 9 or 10	600,000
Win 10 games	0.03	games	
Post-season		Post-season	Additional
exhibition game	0.10	exhibition game	income of
			$100,000

What is the expected net income for the team next season? (*Answer:* $355,000)

10-13 In some casinos, craps is a popular gambling game. One of the many bets available is the 'Hard-way 8'. A $1 bet in this fashion will win the player $4 if in the game the pair of dice come up 4 and 4 before one of the other ways of totalling 8. For a $1 bet, what is the expected result? (*Answer:* 80¢)

10-14 If your interest rate is 8%, what is the expected value of the present worth of the 'extra' insurance payments in Problem 10-9? (*Answer:* $528.7)

10-15 A decision has been made to make certain repairs to the outlet works of a small dam. For a particular 36-inch gate valve, there are three available alternatives:
(a) Leave the valve as it is.
(b) Repair the valve.
(c) Replace the valve.

If the valve is left as it is, the probability of a failure of the valve seats, over the life of the project,

is 60%; the probability of failure of the valve stem is 50%; and of failure of the valve body is 40%.

If the valve is repaired, the probability of a failure of the seats over the life of the project is 40%; of failure of the stem is 30%; and of failure of the body is 20%. If the valve is replaced, the probability, over the life of the project, of a failure of the seats is 30%, of failure of the stem is 20%, and of failure of the body is 10%.

The present worth of cost of future repairs and service disruption of a failure of the seats is $10,000; the present worth of cost of a failure of the stem is $20,000; the present worth of cost of a failure of the body is $30,000. The cost of repairing the valve now is $10,000; and of replacing it is $20,000. If the criterion is to minimize expected costs, which alternative is best?

10-16 A man went to a casino with $500 and placed 100 bets of $5 each, one after another, on the same number on the roulette wheel. There are 38 numbers on the wheel, and the casino pays 35 times the amount bet if the ball drops into the bettor's numbered slot in the roulette wheel. In addition, the bettor receives back the original $5 bet. Estimate how much money the man is expected to win or lose.

10-17 A factory building is located in an area subject to occasional flooding by a nearby river. You have been brought in as a consultant to determine whether flood-proofing of the building is economically justified. The alternatives are as follows:
(1) Do nothing. Damage in a moderate flood is $10,000 and in a severe flood, $25,000.
(2) Alter the factory building at a cost of $15,000 to withstand moderate flooding without damage and to withstand severe flooding with $10,000 worth of damage.
(3) Alter the factory building at a cost of $20,000 to withstand a severe flood without damage.

In any year the probability of flooding is as follows: 0.70, no flooding of the river; 0.20, moderate flooding; and 0.10, severe flooding. If interest is 15% and a 15-year analysis period is used, what do you recommend?

10-18 An industrial park is being planned for a tract of land near the river. To prevent flood damage to the industrial buildings that will be built on this low-lying land, an earthen embankment can be constructed. The height of the embankment will be determined by an economic analysis of the costs and benefits. The following data have been gathered.

Embankment Height above Roadway (m)	Initial Cost
2.0	$100,000
2.5	165,000
3.0	300,000
3.5	400,000
4.0	550,000

Flood Level above Roadway (m)	Average Frequency That Flood Level Will Exceed Height in Col. 1
2.0	Once in 3 years
2.5	Once in 8 years
3.0	Once in 25 years
3.5	Once in 50 years
4.0	Once in 100 years

The embankment can be expected to last 50 years and will require no maintenance. Whenever the flood water flows over the embankment, $300,000 worth of damage occurs. Should the embankment be built? If so, to which of the five heights above the roadway? A 12% rate of return is required.

10-19 Should the following project be undertaken if its life is 10 years and it has no salvage value? The firm uses an interest rate of 12% to evaluate engineering projects.

First Cost	P	Net Revenue	P
$300,000	0.2	$ 70,000	0.3
400,000	0.5	90,000	0.5
600,000	0.3	100,000	0.2

(*Answer:* $45,900, yes)

10-20 A robot has just been installed at a cost of $81,000. It will have no salvage value at the end of its useful life. Given the following estimates and probabilities for the yearly savings and useful life, determine the expected rate of return.

Savings per year	Probability	Useful Life (years)	Probability
$18,000	0.2	12	1/6
20,000	0.7	5	2/3
22,000	0.1	4	1/6

10-21 Five years ago a dam was constructed to impound irrigation water and to provide flood protection for the area below the dam. Last winter a 100-year flood caused extensive damage both to the dam and to the surrounding area. This was not surprising, since the dam was designed for a 50-year flood.

The cost of repairing the dam now will be $250,000. Damage in the valley below amounts to $750,000. If the spillway is redesigned at a cost of $250,000 and the dam is repaired for another $250,000, the dam may be expected to withstand a 100-year flood without sustaining damage. However, the storage capacity of the dam will not be increased and the probability of damage to the surrounding area below the dam will be unchanged. A second dam can be constructed up the river from the existing dam for $1 million. The capacity of the second dam would be more that adequate to provide the desired flood protection. If the second dam is built, a redesign of the existing dam spillway will not be necessary, but the $250,000 worth of repairs must be done.

The development in the area below the dam is expected to be complete in 10 years. A new 100-year flood in the meantime will cause a $1 million loss. After 10 years the loss would be $2 million. In addition, there would be $250,000 of spillway damage if the spillway is not redesigned. A 50-year flood is also likely to cause about $200,000 worth of damage, but the spillway would be adequate. Similarly, a 25-year flood would cause about $50,000 of damage.

There are three alternatives: (1) repair the existing dam for $250,000 but make no other alterations, (2) repair the existing dam ($250,000) and redesign the spillway to take a 100-year flood ($250,000), and (3) repair the existing dam ($250,000) and build the second dam ($1 million). On the basis of an expected annual cash flow analysis and a 7% interest rate, which alternative should be selected? Draw a decision tree to clearly describe the problem.

10-22 The chief uncertainty about a new product is its annual net revenue. So far, $35,000 has been spent on development, but an additional $30,000 is needed to finish development. The firm's interest rate is 10%.

(*a*) What is the expected PW for deciding whether to proceed?

(*b*) Find the P(loss) and the standard deviation for proceeding.

	State		
	Bad	**OK**	**Great**
Probability	0.3	0.5	0.2
Net revenue	−$15,000	$15,000	$20,000
Life, in years	5	5	10

10-23 (*a*) In Problem 10-22 how much is it worth to the firm to terminate the product after 1 year if the net revenues are negative?

(*b*) How much does the ability to terminate early change the *P*(loss) and the standard deviation?

10-24 Find the probability distribution and the expected PW to modify an assembly line. The first cost is $80,000, and its salvage value is $0. The firm's interest rate is 9%. The savings shown in the table depend on whether the assembly line runs 1, 2, or 3 shifts, and on whether the product is made for 3 or 5 years.

Shifts/ day	Savings/ year	Probability	Useful Life (years)	Probability
1	$15,000	0.3	3	0.6
2	30,000	0.5	5	0.4
3	45,000	0.2		

10-25 In Problem 10-24, how much is it worth to the firm to be able to extend the product's life by 3 years, at a cost of $50,000, at the end of the product's initial useful life?

10-26 Al took a mid-term examination in physics and received a mark of 65. The mean was 60 and the standard deviation was 20. Bill received a mark of 14 in mathematics, where the exam mean was 12 and the standard deviation was 4. Which student ranked higher in his class? Explain.

10-27 The Graham Telephone Company may invest in new switching equipment. There are three possible outcomes, having net present worth of $6570, $8590, and $9730. The probability of each outcome is 0.3, 0.5, and 0.2 respectively. Calculate the expected return and risk associated with this proposal. (*Answer:* $E_{PW} = \$8212$, $\sigma_{PW} = \$1158$)

10-28 A new machine will cost $25,000. The machine is expected to last 4 years and have no salvage value. If the interest rate is 12%, determine the return and the risk associated with the purchase.

P	0.3	0.4	0.3
Annual savings	$7000	$8500	$9500

10-29 What is your risk associated with Problem 10-14?

10-30 Measure the risk for Problem 10-19 using the *P*(loss), range of PW values, and standard deviation of the PWs. (*Answer:* $\sigma_{PW} = \$127,900$)

10-31 (*a*) In Problem 10-24, describe the risk using the *P*(loss) and standard deviation of the PWs.

(*b*) How much do the answers change if the possible life extension in Problem 10-25 is allowed?

10-32 An engineer decided to make a careful analysis of the cost of fire insurance for his $200,000 home. From a fire rating bureau he found the following risk of fire loss in any year.

Outcome	Probability
No fire loss	0.986
$ 10,000 fire loss	0.010
40,000 fire loss	0.003
200,000 fire loss	0.001

(*a*) Compute his expected fire loss in any year.

(*b*) He finds that the expected fire loss in any year is less than the $550 annual cost of fire insurance. In fact, an insurance agent explains that this is always true. Nevertheless, the engineer buys fire insurance. Explain why this is or is not a logical decision.

10-33 A firm wants to select one new research and development project. The following table summarizes 6 possibilities. Considering expected return and risk, which projects are good candidates? The firm believes it can earn 5% on a risk-free investment in government securities (labeled as Project *F*).

Project	IRR	Standard Deviation
1	15.8%	6.5%
2	12.0	4.1
3	10.4	6.3
4	12.1	5.1
5	14.2	8.0
6	18.5	10.0
F	5.0	0.0

10-34 A firm is choosing a new product. The following table summarizes 6 new potential products. Considering expected return and risk, which products are good candidates? The firm believes it can earn 4% on a risk-free investment in government securities (labelled as Product F).

Product	IRR	Standard Deviation
1	10.4%	3.2%
2	9.8	2.3
3	6.0	1.6
4	12.1	3.6
5	12.2	8.0
6	13.8	6.5
F	4.0	0.0

CHAPTER 11

Income, Depreciation, and Cash Flow

Is Baseball Going Broke?

In 2001, Major League Baseball brought in record revenues totalling over $3.5 billion. Great business, you say? Not according to Bud Selig, baseball's commissioner.

In December of that year, Selig went before the US Congress to plead poverty, asserting that baseball had suffered a loss of over $200 million for the year. Many journalists (and even more fans) weren't buying it. A few months later, *Forbes* magazine published an article stating that baseball had in fact had an operating profit (that is, earnings before interest, taxes, and depreciation) of $75 million for the year.

Selig angrily denounced the *Forbes* article as 'pure fiction'. In response, *Forbes* suggested that perhaps Selig was 'afraid of being called before Congress and explaining his figures in more detail'.

After Completing This Chapter...

The student should be able to:

- Describe depreciation, deterioration, and obsolescence.
- Distinguish between various types of depreciable property and differentiate between depreciation expenses and other business expenses.
- Use *historical* depreciation methods to calculate the *annual depreciation charge* and *book value* over the asset's life.
- Explain the differences between the historical depreciation methods and the capital cost allowance system (CCA).
- Use CCA to calculate allowable *annual depreciation charge* and *book value* over the asset's life for various asset classes.
- Fully account for *capital gains and losses, loss on disposal of fixed assets,* and *recaptured CCA* due to the disposal of a depreciated business asset.
- Use the *units of production* and *depletion* depreciation methods as needed in engineering economic analysis problems.
- Use spreadsheets to calculate depreciation.

QUESTIONS TO CONSIDER

1. When determining their cash flow, owners of baseball teams deduct stadium depreciation, which can amount to as much as $5 million per year. How would this affect their calculations?

2. In 2003, San Francisco outfielder Barry Bonds was in peak health, coming off a stellar season in which he hit a record 73 home runs. Yet the IRS considered him a 'wasting asset' and allowed the owners of his baseball club to depreciate a large portion of his contract. Explain.

INCOME

'Did we make any money last year?' A valid question since the point of a business is to make money, and after a time it is important to measure whether or not it is doing so. The question refers to a fixed period of time, called an accounting period—usually one year. Whether or not money was made is determined by preparing an *Income Statement,* or *Profit and Loss Statement,* which exhibits the name of the company, the period involved, and the following equation

$$\text{Income} = \text{Revenue} - \text{Costs to obtain revenue}$$

The revenues are usually straightforward, deriving from sales of the products. Costs, however, are subdivided as seen in Figure 11-1.

XYZ Company
Income Statement
For the year ending April 25, 2009

Revenue $

    Sales of product
    Charges for services
                   Total Revenue $

Costs

    Cost of Goods Sold
        Labour wages
        Materials
        Utilities
        Machines **(a portion of the cost)**
        Factory buildings **(a portion of the cost)**
    Selling Costs
        Advertising
        Sale commissions
    Administration Costs
        Administrative salaries
        Office rental
    Financing Costs
        Interest paid on debt

                   Total Costs $

           Net Income before Taxes

FIGURE 11-1 The income statement.

Now some items—wages, materials, and the like—are paid for as they are used—weekly or when they arrive. Other items, like office rent and insurance premiums, are paid for monthy, quarterly, or yearly. Most items are paid within the period of a year, and so it is a reasonable measure to compare revenues for the year with expenses for the year.

There is, however, a group of items called *physical capital assets,* consisting of land, buildings, equipment, and machinery, that last for longer than one year and are usually paid for at the time they are obtained. Therefore, to reasonably represent their contribution to the annual income, some method must be found of allocating the cost of these assets over their useful lives.

Land doesn't wear out, and so it is excluded. All other physical assets are said to depreciate.

We have so far dealt with a variety of economic analysis problems and many techniques for their solution. In the process we have avoided income taxes, which are an important element of most economic analyses. Now, we can move to more realistic—and, unfortunately, more complex—situations.

Our government taxes individuals and businesses to support its processes—lawmaking, domestic and foreign economic policy-making, even the making and issuing of money itself. The omnipresence of taxes requires that they be included in economic analyses, and that means we must understand something about the *way* taxes are imposed. The measure of a business's success is its annual profit (or loss). This measure is arrived at by taking all the revenue that was earned in the year and subtracting from that the expenses incurred in earning that income. For expenses incurred on a weekly, monthly, or hourly basis, this calculation is reasonably straightforward. However, when you consider equipment that you pay for now but that continues to perform for several years, the question arises as to what portion of that initial cost should be allocated to which year's expenses. This is especially important since the government has found that taxing income is one of the better ways of raising money.

Depreciation is the mechanism for allocating the cost of a long-lived property over a number of years for the purpose of calculating annual income. Chapter 11 examines depreciation, and Chapter 12 illustrates how depreciation is used in income tax computations. The goal is decision making on engineering projects, not final tax calculations.

BASIC ASPECTS OF DEPRECIATION

The word *depreciation* is defined as a 'decrease in value'. This is not an entirely satisfactory definition, for *value* has several meanings. In the context of economic analysis, value may refer either to *market value*—that is, the monetary value others place on property—or *value to the owner.* Thus, we now have two definitions of depreciation: a decrease in value to the market or to the owner.

Deterioration and Obsolescence

A machine may depreciate because it is **deteriorating** or wearing out and no longer performing its function as well as when it was new. Many kinds of machinery need increased maintenance as they age and there is a slow but continuing failure of individual parts. In other types of equipment, the quality of output may decline because of wear on components and resulting poorer mating of parts. Anyone who has worked to maintain a car has observed deterioration due to failure of individual parts, such as fan belts, mufflers, and batteries, and the wear on components, such as bearings, piston rings, and alternator brushes.

Depreciation is also caused by **obsolescence.** A machine is described as obsolete when it is no longer needed or useful. A machine may be in excellent working condition, yet may still be obsolete. In the 1970s, mechanical business calculators with hundreds of gears and levers became obsolete. The advance of integrated circuits resulted in a completely different and far superior approach to calculator design. Thus, mechanical calculators rapidly declined or depreciated in value.

If your **car** depreciated in the last year, that means it has declined in market value. It has less value to potential buyers. On the other hand, a manager who says a piece of machinery has depreciated may be describing a machine that has deteriorated because of use and/or because it has become obsolete compared to newer machinery. Both situations indicate the machine has declined in value to the owner.

The accounting profession defines depreciation in yet another way. Although in everyday conversation we are likely to use depreciation to mean a decline in market value or value to the owner, accountants define depreciation as allocating an asset's cost over its **useful** or **depreciable life.** Thus, we now have *three distinct definitions of depreciation:*

1. Decline in market value of an asset.
2. Decline in value of an asset to its owner.
3. Systematic allocation of the cost of an asset over its depreciable life.

Depreciation and Expenses

It is this third (accountant's) definition that is used to compute depreciation for business assets. Business costs are generally either expensed or depreciated. **Expensed** items, such as labour, utilities, materials, and insurance, are part of regular business operations and are 'consumed' over short periods of time (sometimes recurring). These costs do not lose value gradually. For tax purposes they are subtracted from business revenues when they occur. Expensed costs reduce income taxes because businesses are able to *write off* their full amount when they occur.

In contrast, business costs due to **depreciated** assets are not fully written off when they occur. A depreciated asset does lose value gradually and must be written off over an extended period. For instance, consider a plastic injection machine used to produce the beverage cups found at sporting events. The plastic pellets melted into the cup shape lose their value as raw material directly after manufacturing. The raw material cost for production material (plastic pellets) is written off, or expensed, immediately. On the other hand, the plastic mould injection machine itself will lose value over time, and thus its cost (purchase price and installation expenses) are written off (or depreciated) over a period of years. The number of years over which the machine is depreciated is called its **depreciable life** or **recovery period,** which is often different from the asset's useful or most economic life. Depreciable life is determined by the depreciation method used to spread out the cost—many types of depreciated assets operate well beyond their depreciation life.

Depreciation is a **non-cash** cost that requires no exchange of dollars from one hand to another. Companies do not write a cheque to someone to **pay** their depreciation expenses. Rather, it is a business expense that is allowed by the government to offset the loss in value of business assets. Remember, the company has already paid for the asset up front, and depreciation is simply a way to claim this 'business expense' over time. Depreciation is important to the engineering economist because, even though it is a non-cash cost, it represents a cash flow on an *after-tax basis*. Depreciation deductions reduce the taxable income of businesses and thus reduce the amount of taxes paid.

In general, business assets can be depreciated only if they meet the following basic requirements:

The property must be used for business purposes to produce income.

The property must have a useful life that can be determined, and this life must be longer than one year.

The property must be an asset that decays, gets used up, wears out, becomes obsolete, or loses value to the owner from natural causes.

EXAMPLE 11-1

Consider the costs that are incurred by a local pizza business. Identify each cost as either *expensed* or *depreciated* and explain why.

- Cost of pizza dough and toppings
- Cost of new delivery van
- Cost of wages for janitor
- Cost of furnishings in dining room
- Cost of a new baking oven
- Utility costs for soft drink refrigerator

SOLUTION

Cost Item	Type of Cost	Why
Pizza dough and toppings	expensed	Life < 1 year, loses value immediately
New delivery van	depreciated	Meets 3 requirements for depreciation
Wages for janitor	expensed	Life < 1 year, loses value immediately
Furnishings in dining room	depreciated	Meets 3 requirements for depreciation
New baking oven	depreciated	Meets 3 requirements for depreciation
Utilities for soft drink refrigerator	expensed	Life < 1 year, loses value immediately

Types of Property

The rules for depreciation are linked to the classification of business property as either *tangible* or *intangible*. Tangible property is further classified as either *real* or *personal*.

Tangible property can be seen, touched, and felt.

Real property includes land, buildings, and all things growing on, built upon, constructed on, or attached to the land.

Personal property includes equipment, furnishings, vehicles, office machinery, and anything that is tangible, excluding those assets defined as *real property*.

Intangible property is all property that has value to the owner but cannot be directly seen or touched. Examples include patents, copyrights, trademarks, trade names, and franchises.

Many different types of properties that wear out, decay, or lose value can be depreciated as business assets. This wide range includes photocopiers, helicopters, buildings, interior furnishings, production equipment, and computer networks. Almost all tangible property can be depreciated.

One important and notable exception is land, which is *never* depreciated. Land does not wear out, lose value, or have a determinable useful life and thus does not qualify as a depreciable property. Consider the aspect of loss in value. Rather than decreasing in value, most land becomes more valuable as time passes. In addition to the land itself, expenses for clearing, grading, preparing, planting, and landscaping are not generally depreciated because they have no fixed useful life. Other tangible property that *cannot* be depreciated includes factory inventory, containers considered as inventory, equipment used to build capital improvements, and leased property. The leased property exception highlights the fact that only the owner of property may claim depreciation expenses.

Tangible properties used in *both* business and personal activities, such as a car used in a consulting engineering firm that is also used to take one's children to school, can be depreciated. However, in such cases one can only take depreciation deductions in proportion to the use for business purposes. Accurate records indicating the portion of use for business and personal activities are required.

Cost Basis

The cost basis of an asset represents the total cost of acquiring and getting it into working order. It is not just the price. It includes such necessary expenses as engineering, accounting and legal fees, freight, site preparation, installation costs, and commissioning. It is this total cost that must be charged as an expense over the life of the asset.

Fundamentals of Depreciation Calculation

To understand the complexities of depreciation, the first step is to examine the fundamentals of depreciation calculations. Figure 11-2 illustrates the general depreciation problem of allocating the total depreciation charges over the asset's life. The vertical axis is labelled **book value,** and the curve of asset cost or basis minus depreciation charges made starts at the cost basis and declines to the salvage value,

Book value = Cost basis − Depreciation charges made to date

Looked at another way, book value is the remaining unallocated cost of the asset.

In Figure 11-2, *book value* goes from a value of B at time zero in the recovery period to a value of S at time 5. Thus, book value is a *dynamic* variable that changes over an asset's recovery period. The equation used to calculate the book value of an asset over time is:

$$BV_t = \text{Cost basis} - \sum_{i=1}^{t} d_i \qquad (11\text{-}1)$$

FIGURE 11-2 General depreciation.

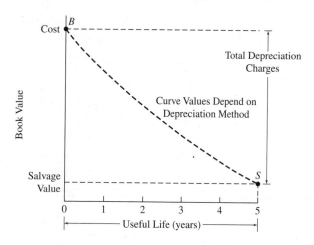

where

$$BV_t = \text{book value of the depreciated asset at the end of time } t$$

Cost basis $= B =$ dollar amount that is being depreciated. This includes the purchase price of the asset, as well as any other costs necessary to make the asset 'ready for use'.

$\sum_{i=1}^{t} d_i =$ the sum of depreciation deduction taken from time 0 to time t, where d_i is the depreciation deduction in year i.

Equation 11-1 shows that year-to-year depreciation charges reduce an asset's book value over its life. The following section describes methods that are or have been allowed under the Income Tax Act for quantifying these yearly depreciation deductions.

DEPRECIATION METHODS

Depreciation accounting, the allocation of the cost for capital expenditures over time, is part of how accountants represent the financial performance of a business. The particular depreciation method chosen is the one that best represents the actual decline in value of the property. In this way it is possible to calculate an annual profit or loss. However, for taxation purposes, governments specify exactly how depreciation is to be calculated. In Canada the method of depreciation for income tax purposes is known as the Capital Cost Allowance (CCA) deductions; in the United States it is called the Modified Accelerated Cost Recovery System (MACRS). In general, accounting depreciation methods can be categorized as follows:

General Depreciation Methods

These methods include the *straight line, sum-of-the-years digits, declining balance,* and *unit-of-production* methods. Each method requires estimates of an asset's useful life and salvage value. Firms elect which method to use for assets, and there is little uniformity in how depreciation expenses are reported.

Tax Reporting Depreciation Methods

Canada.   The Capital Cost Allowance (CCA) is the portion of the capital cost of certain depreciable property that a corporation can deduct from income earned during the year. The CCA method has these features: (1) property class accounts into which property is grouped, (2) declining balance depreciation of the grouped property at a government-specified percentage. (3) reduction of the amount eligible for deduction by 50% in the year of acquisition.

USA.   The Modified Accelerated Cost Recovery System (MACRS) has been in effect since the Tax Reform Act of 1986 (TRA-86). The MACRS method has the following features: (1) Property class lives were created, and all depreciated assets assigned to one particular category. (2) The need to estimate salvage values was eliminated because all assets were *fully* depreciated over their recovery period. (3) The annual depreciation percentages were modified to include a half-year convention for the first and final years. (4) The recovery periods used to calculate annual depreciation *accelerated* the write-off of capital costs more quickly than did the historical methods—thus the name.

In this chapter, our primary purpose is to describe the CCA depreciation method. However, it is useful to first describe the four general depreciation methods. They are still used in some countries, and both the CCA and the MACRS are based on variants of them.

Straight-Line Depreciation

The simplest and best-known depreciation method is **straight-line depreciation.** To calculate the constant **annual depreciation charge,** the total amount to be depreciated, $B - S$, is divided by the depreciable life, in years, N:

$$\textbf{\textit{Annual depreciation charge}} = d_t = \frac{(B - S)}{N} \qquad (11\text{-}2)$$

N is used for the depreciation period because it may be shorter than n, the horizon.

EXAMPLE 11-2

Consider the following:

Cost of the asset, B	$900
Depreciable life, in years, N	5
Salvage value, S	$70

Compute the straight-line depreciation schedule.

SOLUTION

$$\text{Annual depreciation charge} = d_t = \frac{B - S}{N} = \frac{900 - 70}{5} = \$166$$

Year, t	Depreciation for Year t, d_t	Sum of Depreciation Charges Up to Year t, $\sum_{j=1}^{t} d_j$	Book Value at the End of Year t, $BV_t = B - \sum_{j=1}^{t} d_j$
1	$166	$166	$900 - 166 = 734$
2	166	332	$900 - 332 = 568$
3	166	498	$900 - 498 = 402$
4	166	664	$900 - 664 = 236$
5	166	830	$900 - 830 = \ \ 70 = S$

This situation is illustrated in Figure 11-3. Notice that d_t is constant at $166 each year for 5 years, and that the asset has been depreciated down to a book value of $70, which was the estimated salvage value.

FIGURE 11-3 Straight-line depreciation.

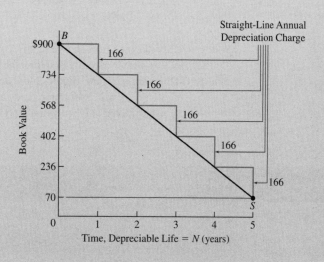

The straight-line (SL) method is often used for intangible property. Veronica's firm bought a patent in April that was not acquired as part of acquiring a business. She paid $6800 for this patent and must use the straight-line method to depreciate it over 17 years with no salvage value. The annual depreciation is $400 (=$6800/17). Since the patent was purchased in April, the deduction must be prorated over the 9 months of ownership. This year the deduction is $300 (= $400 × 9/12), and then next year she can begin taking the full $400 per year.

Sum-of-Years'-Digits Depreciation

Another method for allocating an asset's cost *minus* salvage value *over* its depreciable life is called **sum-of-years'-digits (SOYD) depreciation.** This method results in larger-than-straight-line depreciation charges during an asset's early years and smaller charges as the asset nears the end of its depreciable life. Each year, the depreciation charge equals a fraction of the total amount to be depreciated $(B - S)$. The denominator of the fraction is the sum of the years' digits. For example, if the depreciable life is 5 years, $1 + 2 + 3 + 4 + 5 = 15 =$ SOYD. Then 5/15, 4/15, 3/15, 2/15, and 1/15 are the fractions from Year 1 to Year 5. Each year the depreciation charge shrinks by 1/15 of $B - S$. Because this change is the same every year, SOYD depreciation can be modelled as an arithmetic gradient, G. The equations can also be written as:

$$\begin{pmatrix} \text{Sum-of-years'-digits} \\ \text{depreciation charge for} \\ \text{any year} \end{pmatrix} = \frac{\begin{pmatrix} \text{Remaining depreciable life} \\ \text{at beginning of year} \end{pmatrix}}{\begin{pmatrix} \text{Sum of years' digits} \\ \text{for total depreciable life} \end{pmatrix}} (\text{Total amount depreciated})$$

$$d_t = \frac{N - t + 1}{\text{SOYD}}(B - S) \tag{11-3}$$

where

d_t = depreciation charge in any year t
N = number of years in depreciable life
SOYD = sum of years' digits, calculated as $N(N + 1)/2 =$ SOYD
B = cost of the asset made ready for use
S = estimated salvage value after depreciable life

EXAMPLE 11-3

Compute the SOYD depreciation schedule for the situation in Example 11.2.

Cost of the asset, B	$900
Depreciable life, in years, N	5
Salvage value, S	$70

SOLUTION

$$\text{SOYD} = \frac{5 \times 6}{2} = 15$$

Thus,

$$d_1 = \frac{5 - 1 + 1}{15}(900 - 70) = 277$$

$$d_2 = \frac{5 - 2 + 1}{15}(900 - 70) = 221$$

$$d_3 = \frac{5 - 3 + 1}{15}(900 - 70) = 166$$

$$d_4 = \frac{5 - 4 + 1}{15}(900 - 70) = 111$$

$$d_5 = \frac{5 - 5 + 1}{15}(900 - 70) = 55$$

Year, t	Depreciation for Year t, d_t	Sum of Depreciation Charges Up to Year t, $\sum_{j=1}^{t} d_j$	Book Value at End of Year t, $BV_t = B - \sum_{j=1}^{t} d_j$
1	$277	$277	$900 - 277 = 623$
2	221	498	$900 - 498 = 402$
3	166	664	$900 - 664 = 236$
4	111	775	$900 - 775 = 125$
5	55	830	$900 - 830 = \quad 70 = S$

These data are plotted in Figure 11-4.

FIGURE 11-4 Sum-of-years'-digits depreciation.

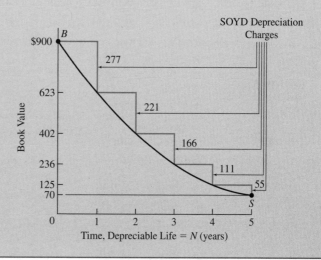

Declining Balance Depreciation

Declining balance depreciation applies a *constant depreciation rate* (D) to the declining book value of the property. The factor D is used to determine the depreciation charge for a given year, d_t, as follows:

D = the fraction of beginning period book value that is allocated

$d_1 = D \times B$

$d_2 = D \times (B - d_1) = D \times B(1 - D)$

$d_3 = D \times (B - d_1 - d_2) = D \times B(1 - D)^2$

Thus for any year,

$$d_n = DB(1 - D)^{n-1} \qquad (11\text{-}4a)$$

and the book value, BV_n, at the end of n years will be

$$BV_n = B(1 - D)^n \qquad (11\text{-}4b)$$

Historically, before the days of CCA and MACRS, companies depreciated assets over their depreciable lives at a straight-line rate of $1/N$, or a declining balance rate of $2/N$. Thus an asset with a 10-year depreciable life would be reduced at 10% of the cost basis each year, or at 20% of the book value each year. Because the declining balance percentage was twice the straight-line, the method came to be known as **double declining balance** or DDB with general equations:

$$\text{Double declining balance } d_t = \frac{2}{N}(\text{Book value}_{t-1}) \qquad (11\text{-}4c)$$

EXAMPLE 11-4

Compute the declining balance depreciation schedule for the situations in Examples 11-2 and 11-3.

Cost of the asset, B	$900
Declining balance rate	40%
Salvage value, S	$ 70

SOLUTION

Year, t	Depreciation for Year t (d_t) from Equation 10-4a	Sum of Depreciation Charges Up to Year t, $\sum_{i=1}^{t} d_i$	Book Value at End of Year t, $BV_t = B - \sum_{i=1}^{t} d_i$
1	40% · 900 = 360	$360	900 − 360 = 540
2	40% · 540 = 216	576	900 − 566 = 334
3	40% · 334 = 130	706	900 − 706 = 194
4	40% · 194 = 78	784	900 − 784 = 116
5	40% · 116 = 46	830	900 − 830 = 70

Figure 11-5 illustrates the situation.

FIGURE 11-5 Declining balance depreciation.

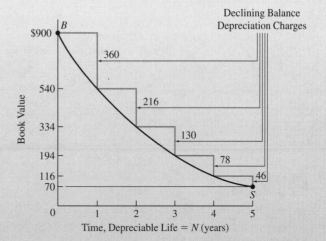

The final salvage value of $70 for Examples 11-2, 11-3, and 11-4 was chosen to match the ending value for the double-declining balance method. This does not normally happen.

UNIT-OF-PRODUCTION DEPRECIATION

At times, there may be situations where the recovery of depreciation on a particular asset is more closely related to use than to time. In these few situations (and they are rare), the **unit-of-production (UOP) depreciation** in any year is:

$$\text{UOP depreciation in any year} = \frac{\text{Production for year}}{\text{Total lifetime production for asset}}(B - S) \qquad (11\text{-}5)$$

This method might be useful for machinery that processes natural resources if the resources will be exhausted before the machinery wears out. It is not considered an acceptable method for general use in depreciating industrial equipment.

EXAMPLE 11-5

For purposes of comparison with previous examples, assume that equipment costing $900 has been purchased for use in a sand and gravel pit. The pit will operate for 5 years, while a nearby airport is being rebuilt and paved. Then the pit will be shut down and the equipment removed and sold for $70. Compute the unit-of-production (UOP) depreciation schedule if the airport reconstruction schedule calls for 40,000 cubic metres of sand and gravel as follows:

Year	Sand and Gravel Required, in m³
1	4,000
2	8,000
3	16,000
4	8,000
5	4,000

SOLUTION

The cost basis, B, is $900. The salvage value, S, is $70. The total lifetime production for the asset is 40,000 cubic metres of sand and gravel. From the airport reconstruction schedule, the first-year UOP depreciation would be:

$$\text{First-year UOP depreciation} = (4000 \text{ m}^3/40{,}000 \text{ m}^3) \times (\$900 - \$70) = \$83$$

Similar calculations for the subsequent 4 years give the complete depreciation schedule:

Year	UOP Depreciation
1	$ 83
2	166
3	332
4	166
5	83
	$830

It should be noted that the actual unit-of-production depreciation charge in any year is based on the actual production for the year rather than on the scheduled production.

DEPRECIATION FOR TAX PURPOSES—CAPITAL COST ALLOWANCE

The Canadian Revenue Agency (CRA) is the agency responsible for collecting income taxes. The Canadian government is continually reviewing and revising the tax laws; and court, and CRA interpretations affect how they are applied. Minor changes are made in these laws every year and major revisions periodically. The following discussion is intended to give the reader an understanding of how the tax laws affect investment decisions. However, Canadian tax law covers many volumes the size of this book. Therefore, with respect to any specific economy study, check with an accountant or tax specialist to ensure that you have included any special tax considerations.

Canadian income tax law permits corporations to depreciate most capital assets by the declining-balance method at a rate specified in the tax legislation. (A capital asset is one with a useful life of more than one year.) With very few exceptions, the taxpayer is given no option as to the depreciation method or the depreciation rate to be used. Since the depreciation allowance for income tax purposes represents an allowable deduction in computing taxable income, it is referred to as capital cost allowance (CCA) in Canada.

The legislation is also very specific about how the capital cost allowance is to be calculated. Basically, the prescribed method is what is considered asset class accounting. That is, all assets of a single class are grouped together into a single ledger account. When additional assets of that class are acquired, their cost is added to the account; when assets are disposed of, the proceeds are deducted from the account. The capital cost allowance for any year is the account total at the end of the year times the capital cost allowance rate for the class. The capital cost allowance rate that applies to each class is the maximum rate. The company can apply any rate between zero and the maximum. The maximum capital cost allowance that a company could take in any one year would just be sufficient to reduce the taxable income to zero. In addition, in the first year of ownership of a depreciable asset, only one-half of the maximum rate can be applied. Table 11-1 lists examples of asset classes and rates.

The 50% Rule or Half-Year Convention

Generally, property acquired that is available for use during the taxation year is eligible for only 50% of the normal maximum CCA for the year. You can claim full CCA for that property in the next taxation year. However, for most classes of assets, disposal of assets during the year is first netted against acquisitions made in the same year. Consequently, the effect of the half-year rule is mitigated when there are major disposals of fixed assets.

Certain properties, Classes 12, 13, 14, 15, 23, 24, 27, 29, and 34, and those acquired through non-arm's-length transfers are exempt from the 50% rule.

Calculating the CCA—Schedule 8

Capital cost allowance is claimed by completing the CRA form called Schedule 8, which is reproduced in Figure 11-6. The use of this form for calculating capital cost allowance is outlined in the following example.

TABLE 11-1 Asset Classes and CCA Rates for 2003

Class Number	Description	CCA Rate
1	Most buildings made of brick, stone, or cement acquired after 1987, including their component parts such as electric wiring, lighting fixtures, plumbing, heating and cooling equipment, elevators, and escalators	4%
3	Most buildings made of brick, stone, or cement acquired before 1988, including their component parts as listed in class 1 above	5%
6	Buildings made of frame, log, stucco on frame, galvanized iron, or corrugated metal that are used in the business of farming or fishing, or that have no footings below-ground; fences and most greenhouses	10%
7	Canoes, boats, and most other vessels, including their furniture, fittings, or equipment	15%
8	Property that is not included in any other class such as furniture, calculators and cash registers (that do not record multiple sales taxes), photocopy and fax machines, printers, display fixtures, refrigeration equipment, machinery, tools costing $200 or more, and outdoor advertising billboards and greenhouses with rigid frames and plastic covers acquired after 1987	20%
9	Aircraft, including furniture, fittings, or equipment attached, and their spare parts	25%
10	Automobiles (except taxis and others used for lease or rent), vans, wagons, trucks, buses, tractors, trailers, drive-in theatres, general-purpose electronic data-processing equipment (for example, personal computers) and systems software, and timber cutting and removing equipment	30%
10.1	Passenger vehicles costing more than $30,000 if acquired after 2000 ($27,000 if acquired in 2000; $26,000 if acquired after 1997 and before 2000; $25,000 if acquired in 1997; $24,000 if acquired after August 31, 1989, and before 1997; and $20,000 if acquired before September 1989)	30%
12	Chinaware, cutlery, linen, uniforms, dies, jigs, moulds or lasts, computer software (except systems software), cutting or shaping parts of a machine, certain property used for earning rental income such as apparel or costumes, and videotape cassettes; certain property costing less than $200 such as kitchen utensils, tools, and medical or dental equipment; certain property acquired after August 8, 1989, and before 1993 for use in a business of selling or providing services such as electronic bar-code scanners, and cash registers used to record multiple sales taxes	100%
13	Property that is leasehold interest (the maximum CCA rate depends on the type of the leasehold and the terms of the lease)	
14	Patents, franchises, concessions, and licences for a limited period—the CCA is limited to whichever is less: the capital cost of the property spread out over the life of the property; or the undepreciated capital cost of the property at the end of the taxation year. Class 14 also includes patents, and licences to use patents for a limited period, that you elect not to include in class 44	Not applicable
16	Automobiles for lease or rent, taxicabs, and coin-operated video games or pinball machines; certain tractors and large trucks acquired after December 6, 1991, that are used to haul freight and that weigh more than 11,788 kilograms	40%
17	Roads, sidewalks, parking-lot or storage areas, telephone, telegraph, or non-electronic data communication switching equipment	8%
38	Most power-operated movable equipment acquired after 1987 used for moving, excavating, placing, or compacting earth, rock, concrete, or asphalt	30%
39	Machinery and equipment acquired after 1987 that is used in Canada primarily to manufacture and process goods for sale or lease	25%
43	Manufacturing and processing machinery and equipment acquired after February 25, 1992, described in class 39 above	30%
44	Patents and licences to use patents for a limited or unlimited period that the corporation acquired after April 26, 1993. However, you can elect not to include such property in class 44 by attaching a letter to the return for the year the corporation acquired the property. In the letter, indicate the property you do not want to include in class 44	25%

Source: Canada Revenue Agency, *2003 T2 Corporate Income Tax Guide*. <www.cra-arc.gc.ca>.

Canada Customs and Revenue Agency

Agence des douanes et du revenu du Canada

CAPITAL COST ALLOWANCE (CCA) (1998 and later taxation years)

Name of corporation

Business Number

Taxation year end | Year | Month | Day

For more information, see the section called "Capital Cost Allowance" in the *T2 Corporation Income Tax Guide*.

Is the corporation electing under regulation 1101(5q)? **101** ☐ 1 Yes ☐ 2 No ☐

1 Class number	2 Undepreciated capital cost at the beginning of the year (undepreciated capital cost at the end of the year from last year's CCA schedule)	3 Cost of acquisitions during the year (new property must be available for use) See note 1 below	4 Net adjustments (show negative amounts in brackets)	5 Proceeds of dispositions during the year (amount not to exceed the capital cost)	6 Undepreciated capital cost (column 2 **plus** column 3 **plus** or **minus** column 4 **minus** column 5)	7 50% rule (1/2 of the amount, if any, by which the net cost of acquisitions exceeds column 5) See note 2 below	8 Reduced undepreciated capital cost (column 6 **minus** column 7)	9 CCA rate %	10 Recapture of capital cost allowance	11 Terminal loss	12 Capital cost allowance (column 8 **multiplied by** column 9; or a lower amount) See note 3 below	13 Undepreciated capital cost at the end of the year (column 6 **minus** column 12)
200	**201**	**203**	**205**	**207**		**211**		**212**	**213**	**215**	**217**	**220**
1.												
2.												
3.												
4.												
5.												
6.												
7.												
8.												
9.												
10.												

Totals

Enter the total of column 10 on line 107 of Schedule 1.
Enter the total of column 11 on line 404 of Schedule 1.
Enter the total of column 12 on line 403 of Schedule 1.

Note 1. Include any property acquired in previous years that has now become available for use. This property would have been previously excluded from column 3. List separately any acquisitions that are not subject to the 50% rule, see Regulation 1100(2) and (2.2).

Note 2. The net cost of acquisitions is the cost of acquisitions plus or minus certain adjustments from column 4.

Note 3. If the taxation year is shorter than 365 days, prorate the CCA claim. See the *T2 Corporation Income Tax Guide* for more information.

T2 SCH 8 (99)
Printed in Canada

(Fran ais au verso)

1388

Canada

FIGURE 11-6 Schedule 8 from *2003 T2 Corporate Income Tax Guide*.
Source: Canada Revenue Agency. <www.cra-arc.gc.ca>.

EXAMPLE 11-6

A firm has 6 vehicles; the make, age, and current book value of each is as follows:

	Description	Book Value (*undepreciated capital cost*)
1998	Chev Van	$22,465
2002	Hyundai XG 350	31,620
2000	Honda Accord	18,732
1980	Ford Bronco	2,419
1995	Dodge 1/2 ton pick-up	11,563
	Total book value	**$86,799**

What depreciation deduction (CCA) is permitted at the end of the current year (year 1)?

SOLUTION

According to tax legislation (see Table 11-1) automotive equipment is a Class 10 asset and can be depreciated at a CCA rate of 30%. Therefore all the vehicles would be grouped into a single CCA schedule as follows:

The Schedule 8 for Year 1

1 Class number	2 Undepreciated capital cost at the beginning of the year (undepreciated capital cost at the end of the year from last year's CCA schedule)	3 Cost of acquisitions during the year (new property must be available for use)	5 Proceeds of dispositions during the year (amount not to exceed the capital cost)	6 Undepreciated capital cost (column 2 **plus** column 3 **plus** or **minus** column 4 **minus** column 5)	7 50% rule (1/2 of the amount, if any, by which the net cost of acquisitions exceeds column 5)	8 Reduced undepreciated capital cost (column 6 **minus** column 7)	9 CCA rate %	12 Capital cost allowance (column 8 **multiplied** by column 9; or a lower amount)	13 Undepreciated capital cost at the end of the year (column 6 **minus** column 12)
10	$86,799	$—	$—	$86,799	$—	$86,799	30%	$26,040	$60,759

The year 1 CCA = $ 26,040

Action. In year 2 the company sells the Honda Accord for $20,000. What CCA is permitted at the end of year 2?

Calculation. When the company disposes of an asset, following the instructions on Schedule 8, the proceeds from the disposal are subtracted from the current total book value of the asset class. The sale of the Honda is treated as follows. The proceeds are entered in column 5 and subtracted from the capital cost in column 2 to produce a reduced total for the class.

The Schedule 8 for Year 2

1 Class number	2 Undepreciated capital cost at the beginning of the year (undepreciated capital cost at the end of the year from last year's CCA schedule)	3 Cost of acquisitions during the year (new property must be available for use)	5 Proceeds of dispositions during the year (amount not to exceed the capital cost)	6 Undepreciated capital cost (column 2 **plus** column 3 **plus** or **minus** column 4 **minus** column 5)	7 50% rule (1/2 of the amount, if any, by which the net cost of acquisitions exceeds column 5)	8 Reduced undepreciated capital cost (column 6 **minus** column 7)	9 CCA rate %	12 Capital cost allowance (column 8 **multiplied** by column 9; or a lower amount)	13 Undepreciated Capital Cost at the end of the year (column 6 **minus** column 12)
10	$60,759	$—	$20,000	$ 40,759	$—	$40,759	30%	$12,228	$28,532

The year 2 CCA = $12,228

Notice that the current book value of the Honda = $18,732 \times (1 - 30\%) = \$13,112$. So the sale resulted in **a recapture of ($20,000 - 13,112) = \$6,888$ capital cost allowance.** However, recaptures and losses are not explicitly calculated for a particular asset: only the class total is adjusted. The result of the Schedule 8 worksheet is that the recaptures or losses are added or subtracted back into income at the same CCA rate as that at which it was taken out. This is covered in more detail in Chapter 12 under 'Disposal of Assets'.

Action.   In year 3 the company buys a Toyota Land Cruiser for $26,000. What CCA is permitted at the end of year 3?

Calculation.   In the year that a company acquires an asset, the **50% rule** permits only one-half the normal CCA. The full capital cost of aquisitions is added in column 3, but then one-half of the net amount for the year (acquisitions minus proceeds) is taken in column 7, and the account total is reduced by that amount to arrive at a reduced undepreciated capital cost (column 8) from which to calculate the CCA.

The Schedule 8 for Year 3

1 Class number	2 Undepreciated capital cost at the beginning of the year (undepreciated capital cost at the end of the year from last year's CCA schedule)	3 Cost of acquisitions during the year (new property must be available for use)	5 Proceeds of dispositions during the year (amount not to exceed the capital cost)	6 Undepreciated capital cost (column 2 **plus** column 3 **plus** or **minus** column 4 **minus** column 5)	7 50% rule (1/2 of the amount, if any, by which the net cost of acquisitions exceeds column 5)	8 Reduced undepreciated capital cost (column 6 **minus** column 7)	9 CCA rate %	12 Capital cost allowance (column 8 **multiplied** by column 9; or a lower amount)	13 Undepreciated Capital Cost at the end of the year (column 6 **minus** column 12)
10	$28,532	$26,000	$0	$54,532	$13,000	$41,532	30%	$12,459	$42,072

The year 3 CCA = $12,459

The effect of the calculation method is that one-half the capital cost, ($26,000/2) = \$13,000$, is added to the previous year's total of $28,532 to arrive at the CCA for second and subsequent years. No further adjustment is necessary.

Action.   In year 4 no capital assets were acquired or disposed of. What CCA is permitted at the end of the year?

Calculation.   With no additions and no subtractions, the year 4 CCA is just a multiplication of the beginning-of-year undepreciated amount by the CCA rate.

The Schedule 8 for Year 4

1 Class number	2 Undepreciated capital cost at the beginning of the year (undepreciated capital cost at the end of the year from last year's CCA schedule)	3 Cost of acquisitions during the year (new property must be available for use)	5 Proceeds of dispositions during the year (amount not to exceed the capital cost)	6 Undepreciated capital cost (column 2 **plus** column 3 **plus** or **minus** column 4 **minus** column 5)	7 50% rule (1/2 of the amount, if any, by which the net cost of acquisitions exceeds column 5)	8 Reduced undepreciated capital cost (column 6 **minus** column 7)	9 CCA rate %	12 Capital cost allowance (column 8 **multiplied** by column 9; or a lower amount)	13 Undepreciated Capital Cost at the end of the year (column 6 **minus** column 12)
10	$42,072	$0	$0	$42,072	$—	$42,072	30%	$12,622	$29,450

The year 4 CCA = **$12,622**

The Schedule 8 for Year 5

1 Class number	2 Undepreciated capital cost at the beginning of the year (undepreciated capital cost at the end of the year from last year's CCA schedule)	3 Cost of acquisitions during the year (new property must be available for use)	5 Proceeds of dispositions during the year (amount not to exceed the capital cost)	6 Undepreciated capital cost (column 2 **plus** column 3 **plus** or **minus** column 4 **minus** column 5)	7 50% rule (1/2 of the amount, if any, by which the net cost of acquisitions exceeds column 5)	8 Reduced undepreciated capital cost (column 6 **minus** column 7)	9 CCA rate %	12 Capital cost allowance (column 8 **multiplied** by column 9; or a lower amount)	13 Undepreciated Capital Cost at the end of the year (column 6 **minus** column 12)
10	$29,450	$25,000	$850	$53,600	$12,075	$41,525	30%	$12,458	$41,143

Action.   In year 5 the company sells the Ford Bronco for $850 and purchases a computer system for $25,000. What CCA is permitted at the end of year 5?

Calculation.   The Income Tax Act (Table 11-1) lists electronic data processing equipment (personal computers) in Asset Class 10, and so the capital cost is lumped in with the automobiles. In this year there is both an acquisition (column 3) and a disposal (column 5), and it is the net amount that is subject to the 50% rule.

The year 5 CCA = $12,458

Action.   In year 6 the company sells all the remaining vehicles for $9,000. What CCA is permitted at the end of year 6?

The Schedule 8 for Year 6

1 Class number	2 Undepreciated capital cost at the beginning of the year (undepreciated capital cost at the end of the year from last year's CCA schedule)	3 Cost of acquisitions during the year (new property must be available for use)	5 Proceeds of dispositions during the year (amount not to exceed the capital cost)	6 Undepreciated capital cost (column 2 **plus** column 3 **plus** or **minus** column 4 **minus** column 5)	7 50% rule (1/2 of the amount, if any, by which the net cost of acquisitions exceeds column 5)	8 Reduced undepreciated capital cost (column 6 **minus** column 7)	9 CCA rate %	12 Capital cost allowance (column 8 **multiplied** by column 9; or a lower amount)	13 Undepreciated Capital Cost at the end of the year (column 6 **minus** column 12)
10	$41,143	$0	$9,000	$32,143	$—	$32,143	30%	$9,643	$22,500

Calculation.   The proceeds from the sale are entered in column 5 and serve to reduce the total undepreciated capital cost of the Class 10 assets owned to $32,143. From this amount the annual CCA is calculated.

The year 6 CCA = $9,643

Observe that at this time the only Class 10 Asset that the company has is a one-year-old computer system that, if we were to consider it alone, has a undepreciated capital cost or book value of:

$$\text{Book value} = \text{Original capital cost minus Accumulated depreciation}$$
$$= \$25,000 - (\$25,000 \times 30\%)$$
$$= \$17,500$$

However, the amount on which the CCA is calculated is $32,143. This illustrates the difference between the asset class account totals and actual book values. The account totals reflect the transactions for the asset class and are a result of assets that are no longer possessed. Generally, as long as the company possesses a single asset in any one class, in this case a computer or a vehicle, the account remains open and the CCA is calculated annually. If the last asset in a class is sold, then the account books for that asset class are closed and the remaining balance (loss or recapture) is added to that year's income.

Since, in engineering economics, we deal with individual assets and not with asset classes, it is necessary to calculate the CCA amounts for the particular asset under consideration. The capital cost allowance for year n (CCA_n), and the undepreciatiated capital cost at the end of year n (UCC_n) can be calculated by the following formulas:

$$CCA_n = \text{Capital cost allowance for year } n$$
$$P = \text{Asset cost basis}$$
$$d = \text{CCA rate}$$
$$CCA_1 = P\left(\frac{d}{2}\right) \qquad \text{for } n = 1 \tag{11-6}$$

$$CCA_n = Pd\left(1 - \frac{d}{2}\right)(1 - d)^{n-2} \qquad \text{for } n \geq 2 \tag{11-7}$$

$$\text{UCC}_n = \text{Undepreciated capital cost at the end of year } n$$

$$\text{UCC}_n = P\left(1 - \frac{d}{2}\right)(1 - d)^{n-1} \tag{11-8}$$

The undepreciated capital cost is also referred to as the book value (BV) of the asset. This is because it is the value shown in the company's books of account.

DEPRECIATION AND ASSET DISPOSAL

In the normal conduct of business, assets are bought and sold. For income tax purposes in Canada, the cost of acquisition and proceeds from disposal are totalled and added to the appropriate column in Schedule 8 and the calculation of annual permissible CCA is fairly automatic. However, in engineering economy studies, it is often necessary to know about a particular asset—how much depreciation is taken, and what the eventual salvage value (SV) is, that is, the after-tax value at the time of disposal. To calculate this, it is necessary to compare the asset's market value, MV (what a willing buyer pays) with the asset's book value, BV, to determine if the depreciation deductions taken match the actual decrease in value. When they do not, an adjustment is made to the amount of taxes paid.

The precise adjustment can be a complex calculation, depending, as it does, on the time of the disposal, the time the money is realized, and the time period for which the tax return is filed. Here we will consider only the straightforward case, assuming that the asset is disposed of at the end of the year, immediately after the depreciation for that year has been taken, and that the adjustment is included in the calculations for that year. For more complex cases, it is recommended that a tax specialist be consulted.

When an asset is disposed of, the important question is which is larger: (1) the book value (what we show in our accounting records after applying the rules set by the government) or (2) the asset's market value, MV (what a willing buyer pays)? If the book value is lower than the market value, too much CCA has been deducted from taxable income. On the other hand, if the book value is higher than the market value, not enough CCA has been deducted. In either case, the current level of taxes owed changes.

Recaptured CCA (See Figure 11-7) occurs when an asset is sold for more than its current book value. If more than the original cost basis is received, only the amount up to the original cost basis is *recaptured depreciation*. **Recaptured depreciation** represents the over-expense in depreciation that has been claimed. In other words we have taken too much expense for the asset's 'loss in value'.

Loss on disposal (See Figure 11-8) occurs when the market value is less than book value. In the accounting records we've exchanged an asset worth its book value for something less—which is a loss. In this case a company has not claimed enough depreciation expense.

Capital gains (See Figure 11-9) occur when the asset is sold for more than its original cost. The excess over the original cost basis is the capital gain. As described in Chapter 12, the tax rate on such gains is sometimes lower than for ordinary income, but there are complicated rules that consider how long the investment has been held. In most engineering economic analyses, capital gains only occur only on land because business and production equipment and facilities almost always lose value over time. Capital gains are much more likely on non-depreciated assets like stocks, bonds, real estate, jewellery, art, and collectibles.

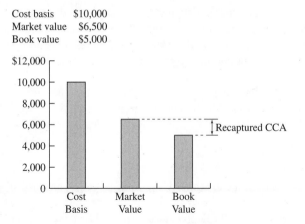

Cost basis $10,000
Market value $6,500
Book value $5,000

If Cost basis > Market value > Book value, there is Recaptured CCA
Recaptured CCA = Market value minus Book value = $1500

FIGURE 11-7 Recaptured CCA.

Cost basis $10,000
Market value $2,250
Book value $5,000

If Book value > Market value, there is a *loss on disposal*.
Loss on disposal = Book value minus Market value = $2750

FIGURE 11-8 Loss on disposal.

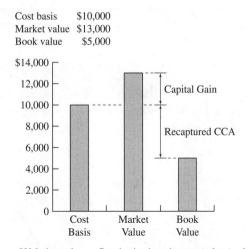

Cost basis $10,000
Market value $13,000
Book value $5,000

If Market value > Cost basis, there is a *capital gain* plus *recaptured depreciation*
Capital gain = Market value minus Cost basis = $3000
Recaptured depreciation = Cost basis minus Book value = $5000

FIGURE 11-9 Capital gain.

NATURAL RESOURCE ALLOWANCES

Depletion refers to the consumption of exhaustible natural resources as a result of their removal. Since depletion covers such things as mineral properties, oil and gas wells, and standing timber, removal may take the form of digging up metallic or non-metallic minerals, producing petroleum or natural gas from wells, or cutting down trees.

In Canada, the federal and provincial governments collect income tax, but each province owns is natural resources and so the tax and royalty regimes surrounding these activities vary across the country. In the United States depletion is recognized for income taxes for the same reason depreciation is—capital investment is being consumed or used up. Thus a portion of the gross income should be considered a return of the capital investment. For mining companies in Canada, the use of an allowance (the lesser of 25% of resource profits or the amount of earned depletion) was discontinued in federal legislation in 1990, although existing mines were grandfathered in.

The following sections illustrate the usual approaches, but for a specific study it will be necessary to consult tax experts to determine which particular treatment applies.

There are two distinct methods of calculating depletion: *percentage depletion* and *cost depletion*.

Percentage Depletion

Percentage depletion is sometimes used for mineral property and some oil or gas wells. The allowance is a certain percentage of the property's gross income during the year. This is an entirely different concept than depreciation. Unlike depreciation, which allocates cost over useful life, the percentage depletion allowance is based on the property's gross income.

Since percentage depletion is computed on the *income* rather than the cost of the property's, the total depletion *may exceed the cost of the property*. In computing the *allowable percentage depletion* on a property in any year, United States regulations specify that the *percentage depletion allowance* cannot exceed 50% of the property's taxable income computed without the depletion deduction. The percentage depletion calculations are illustrated by Ex. 11-7.

EXAMPLE 11-7

A coal mine has a gross income of $250,000 for the year. Mining expenses equal $210,000. Compute the allowable percentage depletion deduction.

SOLUTION

From Table 11-2, coal has a 10% depletion allowance. The percentage depletion deduction is computed from gross mining income. Then the taxable income must be computed. The allowable percentage depletion deduction is limited to the computed percentage depletion or 50% of taxable income, whichever is smaller.

TABLE 11-2 US Percentage Depletion Allowance for Selected Items

Type of Deposit	Percentage
Lead, zinc, nickel, sulphur, uranium	22
Oil and gas (small producers only)	15
Gold, silver, copper, iron ore	15
Coal and sodium chloride	10
Sand, gravel, stone, clam and oyster shells, brick, and tile clay	5
Most other minerals and metallic ores	14

Computed Percentage Depletion

Gross income from mine	$250,000
Depletion percentage	$\times$ 10%
Computed percentage depletion	$25,000

Taxable Income Limitation

Gross income from mine	$250,000
Less: expenses other than depletion	−210,000
Taxable income from mine	40,000
Deduction limitation	× 50%
Taxable income limitation	$20,000

Since the taxable income limitation ($20,000) is less than the computed percentage depletion ($25,000), the allowable percentage depletion deduction is $20,000.

Cost Depletion

The calculation of depreciation relied on the cost of an asset, its depreciable life, and its salvage value to apportion the cost *minus* salvage value *over* the depreciable life. In some cases where the asset is used at fluctuating rates, we might use the unit-of-production (UOP) method of depreciation. For mines, oil wells, and standing timber, fluctuating production rates are the usual situation. Thus, *cost depletion* is computed like unit-of-production depreciation; the calculation uses:

1. Property cost
2. Estimated number of recoverable units (tons of ore, cubic metres of gravel, barrels of oil, million cubic feet of natural gas, thousand board-feet of timber, etc.)
3. Salvage value, if any, of the property

EXAMPLE 11-8

Assume that an oil reservoir is estimated to contain 150,000 barrels (bbls) and the investment cost to develop the reserve is $1,250,000. Then the unit depletion rate would be calculated as:

(cost basis of reserve/number of recoverable units) = 1,250,000/150,000 = $8.33 per bbl.

The depletion allowance for a year when 5,000 bbls of oil were produced would be:

Number of units produced × unit depletion rate = 5,000 × $8.33 = $41,667

As previously stated, in the case of mineral property and some oil and gas wells, the depletion deduction can be based on either cost or percentage depletion, depending upon the rules of the jurisdiction.

SPREADSHEETS AND DEPRECIATION

The spreadsheet functions for straight-line, declining-balance, and sum-of-years'-digits depreciation are listed in Table 11-3. All three functions include parameters for *cost* (initial book value), *salvage* (final salvage value), and *life* (depreciation period). Both DDB and SYD change depreciation amounts every year, and so they include a parameter to pick the *period* (year). Finally, DDB includes a *factor*. The default value is 2 for 200% or double-declining balance, but another commonly used value is 1.5 for 150%.

TABLE 11-3 Spreadsheet Functions for Depreciation

Depreciation Technique	Excel
Straight-line	SLN(cost, salvage, life)
Declining-balance	DDB(cost, salvage, life, period, factor)
Sum-of-years'-digits	SYD(cost, salvage, life, period)

The Income Tax Act uses special terms when referring to CCA. These are the most common ones:

Book Depreciation Terms	**Tax Depreciation Terms**
asset	property
depreciation	capital cost allowance
cost base	capital cost
book value	undepreciated capital cost
salvage value	proceeds from disposition

PROBLEMS

11-1 A depreciable asset costs $10,000 and has an estimated salvage value of $1600 at the end of its 6-year depreciable life. Compute the depreciation schedule for this asset by both SOYD depreciation and DDB depreciation.

11-2 A million-dollar oil drilling rig has a 6-year depreciable life and a $75,000 salvage value at the end of that time. Determine which one of the following methods provides the preferred depreciation schedule: DDB or SOYD. Show the depreciation schedule for the preferred method.

11-3 A new machine tool is being purchased for $16,000 and is expected to have a zero salvage value at the end of its 5-year useful life. Compute the DDB depreciation schedule for this capital asset. Assume that any remaining depreciation is claimed in the last year.

11-4 Some special handling devices can be obtained for $12,000. At the end of 4 years, they can be sold for $3,500. Compute the depreciation schedule for the devices by the following methods:
(a) Straight-line depreciation
(b) Sum-of-years'-digits depreciation
(c) Double-declining balance depreciation
(d) CCA depreciation as a Class 43 asset

11-5 A company treasurer is uncertain which of four depreciation methods the firm should use for office furniture that costs $50,000 and has a zero salvage value at the end of a 10-year depreciable life. Compute the

depreciation schedule for the office furniture by the methods listed:
(a) Straight line
(b) Double-declining balance
(c) Sum-of-years'-digits
(d) Capital Cost Allowance

11-6 The RX Drug Company has just purchased a capsulating machine for $76,000. The plant engineer estimates the machine has a useful life of 5 years and little or no salvage value. He will use zero salvage value in the computations. Compute the depreciation schedule for the machine by:
(a) Straight-line depreciation
(b) Sum-of-years'-digits depreciation
(c) CCA at 30% rate

11-7 The Acme Chemical Company bought $45,000 of research equipment, which it believes will have zero salvage value at the end of its 5-year life. Compute the depreciation schedule for the equipment by each of the following methods:
(a) Straight-line
(b) Sum-of-years'-digits
(c) Double declining balance
(d) CCA as Class 8 assets

11-8 Consider a $6500 piece of machinery, with a 5-year depreciable life and an estimated $1200 salvage value. The projected use of the machinery when it was purchased, and its actual production to date, are shown below.

Year	Projected Production (tons)	Actual Production (tons)
1	3500	3000
2	4000	5000
3	4500	[Not
4	5000	yet
5	5500	known]

Compute the machinery depreciation schedule by each of the following methods:
(a) Straight-line
(b) Sum-of-years'-digits
(c) Double-declining balance
(d) Unit of production (for first 2 years only)
(e) CCA—30% rate

11-9 A large profitable corporation purchased a small jet plane for use by the firm's executives in January. The plane cost $1.5 million and will be kept for 5 years. Compute the CCA depreciation schedule for 5 years.

11-10 For an asset that fits into the CCA 'Property that is not included in any other class ...' designation, show in a table the depreciation and book value over a 10-year life of use. The cost basis of the asset is $10,000.

11-11 A company that manufactures food and beverages in the vending industry has purchased some handling equipment that cost $75,000 and will be depreciated as a Class 43 asset. Show in a table the yearly depreciation amount and book value of the asset over 10 years of depreciation life.

11-12 Consider five depreciation schedules:

Year	A	B	C	D	E
1	$45.00	$35.00	$21.75	$58.00	$43.50
2	36.00	20.00	36.98	34.80	30.45
3	27.00	30.00	25.88	20.88	21.32
4	18.00	30.00	18.12	12.53	14.92
5	9.00	20.00	12.68	7.52	10.44
6	8.36				

They are based on the same initial cost, useful life, and salvage value. Identify each schedule as one of the following

- Straight-line depreciation
- Sum-of-years'-digits depreciation
- 150% declining balance depreciation
- Double-declining balance depreciation
- Unit-of-production depreciation

11-13 The depreciation schedule for an asset with a salvage value of $90 at the end of the recovery period has been computed by several methods. Identify the depreciation method used for each schedule.

Year	A	B	C	D	E
1	$323.3	$212.0	$424.0	$194.0	$212.00
2	258.7	339.2	254.4	194.0	$339.20
3	194.0	203.5	152.6	194.0	$203.52
4	129.3	122.1	91.6	194.0	$122.11
5	64.7	122.1	47.4	194.0	$ 73.27
6	61.1				
	970.0	1060.0	970.0	970.0	950.10

11-14 The depreciation schedule for a computer has been arrived at by several methods. The estimated salvage value of the equipment at the end of its 6-year useful life is $600. Identify the resulting depreciation schedules.

Year	A	B	C	D
1	$2114	$2000	$1,600.00	$1233
2	1762	1500	$2,560.00	1233
3	1410	1125	$1,536.00	1233
4	1057	844	$ 921.60	1233
5	705	633	$ 552.96	1233
6	352	475	$ 331.78	1233

11-15 TELCO Corp has leased some industrial land near its plant. It is building a small warehouse on the site at a cost of $250,000. The building will be ready for use January 1. The lease will expire 15 years after the building is occupied. The warehouse will belong at that time to the landowner, with the result that there will be no salvage value to TELCO. The warehouse is to be depreciated by either CCA at 10% or SL depreciation. If the interest rate is 15%, which depreciation method should be selected?

11-16 A profitable company making earthmoving equipment is considering an investment of $100,000 on equipment that will have 5-year useful life and a $20,000 salvage value. If money is worth 10%, which one of the following three methods of depreciation would be preferable?

(a) Straight-line method
(b) Double-declining-balance method
(c) CCA at 30% rate

11-17 The White Swan Talc Company purchased $120,000 of mining equipment for a small talc mine. The mining engineer's report states the mine contains 40,000 cubic metres of commercial-quality talc. The company plans to mine all the talc in the next 5 years as follows:

Year	Production (m³)
1	15,000
2	11,000
3	4,000
4	6,000
5	4,000

At the end of 5 years, the mine will be exhausted and the mining equipment will be worthless. The company accountant must now decide whether to use sum-of-years'-digits depreciation or unit-of-production depreciation. The company considers 8% to be a reasonable time value of money. Compute the depreciation schedule for each of the two methods. Which method would you recommend that the company adopt? Show the computations to justify your decision.

11-18 For its fabricated metal products, the Able Corp. is buying $10,000 worth of special tools that have a 4-year useful life and no salvage value. Compute the depreciation charge for the *second* year by each of the following methods:

(a) DDB
(b) Sum-of-years'-digits
(c) CCA for Class 12 assets

11-19 Use a spreadsheet to solve 11-8.

11-20 Use a spreadsheet to solve 11-15.

11-21 Use Canadian tax depreciation for each of the assets, 1–3, to calculate the items (a)–(c).

1. A light general-purpose truck used by a delivery business, cost = $17,000.
2. Production equipment used by a Detroit automaker to produce vehicles, cost = $30,000.
3. Cement buildings used by a construction firm, cost $130,000.

(a) The asset class

(b) The depreciation deduction for Year 3
(c) The book value (UCC) of the asset after 6 years

11-22 On July 1, Sarah Engineer paid $600,000 for a commercial building and an additional $150,000 for the land on which it stands. Four years later, also on July 1, she sold the property for $850,000. Compute the CCA depreciation for each of the 5 calendar years during which she had the property and the gain or loss on disposal.

11-23 A group of investors has formed Trump Corporation to buy a small hotel. The asking price is $150,000 for the land and $850,000 for the hotel building. If the purchase takes place in June, compute the CCA depreciation for the first three calendar years. Then assume the hotel is sold in June of the fourth year, and compute the UCC on disposal.

11-24 Use a spreadsheet to solve problem 11-23.

11-25 A company is considering buying a new piece of machinery. A 10% interest rate will be used in the computations. Two models of the machine are available.

	Machine I	Machine II
Initial cost	$80,000	$100,000
End-of-useful-life salvage value, S	20,000	25,000
Annual operating cost	18,000	15,000 first 10 years 20,000 thereafter
Useful life, in years	20	25

(a) Determine which machine should be purchased, based on equivalent uniform annual cost.
(b) What is the capitalized cost of Machine I?
(c) Machine I is purchased and a fund is set up to replace Machine I at the end of 20 years. Compute the required uniform annual deposit.
(d) Machine I will produce an annual saving of material of $28,000. What is the rate of return if Machine I is installed?
(e) What will the book value of Machine I be after 2 years if sum-of-years'-digits depreciation is used?
(f) What will the book value of Machine II be after 3 years if double-declining balance depreciation is used?
(g) Assuming that Machine II is a Class 43 asset, what would the CCA depreciation be in the third year?

11-26 Equipment costing $20,000 that is a CCA 30% asset is disposed of during the second year for $14,000.

Calculate any depreciation recapture, losses, or capital gains associated with disposal of the equipment.

11-27 An asset with a 30% CCA rate costs $50,000 and was purchased on January 1, 2001. Calculate any depreciation recapture, ordinary losses, or capital gains associated with selling the equipment on December 31, 2003, for $15,000, $25,000, and $60,000. Consider two cases of depreciation for the problem: if CCA method is used, and if straight-line depreciation over an 8-year life is used with a $10,000 salvage value.

11-28 When a major highway was to be constructed nearby, a farmer realized that a dry streambed running through his property might be a valuable source of sand and gravel. He shipped samples to a testing laboratory and learned that the material met the requirements for certain low-grade fill material. The farmer contacted the highway construction contractor, who offered 65¢ per cubic metre for 45,000 cubic metres of sand and gravel. The contractor would build a haul road and would use his own equipment. All activity would take place during a single summer.

The farmer hired an engineering student for $2500 to count the truckloads of material hauled away. The farmer estimated that 2 acres of streambed had been stripped of the sand and gravel. The 640-acre farm had cost him $300 per acre, and the farmer felt the property had not changed in value. He knew that there had been no use for the sand and gravel before the construction of the highway, and he could foresee no future use for any of the remaining 50,000 cubic metres of sand and gravel. Determine the farmer's depletion allowance. (*Answer:* $1462.50)

11-29 Mr H. Salt purchased a 1/8 interest in a producing oil well for $45,000. Recoverable oil reserves for the well were estimated at that time at 15,000 barrels, 1/8 of which represented Mr Salt's share of the reserves. During the subsequent year, Mr Salt received $12,000 as his 1/8 share of the gross income from the sale of 1000 barrels of oil. From this amount, he had to pay $3000 as his share of the expense of producing the oil. Compute the depletion allowance that US tax rules would allow Mr Salt for the year. (*Answer:* $3000)

11-30 A heavy construction firm has been awarded a contract to build a large concrete dam. It is expected that a total of 8 years will be needed to complete the work. The firm will buy $600,000 worth of special equipment for the job. During the preparation of the job cost estimate, the following utilization schedule was computed for the special equipment:

Year	Utilization (hr/yr)	Year	Utilization (hr/yr)
1	6000	5	800
2	4000	6	800
3	4000	7	2200
4	1600	8	2200

It is estimated that at the end of the job, the equipment can be sold at auction for $60,000.

(*a*) Compute the sum-of-years-digits' depreciation schedule.

(*b*) Compute the unit-of-production depreciation schedule.

11-31 Some equipment costs $1000, has a 5-year depreciable life, and will have an estimated $50 salvage value at the end of that time. You have been assigned the problem of determining whether to use straight-line or SOYD depreciation. At a 10% interest rate, which is the preferred depreciation method for this profitable corporation? Use a spreadsheet to show your computations of the difference in present worths.

11-32 The FOURX Corp. has purchased $12,000 of experimental equipment. The anticipated salvage value is $400 at the end of its 5-year depreciable life. This profitable corporation is considering two methods of depreciation: sum-of-years'-digits and double-declining balance. If it uses 7% interest in its comparison, which method do you recommend? Show computations to support your recommendation. Use a spreadsheet to develop your solution.

11-33 Given the data in Problem 11-16, use a spreadsheet function to compute the CCA depreciation schedule. Show the total depreciation taken (= sum( )) as well as the PW of the depreciation charges discounted at the MARR%.

11-34 Office equipment whose initial cost is $100,000 has an estimated actual life of 6 years, with an estimated salvage value of $10,000. Prepare tables listing the annual costs of depreciation and the book value at the end of each 6 years, based on the straight-line, sum-of-years'-digits, and CCA depreciation. Use spreadsheet to calculate the depreciation amounts.

11-35 You are equipping an office. The total office equipment will have a first cost of $1,750,000. You expect the equipment will last 10 years. Use a spreadsheet to compute the CCA depreciation schedule.

CHAPTER 12

After-Tax Cash Flows

On with the Wind

For at least three decades, environmental activists have bemoaned North Americans' dependence on oil as our primary source of energy. Not only are there grave concerns about the effect of fossil fuel consumption on the global environment, but the war in Iraq has again highlighted the precariousness of relying on an unstable region of the world for so much of our energy.

One solution to this dilemma is to rely more on renewable sources of energy, such as solar power and wind. The technology for such energy sources has been around for many years, and if good intentions were all it took, we'd be getting much of our electricity from windmills.

But, of course, the transition to greater use of wind power would require a significant investment in infrastructure, especially costly wind turbines. And few investors are willing to commit their money without a solid expectation of a competitive return.

Until fairly recently, cost factors kept wind energy from becoming an attractive investment. As recently as the late 1980s, wind-generated power cost roughly twice as much to produce as energy from conventional sources.

In the past few years, however, wind energy has decreased dramatically in price. The American Wind Energy Association (AWEA) reports that many modern wind plants can now produce power for less than 5 cents per kilowatt hour, making them competitive with conventional sources. Not surprisingly, investment in wind power has also increased substantially.

How did this happen? In part, it was driven by advances in wind turbine technology. But changes in tax policy can also alter investors' behaviour. For instance, in the United States, development of wind power was helped significantly by the Energy Policy Act of 1992, which allowed electricity suppliers a 'production tax credit' of 1.5 cents per kilowatt hour (later adjusted to 1.7 cents to account for inflation).

This tax credit was a key incentive to the American wind power industry, which, like all energy producers, must expend large sums on capital assets. During 2001, for instance, energy producers added almost 1700 megawatts of wind-generating capacity—enough to power nearly half a million homes.

But what the government giveth, the government can take away. When the production tax credit briefly expired at the end of 2001, an estimated $3 billion worth of wind projects were suspended, and hundreds of workers were laid off.

Fortunately for the industry, the credit was subsequently re-introduced, and as a result, the AWEA anticipated that another 1500 to 1800 megawatts of wind-generating capacity would come on line.

After Completing This Chapter...

The student should be able to:

- Calculate *taxes due* or *taxes owed* for both individuals and corporations.
- Understand the incremental nature of the individual and corporate tax rates used for calculating taxes on income.
- Calculate a combined income tax rate for provincial and federal income taxes and select an appropriate tax rate for engineering economic analyses.
- Use an *after-tax table* to find the after-tax cash flows for a prospective investment project.
- Calculate after-tax measures of merit, such as present worth, annual worth, payback period, internal rate of return, and benefit-cost ratio, from developed after-tax cash flows.
- Evaluate investment alternatives on an after-tax basis, including asset disposal.
- Use spreadsheets for solving after-tax economic analysis problems.

QUESTIONS TO CONSIDER

1. The wind energy production tax credit discussed in the vignette has become controversial among some commentators, who see it as 'corporate welfare'. These critics contend that the government should not be in the business of encouraging wind energy—or any other type of energy source, for that matter. Instead, they argue, the market should determine whether alternative energy sources such as wind power succeed or fail. Is this a valid argument?

2. Proponents of the wind production tax credit counter that the energy market has never been genuinely 'free'. They point out that other types of energy production (notably coal, nuclear power, and oil) have obtained large government subsidies over the years. In fact, these proponents argue, carbon-based sources enjoy a tremendous 'hidden subsidy' because the environmental and health effects of these more polluting energy sources are not paid for by coal and oil producers but are borne by society at large. Is this a valid argument?

3. Developing a wind power project takes many years and requires the commitment of large sums of investment capital before the project begins to return a profit. What is the effect on investment when a tax credit is allowed to expire or is extended for periods of only a few years?

4. Using the Internet, can you determine how Canadian tax rates changed throughout the course of the past century? How has this affected the value of tax credits to industry?

An unpleasant fact of life is that the only inevitable things are death and taxes. In this chapter we will examine the structure of income taxes in Canada. There is, of course, a wide variety of taxes ranging from sales taxes (GST and PST) to gasoline taxes, property taxes, provincial and federal income taxes, and so forth. Here we will concentrate our attention on income taxes, because they often have a direct and meaningful effect on the economic viability of an engineering project.

First, we must understand the way in which taxes are imposed. Since the previous chapter concerning depreciation is an integral part of this analysis, it is essential that the principles covered there be well understood. Then, having understood the mechanism of depreciation, we will see how income taxes affect our economic analysis. The various analysis techniques will be used in examples of after-tax calculations.

A Partner in the Business

Probably the most straightforward way to understand income taxes is to consider the government as a partner in every business activity. As a partner, the government shares in the profits from every successful venture. And in a somewhat more complex way, the government shares in the losses of unprofitable ventures. The tax laws are complex, and it is not our purpose to explain them fully. Instead, we will examine the fundamental concepts of the income tax laws—and we must recognize that there are exceptions and variations to almost every statement we shall make!

Calculation of Taxable Income

At the mention of income taxes, one can visualize dozens of elaborate and complex calculations. And there is some truth to that, for there can be all sorts of complexities in the computation of income taxes. Yet some of the difficulty is removed when one defines income taxes as just another type of disbursement. Our economic analysis calculations in prior chapters have dealt with all sorts of disbursements: operating costs, maintenance, labour and materials, and so forth. Now we simply add one more prospective disbursement to the list—income tax.

Taxable Income of Individuals

The amount of federal income tax to be paid depends on taxable income and the income tax rates. Therefore, our first concern is the definition of taxable income. To begin, one must compute his or her *taxable income*:

Personal Income Tax:

Gross income = Wages and salary
+ Interest income
+ Dividend income
+ Rent and royalty income
+ Capital gains
+ Others

Deductions = Retirement plan contribution
+ Union and professional dues
+ Child care expenses
+ Attendant care expenses
+ Business investment losses
+ Moving expenses
+ Alimony and maintenance
+ Interest on money borrowed for investment and others

Gross income − Deductions = Taxable income

To calculate the amount of taxes, it is necessary to multiply the taxable income by the correct tax rates. Federal tax rates are called 'progressive' because the larger the taxable income, the larger the percentage that is required in taxes. Table 12-1 shows the federal tax rates for the year 2004.

TABLE 12-1 Individual Federal Income Tax Structure (2004 rates)

Taxable Income	Tax Rate
First $35,000	16%
Amount between $35,001 and $70,000	22%
Amount between $70,001 and $113,804	26%
Amount above $113,805	29%

From time to time, Parliament might amend the Income Tax Act and change the rates. The breakpoints, amounts at which you switch from one rate to the higher, are changed every year because they are tied to the cost of living. (See also Chapter 14, on inflation.)

From the federal income tax, individuals are allowed to subtract non-refundable tax credits. There is a personal exemption amount, and credits are also given for items such as post-secondary tuition and Canada Pension Plan contributions. The credit is calculated by multiplying the total amount to be credited by 16%.

The total amount of federal income tax is calculated by applying the rates in Table 12-1 to the taxable income, and then subtracting any tax credits. The result is called the **basic federal tax**.

Provincial tax rates vary from province to province, and usually they are of the progressive type. The exception is in Alberta, which uses a flat tax scheme. In 2004 in Alberta the rate was 10% and it applies to taxable income.

Two concepts that we will use extensively in this and subsequent chapters are **average tax rate** and **marginal tax rate**. The average tax rate is just the ratio of total taxes payable to taxable income. Total taxes payable includes provincial and federal taxes as well as all surtaxes. The marginal rate refers to the tax bracket or step that you are in, and it is the rate that will be charged on the next dollar made.

$$\text{Average tax rate} = \frac{\text{total taxes payable}}{\text{taxable income}}$$

Marginal tax rate = the tax rate that applies to the next taxable dollar

Examples 12-1 to 12-3 look at the different average and marginal rates in Alberta, Manitoba, and Ontario.

Personal Tax in Alberta (2004)

EXAMPLE 12-1

An Albertan has a yearly salary of $75,000 and non-refundable tax credits of $8000. Find:

(a) total taxes payable
(b) average tax rate
(c) marginal tax rate

SOLUTION

Data

Taxable Income	**$75,000.00**
Non-refundable tax credit	**$ 8,000.00**

2004 Federal Personal Rates   Tax per Level

		On the first	$ 35,000.00	16.00%	**$5,600.00**
from	$ 35,000.00	to	$ 70,000.00	22.00%	**$ 7,700.00**
from	$ 70,001.00	to	$113,804.00	26.00%	**$ 1,299.74**
above	$113,805.00			29.00%	**$ —**
					$ 14,599.74

Taxable Income	**$ 75,000.00**	Fed. tax =	**$14,599.74**
Non-refundable tax credit	**$ 8,000.00**	× 16% =	$1,280.00
	Total Federal Tax Payable =		**$13,319.74**

2004 Alberta Provincial Tax

       Tax Amount

On all income		10.00%	**$ 7,500.00**
Alberta tax =	**$7,500.00**		
Combined tax =	**$22,099.74**		

Average tax rate = tax payable/taxable income   **29%**

Marginal tax rate = tax payable on next dollar   **36%**

Personal Tax in Manitoba (2004)

EXAMPLE 12-2

A Manitoban has a yearly salary of $75,000 and non-refundable tax credits of $8000.

SOLUTION

Data

Taxable Income	**$75,000**
Non-refundable tax credit	**$ 8,000**

2004 Federal Personal Rates

						Tax per Level
		On the first	$ 35,000.00	16.00%		**$5,600.00**
from	$ 35,000.00	to	$ 70,000.00	22.00%		**$ 7,700.00**
from	$ 70,001.00	to	$113,804.00	26.00%		**$ 1,299.74**
above	$113,805.00			29.00%		$ —
						$ 14,599.74

Taxable Income	**$ 75,000.00**	Fed. tax =	**$14,599.74**
Non-refundable tax credit	**$ 8,000.00**	× 16% =	$1,280.00
	Total Federal Tax Payable =		**$13,319.74**

2004 Manitoba Provincial Tax

					Tax Amount
		On the first	$ 30,544.00	10.90%	**$ 3,329.30**
	$ 30,545.00	to	$ 65,000.00	14.00%	**$ 4,823.70**
	$ 65,001.00			17.4%	**$ 1,739.83**
		Manitoba tax =	**$9,892.82**		
		Combined tax =	**$23,212.56**		

Average tax rate = tax payable/taxable income   **31%**

Marginal tax rate = tax payable on next dollar   **43%**

Personal Tax in Ontario (2004)

EXAMPLE 12-3

An Ontarian has a yearly salary of $75,000 and non-refundable tax credits of $8000.

SOLUTION

Data

Taxable income	**$75,000.00**
Non-refundable tax credit	**$ 8,000.00**

2004 Federal Personal Rates

						Tax per Level
		On the first	$ 35,000.00	16.00%		**$5,600.00**
from	$ 35,000.00	to	$ 70,000.00	22.00%		**$ 7,700.00**
from	$ 70,001.00	to	$113,804.00	26.00%		**$ 1,299.74**
above	$ 113,805.00			29.00%		$ —
						$ 14,599.74

Taxable Income	**$ 75,000.00**	Fed. tax =	**$14,599.74**
Non-refundable tax credit	**$ 8,000.00**	× 16% =	$1,280.00
	Total Federal Tax Payable =		**$13,319.74**

2004 Ontario Provincial Tax

					Tax Amount
		On the first	$33,375.00	6.05%	**$ 2,019.19**
from	$ 33,376.00	to	$66,752.00	9.15%	**$ 3,053.90**
above	$ 66,753.00			11.2%	**$ 920.37**
					$ 5,993.46
Surtax		Surtax on Provincial Tax over			
			$3,685	20.0%	**$ 461.69**
			4864	36.0%	**$ 406.60**

Ontario tax =	**$6,861.75**
Combined tax =	**$20,181.49**

Average tax rate = tax payable/taxable income	**27%**
Marginal tax rate = tax payable on next dollar	**43%**

CORPORATE INCOME TAXES

Engineering economy is usually practised in a corporate environment, and it is corporate income taxes, as opposed to individual income taxes, that are our usual focus.

Just as individuals pay personal tax on the income they have earned, corporations pay tax on the money that they have earned—that is, their net income or profit. Although the calculation is a bit more complex because of credit terms and time lags and difficulties in collection (the cash flow is not always at the same time as the transaction) the corporate accountants apply what are called Generally Accepted Accounting Principals (GAAP) to a corporation's account to try and represent what has actually happened. Depreciation is done according to the rules and forms of capital cost allowance provided by the Canadian Customs and Revenue Agency.

In an effort to be precise, the Income Tax Act defines certain specific accounting concepts. For example, depreciation is called capital cost allowance because it is not really the machine wearing out; rather it is the amount of the original capital cost that the government will allow you to deduct from this year's income to arrive at a taxable income. Special terms are used by CRA to describe property and amounts. The more important ones were listed in Chapter 11 and are repeated here.

Book Depreciation Terms	Tax Depreciation Terms
asset	property
depreciation	capital cost allowance
cost base	capital cost
book value	undepreciated capital cost
salvage value	proceeds from disposition

Unlike personal taxes, corporate taxes are not progressive and do not increase as the amount of income increases. The percentage rate does vary depending upon the type of business and the province, and there is a Small Business Deduction (SBD) that results in a lesser tax being charged on the first $300,000 of taxable income. However, since many engineering economy projects are undertaken by large enterprises, we can be comfortable in the assumption that a single marginal rate, t, is applied to all the taxable income. Start-up firms, special research enterprises, and small businesses are sometimes eligible for

TABLE 12-2 2006 Marginal Tax Rate on Active Business Income over $300,000

	Federal Rate	Provincial Rate	Combined Rate
British Columbia	22.1%	13.5%	**35.6%**
Alberta	22.1	11.5	**33.6**
Saskatchewan	22.1	17.0	**39.1**
Manitoba	22.1	15.0	**37.1**
Ontario	22.1	14.0	**36.1**
Quebec	22.1	8.9	**31.0**
New Brunswick	22.1	13.0	**35.1**
Nova Scotia	22.1	16.0	**38.1**
Prince Edward Island	22.1	16.0	**38.1**
Newfoundland and Labrador	22.1	14.0	**36.1**

INCItME STATEMENT
For ABC Corporation
For the year ending 1999

Operating revenue	OR
Operating costs	−OC
Before-tax cash flow	BTCF
CCA	−CCA
Debt interest	−I
Taxable income	OR − OC − CCA − I
Less income tax (at rate t)	$-t(\text{OR} - \text{OC} - \text{CCA} - \text{I})$
Net profit	$(\text{OR} - \text{OC} - \text{CCA} - \text{I})(1 - t)$

FIGURE 12-1 Simplified income statement formula

a variety of credits and incentives, but those are beyond the scope of this course and can best be explained by a taxation expert.

The terms income and profit are often used interchangeably to describe the amount of money a corporation earns. To avoid confusion, we shall use the word income to describe amounts before the application of income taxes (e.g. net income before tax, taxable income, income before interest and taxes), and profit to mean what is left after income taxes have been subtracted.

Profit is calculated on an accounting statement called the **income statement.** A simplified example is given in figure 12-1. In the right-hand column are formulas that can be used to directly calculate Net Profit and Before-Tax Cash Flow.

Using CCA to Calculate Net Profit

EXAMPLE 12-4

A construction company ($t = 37.62\%$) purchased a new bulldozer for $220,000 (CCA Rate = 30%). The expected annual revenues and costs that will be created by the machine are:

$$\text{Operating revenue (OR)} = \$324{,}000$$

$$\text{Operating cost (OC)} = \$\ 96{,}000$$

Find the net profit for Year 1 and 2.

SOLUTION

Unless specifically stated otherwise, the tax rate, t, given in a question is assumed to be the marginal tax rate, and the income tax payable is the taxable income multiplied by t.

Year 1 CCA = $220,000 × 0.5 × 30% = $33,000 (half-year rule applies)
Year 2 CCA = ($220,000 − 33,000) × 30% = $56,100

Income statements for	Year 1	Year 2
OR	$324,000	$324,000
OC	96,000	96,000
CCA	33,000	56,100
Taxable income	$195,000	$171,900
Less income tax (37.62%)	73,359	64,669
Net profit	**$121,641**	**$107,231**

ACCOUNTING AND ENGINEERING ECONOMY

Engineering economy studies use accounting practices and formulas to determine what will happen, but there are fundamental differences between the two fields. First, accountants are trying to measure what has happened, whereas engineering economists look forward and try to predict what may happen. Second, accountants try to allocate funds in time to attach them to units or products, whereas engineering economy is concerned with **when the cash flows occur**.

The principal accounting statements that deal with cash flows are:

1. Income statement: tells what happened over the past year
2. Cash flow statement: lists the sources and uses of cash

We are concerned with flows, and so a useful comparison is with a pipeline. Figure 12-2 shows money flowing into the firm, and branching off to pay for the different items.

From this diagram we can develop the formulas for calculating the net funds from operations. Beginning at the top, the operating revenue stream is divided into two streams, operating cost and the before-tax cash flow. Thus,

$$\text{Operating revenue (OR)} = \text{Operating cost (OC)} + \text{Before-tax cash flow (BTCF)}$$

To calculate the after-tax cash flow (ATCF), it is first necessary to calculate the taxable income. In the diagram, the BTCF stream is partitioned into interest payments on debt (I), CCA, and taxable income. Thus,

$$\text{BTCF} = \text{Debt interest (I)} + \text{CCA} + \text{Taxable income}$$

rearranging

$$\text{Taxable income} = \text{BTCF} - \text{Debt interest (I)} - \text{CCA}$$

Then, we need to determine the net profit. From the diagram,

$$\text{Taxable income} = \text{Net profit} + \text{Income tax}$$
$$\text{Net profit} = \text{Taxable income} - \text{Income tax}$$

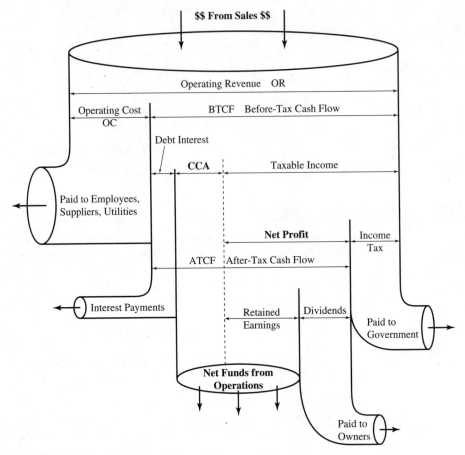

FIGURE 12-2 The operations cash flow pipeline.

where

$$\text{Income tax} = \text{Taxable income} \times \text{Tax rate } (t)$$

So,

$$\textbf{Net profit} = \textbf{Taxable income} \times (1 - t)$$
$$= (\textbf{OR} - \textbf{OC} - \textbf{CCA} - \textbf{I}) \times (1 - t) \qquad (12\text{-}1)$$

This is the same formula that was developed from the income statement.

When the corporation borrows money (called debt capital), it commits itself to making interest payments. The money is received as a loan, and the lender does not own any portion of the corporation. Interest payments are not affected by the profitability (or lack thereof) of the corporation, but the profitability can be very much affected by the interest payments. Borrowing and repayment affect the cash flow in two ways. The interest is a cost of doing business and so is deducted from revenue to arrive at net profit.

In the pipeline diagram, Figure 12-2:

$$\text{After-tax cash flow (ATCF)} = \text{Net profit} + \text{CCA} + \text{Debt interest (I)}$$

Substituting, we get:

$$\text{ATCF} = \text{Taxable income} \times (1 - t) + \text{CCA} + \text{I}$$
$$\text{ATCF} = [\text{BTCF} - \text{I} - \text{CCA}] \times (1 - t) + \text{CCA} + \text{I}$$
$$\text{ATCF} = \text{BTCF}(1 - t) - \text{I} \times (1 - t) - \text{CCA} \times (1 - t) + \text{CCA} + \text{I}$$
$$\text{ATCF} = [\text{OR} - \text{OC}](1 - t) + \text{I} \times t + \text{CCA} \times t$$
$$\mathbf{ATCF = OR(1 - \mathit{t}) - OC(1 - \mathit{t}) + I \times \mathit{t} + CCA \times \mathit{t}} \qquad (12\text{-}2)$$

Corporations are owned by their shareholders. The corporation sells shares, and the shares entitle the buyer to a percentage of ownership of the corporation. The money that the corporation receives as a result of this sale is called equity. **Dividends** are amounts of money, a portion of the profit, that are paid to the shareholders.

To find the net cash from operations,

$$\text{Net cash from Operations} = \text{ATCF} - \text{I} - \text{Dividends}$$
$$\text{Net cash from Operations} = \text{OR}(1 - t) - \text{OC}(1 - t) + \text{I} \times t$$
$$+ \text{CCA} \times t - \text{I} - \text{Dividends}$$
$$\mathbf{Net\ cash\ from\ Operations = OR(1 - \mathit{t}) - OC(1 - \mathit{t}) - I(1 - \mathit{t})}$$
$$\mathbf{+ CCA \times \mathit{t} - Dividends} \qquad (12\text{-}3)$$
$$\mathbf{Net\ cash\ from\ Operations = (1 - \mathit{t})[OR - OC - I] + CCA \times \mathit{t} - Dividends}$$
$$\mathbf{= Net\ profit + CCA - Dividends} \qquad (12\text{-}4)$$

But operations are not the only source or use of cash for a corporation. Sometimes the corporation will buy or sell its production equipment, or borrow money, or repay debt, or issue or repurchase shares. In Figure 12-3, a large tub is shown below the pipeline to catch all the cash flows and to store the excess cash.

The pipes coming into the tub are sources, and the drainpipes are uses. The pipes at the top show that the net funds from operations, new equity, new debt, and proceeds from disposal of assets all add to the pool of funds available to the corporation.

The drains on funds are the repurchase of equity, the repayment of debt, the purchase of assets. Thus, a firm's net cash flow is:

$$
\begin{aligned}
\text{Net cash flow} = \ &\text{Net cash from operations} \\
&+ \text{New equity} \\
&+ \text{New debt} \\
&+ \text{Proceeds from asset disposal} \qquad (12\text{-}5) \\
&- \text{Repurchase of equity} \\
&- \text{Repayment of debt (principal)} \\
&- \text{Purchase of assets}
\end{aligned}
$$

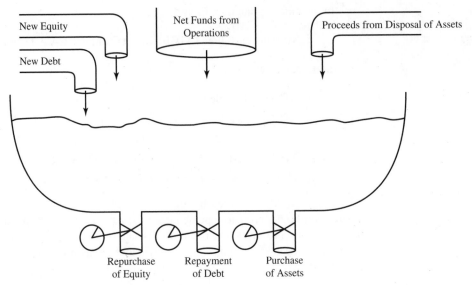

FIGURE 12-3 Sources and uses of cash.

Calculating Net Cash Flow

EXAMPLE 12-5

A construction company ($t = 37.62\%$) bought a new bulldozer for $220,000 (CCA Rate = 30%). The expected annual revenues and costs that will be created by the machine are:

Operating revenue (OR) = $324,000

Operating cost (OC) = $ 96,000

Find the **net cash flow** for Years 1 and 2.

SOLUTION

These are the same data as in Example 12-4, and that solution, net profit calculation, is shown in italics. Below the net profit line, we continue the table to show the actual cash flow. We do this by providing lines for each of the items in equations 12-4 and 12-5.

There is a negative cash flow at the start of Year 1 that occurred when we bought the bulldozer. Negative cash flows (cash outflows) are usually shown in parenthesis like ($220,000).

	Cash Flow at the		
	Beginning of Year 1 (time 0)	End of Year 1	End of Year 2
OR		$324,000	$324,000
OC		96,000	96,000
CCA		33,000	56,100
Taxable income		$195,000	$171,900
Less income tax (37.62%)		73,359	64,669
Net profit		**$121,641**	**$107,231**
Calculation of Net Cash Flow			
Net profit		**$121,641**	**$107,231**
+ CCA		33,000	56,100
− Dividends			
+ New equity			
+ Proceeds from asset disposal			
− Repurchase of equity			
− Repayment of debt			
− Purchase of assets	($220,000)		
Net Cash Flow	**($220,000)**	**$154,641**	**$163,331**

Note: Net Profit and Net Cash Flow can be very different numbers. The result of the CCA deduction is to reduce the taxable income, but not the cash flow, so that the actual effect of a CCA amount d is to increase the tax flow by an amount $t \times$ CCA. This is illustrated in the derivation of equation 12-4, which shows that the CCA is added to the net profit to get the net funds.

Acquiring and Disposing of Assets

In an ongoing business, capital assets are acquired and disposed of regularly, and the accounting system deals with this by adding them to and subtracting them from the asset pool. For an economy study of a particular project or item of equipment, when an asset is acquired, the cost basis is paid and becomes part of the initial investment that must meet the MARR criterion. When an asset is sold or otherwise disposed of, there must be some reconciliation to the cash flow in that year of the project. To do this it is necessary to calculate a net salvage value.

The reality of Canadian tax rules is that any *loss on disposal* or *recaptured CCA* is continuously allocated by the declining balance mechanism as long as the account exists (theoretically on to infinity). This was illustrated in Example 11-6. In the next main section, 'Capital Tax Factors', formulae are derived to accommodate this. Therefore, the usual assumption made about asset disposal is that any difference between the book value and the disposal price will continue to be allocated at the regular CCA rate. This is referred to as the **books open** assumption.

When one is working with spreadsheets, the continuing depreciation is not a convenient assumption because it stretches the calculations far into the future. For these instances, calculating the recaptured CCA or loss and applying them to the final year's income statement is a reasonable procedure, especially when we are dealing with a small adjustment to a distant estimate. This is tantamount to closing the account book on that asset class and is thus known as the **books closed** assumption.

The exception to both of the above assumptions is the situation where there is a *capital gain*. A capital gain occurs when an asset is sold for more than its cost basis. The tax on capital gains is only one-half the marginal tax rate. Capital gains tax must be paid at the time the asset is sold and the gains are realized. (Review Figures 11-7 to 11-9.)

Net Salvage of Land—Capital Gain

EXAMPLE 12-6

Five years ago, anticipating expansion, the XYZ Company bought the lot next to their current factory for $2,300,400. Over the ensuing period they modified their production system and began to make extensive use of outsourcing. Thus, despite increasing sales, they found they used less space, not more. Consequently they sold the lot and, after paying for advertising, legal fees, and commissions, realized a sum of $3,427,958.25. If the company's marginal tax rate is 40%, what is the *net salvage value* of the land?

SOLUTION

$$\text{Capital gain} = \text{Cost basis} - \text{Realized value}$$
$$= \$3.427,958.25 - \$2,300,400 = \$1,127,558.25$$
$$\text{Capital gains tax} = t \times {}^1\!/_2 \times \text{capital gain}$$
$$= 40\% \times {}^1\!/_2 \times \$1,127,558.25 = \$225,511.64$$
$$\textit{Net salvage value} = \textit{Realized value} - \textit{Capital gains tax}$$
$$= \$3.427,958.25 - \$225,511.64 = \$3,202,446.61$$

Net Salvage—Books Closed Assumption

For the books closed situation, if there is a recapture, there is a tax that must be paid; if there is a loss, there is a tax credit to be received. Both situations can be accommodated in a single formula.

$P = $ Original cost basis of a depreciable asset

$t = $ marginal tax rate

$S = $ salvage value (net proceeds from disposal of the asset)

$B_d = $ book value (UCC) at disposal

$DTE = $ disposal tax effect

$DTE = t \times (B_d - S)$

$NSV = $ Net salvage value = after-tax net proceeds from disposal of an asset

Net salvage value NSV $= S + DTE$

$$= S(1 - t) + B_d t \qquad (12\text{-}6)$$

EXAMPLE 12-7

As a result of the outsourcing, the XYZ company auctioned off its production equipment (Class 39—CCA rate 25%) for $320,000; and its fleet of trucks (Class 10—CCA rate 30%) for $176,000. The cost basis of the equipment was $1,500,000, and the current UCC is $415,283. The cost basis of the trucks was $480,000, and the current UCC is $98,000. If the company's marginal tax rate is 40%, what is the *net salvage value (NSV)* of the equipment and vehicles?

SOLUTION

Equipment

$$DTE = t \times (B_d - S)$$
$$= 40\% \times (\$415,283 - \$320,000)$$
$$= \$38,095$$

$$NSV = S + DTE$$
$$= \$320,000 + \$38,095$$
$$= \$358,095$$

With the equipment there was a *loss on disposal*, which resulted in a tax credit of $38,095, and so the actual *net salvage value* is greater than the selling price.

Trucks

$$DTE = t \times (B_d - S)$$
$$= 40\% \times (\$98,000 - \$176,000)$$
$$= -\$70,000$$

$$NSV = S + DTE$$
$$= \$176,000 + (-\$70,000)$$
$$= \$106,000$$

Since trucks sold for more than their book value, there was *recaptured CCA* and consequent tax liability. Thus the actual *net salvage value* is less than the selling price.

Occasionally we find situations where there are both capital gains and recaptured CCA. In these cases, as was outlined in Chapter 11, you can only recapture up to the amount of the original cost basis. Money received above that is capital gain.

EXAMPLE 12-8

The XYZ Company also had a 1965 Aston Martin Sedan in mint condition that it had been storing in a shed on the land. The car had been bought in 1965 by the company's founder, who, unfortunately was stricken with gout and could not drive it. Thus it had remained parked for the last 39 years, and was depreciated on the books from its original purchase price of $32,000 at a CCA rate of 30%. Since the founder was dead, the current board felt they could safely sell the car and did so at a public auction. It was bought by an eccentric car collector for $225,000. If the company's marginal tax rate is 40%, what is the *net salvage value* (*NSV*) of the Aston Martin?

SOLUTION

Use equation 11-8 to calculate the current book value (UCC) of the car:

$$B_{39} = UCC_{39} = \$32,000(1 - 0.3/2)(1 - 0.3/2)^{39-1}$$

$$= \$56.56$$

$$\text{Tax on recapture} = t \times (B_{39} - S)$$

$$= 40\% \times (\$56.56 - \$32,000)$$

$$= \$12,777.38$$

$$\text{Capital gains tax} = t \times {}^{1}\!/_{2} \times (\text{Realized value} - \text{Cost basis})$$

$$= 40\% \times {}^{1}\!/_{2} \times (\$225,000 - \$32,000)$$

$$= \$38,600$$

$$DTE = \text{Tax on recapture} + \text{Capital gains tax}$$

$$= -\$12,777.38 + -\$38,600$$

$$= -\$51,377.38$$

$$NSV = S + DTE$$

$$= \$225,000 - \$51,377.38$$

$$= \$173,622.62$$

All the variants can be easily accommodated in an Excel worksheet. Equation 12-7 below uses Excel functions to calculate a net salvage value for all situations.

Calculating net salvage with books closed assumption and capital gains formula written with excel functions for inclusion in spreadsheet:

P = Original cost basis of depreciable asset

DTE = disposal tax effect

t = marginal tax rate

S = salvage value

B_{disposal} = UCC at disposal or book value at disposal

$$DTE = t \times IF[(S > P), (B_d - P), (B_d - S)] - \tfrac{1}{2} \times t \times MAX\{(S - P), 0\} \quad (12\text{-}7)$$

$$\text{Net salvage value} = S + DTE \quad (12\text{-}8)$$

CAPITAL TAX FACTORS AND BOOKS OPEN

In the normal course of events, as long as a corporation stays in business, the books stay open and the annual CCA calculation has traces of all the assets that have come and gone. If we consider just one single asset, the basic cash flow pattern for an asset depreciated according to the CCA method, including the one-half year rule, is illustrated in Figure 12-4. The purchase of an asset for an amount P generates an infinite series of depreciation deductions which result in positive cash flows of tax credits.

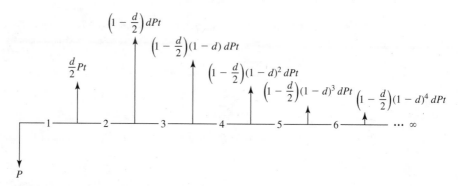

FIGURE 12-4 Cash flow pattern of tax credits due to CCA.

Thus the tax credits form an infinite series. The present worth of the tax credits also forms an infinite series, and we can calculate the sum as follows:

The present worth of the tax credits resulting from an asset of cost basis P is:

$$PW = \frac{\frac{d}{2}Pt}{(1+i)} + \frac{\left(1-\frac{d}{2}\right)dPt}{(1+i)^2} + \frac{\left(1-\frac{d}{2}\right)(1-d)dPt}{(1+i)^3} + \frac{\left(1-\frac{d}{2}\right)(1-d)^2dPt}{(1+d)^4} + \cdots$$

and calculating the sum of the series, this reduces to:

$$PW = P\left[\left(\frac{td}{i+d}\right)\left(\frac{1+i/2}{1+i}\right)\right]$$

and so the present worth of the after-tax cost of an asset is:

$$P\left[1 - \left(\frac{td}{i+d}\right)\left(\frac{1+i/2}{1+i}\right)\right] \tag{12-9}$$

The value in brackets is called the *capital tax factor* and is abbreviated CTF.

The same logic applies when we dispose of an asset. That is, the tax liabilities resulting from the sale of an asset for an amount S produce the pattern shown in Figure 12-5.

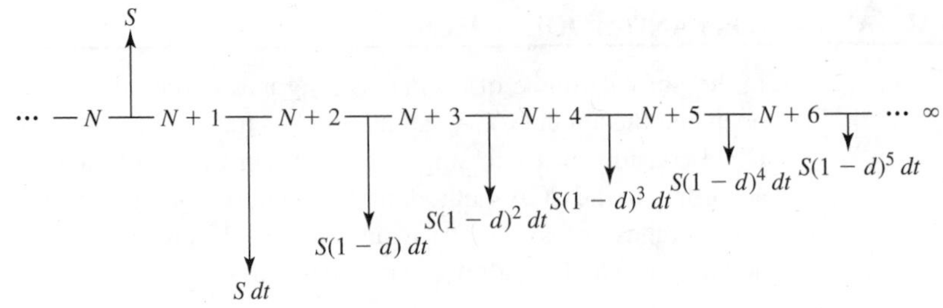

FIGURE 12-5 Salvage value tax effect.

The present worth of these cash flows can likewise be calculated as follows:

$$PW = \frac{Std}{(1+i)} + \frac{S(1-d)td}{(1+i)^2} + \frac{S(1-d)^2td}{(1-i)^3} + \frac{S(1-d)^3td}{(1+i)^4} + \cdots$$

$$= S\left(\frac{td}{i+d}\right)$$

Therefore the present worth of the after-tax cost of salvage at the end of period n is

$$S\left(1 - \frac{td}{i+d}\right) \qquad (12\text{-}10)$$

which is called the *capital salvage factor* and is abbreviated CSF.

EXAMPLE 12-9

A capital expenditure of $246,000 is made for equipment (CCA Class 8, 20% rate). The investment is expected to generate the following before-tax cash flows over the next 10 years.

Year	Before-Tax Cash Flow Amount
1	$ 30,000
2	40,000
3	50,000
4	60,000
5	70,000
6	80,000
7	90,000
8	100,000
9	110,000
10	120,000

At the end of 10 years the equipment is disposed of for $20,000. The marginal tax rate is 35%, and the MARR = 12%. Find the present worth of the investment.

SOLUTION

This problem can be solved in two ways: first, by using the Capital Tax Factors and assuming that the account books stay open; second, by using a computer spreadsheet program and a tabular format. The advantages and shortcomings of each method will be discussed after the example.

Solution Using CTFs:

The before-tax cash flow diagram (values in thousands) is:

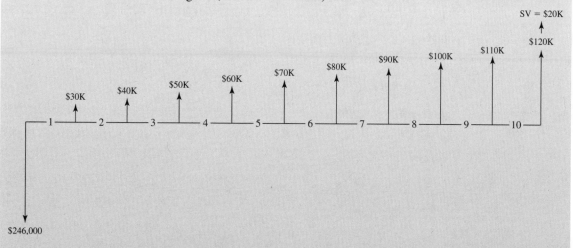

To convert from a before-tax situation to an after-tax situation, it is only necessary to:

- Multiply the cost and revenues (before-tax cash flows) by $(1 - t)$
- Multiply the depreciable capital investment amounts by CTF
- Multiple the proceeds from disposal of capital assets (cash salvage values) by CSF

The corresponding after-tax cash flow diagram therefore is:

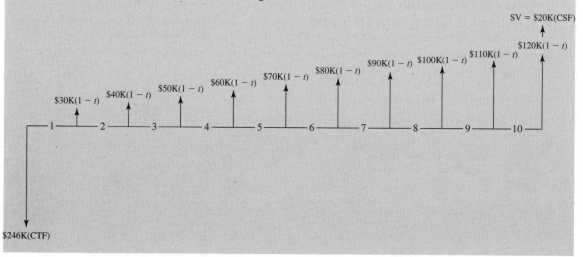

And the calculations are as follows:

$$t = 35\%$$

$$d = 20\%$$

$$i = 12\%$$

$$\text{CTF} = \left[1 - \left(\frac{td}{i+d}\right)\left(\frac{1+i/2}{1+i}\right)\right] = \left[1 - \left(\frac{0.35 \times 0.20}{0.12 + 0.20}\right)\left(\frac{1 + 0.12/2}{1 + 0.12}\right)\right]$$

$$= 0.7930$$

$$\text{CSF} = \left[1 - \left(\frac{td}{i+d}\right)\right]$$

$$= \left[1 - \left(\frac{0.35 \times 0.20}{0.12 + 0.20}\right)\right]$$

$$= 0.7813$$

$$\text{PW} = -\$246\text{K}(\text{CTF}) + \$35\text{K}(1-t)(P/A, i, n) + \$10\text{K}(1-t)(P/G, i, n)$$

$$+ \$20\text{K}(\text{CSF})(P/F, i, n)$$

$$= -246 \times 0.7930 + 35 \times (1 - 0.35) \times (P/A, 12\%, 10)$$

$$+ 10 \times (1 - 0.35) \times (P/G, 12\%, 10) + 20 \times 0.7813 \times (P/F, 12\%, 10)$$

$$= -195,070 + 128,538 + 131,651 + 5,032$$

$$= +\$70,151 \cong \mathbf{\$70,000}$$

Solution in a Spreadsheet

To analyse this in a spreadsheet it is convenient to assume that the account books are closed and a net (after-tax) salvage value can be used. This can be calculated explicity, as follows, or the spreadsheet equation 12-7 can be used.

Data			Calculation of Net Salvage	
$n =$	10			
MARR $= i =$	12%			
$A =$	$ 35,000		UCC at year 10 $=$	$29,716
$DTE =$	$ 10,000		proceeds $S =$	$20,000
equipment $P =$	$246,000		*loss on disposal*	$ 9,716
$S =$	$ 20,000		tax effect $DTE =$	$ 3,401
$d =$	20%		Net salvage $= DTE + S =$	**$23,401**
$t =$	35%			

The spreadsheet tabular format takes the relationships from the corporate cash flow pipeline.

End of YEAR	0	1	2	3	4	5	6	7	8	9	10
BTCF		$35,000	$45,000	$55,000	$65,000	$75,000	$85,000	$95,000	$105,000	$115,000	$125,000
− CCA		$24,600	$44,280	$35,424	$28,339	$22,671	$18,137	$14,510	$ 11,608	$ 9,286	$ 7,429
= Taxable Income		$10,400	$ 720	$19,576	$36,661	$52,329	$66,863	$80,490	$ 93,392	$105,714	$117,571
− Income Tax		$ 3,640	$ 252	$ 6,852	$12,831	$18,315	$23,402	$28,172	$ 32,687	$ 37,000	$ 41,150
= Net Profit		**$ 6,760**	**$ 468**	**$12,724**	**$23,830**	**$34,014**	**$43,461**	**$52,319**	**$ 60,705**	**$ 68,714**	**$ 76,421**
+ CCA		$24,600	$44,280	$35,424	$28,339	$22,671	$18,137	$14,510	$ 11,608	$ 9,286	$ 7,429
= ATCF Operations		$31,360	$44,748	$48,148	$52,169	$56,685	$61,598	$66,828	$ 72,313	$ 78,000	$ 83,850
Cap Investment	$246,000										
+ Net Salvage											23,401
= Net ATCF	$(246,000)	$31,360	$44,748	$48,148	$52,169	$56,685	$61,598	$66,828	$ 72,313	$ 78,000	$107,251

> *Using the Excel NPV function*
> Present worth = **$70,565**

The difference between the two answers ($70,565 − 70,151 = $414) is due to the different assumptions—books open or books closed. For most engineering economy studies that involve long time periods and small salvage values, the difference is not significant.

Both the spreadsheet tabular calculation and the tax factor method have unique advantages. The tax factors are useful when a quick feasibility check is desired and the estimates are based upon either arithmetic or geometric series. But if there is discontinuous cash flows, for example, a major revenue or cost item in a particular year, or when it is necessary to monitor cash and working capital requirements carefully throughout the project, then the spreadsheet provides a more complete picture. The spreadsheet method is also extremely useful when one is doing a 'what-if?' analysis or experimenting with different methods of financing.

WORKING CAPITAL REQUIREMENTS

Supposing you find a product that you can make for 10 cents and sell for 25, you borrow money and buy equipment, you find factory space, you enter into contracts with reputable material suppliers, and you hire a motivated and trained workforce—but still there is more. You need money to operate.

The material suppliers want to be paid when they deliver, or at least within the month. For the workers, payday is Friday. The landlord expects the rent in advance! And even for products that are stamped out by the thousands, there is still the need to inspect them, package them, inventory them, ship them to a retailer, and wait until they are sold to the final consumer and the money comes back to you. There is a time lag between when money is expended in production and when money returns from sales. To cover this time lag it is necessary, at the start of an operation, to inject a sum of money into the operation. This money is referred to as **working capital.**

Often it is only necessary to inject it at the start—in the first few months. Then the returns from sales start coming in and the expenses of today are covered by the cash receipts resulting from the sales of products manufactured two months ago. This situation of cash balancing could, barring seasonal fluctuations, continue for the life of the product. Then, when the product is discontinued and manufacturing ceases, the money continues to come in for several months as the product in the supply chain is used up.

Examples 12-10 and 12-11 illustrate how to deal with working capital requirements.

Initial Working Capital

EXAMPLE 12-10

A company was adding a new product line that needed $80,000 of Class 39 equipment (CCA rate 25%), and initial working capital of $55,000. The product would have production costs of $79,000 a year and annual revenues of $167,000. The product would be manufactured for five years and then discontinued, and then the working capital would be recovered and the equipment sold for $5,000. Find the equivalent uniform annual worth with MARR = 10% and t = 40%.

SOLUTION

The after-tax diagram shows the working capital going in at time zero, and coming back at the end of year 5.

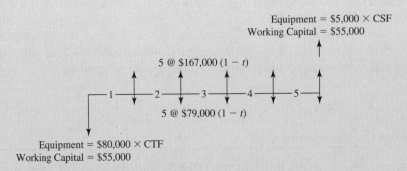

Now we use the annual worth formula on the after-tax cash flows.

$$CTF = [1 - (0.4 \times 0.25/(0.1 + 0.25))(1.05/1.1)] = 0.7273$$

$$CSF = [1 - (0.4 \times 0.25/(0.1 + 0.25))] = 0.7143$$

$$EUAW = -(\$80,000 \times CTF + \$55,000)(A/P, 10\%, 5) + (\$167,000 - \$79,000)(1 - 0.40)$$

$$+ (\$5,000 \times CSF + \$55,000)(A/F, 10\%, 5)$$

$$= -29,858 + 52,800 + 9594$$

$$= 32,536 \cong \mathbf{\$33,000}$$

Spreadsheet Method

Data	
$n =$	5
MARR $= i =$	10%
$R =$	$167,000
$C =$	$ 79,000
WC $=$	$ 55,000
equipment $P =$	$ 80,000
$S =$	$ 5,000
$d =$	25%
$t =$	40%

Disposal tax effect = 6,859.38
Net salvage = 11,859.38

End of Year	0	1	2	3	4	5
Revenue		$167,000	$167,000	$167,000	$167,000	$167,000
− Costs		$ 79,000	$ 79,000	$ 79,000	$ 79,000	$ 79,000
− CCA		$ 10,000	$ 17,500	$ 13,125	$ 9,844	$ 7,383
= Taxable Income		$ 78,000	$ 70,500	$ 74,875	$ 78,156	$ 80,617
− Income Tax		$ 31,200	$ 28,200	$ 29,950	$ 31,263	$ 32,247
= Net Profit		$ 46,800	$ 42,300	$ 44,925	$ 46,894	$ 48,370
+ CCA		$ 10,000	$ 17,500	$ 13,125	$ 9,844	$ 7,383
= ATCF from Operations		$ 56,800	$ 59,800	$ 58,050	$ 56,738	$ 55,753
− Cap investment	$ (80,000)					
+ Net salvage						$ 11,859
Working capital = WC=	**$ (55,000)**					**$ 55,000**
= Net ATCF	$(135,000)	$ 56,800	$ 59,800	$ 58,050	$ 56,738	$122,613

Using the Excel NPV & PMT functions
Present worth = $124,557
    EUAW = **$ 32,858**

Increasing Working Capital Requirement

EXAMPLE 12-11

A company was adding a new product line that required $80,000 worth of Class 39 equipment (CCA rate 25%) and initial working capital of $55,000. *The Working capital requirement will increase at 18% a year.* The product would have production costs of $79,000 a year and annual revenues of $167,000. The product would be manufactured for five years and then discontinued; then the working capital would be recovered and the equipment sold for $5,000. Find the EUAW with MARR = 10% and $t = 40\%$.

SOLUTION

The after-tax diagram shows the initial working capital going in at time zero. Then there are the annual additions to ensure that there is sufficient working capital to meet the increasing requirements. Since the initial working capital remains throughout the project, it is necessary only to add an amount annually to cover the percentage increase.

Working Capital Requirement Increasing at 18%

Year	Working Capital at Beginning of Year	Amount Added at End of Year	Total Available for Next Year
1	$ 55,000	$55,000 × 18% = $9,900	$64,900
2	$ 64,900	($55,000 × 18%)(1+18%) = 11,682	76,582
3	$ 76,582	($55,000 × 18%)(1+18%)^2 = 13,785	90,376
4	$ 90,376	($55,000 × 18%)(1+18%)^3 = 16,266	106,633
5	$106,633		106,633

The increasing working capital thus forms an $n - 1$ long geometric series as shown on the cash flow diagram.

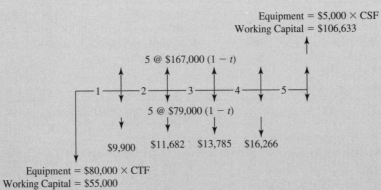

The annual equivalent formula on the after-tax cash flows is the same as in Example 12-11 but with the addition of a geometric series term for 4 periods and an increase amount of working capital recovered at the end of period 5.

CTF = [1 − (0.4 × 0.25/(0.1 + 0.25))(1.05/1.1)] = 0.7273

CSF = [1 − (0.4 × 0.25/(0.1 + 0.25)] = 0.7143

EUAW = − ($80,000 × CTF + $55,000 + ($55,000 × 0.18)(P/A, 18%, 10%, 4))(A/P, 10%, 5)
   + ($167,000 − $79,000)(1 − 0.40) + ($5,000 × CSF + $106,033)(A/F, 10%, 5)

   = −$40,441 + $52,800 + $18,051

   = $30,410 ≅ 30,500

Using the spreadsheet method:

Data

$n =$	5
MARR $= i =$	10%
$R =$	$167,000
$C =$	$ 79,000
WC $=$	$ 55,000
g for WC	**18%**
equipment $P =$	$ 80,000
$S =$	$ 5,000
$d =$	25%
$t =$	40%

End of Year	0	1	2	3	4	5
Revenue		$167,000	$167,000	$167,000	$167,000	$167,000
− Costs		$ 79,000	$ 79,000	$ 79,000	$ 79,000	$ 79,000
− CCA		$ 10,000	$ 17,500	$ 13,125	$ 9,844	$ 7,383
= Taxable income		$ 78,000	$ 70,500	$ 74,875	$ 78,156	$ 80,617
− Income tax		$ 31,200	$ 28,200	$ 29,950	$ 31,263	$ 32,247
= Net profit		$ 46,800	$ 42,300	$ 44,925	$ 46,894	$ 48,370
+ CCA		$ 10,000	$ 17,500	$ 13,125	$ 9,844	$ 7,383
= ATCF from Operations		$ 56,800	$ 59,800	$ 58,050	$ 56,738	$ 55,753
− Cap investment	$ (80,000)					
+ Net salvage						$ 11,859
Working capital = WC =	**$ (55,000)**	**$ (9,900)**	**$ (11,682)**	**$ (13,785)**	**$ (16,266)**	**$106,633**
= Net ATCF	$(135,000)	$ 46,900	$ 48,118	$ 44,265	$ 40,471	$174,245

> *Using the Excel NPV and PMT functions*
> Present worth = $116,496
> EUAW = $ 30,731

LOAN FINANCING

Interest, which is money paid for the use of money, is an expense of doing business. Thus interest is deducted from the before-tax income. The repayment of the loan principal, however, is just the returning of money borrowed and so must come from the after-tax cash flow. Incorporating interest repayment into a capital tax factor formulation is beyond the scope of this text: here we will deal with interest only in the spreadsheet formulation.

Loan with Equal Principal Repayment

EXAMPLE 12-12

A company was adding a new product line that required $80,000 of Class 39 equipment (CCA rate 25%), and initial working capital of $55,000. The product would have production costs of $79,000 a year and annual revenues of $167,000. The product would be manufactured for five years and then discontinued, the working capital would be recovered, and the equipment would be sold for $5,000. To assist in financing the project, the company is borrowing $100,000 at 12% interest. The loan interest is to be paid yearly, and the principal is to be **repaid in five equal annual principal payments.** Fine the equivalent uniform annual worth with MARR = 10% and $t = 40\%$.

SOLUTION

The data are the same as in example 12-11, and so we can use that spreadsheet but with the addition of a line in the income statement for the loan interest, and a line in the Sources and Uses portion for the loans and repayment amounts.

Data

$n =$	5
MARR $= i =$	10%
Loan interest $=$	12%
$R =$	$167,000
$C =$	$ 79,000
WC $=$	$ 55,000
equipment $P =$	$ 80,000
$S =$	$ 5,000
$d =$	25%
$t =$	40%

Using the Excel NPV & PMT functions
Present worth $= \$148,741$
EUAW $= \$ \ 39,238$

End of Year	0	1	2	3	4	5
Revenue		$167,000	$167,000	$167,000	$167,000	$167,000
− Costs		$ 79,000	$ 79,000	$ 79,000	$ 79,000	$ 79,000
− CCA		$ 10,000	$ 17,500	$ 13,125	$ 9,844	$ 7,383
Loan Interest		**$ 12,000**	**$ 9,600**	**$ 7,200**	**$ 4,800**	**$ 2,400**
= Taxable income		$ 78,000	$ 70,500	$ 74,875	$ 78,156	$ 80,617
− Income tax		$ 31,200	$ 28,200	$ 29,950	$ 31,263	$ 32,247
= Net profit		$ 46,800	$ 42,300	$ 44,925	$ 46,894	$ 48,370
+ CCA		$ 10,000	$ 17,500	$ 13,125	$ 9,844	$ 7,383
= ATCF from Operations		$ 56,800	$ 59,800	$ 58,050	$ 56,738	$ 55,753
− Cap investment	$(80,000)					
Loan (repayment)	**$100,000**	**$(20,000)**	**$(20,000)**	**$(20,000)**	**$(20,000)**	**$(20,000)**
+ Net salvage						$ 11,859
Working capital = WC =	$(55,000)					$ 55,000
= **Net ATCF**	$(35,000)	$ 36,800	$ 39,800	$ 38,050	$ 36,738	$102,613

Loan with Equal Annual Repayment

EXAMPLE 12-13

A company was adding a new product line that required $80,000 of Class 39 equipment (CCA rate 25%), and initial working capital of $55,000. The product would have production costs of $79,000 a year and annual revenues of $167,000. The product would be manufactured for five years and then discontinued, the working capital would be recovered, and the equipment would be sold for $5,000. To assist in financing the project, the company is borrowing $100,000 at 12% interest. The loan is to be **repaid in 5 equal annual principal payments.** Find the EUAW with MARR = 10% and $t = 40\%$.

SOLUTION

The Capital Recovery Factor is used to calculate the payment amount.

Data	
$n =$	5
MARR $= i =$	10%
Loan interest =	**12%**
$R =$	$167,000
$C =$	$ 79,000
WC $=$	$ 55,000
equipment $P =$	$ 80,000
$S =$	$ 5,000
$d =$	25%
$t =$	40%

Equal annual payment = $100,000(A/P, 12%, 5)
= **$27,741**

A repayment schedule is then used to calculate the annual principal and interest payments:

Year	Amount Owing at Start of Year	Interest	Principal Repayment	Amount Owing at End of Year
1	$100,000	$12,000	$15,741	$84,259
2	$ 84,259	$10,111	$17,630	$66,629
3	$ 66,629	$ 7,995	$19,745	$46,884
4	$ 46,884	$ 5,626	$22,115	$24,769
5	$ 24,769	$ 2,972	$24,769	$ 0

End of Year	0	1	2	3	4	5
Revenue		$167,000	$167,000	$167,000	$167,000	$167,000
− Costs		$ 79,000	$ 79,000	$ 79,000	$ 79,000	$ 79,000
− CCA		$ 10,000	$ 17,500	$ 13,125	$ 9,844	$ 7,383
Loan interest		**$ 12,000**	**$ 10,111**	**$ 7,995**	**$ 5,626**	**$ 2,972**
= Taxable income		$ 78,000	$ 70,500	$ 74,875	$ 78,156	$ 80,617
− Income tax		$ 31,200	$ 28,200	$ 29,950	$ 31,263	$ 32,247
= Net Profit		$ 46,800	$ 42,300	$ 44,925	$ 46,894	$ 48,370
+ CCA		$ 10,000	$ 17,500	$ 13,125	$ 9,844	$ 7,383
= ATCF from Operations		$ 56,800	$ 59,800	$ 58,050	$ 56,738	$ 55,753
− Cap investment	$(80,000)					
Loan (repayment)	**$ 100,000**	**$ (15,741)**	**$ (17,630)**	**$ (19,745)**	**$ (22,115)**	**$ (24,769)**
+ Net salvage						$ 11,859
Working capital = WC =	$(55,000)					$ 55,000
= **Net ATCF**	$(35,000)	$ 41,059	$ 42,170	$ 38,305	$ 34,623	$ 97,844

> *Using the Excel NPV and PMT functions*
> Present worth = $150,357
> EUAW = **$39,664**

ESTIMATING THE AFTER-TAX RATE OF RETURN

Example 12-13 is the same problem as Example 12-11 except that it is using borrowed money to partially finance the project. The effect of financing was to change EUAW from $30,700 to $39,200. This increase in value results even though the money is borrowed at a rate greater than the MARR value (12% versus 10%)! This is because the interest (which is the cost of borrowing the money) is deductible from revenue before the taxes are calculated, and so the actual rate of borrowing the money is reduced by $(1 - t)$. In this instance, $12\%(1 - 40\%) = 7.2\%$ is the cost of borrowing the money. The equation relating the before- and after-tax cost of debt capital is:

$$i_{dt} = i_d(1 - t) \tag{12-11}$$

where

$$i_{dt} = \text{after-tax cost } t \text{ of debt capital}$$

$$i_d = \text{before-tax cost } t \text{ of debt capital}$$

$$t = \text{marginal tax rate}$$

There is no shortcut to computing the after-tax rate of return from the before-tax rate of return. One possible exception to this statement is when non-depreciable assets and financing are repaid in a lump sum at the end of the project. In this special case, we have

$$\text{After-tax rate of return} = (1 - \text{marginal tax rate}) \times (\text{Before-tax rate of return})$$

This relationship may be helpful for selecting a trial after-tax rate of return when the before-tax rate of return is known. It must be emphasized, however, this relationship is almost always only a rough approximation.

SUMMARY

Since income taxes are part of most situations, no realistic economic analysis can ignore their consequences. Income taxes make the government a partner in many business ventures. Thus the government benefits from all profitable ventures and shares in the losses of unprofitable ventures.

Individuals pay income taxes to both the federal and provincial governments. The personal federal tax is based upon a progressive scale in which the higher the taxable income, the higher the marginal tax rate.

For corporations, taxable income equals gross income minus all ordinary and necessary expenditures (except capital expenditures) and depreciation and depletion charges. The income tax computation (whether for an individual or a corporation) is relatively straightforward; it consists of multiplying the taxable income by the tax rate. The proper rate to use in an economic analysis is the marginal tax rate applicable to the increment of taxable income being considered.

To introduce the effect of income taxes into an economic analysis, the starting point is a before-tax cash flow. Then the depreciation schedule is deducted from appropriate parts of the before-tax cash flow to obtain taxable income. Income taxes are obtained by multiplying taxable income by the proper tax rate. Before-tax cash flow less income taxes equals the after-tax cash flow.

Accounting statements show profit and loss, not the timing of cash flows. Thus it is necessary for an engineering economist to understand the accounting statements, but the information contained in them must be adapted. The economy studies need timing; to convert between systems, use the following equation:

$$\text{Net cash flow} = \text{Net cash from operations}$$

$$+ \text{ New equity}$$

$$+ \text{ New debt}$$

$$+ \text{ Proceeds from asset disposal}$$

$$- \text{ Repurchase of equity}$$

$$- \text{ Repayment of debt (principal and interest)}$$

$$- \text{ Purchase of assets}$$

Working capital also affects the cash flow and cash requirements and must be included.

The process of CCA calculations is such that assets depreciate forever. To make the problems tractable, we must make some assumptions for the tax implications of asset disposal. The books open assumption is easiest for manual calculation, the books closed easiest for spreadsheets. The error introduced by using one or the other is usually minimal.

When one is dealing with non-depreciable assets, there is a nominal relationship between before-tax and after-tax rate of return. It is

$$\text{After-tax rate of return} = (1 - \text{Tax rate})(\text{Before-tax rate of return})$$

There is no simple relationship between before-tax and after-tax rate of return in the more usual case of investments involving depreciable assets.

PROBLEMS

These can be solved by hand, but most can be solved much more easily with a spreadsheet.

12-1 An unmarried taxpayer with no dependents expects a taxable income of $62,000 in a given year. His non-refundable tax credits are expected to be $8,000.
  (a) What will his federal income tax be?
  (b) He is considering an additional activity expected to increase his taxable income. If this increase should be $16,000 and there should be no change in deductions or non-refundable tax credits, what will the increase be in his federal income tax?

12-2 Mary Eve has a $45,000 adjusted gross income and $6000 of non-refundable tax credits. Compute the total tax she would pay as a resident of:
  (a) Alberta
  (b) Ontario

12-3 Bill Jackson, a Manitoban, worked during school and during the first 2 months of his summer vacation. After factoring in his deductions, Bill found that he had a total taxable income of $7800 and non-refundable tax credits of $6000. Bill's employer wants him to work another month during the summer, but Bill had planned to spend the month hiking. If an additional month's work would increase Bill's taxable income by $2600, how much more money would he have after paying the income tax?   (*Answer:* $1900)

12-4 Do Question 12-3 with Bill a resident of Alberta.

12-5 Jane Shay lives in Ontario and operates a management consulting business. The business has been successful and now produces a taxable income of $65,000 a year after all 'ordinary and necessary' expenses and depreciation have been deducted. At present the business is operated as a proprietorship; that is, Jane pays personal federal income tax on the entire $65,000. For tax purposes, it is as if she had a job that paid her a salary of $65,000 a year.

As an alternative, Jane is considering incorporating the business. If she does, she will pay herself a salary of $22,000 a year from the corporation. The corporation will then pay taxes on the remaining $43,000 and retain the balance of the money as a corporate asset. Thus Jane's two alternatives are to operate the business as a proprietorship or as a corporation. Jane is single and has $8000 non-refundable tax credits. Which alternative will result in a smaller total payment of taxes to the government? (*Answer:* Incorporation, $11,889 versus $16,416)

12-6 Do Question 12-5 as if Jane were a resident of Alberta.

12-7 A company wants to set up a new office in a country where the corporate tax rate is as follows: 15% of first $50,000 profits, 25% of next $25,000, 34% of next $25,000, and 39% of everything over $100,000. Executives estimate that they will have gross revenues of $500,000, total costs of $300,000, $30,000 in allowable tax deductions, and a one-time business start-up credit of $8000. What is taxable income for the first year and how much should the company expect to pay in taxes?

12-8 WorldWide oil company purchased two large compressors for $125,000 each. One compressor was installed in the firm's Alberta refinery and is being depreciated by the CCA method. The other compressor was placed in the Oklahoma refinery, where it is being depreciated by sum-of-years'-digits depreciation with zero salvage value. Assume the company pays federal income taxes each year and the tax rate is constant. The corporate accounting department noted that the two compressors are being depreciated differently and wonders whether the corporation will wind up paying more income taxes over the life of the equipment as a result of this. What do you tell them?

12-9 Sole Brother Inc. is a shoe outlet to a major shoe manufacturing industry located in Montreal. Sole Brother uses accounts payable as one of its financing sources. Shoes are delivered to Sole Brother with a 3% discount if payment on the invoice is received within 10 days of delivery. By paying after the 10-day period, Sole is borrowing money and paying (giving up) the 3% discount. Although Sole Brother is not required to pay interest on delayed payments, the shoe manufacturers require that payments not be delayed beyond 45 days after the invoice date. To be sure of paying within 10 days, Sole Brothers decides to pay on the fifth day. Sole has a marginal corporate income tax of 40% (combined provincial and federal). By paying within the 10-day period, Sole is avoiding paying a fairly high price to retain the money owed shoe manufacturers. What would have been the effective annual after-tax interest rate?

12-10 To increase its market share, Sole Brother Inc. decided to borrow $5000 from its banker for the purchase of newspaper advertising for its shoe retail line. The loan is to be paid in four equal annual payments with 15% interest. The loan is discounted 6 points. The 6 'points' is an additional interest charge of 6% of the loan, deducted immediately. This additional interest 6% (5000) = $300 means the actual amount received from the $5000 loan is $4700. The $300 additional interest may be deducted as four $75 additional annual interest payments. What is the after-tax interest rate on this loan?

12-11 Nova Scotia has a corporate tax rate of 16% of taxable income. The federal rate is 22.1%. If a corporation has a taxable income of $150,000, what is the total provincial and federal income tax it must pay?

Compute its combined incremental provincial and federal tax rates.   (*Answers:* $57.150, 38.1%)

12-12 An unmarried woman in British Columbia with a taxable income of about $80,000 has a federal incremental tax rate of 26% and a provincial incremental tax rate of 13.7%. What is her combined incremental tax rate?

12-13 The Lynch Bull investment company suggests that Steven Comstock, a wealthy investor (his incremental income tax rate is 40%), consider the following investment.

Buy corporate bonds on the New York Stock Exchange with a face value (par value) of $100,000 and a 5% coupon rate (the bonds pay 5% of $100,000, which equals $5000 interest per year). These bonds can be purchased at their present market value of $75,000. At the end of each year, Steve will receive the $5000 interest, and at the end of 5 years, when the bonds mature, he will receive $100,000 plus the last $5000 of interest.

Steve will pay for the bonds by borrowing $50,000 at 10% interest for 5 years. The $5000 interest paid on the loan each year will equal the $5000 of interest income from the bonds. As a result Steve will have no net taxable income during the 5 years due to this bond purchase and borrowing money scheme. At the end of 5 years, Steve will receive $100,000 plus $5000 interest from the bonds and will repay the $50,000 loan and pay the last $5000 interest. The net result is that he will have a $25,000 capital gain; that is, he will receive $100,000 from a $75,000 investment. (*Note:* This situation represents an actual recommendation of a brokerage firm.)

(*a*) Compute Steve's after-tax rate of return on this dual bond-plus-loan investment package.

(*b*) What would Steve's after-tax rate of return be if he purchased the bonds for $75,000 cash and *did not* borrow the $50,000?

12-14 Albert Chan decided to buy an old duplex as an investment. After looking for several months, he found a desirable duplex that could be bought for $93,000 cash. He decided that he would rent both sides of the duplex, and determined that the total expected income would be $800 a month. The total annual expenses for property taxes, repairs, gardening, and so forth are estimated at $600 a year. For tax purposes, Al plans to depreciate the building by the capital cost allowance method at a 10% rate and assumes that the building has a 20-year remaining life and no salvage value. Of the total $93,000 cost of the property, $84,000 represents the value of the building and $9000 is the value of the lot. Assume that Al is in the 38% incremental income tax bracket (combined provincial and federal taxes) throughout the 20 years.

In this analysis Al estimates that the income and expenses will remain constant at their present levels. If he buys and holds the property for 20 years, what after-tax rate of return can he expect to receive on his investment, using the following assumptions?

(a) Al believes the building and the lot can be sold at the end of 20 years for the $9000 estimated value of the lot.

(b) A more optimistic estimate of the future value of the lot is that the property can be sold for $100,000 at the end of 20 years.

12-15 Zeon, a large, profitable corporation, is considering adding some automatic equipment to its production facilities. An investment of $120,000 will produce an initial annual benefit of $29,000, but the benefits are expected to decline by $3000 a year, making second-year benefits $26,000, third-year benefits $23,000, and so forth. If the firm uses sum-of-years'-digits depreciation, an 8-year useful life, and $12,000 salvage value, will it obtain the desired 6% after-tax rate of return? Assume that the equipment can be sold for its $12,000 salvage value at the end of the 8 years. Also assume a 46% income tax rate for provincial and federal taxes combined.

12-16 A group of businessmen formed a corporation to lease for 5 years a piece of land at the intersection of two busy streets. The corporation has invested $50,000 in car-washing equipment. They will depreciate the equipment by sum-of-years'-digits depreciation, assuming a $5000 salvage value at the end of the 5-year useful life. The corporation is expected to have a before-tax cash flow, after meeting all expenses of operation (except depreciation), of $20,000 the first year, declining $3000 per year in future years (second year = $17,000, third year = $14,000, etc.). The corporation has other income, and so it is taxed at a combined corporate tax rate of 20%. If the projected income is correct, and the equipment can be sold for $5000 at the end of 5 years, what after-tax rate of return would the corporation receive from this venture? (*Answer:* 14%)

12-17 The effective combined tax rate in an owner-managed corporation is 40%. An outlay of $20,000 for certain new assets is under consideration. It is estimated that for the next 8 years, these assets will be responsible for annual receipts of $9000 and annual disbursements (other than for income taxes) of $4000. After this time, they will be used only for stand-by purposes, and no future excess of receipts over disbursements is estimated.

(a) What is the prospective rate of return before income taxes?

(b) What is the prospective rate of return after taxes if straight-line depreciation can be used to write off these assets for tax purposes in 8 years?

(c) What is the prospective rate of return after taxes if it is assumed that these assets must be written off for tax purposes over the next 20 years, using straight-line depreciation?

12-18 A firm is considering the following investment project:

Year	Before-Tax Cash Flow (thousands)
0	−$1,000
1	+500
2	+340
3	+244
4	+100
5	+100

The project has a 5-year useful life with a $125,000 salvage value, as shown. Double-declining balance depreciation will be used, assuming the $125,000 salvage value. The income tax rate is 34%. If the firm requires a 10% after-tax rate of return, should the project be undertaken?

12-19 The Shellout Corp. has a 34% tax rate and owns a piece of petroleum drilling equipment that costs $100,000 and will be depreciated at a CCA rate of 30%. Shellout will lease the equipment to others and each year receive $30,000 in rent. At the end of 5 years, the firm will sell the equipment for $35,000. What is the after-tax rate of return Shellout will receive from this equipment investment?

12-20 A mining corporation purchased $120,000 of production machinery and depreciated it by SOYD depreciation, a 5-year depreciable life, and zero salvage value. The corporation is a profitable one that has a 34% incremental tax rate. At the end of 5 years the mining company changed its method of operation and sold the production machinery for $40,000. During the 5 years the machinery was used, it reduced mine operating costs by $32,000 a year, before taxes. If the company MARR is 12% after taxes, was the investment in the machinery a satisfactory one?

12-21 An automobile manufacturer is buying some special tools for $100,000. The tools are being depreciated by double-declining balance depreciation using a 4-year depreciable life and a $6250 salvage value. It is expected the tools will actually be kept in service for 6 years and then sold for $6250. The before-tax benefit of owning the tools is as follows:

Year	Before-Tax Cash Flow
1	$30,000
2	30,000
3	35,000
4	40,000
5	10,000
6	10,000
6	$6,250 selling price

Compute the after-tax rate of return for this investment situation, assuming a 46% incremental tax rate. (*Answer:* 11.6%)

12-22 This is a continuation of Problem 12-21. Instead of paying $100,000 cash for the tools, the corporation will pay $20,000 now and borrow the remaining $80,000. The depreciation schedule will remain unchanged. The loan will be repaid by 4 equal end-of-year payments of $25,240. Prepare an expanded cash flow table that takes into account both the special tools and the loan.

(a) Compute the after-tax rate of return for the tools, taking into account the $80,000 loan.

(b) Explain why the rate of return obtained in part (a) is different from the rate of return obtained in Problem 12-21.

Hints: **1.** Interest on the loan is 10%, $25,240 = 80,000 $(A/P, 10\%, 4)$. Each payment is made up of part interest and part principal. Interest portion for any year is 10% of balance due at the beginning of the year.

2. Interest payments are tax deductible (i.e., they reduce taxable income and thus taxes paid). Principal payments are not. Separate each $25,240 payment into interest and principal portions.

3. The Year 0 cash flow is $20,000 (100,000 − 80,000).

4. After-tax cash flow will be before-tax cash flow − interest payment − principal payment − taxes.

12-23 A project will require the investment of $108,000 in equipment (sum-of-years'-digits depreciation with a depreciable life of 8 years and zero salvage value) and $25,000 in raw materials (not depreciable). The annual project income after all expenses except depreciation have been paid is projected to be $24,000. At the end of 8 years the project will be discontinued and the $25,000 investment in raw materials will be recovered. Assume a 34% income tax rate for this corporation. The corporation wants a 15% after-tax rate of return on its investments. Determine by present worth analysis whether this project should be undertaken.

12-24 A profitable incorporated business is considering an investment in equipment having the following before-tax cash flow. The equipment will be depreciated by CCA method with a 25% rate.

Year	Before-Tax Cash Flow
0	−$12,000
1	1,727
2	2,414
3	2,872
4	3,177
5	3,358
6	1,997
$1,000	Salvage value

If the firm wants a 9% after-tax rate of return and its incremental income tax rate is 34%, determine by annual cash flow analysis whether the investment is desirable.

12-25 A salad oil bottling plant can either purchase caps for the glass bottles at 5 cents each or install $500,000 worth of plastic moulding equipment and manufacture the caps at the plant. The manufacturing engineer estimates the material, labour, and other costs would be 3 cents per cap.

(a) If 12 million caps a year are needed and the moulding equipment is installed, what is the payback period?

(b) The plastic moulding equipment would be depreciated by straight-line depreciation using a 5-year useful life and no salvage value. Assuming a 40% income tax rate, what is the after-tax payback period, and what is the after-tax rate of return?

12-26 A firm has invested $14,000 in machinery with a 7-year useful life. The machinery will have no salvage value, as the cost to remove it will equal its scrap value. The uniform annual benefits from the machinery are $3600. For a 47% income tax rate, and sum-of-years'-digits depreciation, compute the after-tax rate of return.

12-27 A firm manufactures padded shipping bags. One hundred bags are packed in a cardboard carton. At present, machine operators fill the cardboard cartons by eye: that is, when the cardboard carton looks full, it is assumed to contain 100 shipping bags. Actual inspection reveals that the cardboard carton may contain anywhere from 98 to 123 bags with an average number of 105.5 bags.

The management has never received complaints from its customers about cartons containing fewer than 100 bags. Nevertheless, management realizes that they are giving away $5^1/_2$% of their output by overfilling the cartons. One solution would be to count the shipping bags to ensure that 100 are packed in each carton. Another solution would be to weigh each filled shipping carton. Underweight cartons would have additional shipping bags added, and overweight cartons would have some shipping bags removed. This would not be a perfect solution because the actual weight of the shipping bags varies slightly. If the weighing is done, it is believed that the average number of bags per carton could be reduced to 102, and almost no cartons would contain fewer than 100 bags. The weighing equipment would cost $18,600. The equipment would be depreciated by straight-line depreciation using a 10-year depreciable life and a $3600 salvage value at the end of 10 years. The $18,600 worth of equipment qualifies for a 10% investment tax credit. One person, hired at a cost of $16,000 per year, would be needed to operate the weighing equipment and to add or remove padded bags from the cardboard cartons. 200,000 cartons would be checked on the weighing equipment each year, with an average removal of 3.5 padded bags per carton with a manufacturing cost of 3 cents a bag. This large profitable corporation has a 50% combined incremental tax rate. Assume a 10-year study period for the analysis and an after-tax MARR of 20%. Compute:

(a) The after-tax present worth of this investment.
(b) The after-tax internal rate of return of this investment.
(c) The after-tax simple payback period of this investment.

12-28 Mr Sam K. Jones, a successful Alberta businessman, is considering erecting a small building on a commercial lot he owns very close to the centre of town. A local furniture company is willing to lease the building for $9000 per year, paid at the end of each year. It is a net lease, which means the furniture company must also pay the property taxes, fire insurance, and all other annual costs. The furniture company will require a 5-year lease with an option to buy the building and land on which it stands for $125,000 at the end of the 5 years. Mr Jones could have the building constructed for $82,000. He could sell the commercial lot now for $30,000, the same price he paid for it. Mr Jones currently has an annual taxable income from other sources of $123,000. He would depreciate the commercial building at a CCA rate of 4%. Mr Jones believes that at the end of the 5-year lease he could easily sell the property for $125,000 ($30,000 for the lot and $90,000 for the building). What is the after-tax present worth of this 5-year venture if Mr Jones uses a 10% after-tax MARR?

12-29 One January, Gerald Adair bought a small house and lot for $99,700. He estimated that $9700 of this amount represented the value of the land. He rented the house for $6500 a year during the 4 years he owned it. Expenses for property taxes, maintenance, and so forth were $500 a year. For tax purposes the house was depreciated at a CCA rate of 10%. At the end of 4 years the property was sold for $105,000 ($90,000 house and $15,000 land). Gerald is married and works as an engineer. He estimates that his incremental combined tax rate is 40%. What after-tax rate of return did he obtain on his investment in the property?

12-30 A corporation with a 34% income tax rate is considering the following investment in research equipment, and has projected the benefits as follows:

Year	Before-Tax Cash Flow
0	−$50,000
1	+2,000
2	+8,000
3	+17,600
4	+13,760
5	+5,760
6	+2,880

Prepare a cash flow table to determine the year-by-year after-tax cash flow assuming CCA rate of 30%.

(a) What is the after-tax rate of return?

(b) What is the before-tax rate of return?

12-31 An engineer is working on the layout of a new research and experimentation facility. Two plant operators will be required. If, however, an additional $100,000 of instrumentation and remote controls were added, the plant could be run by a single operator. The total before-tax cost of each plant operator is projected to be $35,000 per year. The instrumentation and controls will be depreciated at a CCA rate of 20%. If this corporation (the corporate tax rate is 34%) invests in the additional instrumentation and controls, how long will it take for the after-tax benefits to equal the $100,000 cost? In other words, what is the after-tax payback period? (*Answer:* 3.6 years).

12-32 A special power tool for plastic products costs $400 and has a 4-year useful life, no salvage value, and a 2-year before-tax payback period. Assume uniform annual end-of-year benefits.

(a) Compute the before-tax rate of return.

(b) Compute the after-tax rate of return, based on CCA rate of 100% and a 34% corporate income tax rate.

12-33 The Ogi Corporation, a construction company, purchased a pick-up truck for $14,000 and used CCA rate of 30% in the income tax return. During the time the company had the truck, they estimated that it saved $5000 a year. At the end of 4 years, Ogi sold the truck for $3000. The combined income tax rate for Ogi is 45%. Compute the after-tax rate of return for the truck.   (*Answer:* 12%)

12-34 A profitable wood products corporation is considering buying a parcel of land for $50,000, building a small factory at a cost of $200,000 (CCA rate 4%), and equipping it with $150,000 of machinery (CCA rate 30%). Assume the plant is put in service October 1. The before-tax net annual benefit from the project is estimated at $70,000 a year. The analysis period is to be 5 years, and planners assume the total property (land, building, and machinery) will be sold at the end of 5 years, also on October 1, for $328,000 (50,000 for land, 166,470 for building and 111,530 for equipment). Compute the after-tax cash flow at a 34% income tax rate. If the corporation's criterion is a 15% after-tax rate of return, should it proceed with the project?

12-35 A small vessel was purchased by a chemical company for $55,000 and was to be depreciated at a CCA rate of 15% when its requirements changed suddenly, and the chemical company leased the vessel to an oil company for 6 years at $10,000 a year. The lease also provided that the vessel could be purchased at the end of 6 years by the oil company for $35,000. At the end of the 6 years, the oil company exercised its option and bought the vessel. The chemical company has a 34% incremental tax rate. Compute its after-tax rate of return on the vessel.   (*Answer:* 9.6%)

12-36 Xon, a small oil company, bought a new petroleum drilling rig for $1,800,000. Xon will depreciate the drilling rig using a 39% CCA rate. The drilling rig has been leased to a drilling company, which will pay Xon $450,000 a year for 8 years. At the end of 8 years the drilling rig will belong to the drilling company. If Xon has a 34% incremental tax rate and a 10% after-tax MARR, does the investment appear to be satisfactory?

12-37 The profitable Palmer Golf Cart Corp. is considering investing $300,000 in special tools for some of the plastic golf cart components. Executives of the company believe the present golf cart model will continue to be manufactured and sold for 5 years, after which a new cart design will be needed, together with a different set of special tools. The saving in manufacturing costs, owing to the special tools, is estimated to be $150,000 a year for 5 years. Assume CCA rate of 30% for the special tools and a 39% income tax rate.

(a) What is the after-tax payback period for this investment?

(b) If the company wants a 12% after-tax rate of return, is this a desirable investment?

12-38 Michael is contemplating a $10,000 investment in a methane gas generator. He estimates his gross income would be $2000 the first year and increase by $200 each year over the next 10 years. His expenses of $200 the first year would increase by $200 each year over the next 10 years. He would depreciate the generator by CCA rate of 20%. A 10-year-old methane generator has no market value. The income tax rate is 40%. (Remember that recaptured depreciation is taxed at the same 40% rate).

(a) Construct the after-tax cash flow for the 10-year project life.

(b) Determine the after-tax rate of return on this investment. Michael thinks it should be at least 8%.

(c) If Michael could sell the generator for $7000 at the end of the fifth year, would his rate of return be better than if he kept it for 10 years? You don't have to actually find the rate of return, Just do enough calculations to see whether it is higher than that of part (b).

12-39 Katie's Butter and Egg Business is such that she pays an effective tax rate of 40%. Katie is considering the purchase of a new Turbo Churn for $25,000. This churn has an estimated life of 4 years and a salvage value of $5000. The new churn is expected to increase net income by $8000 a year for each of the 4 years of use. If Katie works with an after-tax MARR of 10% and CCA of 30%, should she buy the churn?

12-40 Steve has a house and lot for sale for $70,000. It is estimated that $10,000 is the value of the land and $60,000 is the value of the house. Annthea is purchasing the house on January 1 to rent and plans to own the house for 5 years. After 5 years, it is expected that the house and land can be sold on December 31 for $80,000 ($20,000 for the land and $60,000 for the house). Total annual expenses (maintenance, property taxes, insurance, etc.) are expected to be $3000 a year. The house would be depreciated using a CCA rate of 10%. Annthea wants a 15% after-tax rate of return on her investment. You may assume that Annthea has an incremental income tax rate of 27% in each of the 5 years. Capital gains are taxed at 13.5%. Determine the following:

(a) The annual depreciation.
(b) The capital gain (loss) resulting from the sale of the house.
(c) The annual rent Annthea must charge to produce an after-tax rate of return of 15%.

12-41 Carolyn owns a data processing company. She plans to buy an additional computer for $20,000, use it for 3 years, and sell it for $10,000. She expects that the use of the computer will produce a net income of $8000 a year. The combined federal and provincial incremental tax rate is 45%. Using a CCA rate of 30% and an interest rate of 12%, complete Table 12-41 to determine the net present worth of the after tax cash flow.

12-42 Refer to Problem 12-33. To help pay for the pick-up truck, the Ogi Corp. obtained a $10,000 loan from the truck dealer, payable in four end-of-year payments of $2500 plus 10% interest on the loan balance each year.

TABLE 12-41 Worksheet for Problem 12-41

Year	Before-Tax Cash Flow	Capital Cost Allowance	Taxable Income	Income Tax (45%)	After-Tax Cash Flow	Present Worth (12%)
0	−$20,000					
1	+8,000					
2	+8,000					
3	+8,000					
	+10,000					

Net Present Worth =

(a) Compute the after-tax rate of return for the truck together with the loan. Note that the interest on the loan is tax deductible but the $2500 principal payments are not.
(b) Why is the after-tax rate of return computed in part (a) so much different from the 12.5% obtained in Problem 12-33?

12-43 A store owner, Justin Lang, believes his business has suffered from the lack of adequate customer parking space. Thus, when he was offered an opportunity to buy an old building and lot next to his store, he was interested. He would demolish the old building and make off-street parking for 20 customers' cars. He estimates that the new parking would increase his business and produce an additional before-income tax-profit of $7000 a year. It would cost $2500 to demolish the old building. Mr Lang's accountant advised that both costs (buying the property and demolishing the old building) would be considered to comprise the total value of the land for tax purposes, and it would not be depreciable. Mr Lang would spend an additional $3000 right away to put a light gravel surface on the lot. This expenditure, he believes, may be charged as an operating expense immediately and need not be capitalized. To compute the tax consequences of adding the parking lot, he estimates that his combined incremental income tax rate will average 40%. If Mr Lang wants a 15% after-tax rate of return from this project, how much could he pay to buy the adjoining land with the old building? Assume that the analysis period is 10 years and that the parking lot could always be sold to recover the costs of buying the property and demolishing the old building. (*Answer:* $23,100)

12-44 The management of a private hospital is considering the installation of an automatic telephone switchboard, which would replace a manual switchboard and eliminate the attendant operator's position. The class of service provided by the new equipment is estimated to be at least equal to the present method of operation. To provide telephone service, five operators will work three shifts a day, 365 days a year. Each operator earns $14,000 a year. Company-paid benefits and overhead are 25% of wages. Money costs 8% after income taxes. Combined income taxes are 40%. Annual property taxes and maintenance are $2^1/_2$ and 4% of investment, respectively. Depreciation is 15-year straight-line. Disregarding inflation, how large an investment in the new equipment can be economically justified by savings obtained by eliminating the present equipment and labour costs? The existing equipment has zero salvage value.

12-45 A contractor has to choose one of the following alternatives in performing earthmoving contracts:
(a) Purchase a heavy-duty truck for $13,000. Salvage value is expected to be $3000 at the end of the vehicle's 7-year depreciable life. Maintenance is $1100 a year. Daily operating expenses are $35.
(b) Hire a similar unit for $83 a day. Using a 10% after-tax rate of return, calculate how many days a year the truck must be used to justify its purchase. Base your calculations on straight-line depreciation and a 50% income tax rate. (*Answer:* $91^1/_2$ days)

12-46 The Able Corporation is considering the installation of a small electronic testing device for use in conjunction with a government contract the firm has just won. The testing device will cost $20,000 and have an estimated salvage value of $5000 in 5 years when the government contract is finished. The firm will depreciate the instrument by the sum-of-years'-digits method, using 5 years as the useful life and a $5000 salvage value. Assume that Able pays 50% corporate income taxes and uses 8% *after tax* in economic analysis. What minimum equal annual benefit must Able obtain *before taxes* in each of the 5 years to justify purchasing the electronic testing device? (*Answer:* $5150)

12-47 A house and lot are for sale for $155,000. It is estimated that $45,000 is the value of the land and

$110,000 is the value of the house. If purchased, the house can be rented to provide a net income of $12,000 a year after taking all expenses, except depreciation, into account. The house would be depreciated by straight-line depreciation using a 27.5-year depreciable life and zero salvage value. Mary Silva, the prospective buyer, wants a 10% after-tax rate of return on her investment after considering both annual income taxes and a capital gain when she sells the house and lot. At what price would she have to sell the house at the end of 10 years to achieve her objective, given that the value of the lot is now $75,000? You may assume that Mary has an incremental income tax rate of 27% in each of the 10 years.

12-48 A corporation is considering buying a medium-sized computer that will eliminate a task that must be performed by three shifts a day, 7 days a week, except for one 8-hour shift every week when the operation is shut down for maintenance. At present four people are needed to perform the day and night tasks. Thus the computer will replace four employees. Each employee costs the company $32,000 a year ($24,000 in direct wages plus $8000 a year in other company employee costs). It will cost $18,000 a year to maintain and operate the computer. The computer will be depreciated by sum-of-years'-digits depreciation using a 6-year depreciable life, at which time it will be assumed to have zero salvage value. The corporation has a combined incremental tax rate of 50%. If the firm wants a 15% rate of return after considering income taxes, how much can it afford to pay for the computer?

12-49 A sales engineer has the following alternatives to consider in touring his sales territory.
(a) Buy a new car for $14,500. Salvage value is expected to be about $5000 after 3 years. Maintenance and insurance cost is $1000 in the first year and increases at the rate of $500 a year in subsequent years. Daily operating expenses are $50 a day.
(b) Rent a similar car for $80 a day. Based on a 12% after-tax rate of return, how many days per year must he use the car to justify its purchase? You may assume that this sales engineer is in the 30% incremental tax bracket. Use a CCA rate of 30%.

12-50 A large profitable company, in the 40% tax bracket, is considering the purchase of a new piece of equipment. The new equipment will yield benefits of $10,000 in Year 1, $15,000 in Year 2, $20,000 in Year 3, and $25,000 in Year 4. The CCA rate is 30%. It is expected that the equipment will be sold at the end of the fourth year for 20% of its purchase price. What is the maximum price the company can pay for the equipment if its after-tax MARR is 10%?

12-51 Machine X costs $248,751 and has annual operating and maintenance costs of $9,980. Machine Y costs $264,500 and has annual operating and maintenance cost of $5,120. Both machines are Class 39, which specifies a CCA rate of 25%. The company needs the machines for 11 years, and at the end of year 11 machine X can be sold for $12,257 and machine Y can be sold for 13,033. The Company's MARR is 10%, and its marginal taxation rate is 35%. Do an after-tax analysis to determine which machine should be chosen.

12-52 A small-business corporation is considering whether to replace some equipment in the plant. An analysis indicates there are five alternatives in addition to the do-nothing option, Alt. A. The alternatives have a 5-year useful life with no salvage value. Straight-line depreciation would be used.

Alternatives	Cost (thousands)	Before-Tax Uniform Annual Benefits (thousands)
A	$ 0	$ 0
B	25	7.5
C	10	3
D	5	1.7
E	15	5
F	30	8.7

The corporation has a combined income tax rate of 20%. Prepare a choice table to guide the corporation in selecting the most desirable alternative.

12-53 A corporation with $7 million in annual taxable income is considering two alternatives:

Year	Before-Tax Cash Flow Alt. 1	Alt. 2
0	−$10,000	−$20,000
1–10	4,500	4,500
11–20	0	4,500

Both alternatives will be depreciated by straight-line depreciation assuming a 10-year depreciable life and no salvage value. Neither alternative is to be replaced at the end of its useful life. If the corporation has a tax rate of 34% and a minimum attractive rate of return of 10% *after taxes,* which alternative should it choose? Solve the problem by:
(a) Present worth analysis
(b) Annual cash flow analysis
(c) Rate of return analysis
(d) Future worth analysis
(e) Benefit-cost ratio analysis
(f) Any other method you choose

12-54 Two mutually exclusive alternatives are being considered by a profitable corporation with an annual taxable income between $5 million and $10 million.

Year	Before-Tax Cash Flow Alt. A	Alt. B
0	−$3000	−$5000
1	1000	1000
2	1000	1200
3	1000	1400
4	1000	2600
5	1000	2800

Both alternatives have a 5-year useful and depreciable life and no salvage value. Alternative A would be depreciated by sum-of-years'-digits depreciation, and Alt. B by straight-line depreciation. If the MARR is 10% after taxes, and the tax rate is 34%, which alternative should be chosen? (*Answer:* Alt. B)

12-55 A large profitable corporation is considering two mutually exclusive capital investments:

	Alt. A	Alt. B
Initial cost	$11,000	$33,000
Uniform annual benefit	3,000	9,000
End-of-depreciable-life salvage value	2,000	3,000
Depreciation method	SL	SOYD
End-of-useful-life salvage value obtained	2,000	5,000
Depreciable life, in years	3	4
Useful life, in years	5	5

If the firm's after-tax minimum attractive rate of return is 12% and its incremental income tax rate is 34%, which project should be selected?

CHAPTER 13

Replacement Analysis

The $2 Billion Upgrade

In February 2003, Intel Corporation announced it was planning to spend $2 billion to modernize and update its silicon wafer manufacturing plant in Chandler, Arizona. The upgrade will allow Intel to manufacture 300-millimeter chips, rather than the 200-millimeter size it had been producing at the plant, known as 'Fab 12'. The project is also expected to double manufacturing capacity at the facility, while lowering overall operating costs.

Upgrade work will probably take over a year to be completed. It will include remodelling the interior of the plant, and will require Intel to buy new wafer fabrication tools.

Despite the seemingly huge price tag for the project, Intel states that it will be saving money by upgrading the existing fabrication facility instead of building a new one. The company also noted that, by deciding to remain in its current location, it would be able to retain its current workforce, which is highly skilled.

Moreover, the new, larger wafer will accommodate more chips, which should lower production costs at the plant. The upgraded plant will also use a newer process that allows chips to hold smaller and faster transistors.

After Completing This Chapter...

The student should be able to:

- Recast an equipment reinvestment decision as a *challenger versus defender* analysis.
- Use the *replacement analysis decision map* to select the appropriate economic analysis technique to apply.
- Calculate the *minimum cost life* of economic challengers.
- Incorporate concepts such as *repeatability assumption for replacement analysis* and *marginal cost data for the defender* to select the appropriate economic analysis techniques.
- Perform replacement problems on an after-tax basis, utilizing the *defender sign change procedure* when appropriate.
- Use spreadsheets for solving before-tax and after-tax replacement analysis problems.

QUESTIONS TO CONSIDER

1. Using the Internet, can you determine what it would cost Intel to build a new fab from scratch?
2. Will Intel need to scrap all its current production assets when it does the upgrade?
3. How might Intel determine what to keep in operation and what to purchase new?

Up to this point in our economic analysis we have considered the evaluation and selection of *new* alternatives. Which new car or washing machine should we buy? What new material handling system or ceramic grinder should we install? However, a choice between new alternatives is not always what we must consider—economic analysis weighs more frequently *existing* versus *new* facilities. For most engineers the problem is less likely to be one of building a new plant; rather the goal is more often to keep a present plant operating economically. We are not choosing between new ways to perform the desired task. Instead, we have equipment performing the task, and the question is: should the existing equipment be retained or replaced? This adversarial situation has given rise to the terms **defender** and **challenger.** The defender is the existing equipment; the challenger is the best available replacement equipment. An economic evaluation of the existing defender and the challenger replacement is the domain of **replacement analysis.**

THE REPLACEMENT PROBLEM

The replacement of an existing asset may be reasonable in various situations, including obsolescence, depletion, and deterioration due to aging. In each of these cases, the ability of a previously implemented business asset to produce a desired output is challenged. For cases of obsolescence, depletion, and aging, it may be economical to replace the existing asset. We define each of these situations.

Obsolescence: Occurs when the technology of an asset is surpassed by newer and/or different technologies. Changes in technology cause subsequent changes in the market demand for older assets. As an example, today's personal computers (PCs) with more RAM, faster clock speeds, larger hard drives, and more powerful central processors have made older, less powerful PCs obsolete. Thus, obsolete assets may need to be replaced with newer, more technologically advanced ones.

Depletion: The gradual loss of market value of an asset as it is being consumed or exhausted. Oil wells and stands of timber are examples of such assets. In most cases the asset will be used until it is depleted, at which time a replacement asset will be obtained. Depletion was treated in Chapter 11.

Deterioration due to aging: The general condition of loss in value of some asset due to the aging process. Production machinery and other business assets that were once new eventually become aged. To compensate for a loss in functionality due to the aging process, additional operating and maintenance expenses are usually incurred to maintain the asset at its operating efficiency.

Aging equipment often has a greater risk of break-downs. Planned replacements can be scheduled to minimize the time and cost of disruptions. Unplanned replacements can be very costly or even, as with an airplane engine, catastrophic.

In industry, as in government, expenditures are normally monitored by means of *annual budgets.* One important facet of a budget is the allocation of money for new capital expenditures, either new facilities or replacement and upgrading of existing facilities.

Replacement analysis may, therefore, produce a recommendation that certain equipment be replaced and that money for the replacement be included in the capital expenditures budget. Even if there is no recommendation to replace the equipment at the current time, such a recommendation may be made the following year or subsequently. At *some* point, the existing equipment will be replaced, either when it is no longer necessary or when better equipment is available. Thus, the question is not *if* the defender will be replaced, but *when* it will be replaced. This leads us to the first aspect of the defender-challenger comparison:

> *Shall we replace the defender now, or shall we keep it for one or more additional years?*

REPLACEMENT ANALYSIS DECISION MAP

Figure 13-1 is a basic decision map for conducting a replacement analysis.

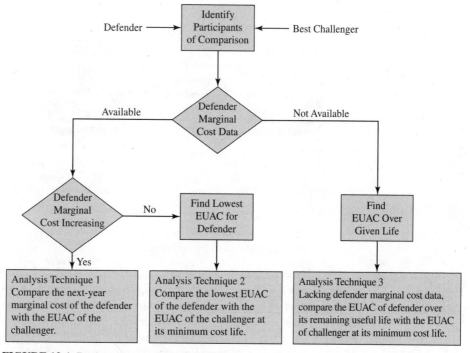

FIGURE 13-1 Replacement analysis decision map.

Looking at the map, we can see there are three *replacement analysis techniques* that are right under different circumstances. The right replacement analysis technique to use in making a replacement comparison of old versus new asset is a function of the data available for the alternatives and how the data behave over time.

WHAT IS THE BASIC COMPARISON?

By looking at the replacement analysis map, we see that the first step is to identify the basic participants of the economic comparison. Again, in replacement analysis we are interested

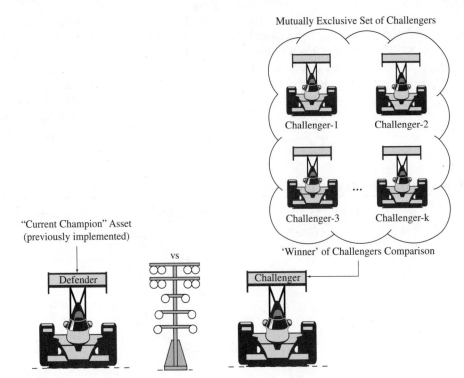

FIGURE 13-2 Defender-challenger comparison.

in comparing our previously implemented asset (the *defender*) with the best current available *challenger.*

> *If the defender proves more economical, it will be kept. If the challenger proves more economical, it will be installed.*

In this comparison the challenger being evaluated against a defender has been selected from a mutually exclusive set of competing challengers. Figure 13-2 illustrates this concept in the context of a drag race between the defender and the challenger. Notice that the challenger that is competing against the defender has emerged from an earlier competition among a set of potential challengers. Any of the methods for evaluating sets of mutually exclusive alternatives discussed previously in this text could be used to identify the 'best' challenger to race against the defender. However, it is important to note that the comparison of these potential challenger alternatives should be made at each alternative's respective *minimum cost life.* This concept is discussed next.

Minimum Cost Life of the Challenger

The **minimum cost life** of any new (or existing) asset is the number of years at which the equivalent uniform annual cost (EUAC) of ownership is minimized. This minimum cost life is often shorter than either the physical or useful life of the asset because of increasing operating and maintenance costs in the later years of asset ownership. The challenger asset selected to 'race' against the defender (in Figure 13-2) is the one having the lowest minimum cost life of all the competing mutually exclusive challengers.

To calculate the minimum cost life of an asset, determine the EUAC that results if the asset is kept for each possible life less than or equal to its useful life. As is illustrated in Example 13-1, the EUAC tends to be high if the asset is kept only a few years, then decreases to some minimum EUAC, and increases again as the asset ages. By identifying the number of years at which the EUAC is a minimum and then keeping the asset for that number of years, we are minimizing the yearly cost of ownership. Example 13-1 illustrates how minimum cost life is calculated for a new asset.

EXAMPLE 13-1

A piece of machinery costs $7500 and has no salvage value after it is installed. The manufacturer's warranty will pay the first year's maintenance and repair costs. In the second year, maintenance costs will be $900, and this item will increase on a $900 arithmetic gradient in subsequent years. Also, operating expenses for the machinery will be $500 the first year and will increase on a $400 arithmetic gradient in the following years. If interest is 8%, compute the useful life of the machinery that results in a minimum EUAC. That is, find its minimum cost life.

SOLUTION

		If Retired at the End of Year n		
Year, n	EUAC of Capital Recovery Costs: $7500(A/P, 8\%, n)$	EUAC of Maintenance and Repair Costs: $900(A/G, 8\%, n)$	EUAC of Operating Costs: $500 + $400(A/G, 8\%, n)$	EUAC Total
1	$8100	$ 0	$ 500	$8600
2	4206	433	692	5331
3	2910	854	880	4644
4	2264	1264	1062	4589 ←
5	1878	1661	1238	4779
6	1622	2048	1410	5081
7	1440	2425	1578	5443
8	1305	2789	1740	5834
9	1200	3142	1896	6239
10	1117	3484	2048	6650
11	1050	3816	2196	7063
12	995	4136	2338	7470
13	948	4446	2476	7871
14	909	4746	2609	8265
15	876	5035	2738	8648

The total EUAC data are plotted in Figure 13-3. From either the tabulation or the figure, we see that the minimum cost life of the machinery is 4 years, with a minimum EUAC of $4589 for each of those 4 years.

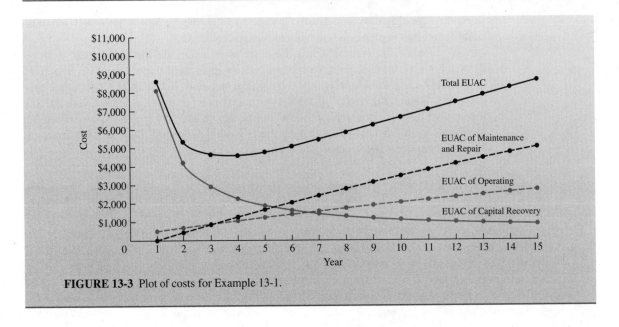

FIGURE 13-3 Plot of costs for Example 13-1.

Looking at Figure 13-3 a bit more closely, we see the effects of each of the individual cost components on total EUAC (capital recovery, maintenance and repair, and operating expense EUACs) and how they behave over time. The total EUAC curve of most assets tends to follow this concave shape—high at the beginning due to capital recovery costs, and high at the end due to increased maintenance and repair and operating expenses. The minimum EUAC occurs somewhere between these high points.

Use of Marginal Cost Data

Once the basic participants in the defender-challenger comparison have been identified (see Figure 13-1), two specific questions regarding marginal costs must be answered: *Do we have marginal cost data for the defender?* and *Are the defender's marginal costs increasing on a year-to-year basis?* Let us first define marginal cost and then discuss why is it important to answer these two questions.

Marginal costs, as opposed to an EUAC, are the year-by-year costs of keeping an asset. Therefore, the 'period' of any yearly marginal cost of ownership is always *1 year.* On the other hand, an EUAC can apply to any number of consecutive years. Thus, the marginal cost of ownership for any year in the life of an asset is the cost for *that year only.* In replacement problems, the total marginal cost for any year can include the capital recovery cost (loss in market value and lost interest for the year), yearly operating and maintenance costs, yearly taxes and insurance, and any other expense that occurs during that year. To calculate the yearly marginal cost of ownership of an asset, it is necessary to have estimates of an asset's market value on a year-to-year basis over its useful life, as well as ordinary yearly expenses. Example 13-2 illustrates how total marginal cost can be calculated for an asset.

EXAMPLE 13-2

A new piece of production machinery has the following costs.

Investment cost	= $25,000
Annual operating and maintenance cost	= $2000 the first year and increasing at $500 per year thereafter
Annual cost for risk of break-down	= $5000 per year for 3 years, then increasing by $1500 per year thereafter
Useful life	= 7 years
MARR	= 15% per year

Calculate the marginal cost of keeping this asset over its useful life.

SOLUTION

From the problem data we can easily find the marginal costs for O&M and risk of break-downs. However, to calculate the marginal capital recovery cost, we need estimates of the year-to-year market value: the prices of the production machinery would bring over its 7-year useful life. Market value estimates were made as follows:

Year	Market Value
1	$18,000
2	13,000
3	9,000
4	6,000
5	4,000
6	3,000
7	2,500

We can now calculate the *marginal cost* (year-to-year cost of ownership) of the production machinery over its 7-year useful life.

Year, n	Loss in Market Value in Year n	Forgone Interest in Year n	O&M Cost in Year n	Cost of Breakdown Risk in Year n	Total Marginal Cost in Year n
1	$25,000 - 18,000 = $7000	$25,000(0.15) = $3750	$2000	$5,000	$17,750
2	$18,000 - 13,000 = 5000	$18,000(0.15) = 2700	2500	5,000	15,200
3	$13,000 - 9,000 = 4000	$13,000(0.15) = 1950	3000	5,000	13,950
4	$9,000 - 6,000 = 3000	$9,000(0.15) = 1350	3500	6,500	14,350
5	$6,000 - 4,000 = 2000	$6,000(0.15) = 900	4000	8,000	14,900
6	$4,000 - 3,000 = 1000	$4,000(0.15) = 600	4500	9,500	15,600
7	$3,000 - 2,500 = 500	$3,000(0.15) = 450	5000	11,000	16,950

Notice that the total marginal cost for each year is made up of loss in market value, forgone interest, O&M cost, and cost for risk of breakdowns. As an example, the Year 5 marginal cost of $14,900 is calculated as $2,000 + 900 + 4,000 + 8,000$.

Do We Have Marginal Cost Data for the Defender?

Our decision map indicates that, in order to decide which replacement technique to use, it is necessary to know whether marginal cost data are available for the defender asset. Usually in engineering economic problems, annual savings and expenses are given for all alternatives. However, as in Example 13-2, it is also necessary to have year-to-year salvage value estimates in order to calculate total marginal costs. If the total marginal costs for the defender can be calculated, and if the data are increasing from year to year, then *replacement analysis technique 1* should be used for comparing the defender to the challenger.

Are These Marginal Costs Increasing?

We have seen that it is important to know whether the marginal cost for the defender is increasing from year to year. This is determined by inspecting the total marginal cost of ownership of the defender over its remaining life. Example 13-3 illustrates the calculation of the total marginal cost for the defender asset.

EXAMPLE 13-3

An asset purchased 5 years ago for $75,000 can be sold today for $15,000. Operating expenses in the past have been $10,000 per year, but these are estimated to increase in the future by $1500 per year each year. It is estimated that the market value of the old asset will decrease by $1000 per year over the next 5 years. If the MARR used by the company is 15%, calculate the total marginal cost of ownership of this old asset (that is, the defender) for each of the next 5 years.

SOLUTION

We calculate the total marginal cost of maintaining the old asset for the next 5-year period as follows:

Year, n	Loss in Market Value in Year n	Forgone Interest in Year n	Operating Cost in Year n	Marginal Cost in Year n
1	$15{,}000 - 14{,}000 = \$1000$	$15{,}000(0.15) = \$2250$	$10{,}000	$13,250
2	$14{,}000 - 13{,}000 = 1000$	$14{,}000(0.15) = 2100$	$11{,}500	14,600
3	$13{,}000 - 12{,}000 = 1000$	$13{,}000(0.15) = 1950$	$13{,}000	15,950
4	$12{,}000 - 11{,}000 = 1000$	$12{,}000(0.15) = 1800$	$14{,}500	17,300
5	$11{,}000 - 10{,}000 = 1000$	$11{,}000(0.15) = 1650$	$16{,}000	18,650

We can see that the marginal costs increase in each subsequent year of ownership. When the condition of increasing marginal costs for the defender has been met, then the defender–challenger comparison should be made by using *replacement analysis technique 1*.

Replacement Analysis Technique 1:
Defender Marginal Costs Can Be Computed and Are Increasing

When our first method of analyzing the defender asset against the best available challenger is used, the basic comparison involves the *marginal cost data of the defender and the minimum cost life data of the challenger.*

When the marginal cost of the defender is increasing from year to year, we will maintain that defender as long as the marginal cost of keeping it one more year is less than the minimum EUAC of the challenger. Thus our decision rule is as follows:

> *Maintain the defender as long as the marginal cost of ownership for one more year is less than the minimum EUAC of the challenger. When the marginal cost of the defender becomes greater than the minimum EUAC of the challenger, then replace the defender with the challenger.*

One can see that this technique assumes that the current best challenger, with its minimum EUAC, will be available and unchanged in the future. However, it is easy to update a replacement analysis when marginal costs for the defender change or when there is a change in the cost and/or performance of available challengers. Example 13-4 illustrates the use of this technique for comparing defender and challenger assets.

EXAMPLE 13-4

Taking the machinery in Example 13-2 as the *challenger* and the machinery in Example 13-3 as the *defender,* use *replacement analysis technique 1* to determine when, if at all, a replacement decision should be made.

SOLUTION

Replacement analysis technique 1 should be used only in the condition of increasing marginal costs for the defender. Since these marginal costs are increasing for the defender (from Example 13-3), we can proceed by comparing defender marginal costs against the minimum EUAC of the challenger asset. In Example 13-2 we calculated only the marginal costs of the challenger; thus it is necessary to calculate the challenger's minimum EUAC. The EUAC of keeping this asset for each year of its useful life is worked out as follows.

Year, n	Challenger Total Marginal Cost in Year n	EUAC of Challenger Ownership If Kept Through Year n	
1	$17,750	$[17,750(P/F,15\%,1)](A/P,15\%,1)$	$= \$17,750$
2	15,200	$[17,750(P/F,15\%,1) + 15,200(P/F,15\%,2)](A/P,15\%,2)$	$= 16,560$
3	13,950	$[17,750(P/F,15\%,1) + \cdots + 13,950(P/F,15\%,3)](A/P,15\%,3)$	$= 15,810$
4	14,350	$[17,750(P/F,15\%,1) + \cdots + 14,350(P/F,15\%,4)](A/P,15\%,4)$	$= 15,520$
5	14,900	$[17,750(P/F,15\%,1) + \cdots + 14,900(P/F,15\%,5)](A/P,15\%,5)$	$= 15,430$
6	15,600	$[17,750(P/F,15\%,1) + \cdots + 15,600(P/F,15\%,6)](A/P,15\%,6)$	$= 15,450$
7	16,950	$[17,750(P/F,15\%,1) + \cdots + 16,950(P/F,15\%,7)](A/P,15\%,7)$	$= 15,580$

A minimum EUAC of $15,430 is attained for the challenger at Year 5, which is the challenger's *minimum cost life*. We proceed by comparing this value against the *marginal* costs of the defender from Example 13-3:

Year, n	Defender Total Marginal Cost in Year n	Challenger Minimum EUAC	Comparison Result and Recommendation
1	$13,250	$15,430	Since $13,250 is *less than* $15,430, keep defender.
2	14,600	15,430	Since $14,600 is *less than* $15,430, keep defender.
3	15,950	15,430	Since $15,950 is *greater than* $15,430, replace defender.
4	17,300		
5	18,650		

On the basis of the data given for the challenger and for the defender, we would keep the defender for 2 more years and then replace it with the challenger because at that point the marginal cost of another year of ownership of the defender would be greater than the minimum EUAC of the challenger.

One may ask, *Why can't replacement analysis technique 1 be used when the marginal costs of the defender do not increase?* To answer this question we must understand that for this technique to be valid, the following basic assumptions must be valid: the best challenger will be available 'with the same minimum EUAC' at any time in the future; and the period of needed service in our business is indefinitely long. In other words, we assume that once the decision has been made to replace, there will be an indefinite replacement of the defender, with continuing 'cycles' of the current best challenger asset. These two assumptions together are much like the repeatability assumptions that allowed us to use the annual cost method from earlier chapters to compare competing alternatives with different useful lives. Taken together, we call these the **replacement repeatability assumptions.** They allow us to greatly simplify the comparison of the defender and the challenger. We state these assumptions formally below.

Replacement Repeatability Assumptions
The two assumptions are as follows:

1. The currently available best challenger will continue to be available in subsequent years and will be unchanged in its economic costs. When the defender is ultimately replaced, it will be replaced with this challenger. Any challengers put into service will also be replaced with the same currently available challenger.
2. The period of needed service of the asset is indefinitely long. Thus the challenger asset, once put into service, will continuously replace itself in repeating, unchanged cycles.

Given that the defender will ultimately be replaced with the current best challenger, we would never want to incur a defender marginal cost greater than the challenger's minimum EUAC. And because the defender's marginal costs are increasing, we can be assured that once the marginal cost of the defender has become greater than the challenger's minimum EUAC, it will continue to be so in the future. However, if the marginal costs do not increase, we have no guarantee that *replacement analysis technique 1* will produce the alternative that is of the greatest economic advantage. One may ask, *Are there ordinary conditions in which the marginal costs are not increasing?* The answer to this question is yes. Consider the new asset in Example 13-2. This new asset has marginal costs that begin at a high of $17,750, then *decrease* over the next years to a low of $13,950, and then *increase* thereafter to $16,950 in Year 7. If this asset were implemented and then evaluated *one year after implementation* as a defender asset, it would not have increasing marginal costs. Thus, defenders in the early stages of their respective implementations would not fit the requirements of *replacement analysis technique 1*. In the situation graphed in Figure 13-3, such defender assets would be in the downward slope of a concave marginal cost curve. Example 13-5 illustrates the error that can be introduced when *replacement analysis technique 1* is applied when defenders do not have consistently increasing marginal cost curves.

EXAMPLE 13-5

Let us look again at the defender and challenger assets in Example 13-4. This time let us arbitrarily change the defender's marginal costs for its 5-year useful life. Now when, if at all, should the defender be replaced with the challenger?

Year, n	Defender Total Marginal Cost in Year n
1	$16,000
2	14,000
3	13,500
4	15,300
5	17,500

SOLUTION

In this case the total marginal costs of the defender are *not* consistently increasing from year to year. However, if we ignore this fact and apply *replacement analysis technique 1*, the recommendation would be to replace the defender now, because the marginal cost of the defender for the first year ($16,000) is greater than the minimum EUAC of the challenger ($15,430). Let us review this decision. One can see that the first-year marginal cost of the defender is greater than the minimum EUAC of the challenger, but in the second to fourth years the marginal costs are less. Thus marginal costs decrease here for 3 years to a minimum and then increase the following 2 years.

Let us calculate the EUAC of keeping the defender asset each of its remaining 5 years, at $i = 15\%$.

Year, n	EUAC of Defender Ownership If Kept n Years	Challenger Minimum EUAC = \$15,430
1	\$16,000	$= 16{,}000(P/F, 15\%, 1)(A/P, 15\%, 1)$
2	15,070	$= [16{,}000(P/F, 15\%, 1) + 14{,}000(P/F, 15\%, 2)](A/P, 15\%, 2)$
3	14,618	$= [16{,}000(P/F, 15\%, 1) + \cdots + 13{,}500(P/F, 15\%, 3)](A/P, 15\%, 3)$
4	14,754	$= [16{,}000(P/F, 15\%, 1) + \cdots + 15{,}300(P/F, 15\%, 4)](A/P, 15\%, 4)$
5	15,162	$= [16{,}000(P/F, 15\%, 1) + \cdots + 17{,}500(P/F, 15\%, 5)](A/P, 15\%, 5)$

The minimum EUAC of the defender for 3 years is \$14,618, which is less than that of the challenger. But this comparison alone is not sufficient to indicate that we should keep the defender for 3 years and then replace it. In looking at a study period of 5 years, we really have six options.

Option 1: Implement the challenger today (at \$15,430/year).

Option 2: Keep the defender for 1 year (at \$16,000) and the challenger for 4 years (at \$15,430/year).

Option 3: Keep the defender for 2 years (at \$15,070/year) and the challenger for 3 years (at \$15,430/year).

Option 4: Keep the defender for 3 more years (at \$14,618/year) and the challenger for 2 years (at \$15,430/year).

Option 5: Keep the defender for 4 years (at \$14,754/year) and the challenger for 1 year (at \$15,430).

Option 6: Keep the defender for 5 years (at \$15,162/year).

The EUAC of each of these options at 15% is:

Option 1: EUAC is \$15,430.

Option 2: EUAC is $[16{,}000 + 15{,}430(P/A, 15\%, 4)](P/F, 15\%, 1)(A/P, 15\%, 5) = \$15{,}578.$

Option 3: EUAC is $[15{,}070(P/A, 15\%, 2) + 15{,}430(P/A, 15\%, 3)(P/F, 15\%, 2)] \times (A/P, 15\%, 5) = \$15{,}255.$

Option 4: EUAC is $[14{,}618(P/A, 15\%, 3) + 15{,}430(P/A, 15\%, 2)(P/F, 15\%, 3)] \times (A/P, 15\%, 5) = \$14{,}877.$

Option 5: EUAC is $[14{,}754(P/A, 15\%, 4) + 15{,}430(P/F, 15\%, 5)](A/P, 15\%, 5) = \$14{,}855.$

Option 6: EUAC is \$15,162.

One can see that Option 5 produces the minimum EUAC. Thus we should keep the defender for 4 more years and then re-evaluate the defender-challenger decision. Notice that if we replace the defender now with the challenger per *replacement decision rule 1*, we would not achieve a minimum EUAC over this 5-year period. Notice also that the lowest cost life of 3 years for the defender is NOT how long we should keep the defender. In Year 5, the defender's marginal cost again exceeds the challenger's minimum EUAC. We emphasize that this analysis is valid only because of the *replacement repeatability assumptions.* These assumptions hold that after the 5-year period, all alternatives have the exact same yearly cash flow of $15,430 (thus we can ignore them), and these identical cash flows continue indefinitely.

Example 13-5 demonstrates that when the marginal costs of the defender are not consistently increasing from year to year, it is necessary to calculate the lowest EUAC of the defender. In the next section we describe this calculation.

Lowest EUAC of the Defender

How long can a defender asset be kept operating? Anyone who has seen or heard old machinery in operation, whether it is a 50-year-old car or 20-year-old piece of production equipment, has realized that almost any machine can be kept operating indefinitely, provided it receives proper maintenance and repair. However, even though one might be able to keep a defender going indefinitely, the cost may prove excessive. So, rather than asking what remaining *operating life* the defender may have, we really want to ask what is the *minimum cost life* of the asset. The minimum cost life of the defender is defined as the number of years of ownership in the future that results in a minimum EUAC for the defender and the challenger combined, as in Example 13-5.

However, for decision making now, the answer to a simpler calculation is enough. Is the defender's lowest or minimum EUAC less than the challenger's minimum EUAC?

EXAMPLE 13-6

An 11-year-old piece of equipment is being considered for replacement. It can be sold for $2000 now, and it is believed that the same salvage value can also be obtained in future years. The current maintenance cost is $500 per year and is expected to increase $100 per year in future years. If the equipment is retained in service, compute the minimum EUAC, based on 10% interest.

SOLUTION

Here the salvage value is not expected to decline from its present $2000. The annual cost of this invested capital is $Si = 2000(0.10) = \$200$. The maintenance is represented by $\$500 + \$100G$. A year-by-year computation of EUAC is as follows:

Year, n	Age of Equipment (years)	EUAC of Invested Capital ($= Si$)	EUAC of Maintenance [$= 500 + 100(A/G, 10\%, n)$]	Total EUAC
1	11	$200	$500	$700
2	12	200	548	748
3	13	200	594	794
4	14	200	638	838
5	15	200	681	881

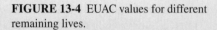

FIGURE 13-4 EUAC values for different remaining lives.

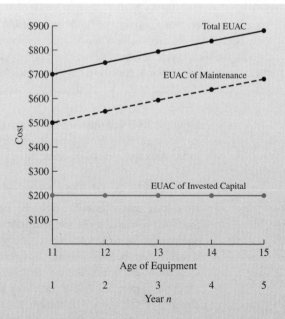

The lowest EUAC is $700. That number can be compared with the challenger's best EUAC to determine whether the equipment should be replaced now.

These data are plotted in Fig. 13-4. We see that the annual cost of continuing to use the equipment is increasing. It is reasonable to assume that if the equipment is not replaced now, it will be reviewed again next year. If the defender's and challenger's cost data do not change, we can compute the marginal cost of extending the defender's service each year in order to decide when to replace.

Example 13-6 represents a common situation. The salvage value is stable, but the maintenance cost is increasing. The marginal cost of extending the defender's life and its total EUAC will continue to increase as time passes, which means that the defender's lowest EUAC will be based on retaining the defender for 1 more year. This is not always the case, as is shown in Example 13-7.

EXAMPLE 13-7

A 5-year-old machine, whose current market value is $5000, is being analyzed to determine its economic life in a replacement analysis. Compute its lowest EUAC at a 10% interest rate. Salvage value and maintenance estimates are given in the following table.

Years of Remaining Life, n	Estimated Salvage Value (S) at End of Year n	Estimated Maintenance Cost for Year	If Retired at End of Year n		
			EUAC of Capital Recovery $(P - S) \times (A/P, 10\%, n) + Si$	EUAC of Maintenance $100(A/G, 10\%, n)$	Total EUAC
0	$P = \$5000$				
1	4000	$ 0	$1100 + 400	$ 0	$1500
2	3500	100	864 + 350	48	1262
3	3000	200	804 + 300	94	1198
4	2500	300	789 + 250	138	1177
5	2000	400	791 + 200	181	1172
6	2000	500	689 + 200	222	1111
7	2000	600	616 + 200	262	1078
8	2000	700	562 + 200	300	1062
9	2000	800	521 + 200	337	1058 ←
10	2000	900	488 + 200	372	1060
11	2000	1000	462 + 200	406	1068

SOLUTION

A minimum EUAC of $1058 is computed at Year 9 for the existing machine.

Looking again at Examples 13-6 and 13-7, we find that the two cases represent the same machine being examined at different points in its life. (In Year 6 and later in Example 13-7 the costs match those of Example 13-6.) Example 13-7 shows that the 5-year-old machine has a minimum EUAC at a life of 9 years.

For older equipment with a negligible or stable salvage value, it is likely that the operating and maintenance costs are increasing. Under these circumstances, the useful life at which EUAC is at a minimum is 1 year.

It is necessary to calculate the defender's minimum EUAC when marginal costs are not consistently increasing. Having made this calculation, we can now use *replacement analysis technique 2* to compare the defender against the challenger.

Replacement Analysis Technique 2: Defender Marginal Cost Can Be Computed and Is Not Increasing

If we use our second method of analyzing the defender asset against the best available challenger, the basic comparison involves the *lowest EUAC of the defender and the EUAC of the challenger at its minimum cost life*.

We calculate the minimum EUAC of the defender and compare this directly with the minimum EUAC of the challenger. Remember that the *replacement repeatability assumptions* allow us to do this. In this comparison we then choose the alternative with the lowest EUAC. In Example 13-5 the comparison involved the defender with the lowest EUAC— $14,618 at a life of 3 years—and the challenger with an economic life of 5 years and EUAC = $15,430. Here we would recommend that the defender be retained for 4 more years because its marginal costs for later years will be increasing above the challenger's EUAC. Then, *replacement analysis technique 1* would apply. Consider Example 13-8.

EXAMPLE 13-8

An economic analysis is to be made to determine whether existing (defender) equipment in an industrial plant should be replaced. A $4000 overhaul must be done now if the equipment is to be retained in service. Maintenance is estimated at $1800 in each of the next 2 years, after which it is expected to increase by $1000 each year. The defender has no present or future salvage value. The equipment described in Example 13-1 is the challenger. Make a replacement analysis to determine whether to keep the defender or replace it by the challenger if 8% interest is used.

SOLUTION

The first step is to determine the lowest EUAC of the defender. The pattern of overhaul and maintenance costs (Figure 13-5) suggests that if the overhaul is done, the equipment should be kept for several years. The computation is as follows:

	If Retired at End of Year n		
Year, n	EUAC of Overhaul $4000(A/P, 8\%, n)$	EUAC of Maintenance $1800 + 1000$ Gradient from Year 3 on	Total EUAC
1	$4320	$1800	$6120
2	2243	1800	4043
3	1552	1800 + 308*	3660 ←
4	1208	1800 + 683†	3691
5	1002	1800 + 1079	3881

*For the first 3 years, the maintenance is $1800, $1800, and $2800. Thus, EUAC = 1800 + 1000(A/F, 8\%, 3) = 1800 + 308.

†EUAC = 1800 + 1000(P/G, 8\%, 3)(P/F, 8\%, 1)(A/P, 8\%, 4) = 1800 + 683.

FIGURE 13-5 Overhaul and maintenance costs for the defender in Example 13-8.

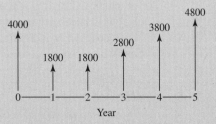

The lowest EUAC of the defender is $3660. In Example 13-1, we determined that the minimum cost life of the challenger is 4 years and that the resulting EUAC is $4589. If we assume the equipment is needed for at least 4 years, the EUAC of the defender ($3660) is less than the EUAC of the challenger ($4589). In this situation the defender should not be replaced yet.

If the defender's and challenger's cost data do not change, we can use *replacement analysis technique 1* to determine when the defender should be replaced. We know from the minimum EUAC calculation that the defender should be kept at least 3 years. Is this the best life? The following table computes the marginal cost to answer this question.

Year, n	Overhaul Cost	Maintenance Cost	Marginal Cost to Extend Service
0	$4000	$ 0	
1	0	1800	$6120 = 4000(1.08) + 1800
2	0	1800	1800
3	0	2800	2800
4	0	3800	3800
5	0	4800	4800

Year 5 is the first year after the year with the lowest EUAC in which the $4800 marginal cost for the defender exceeds the $4589 minimum EUAC for the challenger. Thus, the defender should be kept 4 more years if costs do not change. (Note that if the defender can be overhauled again after 3 years, that might be an even better choice.)

No Defender Marginal Cost Data Available

Earlier we described replacement analysis techniques for comparing the defender and challenger if marginal cost data are available for the old asset. You may recall that yearly salvage value estimates were necessary to calculate these marginal costs. Sometimes, however, it is difficult to obtain end-of-the-year salvage value estimates for the defender asset. A defender based on aging technology with a shrinking market might be such an asset. Or, from a student's problem-solving perspective, perhaps salvage value data are not given as part of the problem data. Without these data, it is impossible to calculate the marginal cost of the defender and thus impossible to use marginal cost data to compare the defender against the challenger. How should we proceed?

Given the *replacement repeatability assumptions,* it is possible for us to calculate the EUAC of the defender over its remaining useful life. Then, knowing that both the defender and challenger will be replaced with indefinite cycles of the challenger, we may use *replacement analysis technique 3* to compare defender and challenger.

Replacement Analysis Technique 3:
When Defender Marginal Cost Data Are Not Available

When our third method of analyzing the defender asset against the best available challenger is used, the basic comparison involves the *EUAC of the defender over its stated useful life, and the minimum EUAC of the challenger.*

We will calculate the EUAC of the defender asset over its remaining useful life and compare this directly against the EUAC of the challenger at its minimum cost life, and then choose the lesser of these two values. However, in making this basic comparison an often complicating factor is deciding what first cost to assign to the challenger and the defender assets.

Defining Defender and Challenger First Costs

Because the defender is already in service, analysts often misunderstand what first cost to assign it. Example 13-9 demonstrates this problem.

EXAMPLE 13-9

A laptop word processor model SK-30 was purchased 2 years ago for $1600; it has been depreciated by straight-line depreciation using a 4-year life and zero salvage value. Because of recent innovations in word processors, the current price of the SK-30 laptop has been reduced from $1600 to $995. An office equipment supply firm has offered a trade-in allowance of $350 for the SK-30 on a new $1200 model EL-40 laptop. Some discussion revealed that without a trade-in, the EL-40 can be bought for $1050, suggesting the originally quoted price of the EL-40 was overstated to allow a larger trade-in allowance. The true current market value of the SK-30 is probably only $200. In a replacement analysis, what value should be assigned to the SK-30 laptop?

SOLUTION

In the example, five different dollar amounts relating to the SK-30 laptop have been outlined:

1. *Original cost:* The laptop cost $1600 2 years ago.
2. *Present cost:* The laptop now sells for $995.
3. *Book value:* The original cost less 2 years of depreciation is $1600 - \frac{2}{4}(1600 - 0) = \$800$.
4. *Trade-in value:* The offer was $350.
5. *Market value:* The estimate was $200.

We know that an economic analysis is based on the current situation, not on the past. We refer to past costs as *sunk* costs to emphasize that, since these costs cannot be altered, they are not relevant. (There is one exception: past costs may affect present or future income taxes.) In the analysis we want to use actual cash flows for each alternative. Here the question is: what value should be used in an economic analysis for the SK-30? The relevant cost is the present market value for the equipment. Neither the original cost, the present cost, the book value, nor the trade-in value is relevant.

At first glance, the trade-in value of an asset would appear to be a suitable present value for the equipment. Often the trade-in price is inflated *along with* the price for the new item. (This practice is so common in new-car showrooms that the term *overtrade* is used to describe the excessive portion of the trade-in allowance. The buyer is also quoted a higher price for the new car.) Distortion of the present value of the defender, or a distorted price for the challenger, can be serious because these distortions do not cancel out in an economic analysis.

Example 13-9 illustrated that of the several different values that can be assigned to the defender, the most appropriate is the present market value. If a trade-in value is obtained, care should be taken to ensure that it actually represents a fair market value.

Determining the value for the installed cost of the challenger asset should be less difficult. In such cases the first cost is usually made up of purchase price, sales tax, installation costs, and other items that occur initially only once owing to the selection of the challenger. These values are usually rather straightforward to obtain if a thorough analysis is conducted. One aspect to consider in assigning a first cost to the challenger is the potential disposition (or market or salvage) value of the defender. One must not arbitrarily subtract the disposition cost of the defender from the first cost of the challenger asset, for this practice can lead to an incorrect analysis.

As described in Example 13-9, the correct first cost to assign to the defender SK-30 laptop is its $200 current market value. This value represents the present economic benefit that we would be *forgoing* to keep the defender. This can be called our *opportunity first cost*. If, instead of assuming that this is an *opportunity cost* to the defender, we assume it is a *cash benefit* to the challenger, a potential error arises. Consider the following case involving the SK-30 and EL-40 laptops. Assume the following data:

	SK-30		**EL-40**
Market value	$200	First cost	$1050
Remaining life	3 years	Useful life	3 years

In this case the remaining life of the defender (SK-30) is 3 years, and so is the useful life of the challenger (EL-40). If we use an *opportunity cost* perspective, then the calculated capital recovery effect of first cost using an annual cost comparison is:

$$\text{Annualized first cost}_{\text{SK-30}} = \$200(A/P, 10\%, 3) = \$80$$

$$\text{Annualized first cost}_{\text{EL-40}} = \$1050(A/P, 10\%, 3) = \$422$$

The *difference* in annualized first cost between the SK-30 and EL-40 is:

$$\text{AFC}_{\text{EL-40}} - \text{AFC}_{\text{SK-30}} = \$422 - \$80 = \$342$$

Now use a *cash flow* perspective to look at the first costs of the defender and challenger. In this case we use the actual cash that changes hands when each alternative is selected. A first cost of zero ($0) cash would be assigned to the defender and a first cost of $850 to the challenger (−$1050 purchase price of the challenger and +$200 in salvage value from defender). We calculate the *difference* due to first cost between the SK-30 and EL-40 to be:

$$\text{Annualized first cost}_{\text{SK-30}} = \$0(A/P, 10\%, 3) = \$0$$

$$\text{Annualized first cost}_{\text{EL-40}} = (\$1050 - 200)(A/P, 10\%, 3) = \$342$$

$$\text{AFC}_{\text{EL-40}} - \text{AFC}_{\text{SK-30}} = \$342 - \$0 = \$342$$

When both the remaining life of the defender and the useful life of the challenger are the same, 3 years in this case, the analysis of the first cost yields an identical (and correct) result. Both the *opportunity cost* and *cash flow* perspectives for considering first cost of the defender and challenger result in a difference of $342 between the two alternatives on an annual cost basis.

Now see what happens when the remaining life of the defender is not equal to the useful life of the challenger. Consider the SK-30 and EL-40 word processors, but assume that the lives have been changed as follows:

	SK-30		**EL-40**
Remaining life	3 years	Useful life	5 years

Looking at this second case from an *opportunity cost* perspective, we use an annual cost comparison to calculate capital recovery effect of the first cost as follows:

$$\text{Annualized first cost}_{\text{SK-30}} = \$200(A/P, 10\%, 3) = \$80$$

$$\text{Annualized first cost}_{\text{EL-40}} = \$1050(A/P, 10\%, 5) = \$277$$

The *difference* in annualized first cost between the SK-30 and EL-40 is:

$$\text{AFC}_{\text{EL-40}} - \text{AFC}_{\text{SK-30}} = \$277 - \$80 = \$197$$

Now using a *cash flow* perspective to look at the first costs of the defender and challenger, we can calculate the *difference* due to first cost between the SK-30 and EL-40.

$$\text{Annualized first cost}_{\text{SK-30}} = \$0(A/P, 10\%, 3) = \$0$$

$$\text{Annualized first cost}_{\text{EL-40}} = (\$1050 - 200)(A/P, 10\%, 5) = \$224$$

$$\text{AFC}_{\text{EL-40}} - \text{AFC}_{\text{SK-30}} = \$224 - \$0 = \$224$$

When the remaining life of the defender (3 years) differs from that of the useful life of the challenger (5 years), analyses of the annualized first costs yield different results. The correct difference of $197 is shown by using the *opportunity cost* approach, and an inaccurate difference of $224 is obtained if the *cash flow* perspective is used. In the opportunity cost case the $200 is spread out over 3 years as a cost to the defender, and yet in the cash flow case the opportunity cost is spread out over 5 years as a benefit to the challenger. Spreading the $200 over 3 years in one case and 5 years in the other case does not produce equivalent annualized amounts. Because of the difference in the lives of the assets, the annualized $200 opportunity cost for the defender cannot be called an equivalent benefit to the challenger.

In the case of unequal lives, the correct method is to assign the current market value of the defender as its time zero opportunity costs, rather than subtracting this amount from the first cost of the challenger. Because the cash flow approach yields an incorrect value when challenger and defender have unequal lives, the *opportunity cost* approach for assigning a first cost to the challenger and defender assets should *always* be used.

Repeatability Assumptions Not Acceptable

Under certain circumstances, the repeatability assumptions described earlier may not apply in a replacement analysis. In these cases replacement analysis techniques 1, 2, and 3 may not be valid methods for comparison. For instance, a decision maker may set the study period instead of assuming that there is an indefinite need for the asset. For example, consider the case of phasing out production after a certain number of years—perhaps a person who is about to retire is closing down a business and selling all the assets. Other examples include production equipment such as moulds and dies that are no longer needed when a new model with new shapes is introduced.

The study period could potentially be set at any number of years relative to the lives of the defender and the challenger, such as equal to the life of the defender, equal to the life of the challenger, less than the life of the defender, greater than the life of the challenger, or somewhere between the lives of the defender and challenger. The essential principle in this case is that in *setting the study period,* the decision maker must use a suitable method as described in earlier chapters. The analyst must be explicit about the economic costs and benefits of the challenger that is assumed for replacement of the defender (when replacement is made), as well as residual or salvage values of the alternatives at the end of the study period. In this case the repeatability replacement assumptions do not apply, and thus the replacement analysis techniques are not necessarily valid. The analysis techniques in the decision map also may not apply when future challengers are not assumed to be identical to the current best challenger. This concept is discussed in the next section.

A Closer Look at Future Challengers

We defined the challenger as the best available alternative to replace the defender. But in time, the best available alternative can change. And given the trend in our technological society, it seems likely that future challengers will be better than the present challenger. If so, the prospect of improved future challengers may affect the present decision between the defender and the challenger.

Figure 13-6 illustrates two possible estimates of future challengers. In many technological areas it seems likely that the equivalent uniform annual costs associated with future challengers will decrease by a constant amount each year. In other fields, however, a rapidly changing technology will produce a sudden and substantially improved challenger—with decreased costs or increased benefits. The uniform decline curve of Figure 13-6 reflects the assumption that each future challenger has a minimum EUAC that is a fixed amount less than the previous year's challenger. This assumption, of course, is only one of many possible assumptions that could be made regarding future challengers.

If future challengers will be better than the present challenger, what influence will this have on an analysis now? The prospect of better future challengers may make it more desirable to retain the defender and to reject the present challenger. By keeping the defender for now, we may be able to replace it later by a better future challenger. Or, to state it another way, the present challenger may be made less desirable by the prospect of improved future challengers. As engineering economic analysts, we must familiarize ourselves with potential

FIGURE 13-6 Two possible ways the EUAC of future challengers may decline.

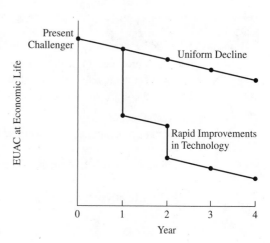

technological advances in assets targeted for replacement. This part of the decision process is much like the search for all available alternatives, from which we select the best. Upon finding out more about what alternatives and technologies are emerging, we will be better able to understand the repercussions of investing in the current best available challenger. Selecting the current best challenger asset can be particularly risky when (1) the costs are very high and/or (2) the useful minimum cost life of that challenger is relatively long (5–10 years or more). When one or both of these conditions exist, it may be better to keep or even augment our defender asset until better future challengers emerge.

There are, of course, many assumptions that *could* be made regarding future challengers. However, if the replacement repeatability assumptions do not hold, the analysis techniques described earlier may not be valid.

AFTER-TAX REPLACEMENT ANALYSIS

As described in Chapter 12, an after-tax analysis adds an expanded perspective to our problems because various very important effects can be included. We saw earlier that examining problems on an after-tax basis provides greater realism and insight. This advantage also exists when one is considering the replacement problems discussed in this chapter. Tax effects have the potential to alter recommendations made in a before-tax-only analysis. After-tax effects may influence calculations in the remaining economic life of defender, in the economic life of challenger, and in the defender-challenger comparisons discussed earlier. Consequently, one should always perform these analyses on an after-tax basis. In this section we illustrate the complicating circumstances that are introduced by an after-tax analysis.

Marginal Costs on an After-Tax Basis

As in the before-tax case, the defender-challenger comparison is sometimes based on the marginal costs of the defender on a year-to-year basis. Marginal costs on an after-tax basis represent the cost that would be incurred through ownership of the asset *in each year.* On an after-tax basis we must consider the effects of ordinary taxes as well as gains and losses due to asset disposal in calculating the after-tax marginal costs. Consider Example 13-10.

EXAMPLE 13-10

Example 13-2 provided the following information, and from it we were able to calculate the marginal cost for each year of the machinery's life.

A new piece of production machinery has the following costs.

Investment cost	= $25,000
Annual operating and maintenance cost	= $2000 the first year and increasing $500 per year thereafter
Annual cost for risk of break-down	= $5000 per year for 3 years, then increasing by $1500 per year thereafter
Useful life in years	7
MARR	= 15%

Calculate the marginal cost of keeping this asset over its useful life.

Year	Market Value
1	$18,000
2	13,000
3	9,000
4	6,000
5	4,000
6	3,000
7	2,500

Now, considering the additional information below, we shall calculate the after-tax marginal cost of this asset.

Machinery is a class 43 asset	30% CCA rate
Marginal tax rate	40%
The interest rate on investments	8%

Use the spreadsheet (*books closed*) assumption.

SOLUTION

The after-tax marginal cost of ownership will involve the following elements: annual loss in after-tax value, forgone interest, tax credit from CCA deduction, and annual operating, maintenance, and break-down costs. We calculate the marginal cost of each of these in Table 13-1.

TABLE 13-1 Marginal Costs of Ownership

	Loss in After-Tax Value and Forgone Interest							
Year, n	Market Value at End on year n	Book Value at End year n	Recapture or Loss on Disposal	Tax Effect of Disposal	After-Tax Market Value	Loss in After-Tax Market Value	Forgone Interest	Tax Credit from CCA Deduction
0	$25,000	$25,000	$ 0	$ 0	$25,000			
1	18,000	21,250	(3,250)	1,300	19,300	$5,700	$1,200	$1,500
2	13,000	14,875	(1,875)	750	13,750	5,550	926	2,550
3	9,000	10,413	(1,413)	565	9,565	4,185	660	1,785
4	6,000	7,289	(1,289)	515	6,516	3,050	459	1,250
5	4,000	5,102	(1,102)	441	4,441	2,075	313	875
6	3,000	3,571	(571)	229	3,229	1,212	213	612
7	2,500	2,500	(0)	0	2,500	729	155	429

After-tax market value = Market value + Tax effect

Forgone interest = After-tax market value at start of year × Before-tax interest rate × $(1 - t)$

Tax credit = If the machine were retained for one year, the CCA would result in a positive cash flow of $(t \times CCA)$

Marginal O&M and Break-down Costs

Year, n	Total O&M and Break-down Costs in Year, n	Marginal Cost in Year n Due to O&M and Break-down
0		
1	$ 7,000	$4,200
2	7,500	4,500
3	8,000	4,800
4	10,000	6,000
5	12,000	7,200
6	14,000	8,400
7	16,000	9,600

After-tax costs $= (1 - t) \times$ before-tax costs

Total Marginal Cost = Loss in Market Value + Forgone Interest − Tax Credit + O&M Cost

Year, n	Loss in After-Tax Market Value	Forgone Interest	Tax Credit from CCA Deduction	Marginal Cost in Year n Due to O&M and Break-down	Total Marginal Cost in Year n
0					
1	$5,700	$1,200	$1,500	$4,200	$9,600
2	5,550	926	2,550	4,500	8,426
3	4,185	660	1,785	4,800	7,860
4	3,050	459	1,250	6,000	8,259
5	2,075	313	875	7,200	8,713
6	1,212	213	612	8,400	9,213
7	729	155	429	9,600	10,055

Notice that the marginal cost for each year is much different from the marginal cost for the same asset on a before-tax basis.

After-Tax Cash Flows for the Challenger

Finding the after-tax cash flows for the challenger asset is straightforward. Here we use the standard after-tax cash flow table and method developed in Chapter 12 to incorporate all the relevant tax effects, which include before-tax cash flows, depreciation, taxes, and gains and losses at disposal. After obtaining the after-tax cash flows for challenger, it is a simple task to calculate the EUAC over the challenger's life.

After-Tax Cash Flows for the Defender

Finding the after-tax cash flows for the defender is not straightforward. The main complicating factor is that the defender is an asset that has been previously placed in service. As such, depreciation has been taken on the cost basis (amount being depreciated) up to the present, and this changes the book value of the asset. Thus, when looking at the first cost of the defender on an after-tax basis (value at time zero), we must consider the current market (salvage) value that can be obtained for the asset, as well as any depreciation recapture or losses at disposal presently associated with the defender.

We have already suggested the use of the *opportunity cost* perspective for assigning a first cost to both the defender and challenger. For the challenger, this first cost will be the after-tax cash flow at time zero. Assuming use of equity financing, the first cost of the challenger is the purchase price plus installation cost. However, for the defender the after-tax first cost will be made up of the *forgone market value* of the asset at the present time plus any *forgone gains or losses* associated with keeping the asset. To develop the after-tax cash flow of the defender at time zero, it is often convenient to use the **defender sign change procedure,** described next.

Defender Sign Change Procedure

The defender sign change procedure allows us to find the after-tax cash flow (ATCF) for the defender asset in after-tax defender-challenger comparisons. To find the time zero ATCF of the defender use the following steps:

1. Assume you are selling the defender now (time 0).
2. Find the ATCF for selling the defender at time 0.
3. Then, because you are actually keeping the defender, not selling it, change all the signs (plus to minus, minus to plus) in the tax table used to develop the ATCF at time 0.
4. Thus the ATCF for selling now becomes the after-tax *opportunity cost* of keeping the defender. Assign this cost to the defender at time 0.

EXAMPLE 13-11

Find the after-tax first cost (today's cost) that should be assigned to a defender asset as described by the following data:

First cost when implemented 5 years ago	= $12,500
Class 43 asset CCA rate	= 30%
Current market value	= $8,000
Remaining useful life in years	10
Annual costs	= $3,000
Annual benefits	= $4,500
Market value after useful life	= $1,500
Marginal tax rate	= 34%

Use the spreadsheet (*books closed*) assumption.

SOLUTION

The defender sign change procedure has four steps.

Steps 1 and 2

Assume that we will sell the defender now (time zero); then:

Current book value (UCC) for equation 11-8

$$UCC_5 = \$12,500(1 - .3/2) \times (1 - 3)^4$$
$$= 2,551.06$$

$$\text{Recaptured CCA} = \$8,000 - \$2,551$$
$$= \$5,449$$

$$\text{Tax effect} = -\$5,449 \times 34\%$$
$$= \$(1,853)$$

$$\text{Net salvage} = \$8,000 - \$1,853$$
$$= \$6,147$$

Thus if we sell at the outset ($n = 0$) we have

Year, n	Before-Tax Cash Flow	Depreciation	Taxable Income	Income Taxes	After-Tax Cash Flow
0	$8,000		$5,449	($1,853)	$6,147

Steps 3 and 4

Change signs. The after-tax cash flow figure is the time zero *opportunity cost* of keeping the defender.

Year, n	Before-Tax Cash Flow	Depreciation	Taxable Income	Income Taxes	After-Tax Cash Flow
0	($8,000)		($5,449)	$1,853	($6,147)

EXAMPLE 13-12

Determine whether the SK-30 laptop of Example 13-9 should be replaced by the EL-40 model. In addition to the data given in Example 13-9, the following estimates have been made:

- The SK-30 maintenance and service contract costs $80 a year.
- The EL-40 will require no maintenance.

- Either laptop is expected to be used for the next 5 years.
- At the end of that time, the SK-30 will have no value, but the EL-40 probably could be sold for $250.
- The EL-40 laptop is faster and easier to use than the SK-30 model. This benefit is expected· to save about $120 a year by reducing the need for part-time employees.

Solve this problem with a MARR equal to 8% after taxes. Both laptop computers will be depreciated by straight-line depreciation using a 4-year depreciable life. The SK-30 is already 2 years old, and so only 2 years of depreciation remain. The analysis period remains at 5 years. Assume a 34% corporate income tax rate.

SOLUTION

Alternative A

Our first option is to keep the SK-30 rather than sell it. Use the following data to compute the after-tax cash flows over the 5-year study period.

	Year	Before-Tax Cash Flow	Straight-Line Depreciation	Taxable Income	34% Income Taxes	After-Tax Cash Flow
(Sell)	0	$200		−$600*	$204*	$404
(Keep)	0	−200		+600	−204*	−404†
	1	−80	$400	−480	+163	+83
	2	−80	400	−480	+163	+83
	3	−80	0	−80	+27	−53
	4	−80	0	−80	+27	−53
	5	−80	0	−80	+27	−53
	5	0 Salvage		0‡	0	0

*If sold for $200, there would be a $600 loss on disposal. If $600 of gains were offset by the loss during the year, there would be no gain to tax, saving 34% × $600 = $204. If the SK-30 is not sold, this loss is not realized, and the resulting income taxes will be a $204 higher than if it had been sold.

†This is the sum of the $200 selling price forgone plus the $204 income tax saving forgone.

‡Gain/Loss = Market value − Book value = 0 − [1600 − (4)400] = 0.

$$EUAC = [404 − 83(P/A, 8\%, 2) + 53(P/A, 8\%, 3)(P/F, 8\%, 2)](A/P, 8\%, 5)$$
$$= \$93$$

Alternative B

Our second option is to purchase an EL-40. Use the following data to compute the after-tax cash flow.

Year	Before-Tax Cash Flow	Straight-Line Depreciation	Taxable Income	34% Income Taxes	After-Tax Cash Flow
0	−$1050				−$1050
1	+120	$200	−$80	+$27	+147
2	+120	200	−80	+27	+147
3	+120	200	−80	+27	+147
4	+120	200	−80	+27	+147
5	+120	0	+120	−41	−79
5	+250		0*	0	+250

*Gain/Loss = Market value − Book value = 250 − [1050 − (4)200] = 0.

Compute the EUAC:

$$EUAC = [1050 - 147(P/A, 8\%, 4) - (79 + 250)(P/F, 8\%, 5)](A/P, 8\%, 5)$$
$$= \$85$$

Based on this after-tax analysis, the EL-40 is the preferred alternative.

EXAMPLE 13-13

Solve Example 13-12 by computing the rate of return on the difference between alternatives. In Example 13-12, the two alternatives were 'keep the SK-30' and 'buy an EL-40.' The difference between the alternatives would be:

Buy an EL-40	Rather than	Keep the SK-30
Alternative B	*minus*	Alternative A

SOLUTION

The after-tax cash flow for the difference between the alternatives may be computed as follows:

Year	A	B	B−A
0	−$404	−$1050	−$646
1	+83	+147	+64
2	+83	+147	+64
3	−53	+147	+200
4	−53	+147	+200
5	−53	+329	+382

The rate of return on the difference between the alternatives is computed as follows:

PW of cost = PW of benefit

$$646 = 64(P/A, i, 2) + 200(P/A, i, 2)(P/F, i, 2) + 382(P/F, i, 5)$$

Try $i = 9\%$:

$$646 \overset{?}{=} 64(1.759) + 200(1.759)(0.8417) + 382(0.6499)$$

$$\overset{?}{=} 656.9$$

Try $i = 10\%$:

$$646 \overset{?}{=} 64(1.736) + 200(1.736)(0.8264) + 382(0.6209)$$

$$\overset{?}{=} 635.2$$

The rate of return IRR $= 9\% + (0.10 - 0.09)\left(\dfrac{656.9 - 646.0}{656.9 - 635.2}\right) = 9.5\%$

The rate of return is greater than the 8% after-tax MARR. The increment of investment is desirable. Buy the EL-40 model.

Example 13-13 illustrates the use of the incremental IRR method in defender-challenger comparisons on an after-tax basis. In the example this analysis method is appropriate because the life of the SK-30 (defender) is equivalent to the economic life of the EL-40 (challenger). Since these lives are the same, a present or future worth analysis also could have been used. However, far more commonly the lives of the defender and the challenger are different. In such cases, if we are willing to make the assumption of a continuing requirement for the asset, then the comparison method to use is the annual cash flow method. As in our previous discussion, this method allows a direct comparison of the annual cash flow of the defender asset over its life against the annual cash flow of the challenger over its minimum cost life. The annual cash flow method is also suitable for comparison when the lives are equal (as in Example 13-12), but more importantly, it is suitable for the case of lives that are different, under the *replacement repeatability assumptions*.

Minimum Cost Life Problems

In this section we illustrate the effect that tax considerations can have on the calculation of the minimum cost life of the defender and the challenger. The calculation of minimum EUAC on an after-tax basis can be affected by both the depreciation method used and by changes in the asset's market value over time, for either defender or challenger. Using an accelerated depreciation method (like CCA) tends to reduce the after-tax costs early in the life of an asset. This effect alters the shape of the total EUAC curve—the concave shape can be shifted and the minimum EUAC changed. Example 13-14 illustrates the effect that taxes can have when either the straight-line or CCA depreciation method is used.

EXAMPLE 13-14

Some new production machinery has a first cost of $100,000 and a useful life of 10 years. Its estimated operating and maintenance (O&M) costs the first year are $10,000, which will increase annually by $4000. The asset's before-tax market value will be $50,000 at the end of the first year and then will decrease by $5000 annually. Calculate the after-tax cash flows using CCA depreciation. This property is a class 43 asset with a 30% CCA rate. The company uses a 6% after-tax MARR and is subject to a combined federal and provincial tax rate of 40%.

SOLUTION

To find the minimum cost life of this new production machinery, we first find the after-tax cash flow effect of the O&M costs and depreciation (Table 13-2). Then we find the ATCFs of disposal if the equipment is sold in each of the 10 years (Table 13-3). Finally in the closing section on spreadsheets, we combine these two ATCFs (in Figure 13-7) and choose the minimum cost life.

In Table 13-2, the O&M expense simply starts at $10,000 and increases at $4000 per year. The taxable income, which is simply the O&M costs minus the depreciation values, is then multiplied by minus the tax rate to determine the impact of this taxable income on taxes. The O&M expense plus taxes is the Table 13-2 portion of the total ATCF.

Regarding the market value data in this problem, it should be pointed out that the initial decrease of $50,000 in year 1 is not uncommon. This is especially true for custom-built equipment for a particular and unique application at a specific plant. Such equipment would not be as valuable to others in the marketplace as to the company for which it was built. Also,

CCA rate = 30%
Cost basis = $100,000
Tax rate = 40%

TABLE 13-2 ATCF for O&M and Depreciation for Example 13-14

Year, t	O&M Expense	CCA Depreciation	Taxable Income	Taxes (at 40%)	O&M ATCF
1	−$10,000	$15,000	−$25,000	$10,000	−$ 0
2	−$14,000	$25,500	−$39,500	$15,800	$ 1,800
3	−$18,000	$17,850	−$35,850	$14,340	−$ 3,660
4	−$22,000	$12,495	−$34,495	$13,798	−$ 8,202
5	−$26,000	$ 8,747	−$34,747	$13,899	−$12,101
6	−$30,000	$ 6,123	−$36,123	$14,449	−$15,551
7	−$34,000	$ 4,286	−$38,286	$15,314	−$18,686
8	−$38,000	$ 3,000	−$41,000	$16,400	−$21,600
9	−$42,000	$ 2,100	−$44,100	$17,640	−$24,360
10	−$46,000	$ 1,470	−$47,470	$18,988	−$27,012

the $100,000 first cost (cost basis) could have included costs of installation, facility modifications, or removal of old equipment. The $50,000 is realistic for the market value of one-year-old equipment.

The next step is to determine the ATCFs that would occur in each possible year of disposal. (The ATCF for year 0 is easy; it is −$100,000.) For example, as shown in Table 13-3, in Year 1 there is a $35,000 loss as the book value exceeds the market value. The tax savings from this loss are added to the salvage (market) value to determine the ATCF (*If the asset is disposed during Year 1*).

This table assumes that depreciation is taken during the year of disposal and then calculates the recaptured depreciation (gain) or loss on the book value at the end of the year.

TABLE 13-3 ATCF in Year of Disposal for Example 13-14

Year, t	Market Value	Book Value UCC	Recapture or Loss	Tax Effect	ATCF If Disposed of
1	$50,000	$85,000	−$35,000	−$14,000	$64,000
2	45,000	59,500	−14,500	−5,800	50,800
3	40,000	41,650	−1,650	−660	40,660
4	35,000	29,155	5,845	2,338	32,662
5	30,000	20,409	9,592	3,837	26,163
6	25,000	14,286	10,714	4,286	20,714
7	20,000	10,000	10,000	4,000	16,000
8	15,000	7,000	8,000	3,200	11,800
9	10,000	4,900	5,100	2,040	7,960
10	5,000	3,430	1,570	628	4,372

SPREADSHEETS AND REPLACEMENT ANALYSIS

Spreadsheets are obviously useful in nearly all after-tax calculations. However, they are absolutely required for optimal life calculations in after-tax situations. Because CCA is the tax law, the after-tax cash flows are different in every year. Thus, the NPV function and the PMT function are both needed to find the minimum EUAC after taxes. Figure 13-7 illustrates the calculation of the minimum cost life for Example 13-14.

In Figure 13-7, the NPV finds the present worth of the irregular cash flows from Period 1 through Period t for $t = 1$ to life. The PV function is used to find the PW of the salvage value. Then PMT can be used to find the EUAC over each potential life. Before-tax replacement analysis can also be done this way. The spreadsheet block function NPV is used to find the PW of cash flows from Period 1 to Period t. Note that the cell for Period 1 is an absolute address and the cell for period t is a relative address. This allows the formula to be copied.

	A	B	C	D	E	F
1		Table 13-2	Table 13-3	6% Interest Rate		
2		O&M & Depr.				
3	Year	ATCF	ATCF	NPV	EAC	
4	0			−100,000		
5	1	0	64,000	−39,623	42,000	=PMT($D$1,B5,E5)
6	2	1,800	50,800	−53,186	29,010	
7	3	−3,660	40,660	−67,332	25,190	
8	4	−8,202	32,662	−82,096	23,692	
9	5	−12,101	26,163	−97,460	23,137	optimal life
10	6	−15,551	20,714	−115,049	23,397	
11	7	−18,686	16,000	−136,553	24,461	
12	8	−21,600	11,800	−155,103	24,977	
13	9	−24,360	7,960	−171,546	25,221	
14	10	−27,012	4,372	−186,476	25,336	=PMT($D$1,B14,E14)
15						
16			=NPV($D$1,B5:B14)+$D$4+PV($D$1,A14,0,−C14)			
17			=NPV(*i*, B column) + year 0 + present value of a future salvage			

FIGURE 13-7 Spreadsheet for life with minimum after-tax cost.

SUMMARY

In selecting equipment for a new plant the question is, which of the machines available on the market will be more economical? But when a given piece of equipment is now performing the desired task, the analysis is more complicated. The existing equipment (called the **defender**) is already in place, so the question is, shall we replace it now, or shall we keep it for one or more years? When a replacement is indicated, it will be by the best available replacement equipment (called the **challenger**). When we already have equipment, there may be a tendency to use past costs in the replacement analysis. But only present and future costs are relevant.

This chapter has presented three distinctly different **replacement analysis techniques,** which are all relevant and appropriate depending upon the conditions of the cash flows for the defender and the challenger. In all cases of analysis the simplifying **replacement repeatability assumptions** are accepted. These state that the defender will ultimately be replaced by the current best challenger (as will any challengers implemented in the future) and that we have an indefinite need for the service of the asset in question.

In the usual case, marginal cost data are both available and are increasing every year, and thus **replacement analysis technique 1** allows a comparison of *the marginal cost data of the defender, with the minimum EUAC of the challenger.* In this case we should keep the defender as long as its marginal cost is less than the minimum EUAC of the challenger.

When marginal cost data are available for the defender but are not increasing every year, **replacement analysis technique 2** calls for a comparison of *the lowest EUAC of the defender with the minimum EUAC of the challenger.*

If the defender's lowest EUAC is smaller, we do not replace it yet. If the challenger's EUAC is less, we would select this asset in place of the defender today. If the cost data for the challenger and the defender do not change, we will replace the defender after the life that minimizes its EUAC when its marginal cost data exceed the minimum EUAC for the challenger.

In the case of no marginal cost data being available for the defender, **replacement analysis technique 3** prescribes a comparison of *the EUAC of the defender over its stated life with the minimum EUAC of the challenger.*

As in the case of replacement analysis technique 2, we would select the alternative that has the smallest EUAC. An important concept when calculating the EUAC of both defender and challenger is the first cost to be assigned to each alternative for calculation purposes. When the lives of the two alternatives are equivalent, either an **opportunity cost** or a **cash flow approach** may be used. However, in the more common case of different useful lives, only the opportunity cost approach accurately assigns an investment cost to the defender and challenger assets.

It is important when performing engineering economic analyses to include the effects of taxes. In minimum economic life and marginal cost calculations, and in finding cash flows over the life of a defender or challenger, the effects of taxes can be significant. The replacement analysis techniques described on a before-tax basis are also used for the after-tax case—the difference being that after-tax cash flows are used in place of before-tax cash flows. Effects on an after-tax basis include opportunity gains and losses at time zero, income taxes and depreciation over the assets' lives, and gains and losses at disposition time. The *sign-change procedure* can be used to determine opportunity gains and losses when one is assigning a first cost to the defender in after-tax problems.

Replacement analyses are vastly important, yet often ignored by companies as they invest in equipment and facilities. Investments in business and personal assets should not be forgotten once an initial economic evaluation has produced a 'buy' recommendation. It is important to continue to evaluate assets over their respective life cycles to ensure that monies invested are continuing to yield the greatest benefit for the investor. Replacement analyses help us to ensure this.

PROBLEMS

13-1 Usually there are two alternatives in a replacement analysis. One alternative is to replace the defender now. Which one of the following is the second alternative?

(1) Keep the defender for its remaining useful life.

(2) Keep the defender for another year and then re-examine the situation.

(3) Keep the defender until there is an improved challenger that is better than the present challenger.

(4) The answer to this question depends on the data available for the defender and challenger as well as the assumptions made regarding the period of needed service and future challengers.

13-2 The economic life of the defender can be obtained if certain estimates about the defender can be made. Assuming those estimates prove to be exactly correct, one can accurately predict the year when the defender should be replaced, even if nothing is known about the challenger. Is this assumption true or false? Explain.

13-3 A proposal has been made to replace a large heat exchanger (3 years ago, the initial cost was $85,000) with a new, more efficient unit at a cost of $120,000. The existing heat exchanger is being depreciated by the CCA method. Its present book value is $35,400, but it has no current value, since its scrap value just equals the cost to remove it from the plant.

In preparing the before-tax economic analysis to see whether the existing heat exchanger should be replaced, the proper treatment of the $35,400 book value becomes an issue. Three possibilities are that the $35,400 book value of the old heat exchanger is:
(1) *Added* to the cost of the new exchanger in the economic analysis.
(2) *Subtracted* from the cost of the new exchanger in the economic analysis.
(3) *Ignored* in this before-tax economic analysis.
Which of the three possibilities is correct?

13-4 A machine tool, which has been used in a plant for 10 years, is being considered for replacement. It cost $9500 and was depreciated by CCA depreciation. An equipment dealer says that the machine has no resale value. Maintenance on the machine tool has been a problem, with an $800 cost this year. Future annual maintenance costs are expected to be higher. What is the economic life of this machine tool if it is kept in service?

13-5 A new $40,000 bottling machine has just been installed in a plant. It will have no salvage value when it is removed. The plant manager has asked you to estimate the economic service life for the machine, ignoring income taxes. He estimates that the annual maintenance cost will be constant at $2500 per year. What service life will result in the lowest equivalent uniform annual cost?

13-6 Which one of the following is the proper dollar value of defender equipment to use in replacement analysis?
(1) Original cost.
(2) Present market value.
(3) Present trade-in value.
(4) Present book value.
(5) Present replacement cost, if different from original cost.

13-7 You have the following options for a major equipment unit:
(a) Buy new.
(b) Trade in and buy a similar, rebuilt equipment from the manufacturer.
(c) Have the manufacturer rebuild your equipment with all new available options.
(d) Have the manufacturer rebuild your equipment to the original specifications.
(e) Buy used equipment.
State the advantages and disadvantages of each option with respect to after-tax benefits.

13-8 Consider following data for a defender asset. What is the correct replacement analysis technique for comparing this asset to a competing challenger? How is this method used? That is, what comparison is made, and how do we choose?

Year, n	BTCF in Year n (marginal costs)
1	−$2000
2	−1750
3	−1500
4	−1250
5	−1000
6	−1000
7	−1000
8	−1500
9	−2000
10	−3000

13-9 The Clap Chemical Company needs a large insulated stainless steel tank for the expansion of its plant. Clap has located such a tank at a recently closed brewery. The brewery has offered to sell the tank for $15,000 delivered to the chemical plant. The price is so low that Clap believes it can sell the tank at any future time and recover its $15,000 investment.

The outside of the tank is covered with heavy insulation that requires considerable maintenance with estimated costs as follows:

Year	Insulation Maintenance Cost
0	$2000
1	500
2	1000
3	1500
4	2000
5	2500

(a) If before-tax MARR is 15%, what life of the insulated tank has the lowest EUAC?
(b) Is it likely that the insulated tank will be replaced by another tank at the end of the period with the lowest EUAC? Explain.

13-10 The plant manager has just purchased a piece of unusual machinery for $10,000. Its resale value at the end of 1 year is estimated to be $3000, because the device is sought by antique collectors, resale value is rising at the rate of $500 per year.

The maintenance cost is expected to be $300 per year for each of the first 3 years, and then it is expected to double each year after that. Thus the fourth-year maintenance will be $600; the fifth-year maintenance, $1200, and so on. If before-tax MARR is 15%, what life of this machinery has the lowest EUAC?

13-11 In a replacement analysis problem, the following facts are known:

Initial cost	$12,000
Annual maintenance	None for the first 3 years
	$2000 at the end of the fourth year
	$2000 at the end of the fifth year
	Increasing $2500 per year after the fifth year ($4500 at the end of the sixth year, $7000 at the end of the seventh year, etc.)

Actual salvage value in any year is zero. Assume a 10% interest rate and ignore income taxes. Compute the life for this challenger having the lowest EUAC. (*Answer:* 5 years)

13-12 An injection moulding machine has a first cost of $1,050,000 and a salvage value of $225,000 in any year the machine is sold. The maintenance and operating cost is $235,000 with an annual gradient of $75,000. The MARR is 10%. What is the most economic life?

13-13 Mario's father read that, at the end of each year, a car is worth 25% less than it was at the beginning of the year. After a car is three years old, the rate of decline falls to 15%. Maintenance and operating costs, on the other hand, increase as the age of the car increases. Because of the manufacturer's warranty, first-year maintenance is very low.

Age of Car (years)	Maintenance Expense
1	$ 50
2	150
3	180
4	200
5	300
6	390
7	500

Mario decided this is a good economic analysis problem. His dad wants to keep his annual cost of car ownership low. The car Mario's dad prefers costs $11,200

new. Should he buy a new or a used car and, if used, when would you suggest he buy it, and how long should it be kept? Give a practical, rather than a theoretical, solution. (*Answer:* Buy a 3-year-old car and keep it 3 years.)

13-14 A professor of engineering economics owns a 1996 car. In the past 12 months, he has paid $2000 to replace the transmission, bought two new tires for $160, and installed a new tape deck for $110. He wants to keep the car for 2 more years because he invested money 3 years ago in a 5-year certificate of deposit, which is earmarked to pay for his dream machine, a red European sports car. Today the old car's engine failed. The professor has two alternatives. He can have the engine overhauled at a cost of $1800 and then most likely have to pay another $800 per year for the next 2 years for maintenance. The car will have no salvage value at that time. Alternatively, a colleague offered to make the professor a $5000 loan to buy another used car. He must pay the loan back in two equal instalments of $2500 due at the end of Year 1 and Year 2, and at the end of the second year he must give the colleague the car. The 'new' used car has an expected annual maintenance cost of $300. If the professor selects this alternative, he can sell his current vehicle to a junkyard for $1500. Interest is 5%. Using present worth analysis, decide which alternative he should select and explain why.

13-15 The Ajax Corporation purchased a railroad tank car 8 years ago for $60,000. It is being depreciated by SOYD depreciation, assuming a 10-year depreciable life and a $7000 salvage value. The tank car needs to be reconditioned now at a cost of $35,000. If this is done, it is estimated the equipment will last for 10 more years and have a $10,000 salvage value at the end of the 10 years.

On the other hand, the existing tank car could be sold now for $10,000 and a new tank car purchased for $85,000. The new tank car would be depreciated by CCA depreciation. Its estimated actual salvage value would be $15,000. In addition, the new tank car would save $7000 per year in maintenance costs, compared to the reconditioned tank car.

Using a 15% before-tax rate of return, determine whether the existing tank car should be reconditioned or a new one purchased. (*Note:* The problem statement provides more data than are needed, which is typical of real situations.) (*Answer:* Recondition the old tank car.)

13-16 The Quick Manufacturing Company, a large profitable corporation, is considering the replacement of a production machine tool. A new machine would cost $3700, have a 4-year useful and depreciable life, and have no salvage value. For tax purposes, sum-of-years'-digits depreciation would be used. The existing machine tool was purchased 4 years ago at a cost of $4000 and has been depreciated by straight-line depreciation assuming an 8-year life and no salvage value. The tool could be sold now to a used equipment dealer for $1000 or be kept in service for another 4 years. It would then have no salvage value. The new machine tool would save about $900 per year in operating costs compared to the existing machine. Assume a 40% combined federal and provincial tax rate.

  (a) Compute the before-tax rate of return on the replacement proposal of installing the new machine rather than keeping the existing machine.

  (b) Compute the after-tax rate of return on this replacement proposal.   (*Answer: (a)* 12.6%)

13-17 The local telephone company purchased four special pole hole diggers 8 years ago for $14,000 each. They have been in constant use to the present. Owing to an increased workload, additional machines will soon be required. Recently it was announced that an improved model of the digger has been put on the market. The new machines have a higher production rate and lower maintenance expense than the old machines, but will cost $32,000 each. The service life of the new machines is estimated to be 8 years, with salvage value estimated at $750 each. The four original diggers have an immediate salvage of $2000 each and an estimated salvage value of $500 each 8 years hence. The estimated average annual maintenance expense associated with the old machines is approximately $1500 each, compared with $600 each for the new machines.

  A field study and trial show that the workload would require three additional new machines if the old machines were continued in service. However, if the old machines were all retired from service, the present workload plus the estimated increased load could be carried by six new machines with an annual savings of $12,000 in operation costs. A training program to teach employees to run the machines will be necessary at an estimated cost of $700 per machine. If the MARR is 9% before taxes, what should the company do?

13-18 Five years ago, Thomas Martin bought and implemented production machinery that had a first cost of $25,000. At the time of the initial purchase it was estimated that yearly costs would be $1250, increasing by $500 in each year that followed. It was also estimated that the market value of this machinery would be only 90% of the previous year's value. It is currently projected that this machine will be useful in operations for 5 more years. There is a new machine available now that has a first cost of $27,900 and no yearly costs over its 5 year minimum cost life. If Thomas Martin uses an 8% before-tax MARR, when, if at all, should he replace the existing machinery with the new unit?

13-19 Consider Problem 13-18 involving Thomas Martin. Suggest when, if at all, the old should be replaced with the new, if the values for the old machine are as follows. The old machine retains only 70% of its value in the market from year to year. The yearly costs of the old machine were $3000 in Year 1 and increase at 10% thereafter.

13-20 Mary O'Leary's company ships fine wool garments from County Cork, Ireland. Five years ago she purchased some new automated packing equipment having a first cost of $125,000 and a CCA rate of 30%. The annual costs for operating, maintenance, and insurance, as well as market value data for each year of the equipment's 10-year useful life are as follows:

Year, n	Annual Costs in Year n for			Market Value in Year n
	Operating	Maintenance	Insurance	
1	$16,000	$ 5,000	$17,000	$80,000
2	20,000	10,000	16,000	78,000
3	24,000	15,000	15,000	76,000
4	28,000	20,000	14,000	74,000
5	32,000	25,000	12,000	72,000
6	36,000	30,000	11,000	70,000
7	40,000	35,000	10,000	68,000
8	44,000	40,000	10,000	66,000
9	48,000	45,000	10,000	64,000
10	52,000	50,000	10,000	62,000

Now Mary is looking at the remaining 5 years of her investment in this equipment, which she had initially evaluated on the basis of an after-tax MARR of 25% and a tax rate of 35% on ordinary equipment. Assuming that the replacement repeatability assumptions are valid, answer the following questions.

  (a) What is the before-tax marginal cost for the remaining 5 years?

(b) When, if at all, should Mary replace this packing equipment if a new challenger, with a minimum EUAC of $110,000, has been identified: use the data from the table and the decision map from Figure 13-1.

13-21 Big-J Construction Company, Inc. (Big-J CC) is conducting routine periodic reviews of existing field equipment. The owner of Big-J CC has asked for a replacement evaluation of a paving machine now in use. A newer, more efficient machine is being considered. The old machine was purchased 3 years ago for $200,000, and yearly operating and maintenance costs are as follows. Big-J CC uses a MARR of 20%; the current market value of the paver is $120,000.

Estimates of Operating & Maintenance Cost and Market Value for Next 7 Years (old paver)

Year, n	Operating Cost in Year n	Maintenance Cost in Year n	Market Value If Sold in Year n
1	$15,000	$ 9,000	$85,000
2	15,000	10,000	65,000
3	17,000	12,000	50,000
4	20,000	18,000	40,000
5	25,000	20,000	35,000
6	30,000	25,000	30,000
7	35,000	30,000	25,000

Data for the new paving machine have been analyzed. Its most economic life is at 8 years with a minimum EUAC of $62,000. Make a recommendation to Big-J CC regarding the paving machine in question.

13-22 VMIC Corp. has asked you to look at the following data. The interest rate is 10%. After considering the data, answer the questions below.

Year, n	Marginal Cost Data Defender	EUAC If Kept n Years Challenger
1	$2500	$4500
2	2400	3600
3	2300	3000
4	2550	2600
5	2900	2700
6	3400	3500
7	4000	4000

(a) What is the lowest EUAC of the *defender*?
(b) What is the minimum cost life of the *challenger*?
(c) When, if at all, should we replace the *defender* with the *challenger*?

13-23 SHOJ Enterprises has asked you to look at the following data. The interest rate is 10%. After considering the data, answer the questions below.

Year, n	Marginal Cost Data Defender	EUAC If Kept n Years Challenger
1	$3000	$4500
2	3150	4000
3	3400	3300
4	3800	4100
5	4250	4400
6	4950	6000

(a) What is the lowest EUAC of the *defender*?
(b) What is the economic life of the *challenger*?
(c) When, if at all, should we replace the *defender* with the *challenger*?

13-24 As proprietor of your own business, you are considering the option of purchasing a new high-efficiency machine to replace older machines currently in use. You believe that the new technology can be used to replace four of the older machines, each with a current market value of $600. The new machine will cost $5000 and will save the equivalent of 10,000 kWh of electricity per year over the older machines. After a period of 10 years, neither option (new or old) will have any market value. If you use a before-tax MARR of 25% and pay $0.075 per kilowatt hour, would you replace the old machines today with the new one?

13-25 Fifteen years ago the Acme Manufacturing Company bought a propane-powered forklift truck for $4800. The company depreciated the forklift, using straight-line depreciation, a 12-year life, and zero salvage value. Over the years, the forklift has been a good piece of equipment, but lately the maintenance cost has risen sharply. Estimated end-of-year maintenance costs for the next 10 years are as follows:

Year	Maintenance Cost
1	$400
2	600
3	800
4	1000
5–10	1400/year

The old forklift has no present or future net salvage value since its scrap metal value just equals the cost of hauling it away. A replacement is now being considered for the old forklift. A modern unit can be purchased for $6500. It has an economic life equal to its 10-year depreciable life. Straight-line depreciation will be employed, with zero salvage value at the end of the 10-year depreciable life. At any time the new forklift can be sold for its book value. Maintenance on the new forklift is estimated to be a constant $50 per year for the next 10 years, after which maintenance is expected to increase sharply. Should Acme Manufacturing keep its old forklift truck for the present, or replace it now with a new one? The firm expects an 8% after-tax rate of return on its investments. Assume a 40% combined income tax rate.

(*Answer:* Keep the old forklift truck.)

13-26 A firm is concerned about the condition of some of its plant machinery. Bill James, a newly hired engineer, was assigned the task of reviewing the situation and determining what alternatives are available. After a careful analysis, Bill reports that there are five feasible, mutually exclusive alternatives.

Alternative A: Spend $44,000 now repairing various items. The $44,000 can be charged as a current operating expense (rather than capitalized) and deducted from other taxable income immediately. These repairs are anticipated to keep the plant functioning for the next 7 years with operating costs remaining at present levels.

Alternative B: Spend $49,000 to buy general-purpose equipment. Depreciation would be straight-line, with the depreciable life equal to the 7-year useful life of the equipment. The equipment will have no end-of-useful-life salvage value. The new equipment will reduce operating costs $6000 per year below the present level.

Alternative C: Spend $56,000 to buy new specialized equipment. This equipment would be depreciated by sum-of-years'-digits depreciation over its 7-year useful life. This equipment would reduce operating costs $12,000 per year below the present level. It will have no end-of-useful-life salvage value.

Alternative D: This alternative is the same as Alternative B, except that this particular equipment would reduce operating costs $7000 per year below the present level.

Alternative E: This is the 'do-nothing' alternative. If nothing is done, future annual operating costs are expected to be $8000 above the present level.

This profitable firm pays 40% corporate income taxes. In their economic analysis, they require a 10% after-tax rate of return. Which of the five alternatives should the firm adopt?

13-27 Machine A has been completely overhauled for $9000 and is expected to last another 12 years. The $9000 was treated as an expense for tax purposes last year. Machine A can be sold now for $30,000 net after selling expenses but will have no salvage value 12 years hence. It was bought new 9 years ago for $54,000 and has been depreciated since then by straight-line depreciation using a 12-year depreciable life.

Because less output is now required, Machine A can now be replaced with a smaller machine: Machine B costs $42,000, has an anticipated life of 12 years, and would reduce operating costs $2500 per year. It would be depreciated by straight-line depreciation with a 12-year depreciable life and no salvage value.

The income tax rate is 40%. Compare the after-tax annual cost of the two machines and decide whether Machine A should be retained or replaced by Machine B. Use a 10% after-tax rate of return in the calculations.

13-28 Fred's Rodent Control Corporation has been using a low-frequency sonar device to locate subterranean pests. This device was purchased 5 years ago for $18,000. The device has been depreciated using SOYD depreciation with an 8-year depreciable life and a salvage value of $3600. At present, it could be sold to the cat next door for $7000. If it is kept for the next 3 years, its market value is expected to drop to $1600.

A new lightweight subsurface heat-sensing searcher (SHSS) that is available for $10,000 would improve the annual net income by $500 for each of the next 3 years. The SHSS would be depreciated at a CCA rate of 30%. At the end of 3 years, the SHSS should have a market value of $4000. Fred's Rodent Control is a profitable enterprise subject to a 40% tax rate.

(*a*) Construct the after-tax cash flow for the old sonar unit for the next 3 years.

(*b*) Construct the after-tax cash flow for the SHSS unit for the next 3 years.

(*c*) Construct the after-tax cash flow for the difference between the SHSS unit and the old sonar unit for the next 3 years.

(*d*) Should Fred buy the new SHSS unit if his MARR is 20%? You do not have to calculate the incremental rate of return; just show how you reach your decision.

13-29 (*a*) A new employee at CLL Engineering Consulting Inc., you are asked to join a team performing an economic analysis for a client. Your team seems stumped on how to assign an after-tax first cost to the defender and challenger assets under consideration. Your task is to take the following data and find the ATCF for each alternative. There is no need for a complete analysis—your colleagues can handle that responsibility—they need help only with the time zero ATCFs. CLL Inc. has a combined income tax rate of 45% on ordinary income, depreciation recapture, and losses.

Defender: This asset was placed in service 7 years ago. At that time the $50,000 cost basis was set up on a straight-line depreciation schedule with an estimated salvage value of $15,000 over its 10-year ADR life. This asset has a present market value of $30,000.

Challenger: The new asset being considered has a first cost of $85,000 and will be depreciated with a CCA rate of 25%. This asset qualifies for a 10% investment tax credit.

(*b*) How would your calculations change if the present market value of the *defender* is $25,500?

(*c*) How would your calculations change if the present market value of the *defender* is $18,000?

13-30 Foghorn Leghorn is considering the replacement of an old egg-sorting machine used with his Foggy's Farm Fresh Eggs business. The old egg machine is not running quite the way it was originally designed to run and will require an additional investment now of $2500 (expensed at time zero) to get it back in working shape. This old machine was purchased 6 years ago for $5000 and has been depreciated by the straight-line method at $500 per year. Six years ago the estimated salvage value for tax purposes was $1000. Operating expenses for the old machine are projected at $600 this next year and are increasing by $150 per year each year thereafter. Foggy projects that with refurbishing, machine will last another 3 years. Foggy believes that he could sell the old machine as is today for $1000 to his friend Fido for sorting bones. He also believes he could sell it 3 years from now at the barnyard flea market for $500.

The new egg-sorting machine, a deluxe model, has a purchase price of $10,000 and will last 6 years, at which time it will have a salvage value of $1000. The new machine qualifies for a CCA rate of 30% and will have operating expenses of $100 the first year, increasing by $50 per year thereafter. Foghorn uses an after-tax MARR of 18% and a tax rate of 35% on original income.

(*a*) What was the depreciation life used with the defender asset (the old egg sorter)?

(*b*) Calculate the after-tax cash flows for both the defender and challenger assets.

(*c*) Use the annual cash flow method to offer a recommendation to Foggy. What assumptions did you make in this analysis?

13-31 BC Junction purchased some embroidering equipment for their Denver facility 3 years ago for $15,000. This equipment qualified for a CCA rate of 30%. Maintenance costs are estimated to be $1000 this next year and will increase by $1000 per year thereafter. The market (salvage) value for the equipment is $10,000 at the end of this year and declines by $1000 per year in the future. If BC Junction has an after-tax MARR of 30%, a marginal tax rate of 45% on ordinary income, and depreciation recapture and losses, what after-tax life of this previously purchased equipment has the lowest EUAC? Use a spreadsheet to develop your solution.

13-32 Reconsider the acquisition of packing equipment for Mary O'Leary's business, as described in Problem 13-20. Given the data tabulated there, and again using an after-tax MARR of 25% and a tax rate of 35% on ordinary income to evaluate the investment, determine the after-tax lowest EUAC of the equipment. Use a spreadsheet to develop your solution.

CHAPTER 14

Inflation and Price Change

The Athabasca Oil Sands

For centuries, people have known about the sticky bitumen that lined the banks of the Athabasca River in northern Alberta. Even before the coming of the European fur traders, Native people used it to seal canoes. Over the years, many people dreamed of producing usable oil from the bitumen, but the sand and oil are not easily separated, and the recovery was not viewed as economically viable. In the 1950s, the world price of oil was around $3 a barrel and the estimated cost of mining and separating oil from the sands was over $30 a barrel.

In 1964, the Sun Oil Company, with government support, formed the Great Canadian Oil Sands Company, and in 1967 it started to mine and process shallow deposits of oil sands. The target production was 31,000 barrels per day, and the initial production cost would be in the area of $25 a barrel. The world price was then about $3.50, but the planners were predicting that production costs would decline and market prices would increase.

Forty years later, in 2004, two major firms in the mineable oil sands supply one-third of Canada's oil, other firms extract bitumen from deeper deposits by steam heating, and $40 billion worth of new oil sands projects are on the books. The successor to Great Canadian Oil Sands, Suncor Energy, has 4500 employees and is producing 130,000 barrels of oil per day. The production costs are in the range of $12 a barrel, and the world price of oil is near $50.

—Peter Flynn

After Completing This Chapter...

The student should be able to:

- Describe inflation, explain how it happens, and list its effects on purchasing power.
- Define real and actual dollars and interest rates.
- Conduct constant dollar and the current dollar analyses.
- Define and use composite and commodity-specific price indexes.
- Develop and use cash flows that inflate at different interest rates and cash flows subject to different interest rates per period.
- Incorporate the effects of inflation in before-tax and after-tax calculations.
- Develop spreadsheets to solve engineering economy problems that incorporate the effects of inflation and price change.

QUESTIONS TO CONSIDER

1. Market analysts have estimated that at a world price of $25 per barrel, oil sands projects will provide about a 10 per cent rate of return. How will inflation affect these estimates?
2. The prices of some items—for example, gas at the pumps, and houses—increase over time while others, such as calculators and computers, decline in price. Given these variations, how can we know if inflation is occurring, and how could we measure it.
3. In 1967, the Canadian consumer price index was 21.5; in 2004, the CPI was 123.2. What is the 2004 production cost in 1967 dollars?

Thus far we have assumed that the dollars in our analyses were unaffected by inflation or price change. However, this assumption is not always valid or realistic. If inflation occurs in the general economy, or if there are price changes in economic costs and benefits, the impact can be substantial on both before- and after-tax analyses. In this chapter we develop several key concepts and illustrate how inflation and price changes may be incorporated into our problems.

MEANING AND EFFECT OF INFLATION

Inflation is an important concept because the purchasing power of money used in most world economies rarely stays constant. Rather, over time the amount of goods and services that can be bought with a fixed amount of money tends to change. Inflation causes money to lose **purchasing power.** That is, when prices are inflated we can buy less with the same amount of money. **Inflation makes future dollars less valuable than present dollars.** Think about examples in your own life, or for an even starker comparison, ask your grandparents how much a loaf of bread or a new car cost 50 years ago. Then compare these prices with what you would pay today for the same items. This exercise will reveal the effect of inflation: as time passes, goods and services cost more, and more of the same monetary units are needed to purchase the same goods and services.

Because of inflation, dollars in one period of time are not equivalent to dollars in another. We know from our previous study that engineering economic analysis requires that comparisons be made on an **equivalent** basis. So, it is important for us to be able to incorporate the effects of inflation in our analysis of alternatives.

When the purchasing power of a monetary unit *increases* rather than decreases as time passes, the result is **deflation.** Deflation, very rare in the modern world, nonetheless can exist. Deflation has the opposite effect of inflation—one can buy **more** with money in future years than can be bought today. This deflation makes future dollars more valuable than current dollars.

How Does Inflation Happen?

Economists do not agree on all the sources of inflation, but most believe that the following effects influence inflation either in isolation or in combination.

Money supply: The amount of money in our national economy is thought to have an effect on its purchasing power. If there is too much money in the system (the Bank of Canada controls the flow of money) in relation to goods and services to purchase with that money, the value of dollars tends to decrease. When there are fewer dollars in the system, they become more valuable. The Bank of Canada, through its influence on the money supply, seeks to increase the volume of money in the system at the same rate that the economy is growing.

Exchange rates: The strength of the dollar in world markets affects the profitability of international companies in those markets. Prices may be adjusted to compensate for the relative strength or weakness of the dollar in the world market. As corporations' profits are weakened or eliminated in some markets owing to fluctuations in exchange rates, prices may be raised in other markets to compensate.

Cost-push inflation: This type of inflation develops as producers of goods and services 'push' their increasing operating costs along to the customer through higher prices. These operating costs include fabrication and manufacturing, marketing, and sales.

Demand-pull inflation: This effect is realized when consumers spend money freely on goods and services. Often 'free spending' is at the expense of consumer saving. As more and more people demand certain goods and services, the prices of those goods and services will rise (demand exceeding supply).

A further consideration in analyzing how inflation works is the usually different rates at which prices and wages rise. Do workers benefit if, as their wages increase, the prices of goods and services increase as well? To determine the net effect of differing rates of inflation, we must be able to make comparisons and understand costs and benefits from an equivalent perspective. In this chapter we will learn how to make such comparisons.

Definitions for Considering Inflation in Engineering Economy

The following definitions are used throughout this chapter to illustrate how inflation and price change affect two quantities: interest rates and cash flows.

Inflation rate (f): As suggested earlier, the inflation rate captures the effect of goods and services costing more—a decrease in the purchasing power of dollars. More money is required to buy a good or service whose price has inflated. The inflation rate is measured as the annual rate of increase in the number of dollars needed to pay for the same amount of goods and services.

Real interest rate (i'): This interest rate measures the 'real' growth of our money excluding the effect of inflation. Because it does not include inflation, it is sometimes called the *inflation-free interest rate.*

Market interest rate (i): This is the rate of interest that one obtains in the general marketplace. For instance, the interest rates on savings accounts, chequing accounts, and term deposits quoted at the bank are all market rates. The lending interest rate for cars and boats is also a market rate. This rate is sometimes called the *combined interest* rate because it incorporates the effect of both real money growth **and** inflation. We can view i as follows:

Market interest rate	**has in it**	**'Real' growth of money**	**and**	**Effect of inflation**

The mathematical relationship between the inflation, real interest rates, and market interest rates is given as:

$$i = i' + f + (i')(f) \qquad (14\text{-}1)$$

EXAMPLE 14-1

Suppose Tiger Woods wants to invest some recent golf winnings in his hometown bank for one year. Currently, the bank is paying a rate of 5.5% *compounded annually*. Assume inflation is expected to be 2% a year or 8% a year for the year of Tiger's investment. In each case identify i, f, and i'.

SOLUTION

If Inflation Is 2% a Year

From the preceding definitions the interest rate that the bank is paying is the *market rate* (i). The *inflation rate* (f) is given in the problem statement. What is left, then, is to find the *real interest rate* (i').

$$i = 5.5\% \qquad f = 2\% \qquad i' = ?$$

Solving for i' in Equation 14-1, we have

$$i = i' + f + (i')(f)$$
$$i - f = i + (i')(f)$$
$$i - f = i'(1 + f)$$
$$i' = (i - f)/(1 + f)$$
$$= (0.055 - 0.02)/(1 + 0.02)$$
$$= 0.034 \quad \text{or} \quad \textbf{3.4\% per year}$$

This means that Tiger Woods will have 3.4% **more** purchasing power with the dollars invested in that account than he had a year ago. At the end of the year he can buy 3.4% more goods and services than he could have at the beginning of the year. As an example of the growth of his money, assume he was purchasing golf balls that cost $5 each and that he had invested $1000 in his hometown bank account.

At the *beginning* of the year he could purchase:

$$\text{Number of balls purchased today} = \frac{\text{Dollars today available to buy balls}}{\text{Cost of balls today}}$$

$$= 1000/\$5 = 200 \text{ golf balls}$$

At the *end* of the year he could purchase:

$$\text{Number of balls purchased at end of year} = \frac{\text{Dollars available for purchase at end of year}}{\text{Cost per ball at end of year}}$$

In this case:

$$\text{Dollars available at end of year} = (\$1000)(F/P, 5.5\%, 1) = \$1055$$

$$\text{Ball cost at end of year (inflated at 2\%)} = (\$5)(1 + 0.02)^1 = \$5.10$$

Thus:

$$\text{Number of balls purchased at end of year} = \$1055/5.10 = 207 \text{ golf balls}$$

Tiger Woods can, after one year, purchase 3.4% more golf balls than he could before. In this case 207 balls is about 3.4% more than 200 balls.

If Inflation Is 8%

As for the lower inflation rate, we would solve for i':

$$i' = (i - f)/(1 + f)$$

$$= (0.055 - 0.08)/(1 + 0.08)$$

$$= -0.023 \quad \text{or} \quad \textbf{-2.3\% per year}$$

In this case we can see that the real growth in money has *decreased* by 2.3%, so that Tiger can now purchase 2.3% fewer balls with the money he had invested in the bank. Even though he has more money at the end of the year, it is worth less, so he can purchase less.

Regardless of how inflation behaves over the year, the bank will pay Tiger $1055 at the end of the year. However, as we have seen, inflation can greatly affect the 'real' growth of dollars over time.

Let us continue the discussion of the effects of inflation by focusing now on cash flows in our problems. We define dollars of two 'types' in our analysis:

Actual dollars (A$): These are the dollars that we ordinarily think of when we think of money. These dollars circulate in our economy and are used for investments and payments. We can touch these dollars and often keep them in our purses and wallets—they are 'actual' and exist physically. Sometimes they are called *inflated dollars* because they carry any inflation that has reduced their worth.

Real dollars (R$): These dollars are a bit harder to define. They are always expressed in terms of some constant purchasing power 'base' year, for example, 2002-based dollars. Real dollars are sometimes called *constant dollars* or *constant purchasing power dollars*, and because they do not carry the effects of inflation, they are also known as *inflation-free dollars*.

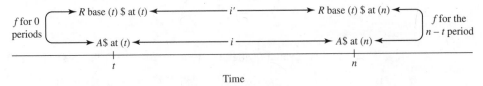

FIGURE 14-1 Relationship between i, f, i', $A\$$, and $R\$$.

Having defined *market, inflation,* and *real interest rates* as well as *actual* and *real dollars,* let us describe how these quantities are related. Figure 14-1 illustrates the relationship between these quantities.

Figure 14-1 illustrates the following principles:

When dealing with actual dollars ($A\$$), use a market interest rate (i), and when discounting $A\$$ over time, also use i.

When dealing with real dollars ($R\$$), use a real interest rate (i'), and when discounting $R\$$ over time, also use i'.

Figure 14-1 shows the relationships between $A\$$ and $R\$$ that occur **at the same period of time.** Actual and real dollars are related by the *inflation rate,* in this case, over the period of years defined by $n - t$. To translate between dollars of one type to dollars of the other ($A\$$ to $R\$$ or $R\$$ to $A\$$), use the inflation rate for the right number of periods. The following example illustrates many of these relationships.

EXAMPLE 14-2

A university is considering replacing its stadium with a new facility. When the present building was completed in 1950, the total cost was $1.2 million. At that time a wealthy alumnus made the university a gift of $1.2 million to be used for a future replacement. University administrators are now considering building the new facility in the year 2005.
Assume the following:

- Inflation is 6.0% per year from 1950 to 2005.
- In 1950 the university invested the gift at a market interest rate of 8.0% per year.

(a) Define i, i', f, $A\$$, and $R\$$ from the problem.
(b) How many actual dollars in the year 2005 will the gift be worth?
(c) How much would the actual dollars in 2005 be in terms of 1950 *purchasing power?*

SOLUTION TO a

Since 6.0% is the inflation rate (f) and 8.0% is the market interest rate (i), we can write

$$i' = (0.08 - 0.06)/(1 + 0.06) = 0.01887, \quad \text{or} \quad 1.887\%$$

Therefore, $1,200,000 was the cost of the building in 1950. These were the actual dollars ($A\$$) spent in 1950.

SOLUTION TO b

From Figure 14-1 we are going from *actual dollars at t, in 1950,* to *actual dollars at n, in 2005.* To do so, we use the *market interest rate* and compound this amount forward 55 years, as illustrated in Figure 14-2.

$$\text{Actual dollars in 2005} = \text{Actual dollars in 1950 } (F/P, i, 55 \text{ years})$$

$$= \$1,200,000(F/P, 8\%, 55)$$

$$= \$82,701,600$$

FIGURE 14-2 Discounting A\$ in 1950 to A\$ in 2005.

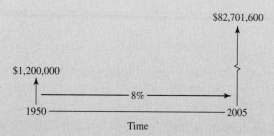

SOLUTION TO c

Now we want to determine the number of *real dollars—based in 1950—that in the year 2005 are equivalent to the $82.7 million from the solution to part **b**.* Let us solve this problem two ways.

1. In this approach let us directly translate the *actual dollars in the year 2005 to real 1950-based dollars in the year 2005.* From Figure 14-1 we can use the inflation rate to **strip 55 years of inflation** from the actual dollars. We do this by using the P/F factor for 55 years at the inflation rate. We are not physically moving the dollars in time in this case; rather we are simply removing inflation from these dollars one year at a time—the P/F factor does that for us. This is illustrated in the following equation and Figure 14-3.

$$\text{Real 1950-based dollars in 2005} = (\text{Actual dollars in 2005})(P/F, f, 55)$$

$$= (\$82,701,600)(P/F, 6\%, 55)$$

$$= \$3,357,000$$

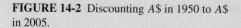

FIGURE 14-3 Translation of A\$ in 2005 to R 1950-based dollars in 2005.

2. In this method we start by recognizing that the *actual dollars in 1950* are exactly equivalent to *real 1950-based dollars that exist in 1950*. By definition, dollars that have *today* as the purchasing power base are the same as *actual dollars today*. Thus, actual dollars in 2004 are the same as real 2004-based dollars that circulated in 2004. So in this example, the $1.2 million can also be said to be *real 1950-based dollars that circulated in 1950*. So, let us translate those real dollars from 1950 to the year 2005. Since they are *real dollars,* we use the *real interest rate.*

$$\text{Real 1950-based dollars in 2005} = (\text{Real 1950-based dollars in 1950})(P/F, i', 55)$$

$$= (\$1,200,000)(F/P, 1.887\%, 55)$$

$$= \$3,355,000$$

FIGURE 14-4 Translation of R 1950-based dollars in 1950 to R 1950-based dollars in 2005.

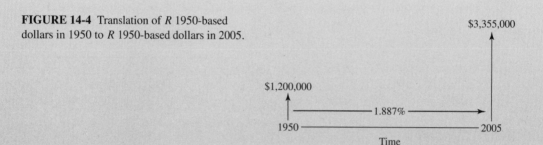

(*Note:* The difference between the answers to parts **1** and **2** is due to our having rounded the market interest rate off to 1.887% rather than carrying it out to more significant digits. The difference due to this rounding is less than 1%. If we were to carry the calculation of i' to enough digits, the answers to the two parts would be identical.)

From Example 14-2 we can see the relationship between dollars of different purchasing power bases, the choice of interest rates to use for moving dollars in time, and stripping out or adding in inflation. In that example the $1.2 million initially invested grew in the 55-year period to over $82.7 million, which becomes the amount available to pay for construction of the new complex. Does this mean that the new stadium will be 82.7/1.2 or about 70 times 'better' than the one built in 1950? The answer is no, because in the year 2005 the purchasing power of a dollar is less than it was in the year 1950. Assuming that construction costs increased at the rate of 6% per year given in the problem, then the amount available for the project *in terms of 1950-based dollars* is almost $3.4 million. This means that the new stadium will be about 3.4/1.2 or approximately 2.8 times 'better' than the original one using *real dollars*—not the 70 times ratio if *actual dollars* are used.

EXAMPLE 14-3

In 1924 Mr. O'Leary buried $1000 worth of quarters in his backyard. Over the years he had always thought that the money would be a nice nest egg to give to his first grandchild. His first granddaughter, Gabrielle, arrived in 1994. At the time of her birth Mr. O'Leary was taking an economic analysis course in his spare time. He had learned that over the years 1924 to 1994, inflation averaged 4.5%, the stock market increased an average of 15% per year, and investments in guaranteed government bonds averaged a 6.5% return per year. What was the relative purchasing power of the jar of quarters that Mr. O'Leary gave to his granddaughter Gabrielle at the time of her birth? What might have been a better choice for his backyard investment?

SOLUTION

Mr. O'Leary's $1000 dollars are *actual dollars* in both 1924 **and** in 1994.

To obtain the *real 1924-based dollar equivalent* of the $1000 that Gabrielle received in 1994, we would **strip 70 years of inflation out of those dollars.** As it turned out, Gabrielle's grandfather gave her $45.90 worth of 1924 purchasing power at the time of her birth. Because inflation has 'stolen' purchasing power from his stash of quarters during the 70-year period, Mr. O'Leary gave his grandaughter much less than the amount he first spaded underground. This loss of purchasing power caused by inflation over time is calculated as follows:

$$\text{Real 1924-based dollars in 1994} = (\text{Actual dollars in 1994})(P/F, f, 1994 - 1924)$$

$$= \$1000(P/F, 4.5\%, 70) = \$45.90$$

On the other hand, if Mr. O'Leary had put his $1000 in the stock market in 1924, he would have made baby Gabrielle an instant multi-millionaire by giving her $17,735,000. We calculate this as follows:

$$\text{Actual dollars in 1994} = (\text{Actual dollars in 1924})(F/P, i, 1994 - 1924)$$

$$= \$1000(F/P, 15\%, 70) = \$17,735,000$$

At the time of Gabrielle's birth, that $17.7 million translates to $814,069 in 1924 purchasing power. This is quite a bit different from the $45.90 in 1924 purchasing power calculated for the unearthed jar of quarters.

$$\text{Real 1924-based dollars in 1994} = \$17,735,000(P/F, 4.5\%, 70) = \$814,069$$

Mr. O'Leary was never a risk taker, and so it is doubtful he would have chosen the stock market for his future grandchild's nest egg. If he had chosen guaranteed government bonds instead of his backyard, by 1994 the investment would have grown to $59,076 (actual dollars)—the equivalent of $2712 in 1924 purchasing power.

$$\text{Actual dollars in 1994} = (\text{Actual dollars in 1924})(F/P, i, 1994 - 1924)$$

$$= \$1000(F/P, 6\%, 70) = \$59,076$$

$$\text{Real 1924-based dollars in 1994} = \$59,076(P/F, 4.5\%, 70) = \$2712$$

Obviously, either option would have been better than the choice Mr. O'Leary made. This example illustrates the effects of inflation and purchasing power, as well as the power of compounded interest over time. However, in Mr. O'Leary's defence, if Canada had experienced 70 years of *deflation* instead of *inflation*, he might have had the last laugh!

There are in general two ways to approach an economic analysis problem after the potential effects of inflation have been recognized. The first is to conduct the analysis by systematically including the effects of inflation; the second is to ignore the effects of inflation in conducting the analysis. Each case requires a different approach.

> *Incorporating inflation in the analysis:* Use a **market interest rate** and **actual dollars** that include inflation.

> *Ignoring inflation in the analysis:* Use **real dollars** and a **real interest rate** that does not reflect inflation.

ANALYSIS IN CONSTANT DOLLARS VERSUS THEN-CURRENT DOLLARS

Performing an analysis requires that we distinguish cash flows as being either constant dollars (real dollars, expressed in terms of some purchasing power base) or then-current dollars (actual dollars that are then-current when they occur). As previously stated, constant (real) dollars require the use of a *real interest rate* for discounting, and then-current dollars require a *market (or combined) interest rate*. These two types of dollar must not be mixed when one is performing an analysis. If both types are stated in the problem, one must convert either the constant dollars to then-current dollars or the then-current dollars to constant dollars, so that a consistent comparison can be made.

EXAMPLE 14-4

The Waygate Corporation is interested in evaluating a major new video display technology (VDT). Two competing computer innovation companies have approached Waygate to develop the technology for the new VDT. Waygate believes that both companies will be able to deliver equivalent products at the end of a 5-year period. From the yearly development costs of the VDT for each of the two companies as given, determine which Waygate should choose if the corporate MARR (investment market rate) is 25% and general price inflation is assumed to be 3.5% per year over the next 5 years.

> *Company Alpha costs:* Development costs will be $150,000 the first year and will increase at a rate of 5% over the 5-year period.

> *Company Beta costs:* Development costs will be a constant $150,000 per year in terms of today's dollars over the 5-year period.

SOLUTION

The costs for each of the two alternatives are as follows:

Year	Then-Current Costs Stated by Alpha	Constant Dollar Costs Stated by Beta
1	$150,000 \times (1.05)^0 = \$150,000$	$150,000
2	$150,000 \times (1.05)^1 = 157,500$	150,000
3	$150,000 \times (1.05)^2 = 165,375$	150,000
4	$150,000 \times (1.05)^3 = 173,644$	150,000
5	$150,000 \times (1.05)^4 = 182,326$	150,000

We inflate (or escalate) the stated yearly cost given by Company Alpha by 5% per year to obtain the then-current (actual) dollars each year. Company Beta's costs are given in terms of today-based constant dollars.

Using a Constant Dollar Analysis

Here we must convert the then-current costs given by Company Alpha to constant today-based dollars. We do this by stripping the right number of years of general inflation from each year's cost using $(P/F, f, n)$ or $(1 + f)^{-n}$.

Year	Constant Dollar Costs Stated by Alpha	Constant Dollar Costs Stated by Beta
1	$150,000 \times (1.035)^{-1} = \$144,928$	$150,000
2	$157,500 \times (1.035)^{-2} = 147,028$	150,000
3	$165,375 \times (1.035)^{-3} = 149,159$	150,000
4	$173,644 \times (1.035)^{-4} = 151,321$	150,000
5	$182,326 \times (1.035)^{-5} = 153,514$	150,000

We use the *real interest rate* (i') calculated from Equation 14-1 to calculate the present worth of costs for each alternative:

$$i' = (i - f)/(1 + f) = (0.25 - 0.035)/(1 + 0.035) = 0.208$$

PW of cost (Alpha) $= \$144,928(P/F, 20.8\%, 1) + \$147,028(P/F, 20.8\%, 2)$

$+ \$149,159(P/F, 20.8\%, 3) + \$151,321(P/F, 20.8\%, 4)$

$+ \$153,514(P/F, 20.8\%, 5) = \$436,000$

PW of cost (Beta) $= \$150,000(P/A, 20.8\%, 5) = \$150,000(2.9387) = \$441,000$

Using a Then-Current Dollar Analysis

Here we must convert the constant dollar costs of Company Beta to then-current dollars. We do this by using $(F/P, f, n)$ or $(1 + f)^n$ to "add in" the appropriate number of years of general inflation to each year's cost.

Year	Then-Current Costs Stated by Alpha	Then-Current Costs Stated by Beta
1	$150,000 \times (1.05)^0 = \$150,000$	$150,000 \times (1.035)^1 = \$155,250$
2	$150,000 \times (1.05)^1 = 157,500$	$150,000 \times (1.035)^2 = 160,684$
3	$150,000 \times (1.05)^2 = 165,375$	$150,000 \times (1.035)^3 = 166,308$
4	$150,000 \times (1.05)^3 = 173,644$	$150,000 \times (1.035)^4 = 172,128$
5	$150,000 \times (1.05)^4 = 182,326$	$150,000 \times (1.035)^5 = 178,153$

Using the *market interest rate* (i), calculate the present worth of costs for each alternative.

$$PW \text{ of cost (Alpha)} = \$150,000(P/F, 25\%, 1) + \$157,500(P/F, 25\%, 2)$$

$$+ \$165,375(P/F, 25\%, 3) + \$173,644(P/F, 25\%, 4)$$

$$+ \$182,326(P/F, 25\%, 5) = \$436,000$$

$$PW \text{ of cost (Beta)} = \$155,250(P/F, 25\%, 1) + \$160,684(P/F, 25\%, 2)$$

$$+ \$166,308(P/F, 25\%, 3) + \$172,128(P/F, 25\%, 4)$$

$$+ \$178,153(P/F, 25\%, 5) = \$441,000$$

Using either a constant dollar or then-current dollar analysis, Waygate should choose Company Alpha's offer, which has the lower present worth of costs. There may, of course, be intangible elements in the decision that also should be considered.

PRICE CHANGE WITH INDEXES

We have already described the effects that inflation can have on money over time. Also, several definitions and relationships regarding dollars and interest rates have been given. We have seen that it is not correct to compare the benefits of an investment in 2004-based dollars with costs in 2006-based dollars. This is like comparing apples and oranges. Such comparisons of benefits and costs can be meaningful only if a standard purchasing power base of money is used. An often asked question is: how do I know what inflation rate to use in my studies? This is a valid question. What **can** we use to measure price changes over time?

What Is a Price Index?

Price indexes are used as a means to describe the relative price fluctuation of goods and services in our national economy. They provide a *historical* record of the behaviour of these quantities over time. Price indexes are tracked for *specific commodities* as well as for *bundles (composites) of commodities.* Thus, price indexes can be used to measure historical price changes for individual cost items (like labour and material) as well as general costs (like consumer products). In understanding the **past** price fluctuations, we have more information for predicting the **future** behaviour of those cash flows.

Table 14-1 lists the historic prices of sending a first-class letter in the United States via the Postal Service from 1970 to 2003. The cost is given both in terms of dollars (cents) and as measured by a fictitious price index that we could call the letter cost index (LCI).

TABLE 14-1 Historic Prices of US First-Class Mail, 1970–2003, and Letter Cost Index

Year, n	Cost of First-Class Mail	LCI	Annual Increase for n	Year, n	Cost of First-Class Mail	LCI	Annual Increase for n
1970	$0.06	100	0.00%	1987	$0.22	367	0.00%
1971	0.08	133	33.33	1988	0.25	417	13.64
1972	0.08	133	0.00	1989	0.25	417	0.00
1973	0.08	133	0.00	1990	0.25	417	0.00
1974	0.10	166	25.00	1991	0.29	483	16.00
1975	0.13	216	30.00	1992	0.29	483	0.00
1976	0.13	216	0.00	1993	0.29	483	0.00
1977	0.13	216	0.00	1994	0.29	483	0.00
1978	0.15	250	15.74	1995	0.32	533	10.34
1979	0.15	250	0.00	1996	0.32	533	0.00
1980	0.15	250	0.00	1997	0.32	533	0.00
1981	0.20	333	33.33	1998	0.33	550	3.13
1982	0.20	333	0.00	1999	0.33	550	0.00
1983	0.20	333	0.00	2000	0.33	550	0.00
1984	0.20	333	0.00	2001	0.34	567	3.03
1985	0.22	367	10.00	2002	0.37	617	8.82
1986	0.22	367	0.00	2003	0.37	617	0.00

Notice two important aspects of the LCI. First, as with all cost or price indexes, the numbers used to express the change in price over time are based on some **base year.** With price (or cost) indexes the base year is always assigned a value of 100. Our LCI has a base year of 1970—thus for 1970, LCI = 100. The letter cost index value given in subsequent years is stated in relation to 1970 as the base year. A second aspect to notice is that the LCI changes only when the cost of first-class postage changes. In years when this quantity does not change (in other words, there was no price increase), the LCI is not affected. These general observations apply to all price indexes.

In general, engineering economists are the 'users' of cost indexes such as our hypothetical LCI. That is, cost indexes are calculated or tabulated by some other party, and our interest is in assessing what the index tells us about the historical prices and how these may affect our estimate of future costs. However, since we are 'users' of indexes, rather than 'compilers', it may be of interest to illustrate how the LCI in Table 14-1 was calculated.

In Table 14-1, the LCI is assigned a value of 100 because 1970 serves as our base year. In the following years the LCI is calculated from year to year on the basis of the annual percentage increase in first-class mail. Equation 14-2 illustrates the arithmetic used.

$$\text{LCI year, } n = ([(\text{cost }(n) - \text{cost } 1970)/\text{cost } 1970] \times 100\%) + 100 \qquad (14\text{-}2)$$

For example, consider the LCI for the year 1980. We calculate the LCI as follows.

$$\text{LCI year } 1980 = ([(0.15 - 0.06)/0.06] \times 100) + 100 = 250$$

As mentioned, engineering economists are often in the business of using cost indexes to project future cash flows. Therefore, our first job is to use a cost index to **calculate** annual cost increases for the items tracked by the index. To calculate the *year-to-year* percentage increase (or *inflation*) of prices tracked by an index, we can use Equation 14-3.

Annual percentage increase, $n = ([\text{Index}(n) - \text{Index}(n-1)]/\text{Index}(n-1)) \times 100\%$ (14-3)

To illustrate the use of this equation, let us look at the percentage change from 1977 to 1978 for the LCI just given.

Annual percentage increase (1978) $= [250 - 216]/216 \times 100\% = 15.74\%$

For 1978 the price of mailing a first-class letter increased by 15.74% over the previous year. This is the value tabulated in Table 14-1.

An engineering economist often wants to know how a particular cost quantity changes over time. Often we are interested in calculating the *average* rate of price increase or inflation in some quantity, such as the cost of postage, over a period of time. For instance, we might want to know the average yearly increase in postal prices from 1970 to 2003. How do we calculate this quantity? Can we use Equation 14-2? If we were to use Equation 14-2 to calculate the percentage change from 1970 to 2003, we would obtain the following:

% Increase (1971 to 1991) $= (617 - 100)/100 \times 100\% = 517\%$

But how do we use this calculation to obtain the **average** rate of increase over those years? Should we divide 517% by 33 years ($517/33 = 15.67\%$)? Of course not! As was established in earlier chapters, the concept of *compounding* precludes such a simple division. To do so would be to treat the interest rate as simple interest—where compounding is not in effect. So the question remains: how do we calculate an *equivalent average rate of increase* in postage rates over a period of time? Let us start by thinking about the LCI. We have a number (index value) of 100 in year 1970 and another number, 617, in year 2003, and we want to know the interest rate that relates these two numbers. If we think of the index numbers as cash flows, it is easy to see that we have a simple internal rate of return problem. Given this approach, let us calculate the *average rate of increase* in postage rates for the years under consideration.

$$P = 100 \qquad F = 617 \qquad n = 33 \text{ years} \qquad i = ?$$

Using $F = P(1+i)^n$ $617 = 100(1+i)^{33}$ $i = (617/100)^{1/33} - 1$ $i = 0.0567 = 5.7\%$

In the same way, we can use a cost index to calculate the average rate of increase over any period of years. Understanding of how costs have behaved historically should provide insight into how they may behave in the future.

Composite versus Commodity Indexes

Cost indexes, in general, come in two types: commodity-specific indexes and composite indexes. Each type of index is useful to the engineering economist. Commodity-specific indexes measure the historical change in price for specific items—such as green beans or iron ore. Common commodities that are tracked by price indexes and used in engineering economic analysis include *utility commodities, labour costs,* and *purchase prices.* Commodity indexes, like our letter cost index, are useful when an economic analysis includes individual cost items that are tracked by such indexes. For example, if we need to estimate the direct-labour cost portion of a construction project, we could use an index that tracks the inflation, or escalation, of this particular cost over time. Statistics Canada and the US Department of Economic Analysis and Bureau of Labor Statistics track many cost quantities. For our example construction project, we would refer to the appropriate labour index and investigate how this cost item had behaved in the past. This should give us valuable information about how to estimate this cost in the future.

EXAMPLE 14-5

Congratulations! Your sister just had a baby girl named Veronica. When you hear this happy news, you fondly remember the fine pony you had as a young child, and you decide you would like to give your new niece a pony for her fifth birthday. You would like to know how much you must put into your savings account (earning 4% interest per year) today to buy a pony and saddle for Veronica 5 years from now. But you have no idea how much a pony and saddle might cost in 5 years.

SOLUTION

Divide the problem into the following steps:

Step 1   Use commodity indexes to measure past price changes for both ponies and saddles.

Step 2   Call several dealers to get prices on the current costs of ponies and saddles.

Step 3   Use the *average price increase values* for each commodity (pony and saddle) to estimate what the cost of each will be in 5 years.

Step 4   Calculate how much you would have to put into your bank account today to cover those costs in 5 years.

Step 1

A trip to the library reveals that, indeed, there are price indexes for the two items that you would like to buy in 5 years: a pony and a saddle. From the indexes you find that in the last 10 years the little pony price index (LPPI) has gone from 213 to 541, and the leather saddle index (LSI) from 1046 to 1229. For each commodity you calculate the *average rate of price increase* for the past 10 years as:

$$\text{Average price change (LPPI) is } 541 = 213(1+i)^{10}$$

$$\text{solving for } i \text{ for ponies} = 9.8\% \text{ per year}$$

$$\text{Average price change (LSI) is } 1229 = 1046(1+i)^{10}$$

$$\text{solving for } i \text{ for saddles} = 1.6\% \text{ per year}$$

Step 2

By calling a stable and a tack shop, you find that the current prices of a registered pony and a leather pony saddle are $600 and $350 respectively.

Step 3

Using the current prices for both the pony and saddle, *inflate* these costs at the respective *price change rates* calculated in Step 1. The assumption you are making is that the prices for ponies and saddles will change in the next 5 years at the same rate as the average of the last 10 years. The cost for the two items, and the total, in 5 years will be:

$$\text{Cost of pony in 5 years} = 600(1 + 0.098)^5 = \$958$$

$$\text{Cost of saddle in 5 years} = 350(1 + 0.016)^5 = \$378$$

$$\text{Total cost in 5 years} = \$958 + \$378 = \$1336$$

Step 4

Your final step is to calculate the amount that must be set aside today at 4% interest for 5 years to accumulate the future cost.

$$\text{Amount invested now} = 1336(1 + 0.04)^{-5} = \$1098$$

So it will require **$1098** today to make Veronica a very happy pony rider on her fifth birthday!

Composite cost indexes do not track historical prices for individual items. Instead, they measure the historical prices of *groups* or *bundles* of assets. Thus a **composite index** measures an overall price change that is a composite of several effects. Examples of composite indexes are the *Consumer Price Index* (CPI) and the *Producer Price Index* (PPI). The CPI measures the effect of prices as experienced by consumers, and the PPI measures prices as felt by producers of goods.

The CPI, an index calculated by Statistics Canada, tracks the cost of a standard *bundle of consumer goods* from year to year. This 'consumer bundle' or 'basket of consumer goods' is made up of common consumer expenses, including housing, clothing, food, transportation, and entertainment. Because of its emphasis on consumer goods, people often use the CPI as a substitute measure for general inflation in the economy. There are several problems with the use of the CPI in this manner, one being the assumption that all consumers purchase the same 'basket of consumer goods' year after year. However, even with its deficiencies, the CPI enjoys popular identification as an 'inflation' indicator. Table 14-2 gives the yearly index values and annual percentage increase in the CPI for the past 30 years.

Composite indexes can be used in much the same way as commodity-specific indexes, described earlier. That is, we can pick a single value from the table if we are interested in measuring the historic price for a single year, or we can calculate an *average inflation rate* or *average rate of price increase* as measured by the index over several years.

How to Use Price Indexes in Engineering Economic Analysis

One may question the usefulness of *historical* data (as provided by price indexes) when engineering economic analysis deals with economic effects projected to occur in the *future*. However, both commodity-specific and composite indexes are indeed useful in many analyses. Engineering economic analysis is concerned with making estimates of future events: the outcomes of yearly costs and benefits, interest rates, salvage values, and tax rates are all

TABLE 14-2 CPI Index Values and Yearly Percentage Increases, 1974–2003

Year	CPI Value*	CPI Increase	Year	CPI Value*	CPI Increase
1974	31.1	10.7%	1989	89.0	5.0%
1975	34.5	10.9	1990	93.3	4.8
1976	37.1	7.5	1991	98.5	5.6
1977	40.0	7.8	1992	100.0	1.5
1978	43.6	9.0	1993	101.8	1.8
1979	47.6	9.2	1994	102.0	0.2
1980	52.4	10.1	1995	104.2	2.2
1981	58.9	12.4	1996	105.9	1.6
1982	65.3	10.9	1997	107.6	1.6
1983	69.1	5.8	1998	108.6	0.9
1984	72.1	4.3	1999	110.5	1.7
1985	75.0	4.0	2000	113.5	2.7
1986	78.1	4.1	2001	116.4	2.6
1987	81.5	4.4	2002	119.0	2.2
1988	84.8	4.0	2003	122.3	2.8

*1992 = 100
Source: Statistics Canada

examples of such estimates. Associated with these estimates are varying degrees of uncertainty.

The challenge for the engineering economist is to reduce this uncertainty for each estimate. Historical data provide a snapshot of how the quantities of interest have behaved in the past. Knowing this past (historical) behaviour should provide insight on how to estimate their behaviour in the future, as well as to reduce the uncertainty of that estimate. This is where the data that price indexes provide come into play. Although it is very dangerous to extrapolate past data into the future in the short run, price index data can be useful in making estimates (especially when considered from a long-term perspective). In this way the engineering economist can use *average historical percentage increases (or decreases)* from commodity-specific and composite indexes, along with data from market analyses and other sources, to estimate how economic quantities may behave in the future.

One may wonder how both commodity-specific and composite price indexes may be used in engineering economic analyses. The answer to that question is reasonably straightforward. As we have established, price indexes can be useful for estimates of future outcomes. The following principle applies to commodity-specific and composite price indexes and such estimates:

> When the estimated quantities are items that are tracked by commodity specific indexes, then those indexes should be used to calculate *average historical percentage increases (or decreases)*.

If no commodity-specific indexes are kept, one should use a suitable composite index to make this calculation. For example, to estimate electric usage costs for a turret lathe over a 5-year period, one would first want to refer to a commodity-specific index that tracks this quantity. If such an index does not exist, one might use a specific index for a very closely related commodity—perhaps, in this case, an index of electric usage costs of screw lathes. In the absence of such substitute or related commodity indexes, one could use suitable

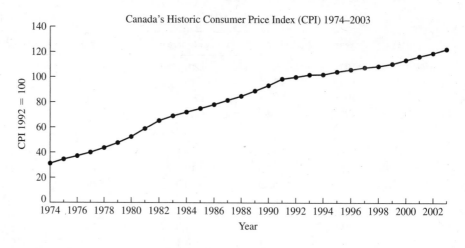

FIGURE 14-5 CPI historic inflation rate over 28 years.

composite indexes: there may be a composite index that tracks electric usage costs for industrial metal-cutting machinery. Or, as before, a related composite index could be used. The point is that one should try to identify and use a price index that is most closely related to the quantity being estimated in the analysis.

CASH FLOWS THAT INFLATE AT DIFFERENT RATES

Engineering economic analysis requires the estimation of various parameters. It is not uncommon that over time, these parameters will *inflate* or *increase* (or even decrease) at different rates. For instance, one parameter might *increase* 5% per year, another 15% per year, and a third *decrease* 3.5% per year. This phenomenon is important because of the various items of different types that are sometimes included in engineering economic analyses. Since we are looking at the behaviour of cash flows over time, we must have a way of handling this effect.

EXAMPLE 14-6

On your first assignment as an engineer, your boss asks you to develop the utility cost portion of an estimate for the cost of a new manufacturing facility. After some research you define the problem as finding the present worth of utility costs given the following data:

- Your company uses a minimum attractive rate of return (MARR) = 35% as i.
- The project has a useful life of 25 years.
- The utilities to be estimated are electricity, water, and natural gas.
- The 35-year historical data reveal:

  Electricity costs increase at 8.5% per year
  Water costs increase at 5.5% per year
  Natural gas costs increase at 6.5% per year

- First-year estimates of the utility costs (in today's dollars) are as follows:

  Electricity will cost $55,000
  Water will cost $18,000
  Natural gas will cost $38,000

SOLUTION

For this problem we will take each of the utilities used in our manufacturing facility and inflate them independently at their various historical annual rates. Once we have these actual dollar amounts ($A\$$), we can total them and then discount each year's total at 35% for the right number of periods back to the present.

Year	Electricity		Water		Natural Gas		Total
1	$55,000(1.085)^0$ =	$ 55,000	$18,000(1.055)^0$ =	$18,000	$38,000(1.065)^0$ =	$ 38,000	$111,000
2	$55,000(1.085)^1$ =	59,675	$18,000(1.055)^1$ =	18,990	$38,000(1.065)^1$ =	40,470	119,135
3	$55,000(1.085)^2$ =	64,747	$18,000(1.055)^2$ =	20,034	$38,000(1.065)^2$ =	43,101	127,882
4	$55,000(1.085)^3$ =	70,251	$18,000(1.055)^3$ =	21,136	$38,000(1.065)^3$ =	45,902	137,289
5	$55,000(1.085)^4$ =	76,222	$18,000(1.055)^4$ =	22,299	$38,000(1.065)^4$ =	48,886	147,407
6	$55,000(1.085)^5$ =	82,701	$18,000(1.055)^5$ =	23,525	$38,000(1.065)^5$ =	52,063	158,290
7	$55,000(1.085)^6$ =	89,731	$18,000(1.055)^6$ =	24,819	$38,000(1.065)^6$ =	55,447	169,997
8	$55,000(1.085)^7$ =	97,358	$18,000(1.055)^7$ =	26,184	$38,000(1.065)^7$ =	59,051	182,594
.	.		.		.		.
.	.		.		.		.
.	.		.		.		.
24	$55,000(1.085)^{23}$ =	359,126	$18,000(1.055)^{23}$ =	61,671	$38,000(1.065)^{23}$ =	161,743	582,539
25	$55,000(1.085)^{24}$ =	389,652	$18,000(1.055)^{24}$ =	65,063	$38,000(1.065)^{24}$ =	172,256	626,970

The present worth of the total yearly utility costs is:

$$PW = \$111,000(P/F, 35\%, 1) + \$119,135(P/F, 35\%, 2) + \cdots + \$626,970(P/F, 35\%, 25)$$

$$= \$5,540,000$$

In Example 14-6 several parameters changed at different rates over the period of the economic study. By using the respective individual inflation rates, we obtained the **actual dollar** amounts for each parameter in each year. Then, we used a market interest rate to discount these actual dollar amounts. Problems of this type can be handled by inflating the various parameters at their respective estimated inflation rates, combining these cash flows if appropriate, and then treating them as actual dollars that occur in those years.

DIFFERENT INFLATION RATES PER PERIOD

In this section we address the situation of inflation rates for the various cash flows in an analysis that are changing over the life of the study period. Rather than different inflation rates for different cash flows, in Example 14-7 the *interest rate* for the same cash flow is changing over time. A method for handling this situation is much like that of the preceding section. We can simply apply the inflation rates in the years in which they are projected to occur. We would do this for each of the individual cash flows over the entirety of the study period. Once we have all these actual dollar amounts, we can use the market interest rate and decision criteria to apply any of the measures of merit developed earlier.

EXAMPLE 14-7

While working as a clerk at Piggly Wiggly, Elvis has learned much about the cost of different vegetables. The kitchen manager at Heartbreak Hotel called recently, requesting Elvis to estimate the raw material cost over the next 5 years of introducing succotash (lima beans and corn) to the buffet line. To develop his estimate, Elvis has used his advanced knowledge of soil growing conditions, world demand, and government subsidy programs for these two crops. He has estimated the following data:

- Costs for lima beans will inflate at 3% per year for the next 3 years and then at 4% for the following 2 years.
- Costs for corn will inflate at 8% per year for the next 2 years and then will decrease 2% in the following 3 years.

The kitchen manager wants to know the equivalent annual cost of providing succotash on the buffet line over the 5-year period. His before-tax MARR is 20%. The manager estimates that he will need an average of 50 pounds each of beans and corn each day. The hotel kitchen operates 6 days a week, 52 weeks a year. Current costs are $0.35/lb for lima beans and $0.80/lb for corn.

SOLUTION

Today's cost for one year's supply of vegetables is:

Lima beans	0.35 $/lb × 50 lb/day × 6 day/wk × 52 wk/yr =	$ 5460/yr
Corn	0.80 $/lb × 50 lb/day × 6 day/wk × 52 wk/yr =	12,480/yr

Year	Lima Beans	Corn	Total
0	$5,460	$12,480	
1	5,460(1.03) = 5,624	12,480(1.08) = 13,478	$19,102
2	5,624(1.03) = 5,793	13,478(1.08) = 14,556	20,349
3	5,793(1.03) = 5,967	$14,556(1.02)^{-1}$ = 14,271	20,238
4	5,967(1.04) = 6,206	$14,271(1.02)^{-1}$ = 13,991	20,197
5	6,206(1.04) = 6,454	$13,991(1.02)^{-1}$ = 13,717	20,171

$$\text{EUAC} = [19,102(P/F, 20\%, 1) + 20,349(P/F, 20\%, 2) + 20,238(P/F, 20\%, 3)$$

$$+ 20,197(P/F, 20\%, 4) + 20,171(P/F, 20\%, 5)](A/P, 20\%, 5)$$

$$= \$19,900 \text{ per year}$$

In Example 14-7, both today's cost of each vegetable and the respective inflation rates were used to calculate the yearly costs of purchasing the desired quantities over the 5-year period. As in Example 14-6, we obtained a total marginal cost (in terms of actual dollars) by combining the two individual yearly costs. We then calculated the equivalent uniform annual cost (EUAC) using the given market interest rate.

Example 14-8 provides another example of how the effect of changes in inflation rates over time can affect an analysis.

EXAMPLE 14-8

If general price inflation is estimated to be 5% for the next 5 years, 7.5% for the 3 years after that, and 3% the following 5 years, at what market interest rate (i) would you have to invest your money to maintain a real purchasing power growth rate (i') of 10% during those years?

SOLUTION

In Years 1–5 you must invest at $0.10 + 0.050 + (0.10)(0.050) = 0.1150 = 11.50\%$ per year.
In Years 6–8 you must invest at $0.10 + 0.075 + (0.10)(0.075) = 0.1825 = 18.25\%$ per year.
In Years 9–13 you must invest at $0.10 + 0.030 + (0.10)(0.030) = 0.1330 = 13.30\%$ per year.

(*Note:* This example illustrates a common problem in timing: the market interest rate necessary to maintain the real purchasing power of the investment must be calculated at a time when the inflation rate is only an estimate. Most interest-bearing investments have fixed, up-front rates that the investor understands well when making an investment. Variable rate investments are the exception. On the other hand, inflation is not quantified, and its effect on our real return is not measured until the end of the year. Therefore, to achieve the conditions required in the example, one would either have to anticipate inflation and adjust one's investments accordingly, or accept the fact that the real investment return (i') may not turn out to be what was originally required.)

INFLATION EFFECT ON AFTER-TAX CALCULATIONS

Earlier we noted the impact of inflation on before-tax calculations. We found that if the subsequent benefits brought constant quantities of dollars, then inflation will diminish the true value of the future benefits and, hence, the real rate of return. If, however, the future benefits keep up with the rate of inflation, the rate of return will not be adversely affected by the inflation. Unfortunately, we are not so lucky when we consider a situation with income taxes, as illustrated by Example 14-9.

EXAMPLE 14-9

A $12,000 investment will return annual benefits for 6 years with no salvage value at the end of the period. Assume straight-line depreciation and a 46% income tax rate. The problem is to solve for both before- and after-tax rates of return, the latter for two situations:

1. *No inflation:* the annual benefits are constant at $2918 per year.
2. *Inflation equal to 5%:* the benefits from the investment increase at this same rate, so that they continue to be the equivalent of $2918 in Year-0-based dollars.

The benefit schedule for the two situations is as follows:

Year	Annual Benefit for Both Situations (Year-0-based dollars)	No Inflation, Actual Dollars Received	5% Inflation Factor*	5% Inflation, Actual Dollars Received
1	$2918	$2918	$(1.05)^1$	$3064
2	2918	2918	$(1.05)^2$	3217
3	2918	2918	$(1.05)^3$	3378
4	2918	2918	$(1.05)^4$	3547
5	2918	2918	$(1.05)^5$	3724
6	2918	2918	$(1.05)^6$	3910

*May be read from the 5% compound interest table as $(F/P, 5\%, n)$.

SOLUTIONS

Before-Tax Rate of Return

Since both situations (no inflation and 5% inflation) have an annual benefit, stated in Year-0-based dollars of $2918, they have the same before-tax rate of return.

$$\text{PW of cost} = \text{PW of benefit}$$

$$12,000 = 2918(P/A, i, 6) \qquad (P/A, i, 6) = \frac{12,000}{2918} = 4.11$$

From compound interest tables: before-tax rate of return equals 12%.

After-Tax Rate of Return, No Inflation

Year	Before-Tax Cash Flow	Straight-Line Depreciation	Taxable Income	46% Income Taxes	Actual Dollars, and Year-0-Based Dollars, After-Tax Cash Flow
0	−$12,000				−$12,000
1–6	+2,918	$2000	$918	−$422	+2,496

PW of cost = PW of benefit

$$12,000 = 2496(P/A, i, 6) \qquad (P/A, i, 6) = \frac{12,000}{2496} = 4.81$$

From compound interest tables: after-tax rate of return equals 6.7%.

After-Tax Rate of Return, 5% Inflation

Year	Before-Tax Cash Flow	Straight-Line Depreciation	Taxable Income	46% Income Taxes	Actual Dollars, After-Tax Cash Flow
0	−$12,000				−$12,000
1	+3,064	$2000	$1064	−$489	+2,575
2	+3,217	2000	1217	−560	+2,657
3	+3,378	2000	1378	−634	+2,744
4	+3,547	2000	1547	−712	+2,835
5	+3,724	2000	1724	−793	+2,931
6	+3,910	2000	1910	−879	+3,031

Converting to Year-0-Based Dollars and Solving for Rate of Return

Year	Actual Dollars, After-Tax Cash Flow	Conversion Factor	Year-0-Based Dollars, After-Tax Cash Flow	Present Worth at 5%	Present Worth at 4%
0	−$12,000		−$12,000	−$12,000	−$12,000
1	+2,575	$\times (1.05)^{-1} =$	+2,452	+2,335	+2,358
2	+2,657	$\times (1.05)^{-2} =$	+2,410	+2,186	+2,228
3	+2,744	$\times (1.05)^{-3} =$	+2,370	+2,047	+2,107
4	+2,835	$\times (1.05)^{-4} =$	+2,332	+1,919	+1,993
5	+2,931	$\times (1.05)^{-5} =$	+2,297	+1,800	+1,888
6	+3,031	$\times (1.05)^{-6} =$	+2,262	+1,688	+1,788
				−25	+362

Linear interpolation between 4 and 5%:

$$\text{After-tax rate of return} = 4\% + 1\%[362/(362 + 25)] = 4.9\%$$

From Example 14-9, we see that the before-tax rate of return for both situations (no inflation and 5% inflation) is the same. Equal before-tax rates of return are expected because the benefits in the inflation situation increased in proportion to the inflation. This example shows that when future benefits fluctuate with changes in inflation or deflation, the effects do not alter the year-0-based dollar estimates. Thus, no special calculations are needed in before-tax calculations when future benefits are expected to respond to inflation or deflation rates.

The after-tax calculations illustrate a different result. The two situations, with equal before-tax rates of return, do not produce equal after-tax rates of return:

Situation	Before-Tax Rate of Return	After-Tax Rate of Return
No inflation	12%	6.7%
5% inflation	12%	4.9%

Thus, 5% inflation results in a smaller after-tax rate of return, even though the benefits increase at the same rate as the inflation. A review of the cash flow table reveals that while benefits increase, the depreciation schedule does not. Thus, the inflation results in increased taxable income and, hence, larger income tax payments; but there are not sufficient increases in benefits to offset these additional disbursements.

The result is that while the after-tax cash flow in actual dollars increases, the augmented amount is not high enough to offset *both* inflation and increased income taxes. This effect is readily apparent when the equivalent year-0-based-dollar after-tax cash flow is examined. With inflation, the year-0-based-dollar after-tax cash flow is smaller than the year-0-based-dollar after-tax cash flow without inflation. Of course, inflation might cause equipment to have a salvage value that was not forecast, or a larger one than had been projected. This effect would tend to reduce the unfavourable consequences of inflation on the after-tax rate of return.

USING SPREADSHEETS FOR INFLATION CALCULATIONS

Spreadsheets are the perfect tool for incorporating a consideration of inflation into analyses of economic problems. For example, next year's labour costs are likely to be estimated as equal to this year's costs times $(1 + f)$, where f is the inflation rate. Thus each year's value is different, and so we can't use factors for uniform flows, A. Also the formulas that link different years are easy to write. The result is problems that are very tedious to do by hand, but easy by spreadsheet.

Example 14-10 illustrates two different ways to write the equation for inflating costs. Example 14-11 illustrates that inflation reduces the after-tax rate of return because inflation makes the depreciation deduction less valuable.

EXAMPLE 14-10

Two costs for construction of a small, remote mine are for labour and transportation. Labour costs are expected to be $350,000 the first year, with inflation of 6% annually. Unit transportation costs are expected to inflate at 5% annually, but the volume of material being moved changes each year. In time-0 dollars, the transportation costs are estimated to be $40,000, $60,000, $50,000, and $30,000 in Years 1 to 4. The inflation rate for the value of the dollar is 3%. If the firm uses an i' of 7%, what is the equivalent annual cost for this 4-year project?

SOLUTION

The data for labour costs can be stated so that no inflation needs to be applied in Year 1: the cost is $350,000. In contrast, the transportation costs for Year 1 are determined by multiplying $40,000 by 1.05 ($= 1 + f$).

Also in later years the labour cost$_t$ = labour cost$_{t-1}(1 + f)$, while each transportation cost must be computed as the time-0 value times $(1 + f)^t$. In Figure 14-6, the numbers in the Year 0 (or real) dollar column equal the values in the actual dollars column divided by $(1.03)^t$.

	A	B	C	D	E	F	G	H
1							7% Inflation-free interest	
2	Inflation rate	6%		5%		3%		
3			Transportation Costs		Total	Total		
4	Year	Labour Costs	Year 0 $s	Actual $s	Actual $s	Real $s		
5	1	120,000	40,000	42,000	162,000	157,282	= E5/(1+$F$2)^A5	
6	2	127,200	60,000	66,150	193,350	182,251		
7	3	134,832	50,000	57,881	192,713	176,360		
8	4	142,922	30,000	36,465	179,387	159,383		
9						$571,732	= NPV(F1,F5:F8)	
10					=B8+D8	$168,791	= −PMT(F1,4,F9)	
11	=B7*(1+$B$2)		=C8*(1+$D$2)^A8					

FIGURE 14-6 Spreadsheet for inflation.

The equivalent annual cost equals $168,791.

EXAMPLE 14-11

For the data of Example 14-9, calculate the IRR with and without inflation with CCA depreciation.

SOLUTION

Most of the formulas for this spreadsheet are given in rows 11 and 12 for the data in Year 6. The CCA is calculated with formulas 11-6 and 11-7. It could also be easily calculated by adding a column called 'book value' and applying the CCA percentage. Since CCA is not influenced by inflation, the depreciation deduction is less valuable as inflation increases. The tax paid equals the tax rate times the taxable income, which equals dollars received minus the depreciation charge. Then ATCF (after-tax cash flow) equals the before-tax cash flow minus the tax paid.

In Figure 14-7, notice that in Year 2 the depreciation charge is large enough to cause this project to pay 'negative' tax. For a firm, this means that the deduction on this project will be used to offset income from other projects.

	A	B	C	D	E	F	G	H
1		P = $12,000						
2		CCA = 40%						
3		t = 46%						
4		S = 0			f = 0%			
5			Actual $s	Actual	Actual $s	Net	Actual $s	Real $s	
6	Year	Received	CCA	Tax	Salvage	ATCF	ATCF	
7	0	−$12,000				−$12,000	−$12,000	= F7/(1+$B$5)^$A7
8	1	$2,918	−$2,400	$238		$2,680	$2,680	= F8/(1+$B$5)^$A8
9	2	$2,918	−$3,840	−$424		$3,342	$3,342	
10	3	$2,918	−$2,304	$282		$2,636	$2,636	
11	4	$2,918	−$1,382	$706		$2,212	$2,212	
12	5	$2,918	−$829	$961		$1,957	$1,957	
13	6	$2,918	−$498	$1,113	$343	$2,148	$2,148	
14	Formulas			= (B13+C13)*$B$3		= B13−D13+E13		
15	for year 6	= 2918*(1+$B$5)^$A13		= $B$7*$B$2*(1−$B$2/2)*(1−$B$2)^($A13−2)				
16							7.26% = IRR	
17	Net salvage calculation from Equation 12-7, 12-8, and 11-8							
18	UCC$_6$ =	$746	= $B$1*(1−$B$2/2)*(1−$B$2)^(6−1)					
19	Net salvage							
20	end of yr 6 = $343							
21	= $B$4+$B$3*IF($B$4>$B$1,$C$18−$B$1,$C$18−$B$4)−1/2*$B$3*MAX(($B$4−$B$1),0)							

FIGURE 14-7a After-tax IRR with CCA and 0% inflation.

	A	B	C	D	E	F	G	H
1		P = $12,000						
2		CCA = 40%						
3		t = 46%						
4		S = 0			f = 5%			
5			Actual $s	Actual	Actual $s	Net	Actual $s	Real $s	
6	Year	Received	CCA	Tax	Salvage	ATCF	ATCF	
7	0	−$12,000				−$12,000	−$12,000	= F7/(1+$B$5)^$A7
8	1	$3,064	−$2,400	$305		$2,759	$2,627	= F8/(1+$B$5)^$A8
9	2	$3,217	−$3,840	−$287		$3,504	$3,178	
10	3	$3,378	−$2,304	$494		$2,884	$2,491	
11	4	$3,547	−$1,382	$996		$2,551	$2,099	
12	5	$3,724	−$829	$1,332		$2,393	$1,875	
13	6	$3,910	−$498	$1,570	$343	$2,684	$2,003	
14	Formulas			= (B13+C13)*$B$3		= B13−D13+E13		
15	for year 6	= 2918*(1+$B$5)^$A13		= $B$7*$B$2*(1−$B$2/2)*(1−$B$2)^$A1				
16							5.64% = IRR	
17	Net salvage calculation from Equation 12-7, 12-8, and 11-8							
18	UCC$_6$ =	$746	= $B$1*(1−$B$2/2)*(1−$B$2)^(6−1)					
19	Net salvage							
20	end of yr 6 = $343							
21	= $B$4+$B$3*IF($B$4>$B$1,$C$18−$B$1,$C$18−$B$4)−1/2*$B$3*MAX(($B$4−$B$1),0)							

FIGURE 14-7b After-tax IRR with CCA and 5% inflation.

SUMMARY

Inflation is characterized by rising prices for goods and services, whereas deflation produces a fall in prices. An inflationary trend makes future dollars have less **purchasing power** than present dollars. Inflation benefits a long-term borrower of money because payment of debt in the future is made with dollars that have reduced purchasing power. This advantage to borrowers is at the expense of lenders.

Deflation has the opposite effect from inflation. If money is borrowed over a period of time in which deflation is occurring, then debt will be repaid with dollars that have **more** purchasing power than those originally borrowed. This condition is advantageous to lenders at the expense of borrowers. Inflation and deflation have opposite effects on the purchasing power of a monetary unit over time.

To distinguish and account for the effect of inflation in our engineering economic analysis, we define *inflation, real,* and *market* interest rates. These interest rates are related by the following expression:

$$i = i' + f + i'f$$

Each rate applies in a different circumstances, and it is important to apply the correct rate for the circumstances. Cash flows are expressed in terms of either *actual* or *real dollars.* The *market interest* rate should be used with *actual dollars,* and the *real interest rate* should be used with *real dollars.*

The different cash flows in our analysis may inflate or change at different interest rates when we look over the life cycle of the investment. Also, a single cash flow may inflate or deflate at different rates over time. These two circumstances are handled easily by applying the proper inflation rate to each cash flow over the study period to obtain the actual dollar amounts occurring in each year. After the actual dollar quantities have been calculated, the analysis proceeds, as in earlier chapters, using the market interest rate to calculate the measure of merit of interest.

Historical price change for single commodities and bundles of commodities are tracked with price indexes. The Consumer Price Index (CPI) is an example of a composite index formed by a bundle of consumer goods. The CPI serves as a surrogate for general inflation in our economy. Indexes can be used to calculate the *average annual increase* (or decrease) of the costs and benefits in our analysis. The historical data provide valuable information about how economic quantities may behave in the future over the long run.

The effect of inflation on the computed rate of return for an investment depends on how future benefits respond to the inflation. If benefits produce constant dollars, which are not increased by inflation, the effect of inflation is to reduce the before-tax rate of return on the investment. If, on the other hand, the dollar benefits increase to keep up with the inflation, the before-tax rate of return will not be adversely affected by the inflation. This outcome is not found when an after-tax analysis is made. Even if the future benefits increase to match the inflation rate, the allowable depreciation schedule does not increase. The result will be increased taxable income and income tax payments, which reduce the available after-tax benefits and, therefore, the after-tax rate of return. The important conclusion is that estimates of future inflation or deflation may be important in evaluating capital expenditure proposals.

PROBLEMS

14-1 Define inflation in terms of the purchasing power of dollars.

14-2 Define and describe the relationships between the following: actual and real dollars; and inflation, real and market (combined) interest rates.

14-3 How does inflation happen—describe a few circumstances that cause prices in an economy to increase.

14-4 Is it necessary for inflation to be accounted for in an engineering economy study? What are the two approaches for handling inflation in such analyses?

14-5 What is the Consumer Price Index (CPI)? What is the difference between commodity specific and composite price indexes? Can each be used in engineering economic analysis?

14-6 In Chapters 5 'Present Worth Analysis' and 6 'Annual Cash Flow Analysis', it is assumed that prices are stable and a machine purchased today for $5000 can be replaced for the same amount many years hence. In fact, prices have generally been rising, so the stable price assumption tends to be incorrect. Under what circumstances is it correct to use the 'stable price' assumption when prices are actually changing?

14-7 An economist has predicted that there will be a 7% per year inflation of prices during the next 10 years. If this proves to be correct, how much will an item that presently sells for $10 bring a decade hence? (*Answer:* $19.67)

14-8 A man bought a 5% tax-free provincial bond. It cost $1000 and will pay $50 interest each year for 20 years. The bond will mature at the end of the 20 years and return the original $1000. If there is 2% annual inflation during this period, what rate of return will the investor receive after the effect of inflation has been accounted for?

14-9 A woman wishes to set aside some money for her daughter's university education. Her goal is to have a bank savings university containing an amount equivalent to $20,000 with today's purchasing power of the dollar, on the girl's 18th birthday. The estimated inflation rate is 8%. If the bank pays 5% compounded annually, what lump sum of money should he deposit in the bank savings account on the child's 4th birthday? (*Answer:* $29,670)

14-10 An economist has predicted that for the next 5 years, annual inflation will be 8%, and then there will be 5 years at a 6% inflation rate. This is equivalent to what average price change per year for the entire 10-year period?

14-11 A newspaper reports that in the last 5 years, prices have increased a total of 50%. This is equivalent to what annual inflation rate, compounded annually? (*Answer:* 8.45%)

14-12 A South American country has had a high rate of inflation. Recently, its exchange rate was 15 cruzados per US dollar; that is, one dollar will buy 15 cruzados in the foreign exchange market. It is likely that the country will continue to experience a 25% inflation rate and that the United States will continue at a 7% inflation rate. Assume that the exchange rate will vary the same as the inflation rate. In this situation, how many cruzados will one US dollar buy 5 years from now? (*Answer:* 32.6)

14-13 An automobile manufacturer has a car that gets 10 kilometres per litre of gasoline. It is estimated that gasoline prices will increase at a 12% per year rate, compounded annually, for the next 8 years. This manufacturer believes that the fuel consumption for its new cars should decline as fuel prices increase, so that the fuel cost will remain constant. To achieve this, what must the fuel rating, in kilometres per litre, of the cars be 8 years hence?

14-14 An economist has predicted that during the next 6 years, prices in the United States will increase 55%. He expects a further increase of 25% in the subsequent 4 years, so that prices at the end of 10 years will have increased to 180% of the present level. Compute the inflation rate, f, for the entire 10-year period.

14-15 Sally Johnson lent a friend $10,000 at 15% interest, compounded annually. She is to repay the loan in five equal end-of-year payments. Sally estimates the inflation rate during this period is 12%. After taking inflation into account, what rate of return is Sally receiving on the loan? Compute your answer to the nearest 0.1%. (*Answer:* 2.7%)

14-16 Dale saw that the campus bookstore is having a special on pads of computation paper normally priced at $3 a pad, and now on sale for $2.50 a pad. This sale is unusual and Dale assumes the paper will not be put on sale again. On the other hand, he expects that there will be no increase in the $3 regular price, even though the inflation rate is 2% every 3 months. Dale believes that competition in the paper industry will keep wholesale and retail prices constant. He uses a pad of computation paper every 3 months. He considers 19.25% a suitable minimum attractive rate of return. Dale will buy one pad of paper for his immediate needs. How many extra pads of computation paper should he buy?   (*Answer:* 4)

14-17 An investor wants a real rate of return i' (rate of return without inflation) of 10% per year on any projects in which he invests. If the expected annual inflation rate for the next several years is 6%, what interest rate i should be used in project analysis calculations?

14-18 (*a*) Compute the equivalent annual inflation rate, based on the Consumer Price Index, for the period from 1981 to 1986.

(*b*) Using the equivalent annual inflation rate computed in part (*a*), estimate the Consumer Price Index in 1996, working from the 1987 Consumer Price Index.

14-19 How much will a $20,000 car cost 10 years from now if inflation continues at an annual rate of 4% for the next decade?

14-20 You are considering the purchase, for $15,000, of an annuity that will pay $2500 per year for the next 10 years. You want to have a real rate of return of 5%, and you estimate inflation will average 6% per year over the next 10 years. Should you buy the annuity?

14-21 Inflation is a reality for the general economy for the foreseeable future. Given this assumption, calculate the number of years it will take for the purchasing power of today's dollars to equal *one-fifth* of their present value. Assume that inflation will average 6% per year.

14-22 A homebuilder's advertising has the caption 'Inflation to Continue for Many Years'. The advertisement continues with the explanation that if one buys a home now for $97,000, and inflation continues at a 7% annual rate, the home will be worth $268,000 in 15 years. According to the advertisement, by purchasing a new home now, the buyer will realize a profit of $171,000 in 15 years. Do you find this logic persuasive? Explain.

14-23 Sally Seashell bought a lot at the Salty Sea for $18,000 cash. She does not plan to build on the lot, but instead will hold it as an investment for 10 years. She wants a 10% after-tax rate of return after taking the 6% annual inflation rate into account. If income taxes amount to 15% of the capital gain, at what price must she sell the lot at the end of the 10 years?   (*Answer:* $95,188)

14-24 A group of students decided to lease and run a gasoline service station. The lease is for 10 years. Almost immediately the students were confronted with the need to alter the gasoline pumps to read in litres. The Dayton Company has a conversion kit available for $900 that may be expected to last 10 years. The firm also sells a $500 conversion kit that has a 5-year useful life. The students believe that any money not invested in the conversion kits may be invested elsewhere at a 10% interest rate. Income tax consequences are to be ignored in this problem.

(*a*) Assuming that future replacement kits cost the same as today, which alternative should be selected?

(*b*) If one assumes a 7% inflation rate, which alternative should be selected?

14-25 Pollution control equipment must be purchased to remove the suspended organic material from liquid being discharged from a vegetable packing plant. Two different pieces of equipment are available that would accomplish the task. A Filterco unit costs $7000 and has a 5-year useful life. A Duro unit, on the other hand, now costs $10,000 but will have a 10-year useful life.

With inflation, equipment costs are rising at 8% per year, compounded annually, so that when the Filterco unit needed to be replaced, the cost would be much more than $7000. Using a 10-year analysis period, and a 20% minimum attractive rate of return, before taxes, calculate which piece of pollution control equipment should be purchased.

14-26 The City of Columbia is trying to attract a new manufacturing business to the area. It has offered to install and operate a water pumping plant to provide service to the proposed plant site. This would cost $50,000 now, plus $5000 per year in operating costs for the next 10 years, all measured in year-0 dollars.

To reimburse the city, the new business must pay a fixed uniform annual fee, A, at the end of each year for 10 years. In addition, it is to pay the city $50,000 at the end of 10 years. It has been agreed that the city should receive a 3% rate of return, after taking an inflation rate, f, of 7% into account.

Determine the amount of the uniform annual fee. (*Answer:* $12,100)

14-27 Sam bought a house for $150,000 with some creative financing. The bank, which agreed to lend Sam $120,000 for 6 years at 15% interest, took a first mortgage on the house. The Joneses, who sold Sam the house, agreed to lend him the remaining $30,000 for 6 years at 12% interest. They received a second mortgage on the house. Thus Sam became the owner without putting up any cash. He pays $1500 a month on the first mortgage and $300 a month on the second mortgage. In both cases these are 'interest-only' loans, and the principal is due at the end of the loan.

Sam rented the house to Justin and Shannon, but after paying the taxes, insurance, and so on, he had only $800 left, and so was forced to put up $1000 a month of his own money to make the monthly payments on the mortgages. At the end of 3 years, Sam sold the house for $205,000. After paying off the two loans and the real estate broker, he had $40,365 left. After an 8% inflation rate is taken into account, what was his before-tax rate of return?

14-28 General price inflation is estimated to be 3% for the next 5 years, 5% the 5 years after that, and 8% the following 5 years. If you invest $10,000 at 10% for those 15 years, what is the future worth of your investment in terms of actual dollars at that time and in terms of real base-zero dollars at that time?

14-29 Ima Luckygirl recently found out that her grandfather has passed away and left her his Rocky Mountain Gold savings account. The account was originally opened 50 years ago when Ima's grandfather deposited $2500. He had not added to or subtracted from the account since then. If the account has earned an average rate of 10% per year and inflation has been 4% per year, answer the following:
  (a) How much money is currently in the account in *actual dollars*?
  (b) Express the answer to part (a) in terms of the purchasing power of dollars from 50 years ago.

14-30 Auntie Frannie wants to help pay for her twin nephews to attend a private school. She intends to send a cheque for $2000 at the end of each of the next 8 years to apply to the cost of schooling.
  (a) If general price inflation, as well as tuition price inflation, is expected to average 5% per year for those 8 years, calculate the present worth of the gifts. Assume that the real interest rate will be 3% per year.
  (b) If Auntie Frannie wants her gifts to keep pace with inflation, what would be the present worth of her gifts? Again assume inflation is 5% and the real interest rate is 3%.

14-31 As a recent graduate, you are considering employment offers from three different companies. However, in an effort to confuse you and perhaps make their offers seem better, each company has used a different *purchasing power base* for expressing your annual

salary over the next 5 years. If you expect inflation to be 6% for the next 5 years and your personal (real) MARR is 8%, which plan would you choose?

> *Company A:* A constant $50,000 per year in terms of today's purchasing power.
>
> *Company B:* $45,000 the first year, with increases of $2500 per year thereafter.
>
> *Company C:* A constant $65,000 per year in terms of Year-5-based purchasing power.

14-32 Calculate the future equivalent in Year 15 of:
  (a) Dollars having today's purchasing power.
  (b) Then-current purchasing power dollars, of $10,000 today. Use a market interest rate of 15% and an inflation rate of 8%.

14-33 A firm is having a large piece of equipment overhauled. It expects that the machine will be needed for the next 12 years. The firm has an 8% minimum attractive rate of return. The contractor has suggested three alternatives:
  (1) A complete overhaul for $6000 that should permit 12 years of operation.
  (2) A major overhaul for $4500 that can be expected to provide 8 years of service. At the end of 8 years, a minor overhaul would be needed.
  (3) A minor overhaul now. At the end of 4 and 8 years, additional minor overhauls would be needed.

If minor overhauls cost $2500, which alternative should the firm select? If minor overhauls, which now cost $2500, increase in cost at 5% per year, but other costs remain unchanged, which alternative should the firm select? (*Answers:* Alt. (3); Alt. (1))

14-34 A couple in Regina, Saskatchewan, must decide whether it is more economical to buy a home or to continue to rent during an inflationary period. Presently the couple rents a one-bedroom duplex for $450 a month plus $139 a month in basic utilities (heating and electricity). These costs tend to increase with inflation, and with the projected inflation rate of 5%, the couple's monthly costs per year over a 10-year planning horizon are as follows.

$n =$	1	2	3	4	5	6	7	8	9	10
Rent	450	473	496	521	547	574	603	633	665	698
Utilities	139	146	153	161	169	177	186	196	205	216

The couple would like to live on the north side of the town, where an average home of 150 m^2 of heating area costs $75,000. A local mortgage company will provide a loan for the property provided the couple makes a down payment of 5% plus estimated closing costs of 1% cash for the home. The couple prefers a 30-year fixed-rate mortgage with an 8% interest rate. It is estimated that the basic utilities for the home, inflating at 5%, will cost $160 a month; insurance and maintenance also inflating at 5% will cost $50 a month. The home will appreciate in value about 6% a year. Assuming a nominal interest rate of 15.5%, which alternative will be more attractive to the couple on the basis of the present worth analysis? (*Note:* Realtor's sales commission here is 5%.)

14-35 Given the following data, calculate the present worth of the investment.

> First cost = $60,000   Project life = 10 years
>
> Salvage value = $15,000   MARR = 25%

General price inflation = 4% per year
> Annual cost 1 = $4500 in Year 1 and
> inflating at 2.5% per year
> Annual cost 2 = $7000 in Year 1 and
> inflating at 10.0% per year
> Annual cost 3 = $10,000 in Year 1 and
> inflating at 6.5% per year
> Annual cost 4 = $8500 in Year 1 and
> inflating at −2.5% per year

14-36 Here is some information about a professors' salary index (PSI).

Year	PSI	Change in PSI
1991	82	3.22%
1992	89	8.50
1993	100	*a*
1994	*b*	4.00
1995	107	*c*
1996	116	*d*
1997	*e*	5.17
1998	132	7.58

(*a*) Calculate the unknown quantities *a*, *b*, *c*, *d*, *e* in the table. Review Equation 14-3.

(*b*) What is the *base year* of the PSI? How did you determine it?

(*c*) Given the data for the PSI, calculate the *average annual price increase* in salaries paid to professors for between 1991 and 1995 and between 1992 and 1998.

14-37 From the data in Table 14-1 in the text, calculate the *average annual inflation rate* of first-class postage as measured by the LCI for the following years:

(*a*) End of 1970 to end of 1979

(*b*) End of 1980 to end of 1989

(*c*) End of 1990 to end of 1999

14-38 From the data in Table 14-1 in the text, calculate the *overall rate change* of first-class postage as measured by the LCI for the following decades:

(*a*) The 1970s (1970–1979)

(*b*) The 1980s (1980–1989)

(*c*) The 1990s (1990–1999)

14-39 From the data in Table 14-2 in the text calculate the *average annual inflation rate* as measured by the CPI for the following years:

(*a*) End of 1974 to end of 1982

(*b*) End of 1980 to end of 1989

(*c*) End of 1985 to end of 2002

14-40 Homeowner Henry is building a fireplace for the house he is constructing. He estimates that his fireplace will require 800 bricks. Answer the following:

(*a*) If the cost of a chimney brick in 1978 was $2.10, calculate the material cost of Henry's project in 1998. The chimney brick index (CBI) was 442 in 1970 and is expected to be 618 in 1998.

(*b*) Estimate the cost of materials for a similar fireplace to be built in the year 2008. What assumption did you make?

14-41 If a composite price index for the cost of vegetarian foods called *eggs, artichokes and tofu* (EAT) was at a value of 330 ten years ago, and has averaged an increase of 12% a year after that, calculate the current value of the index.

14-42 As the owner of Beanie Bob's Basement Brewery, you are interested in a construction project to increase production to offset competition from your crosstown rival, Bad Brad's Brewery and Poolhall. Construction cost percentage increases, as well as current cost estimates, for construction costs are given in the table below over a 3-year period. Use a market interest rate of 25%, and assume that general price inflation is 5% over the 3-year period.

Item	Cost If Incurred Today	Cost Percentage Increase		
		Year 1	Year 2	Year 3
Structural metal and concrete	$120,000	4.3%	3.2%	6.6%
Roofing materials	14,000	2.0	2.5	3.0
Heating and plumbing equipment and fixtures	35,000	1.6	2.1	3.6
Insulation material	9,000	5.8	6.0	7.5
Labour	85,000	5.0	4.5	4.5

(a) What would the costs be for labour in Years 1, 2, and 3?

(b) What is the *average percentage increase* of labour cost over the 3-year period? .

(c) What is the present worth of the insulation cost of this project?

(d) Calculate the future worth of the labour and insulation material cost portion of the project.

(e) Calculate the present worth of the total construction project for Beanie Bob.

14-43 Philippe wants to race in the Tour de France 10 years from now. He wants to know what the cost of a custom-built racing bicycle will be 10 years from today. Calculate the cost given the following data.

Item	Current Cost	Cost Will Inflate x% per Year
Frame	$800	2 %
Wheels	350	10
Gearing system	200	5
Braking system	150	3
Saddle	70	2.5
Finishes	125	8

14-44 Owing to cost structures, trade policies, and corporate changes, the costs for three big automakers are estimated to vary over the next 3-year period. These changes will be reflected in the purchase prices of their vehicles. Mary Clare will graduate in 3-years and will be buying a new car—she is considering one model from each company. Which car should Mary Clare purchase 3 years from now, assuming everything but purchase price is equivalent?

Automaker	Current Price	Price Will Inflate x% per Year
X	$27,500	4 %
Y	30,000	1.5
Z	25,000	8

14-45 Granny Viola has been saving money in the Bread & Butter mutual fund for 15 years. She has been a steady contributor over those years and has a pattern of putting $100 into the account every 3 months. If her original investment 15 years ago was $500 and interest in the account has varied as shown, what is the current value of her savings?

Years	Interest Earned in the Account
1–5	12% compounded quarterly
6–10	16 compounded quarterly
10–15	8 compounded quarterly

14-46 Andrew just bought a new boat for $15,000 to use on the river near his home. He has received delivery of the boat and agreed to the terms of the following loan: all principal and interest is due in 3 years (balloon loan), first year annual interest (on the purchase price) is set at 5%, and this is to be adjusted up 1.5% per year for each of the following years of the loan. How much does Andrew owe if he is to pay off the loan in 3 years?

14-47 You were recently looking at the historical prices paid for homes in a neighbourhood that you are interested in. The data that you found, average price paid, is given below. Calculate on a year-to-year basis how home prices in this neighbourhood have inflated (*a–e* in the table below).

Year	Average Home Price	Inflation Rate for That Year
5 years ago	$165,000	(a)
4 years ago	167,000	(b)
3 years ago	172,000	(c)
2 years ago	180,000	(d)
last year	183,000	(e)
this year	190,000	(f, see below)

(f) What is your estimate of the inflation rate for this year?

14-48 The tax laws provide for the depreciation of equipment based on original cost. Yet owing to substantial inflation, the replacement cost of equipment is often much greater than the original cost. What effect, if any, does this have on a firm's ability to buy new equipment to replace old equipment?

14-49 Emma Johnson inherited $85,000 from her father. She is considering investing the money in a house, which she will then rent to tenants. The $85,000 cost of the property consists of $17,500 for the land and $67,500 for the house. Emma believes she can rent the house and have $8000 a year net income left after paying the property taxes and other expenses. The house will be depreciated by straight-line depreciation using a 45-year depreciable life.

(a) If the property is sold at the end of 5 years for its book value at that time, what after-tax rate of return will Emma receive? Assume that her marginal personal income tax rate is 34%.

(b) Now assume there is 7% per year inflation, compounded annually. Emma will raise the rent 7% per year to match the inflation rate, so that after higher property taxes and other expenses are taken into account, the annual net income will go up 7% per year. Assume Emma's marginal income tax rate remains at 34% for all ordinary taxable income related to the property. The value of the property is now projected to increase from its present $85,000 at a rate of 10% per year, compounded annually.

If the property is sold at the end of 5 years, compute the rate of return on the after-tax cash flow in actual dollars. Also compute the rate of return on the after-tax cash flow in year-0 dollars.

14-50 Tom Ward put $10,000 in a 5-year guaranteed investment certificate that pays 12% interest per year. At the end of the 5 years the certificate will mature and he will receive his $10,000 back. Tom has substantial income from other sources and estimates that his marginal income tax rate is 42%. If the inflation rate is 7% per year, find his

(a) before-tax rate of return, ignoring inflation
(b) after-tax rate of return, ignoring inflation
(c) after-tax rate of return, after taking inflation into account

14-51 Annthea has a total taxable income of $60,000 this year and pays federal tax according to the rates in Table 12-1. If inflation continues for the next 20 years at a 7% rate, compounded annually, she wonders what her taxable income must be in the future to provide the same purchasing power, after taxes, as her present taxable income. Assuming the federal income tax rate table is unchanged, what must her taxable income be 20 years from now?

14-52 A small research device is purchased for $10,000 and depreciated by CCA depreciation. The net benefits from the device, before deducting depreciation, are $2000 at the end of the first year, increasing $1000 per year after that (second year equals $3000, third year equals $4000, etc.), until the device is hauled to the junkyard at the end of 7 years. During the 7-year period there is an inflation rate f of 7%.

This profitable corporation has a 50% combined federal and provincial income tax rate. If it requires a 12% after-tax rate of return on its investment, after taking inflation into account, should the device have been purchased?

14-53 When there is little or no inflation, a homeowner can expect to rent an unfurnished home for 12% of the market value of the property (home and land) per year. About $^1/_8$ of the rental income is paid out for property taxes, insurance, and other operating expenses. Thus the net annual income to the owner is 10.5% of the market value of the property. Since prices are relatively stable, the future selling price of the property often equals the original price paid by the owner.

For a $150,000 property (where the land is estimated at $46,500 of the $150,000), compute the after-tax rate of return, assuming the selling price 59 months later (in December) equals the original purchase price. Use CCA depreciation beginning January 1. Also, assume a 35% income tax rate.

14-54 (This is a continuation of Problem 14-53.) As inflation has increased throughout the world, the rental income of homes has decreased and a net annual rental income of 8% of the market value of the property is common. On the other hand, the market value of homes tends to rise about 2% per year more than the inflation rate. As a result, both annual net rental income and the resale value of the property rise faster than the inflation rate. Consider the following situation.

A $150,000 property (with the house valued at $103,500 and the land at $46,500) is purchased for cash in Year 0. The market value of the property increases at a 12% annual rate. The annual rental income is 8% of the beginning-of-year market value of the property. Thus the rental income also increases each year. The general inflation rate f is 10%.

The individual who purchased the property has an average income tax rate of 35%.

(a) Use CCA depreciation, beginning January 1, to compute the actual dollar after-tax rate of return for the owner, assuming he sells the property 59 months later (in December).

(b) Similarly, compute the after-tax rate of return for the owner, after taking the general inflation rate into account, assuming he sells the property 59 months later.

14-55 Consider two mutually exclusive alternatives stated in year-0 dollars. Both alternatives have a 3-year life with no salvage value. Assume the annual inflation rate is 5%, an income tax rate of 25%, and straight-line depreciation. The minimum attractive rate of return (MARR) is 7%. Use rate of return analysis to determine which alternative is preferable.

Year	A	B
0	−$420	−$300
1	200	150
2	200	150
3	200	150

CHAPTER 15

Selection of a Minimum Attractive Rate of Return

BP Goes to Russia

In the early 1990s, western investors flocked to Russia, hoping to reap a fortune as markets opened up after the fall of the Soviet Union. Many didn't stay long once they discovered what it was like to do business in a country where contracts were often impossible to enforce and bribery was the norm. In 1998, when Russia devalued its currency and defaulted on debt obligations, most of the remaining investors fled in panic.

Despite this dismal business outlook, British Petroleum (BP) announced in early 2003 that it was planning to pay $6.75 billion for a 50% interest in Tyumen Oil Company, Russia's fourth-largest producer of oil. BP's decision is particularly striking in view of the company's own history in Russia: in 1997 it bought a share in a small Russian oil company, only to lose part of its investment a few years later after a bitter court battle. Moreover, Russia's state-owned pipeline infrastructure is outdated and inadequate—and the government has been slow to allow private companies to build and operate their own pipelines.

Given all these drawbacks, BP would seem to be taking a big gamble by investing in Tyumen.

After Completing This Chapter...

The student should be able to:

- Define various sources of capital and the costs of those funds to the firm.
- Select a firm's MARR based on the opportunity cost approach for analyzing investments.
- Adjust the firm's MARR to account for risk and uncertainty.
- Discuss the impact of inflation and the cost of borrowed money.
- Use spreadsheets to develop cumulative investments and the opportunity cost of capital.

QUESTIONS TO CONSIDER

1. Many of BP's older sources of oil are now reaching the end of their useful life. How might this have affected the firm's decision to invest in Tyumen Oil?
2. Outside of Tyumen, there are few other oil companies left in Russia for competitors to buy. How might this have affected BP's decision?
3. Where else are the world's largest available oil reserves located? How does the business climate in these areas compare to Russia's?
4. Most of the largest oil-producing nations are members of OPEC (the Organization of Petroleum Exporting Countries), but Russia is not. Is this good or bad for BP?

The preceding chapters have said very little about what interest rate or minimum attractive rate of return is suitable for use in a particular situation. Since this problem is quite complex, there is no single answer that is always right. A discussion of a suitable interest rate to use must inevitably begin with an examination of the sources of capital, followed by a look at the prospective investment opportunities and risk. Only in this way can an interest rate or minimum attractive rate of return be chosen intelligently.

SOURCES OF CAPITAL

In broad terms there are four sources of capital available to a firm: money generated from the operation of the firm, borrowed money, sale of mortgage bonds, and sale of capital stock.

Money Generated from the Operation of the Firm

A major source of capital investment money is through the retention of profits resulting from the operation of the firm. Since only about half of the profits of industrial firms are paid out to shareholders, the half that is retained is an important source of funds for all purposes, including capital investments. In addition to profit, there is money generated in the business equal to the annual depreciation charges on existing capital assets if the firm is profitable. In other words, a profitable firm will generate money equal to its depreciation charges *plus* its retained profits. Even a firm that earns zero profit will still generate money from operations equal to its depreciation charges. (A firm with a loss, of course, will have still fewer funds.)

External Sources of Money

When a firm requires money for a few weeks or months, it usually borrows from banks. Longer-term unsecured loans (of, say, 1–4 years) may also be arranged through banks. While banks undoubtedly finance a lot of capital expenditures, regular bank loans cannot be considered a source of permanent financing.

Longer-term secured loans may be obtained from banks, insurance companies, pension funds, or even the public. The security for the loan is frequently a mortgage on specific property of the firm. When sold to the public, this financing is by mortgage bonds. The sale of shares in the firm is still another source of money. While bank loans and bonds represent debt that has a maturity date, shares are considered a permanent addition to the ownership of the firm.

Choice of Source of Funds

Choosing the source of funds for capital expenditures is a decision for the firm's top executives, and it may require approval of the board of directors. When internal operations generate adequate funds for the desired capital expenditures, external sources of money are not likely to be used. But when the internal sources are inadequate, external sources must be employed or the capital expenditures will have to be deferred or cancelled.

COST OF FUNDS

Cost of Borrowed Money

A first step in deciding on a minimum attractive rate of return might be to determine the interest rate at which money can be borrowed. Longer-term secured loans may be obtained from banks, insurance companies, or the variety of places in which substantial amounts of money accumulates (for example, the oil-producing nations).

A large, profitable corporation might be able to borrow money at the **prime rate,** that is, the interest rate that banks charge their best and most sought-after customers. All other firms are charged an interest rate that is higher by one-half to several percentage points. In addition to the financial strength of the borrower and his ability to repay the loan, the interest rate will depend on the duration of the loan.

Cost of Capital

Another relevant interest rate is the **cost of capital.** The general assumption concerning the cost of capital is that all the money the firm uses for investments is drawn from all the components of the overall capitalization of the firm. The mechanics of the computation is given in Example 15-1.

EXAMPLE 15-1

For a particular firm, the purchasers of common shares require an 11% rate of return, mortgage bonds are sold at a 7% interest rate, and bank loans are available at 9%. Compute the cost of capital for the following capital structure:

		Rate of Return	Annual Amount
$ 20 million	Bank loan	9%	$1.8 million
20	Mortgage bonds	7	1.4
60	Common shares and retained earnings	11	6.6
$100 million			$9.8 million

SOLUTION

Interest payments on debt, like bank loans and mortgage bonds, are tax-deductible business expenses. Thus:

$$\text{After-tax interest cost} = (\text{Before-tax interest cost}) \times (1 - \text{Tax rate})$$

If we assume that the firm pays 40% income tax, the computations become:

Bank loan   After-tax interest cost $= 9\%(1 - 0.40) = 5.4\%$
Mortgage bonds   After-tax interest cost $= 7\%(1 - 0.40) = 4.2\%$

Dividends paid on the ownership in the firm (common shares + retained earnings) are not tax-deductible. Combining the three components, the after-tax interest cost for the $100 million of capital is:

$$\$20 \text{ million } (5.4\%) + \$20 \text{ million } (4.2\%) + \$60 \text{ million } (11\%) = \$8.52 \text{ million}$$

$$\text{Cost of capital} = \frac{\$8.52 \text{ million}}{\$100 \text{ million}} = 8.52\%$$

In an actual situation, the cost of capital is quite difficult to compute. The fluctuation in the price of common shares, for example, makes it difficult to pick a cost, and because of the fluctuating prospects of the firm, it is even more difficult to estimate the future benefits that purchasers of the shares might expect to receive. Given the fluctuating costs and prospects of future benefits, what rate of return do shareholders require? There is no precise answer, but we can obtain an approximate answer. Similar assumptions must be made for the other components of a firm's capitalization.

INVESTMENT OPPORTUNITIES

An industrial firm can invest its money in many more places than are available to an individual. A firm has larger amounts of money, and this alone makes certain kinds of investment possible that are unavailable to individual investors, with their more limited investment funds. The Canadian government, for example, borrows money for short terms of 90 or 180 days by issuing certificates called Treasury bills that frequently yield a greater interest rate than savings accounts. The customary minimum purchase is $25,000.

More important, however, is the fact that a firm conducts a business, which itself offers many investment opportunities. While exceptions can be found, a good generalization is that the opportunities for investment of money within the firm are superior to the investment opportunities outside the firm. Consider the available investment opportunities for a particular firm as outlined in Table 15-1. Figure 15-1 plots these projects by rate of return versus investment. The cumulative investment required for all projects at or above a given rate of return is given in Figure 15-2.

Figures 15-1 and 15-2 illustrate that a firm may have a broad range of investment opportunities available at varying rates of return and with varying lives and uncertainties. It may take some study and searching to identify the better investment projects available to a firm. If this is done, the available projects will almost certainly exceed the money the firm budgets for capital investment projects.

Opportunity Cost

We see that there are two aspects of investing that are basically independent. One factor is the source and quantity of money available for capital investment projects. The other aspect is the investment opportunities themselves that are available to the firm.

These two situations are usually out of balance, with investment opportunities exceeding the available money supply. Thus some investment opportunities can be selected and many

TABLE 15-1 A Firm's Available Investment Opportunities

Project Number	Project	Cost (× 10³)	Estimated Rate of Return
	Investment Related to Current Operations		
1	New equipment to reduce labour costs	$150	30%
2	Other new equipment to reduce labour costs	50	45
3	Overhaul particular machine to reduce material costs	50	38
4	New test equipment to reduce defective products produced	100	40
	New Operations		
5	Manufacture parts that previously had been purchased	200	35
6	Further processing of products previously sold in semi-finished form	100	28
7	Further processing of other products	200	18
	New Production Facilities		
8	Relocate production to new plant	250	25
	External Investments		
9	Investment in a different industry	300	20
10	Other investment in a different industry	300	10
11	Overseas investment	400	15
12	Purchase of Treasury bills	Unlimited	8

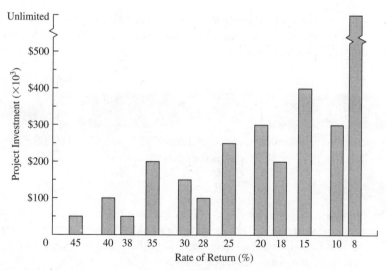

FIGURE 15-1 Rate of return versus project investment.

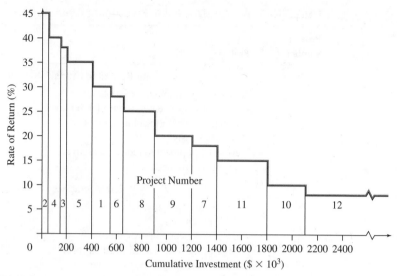

FIGURE 15-2 Cumulative investment required for all projects at or above a given rate of return.

must be rejected. Obviously, we want to ensure that *all the selected projects are better than the best rejected project.* To do this, we must know something about the rate of return on the best rejected project. The best rejected project is the best opportunity forgone, and this in turn is called the **opportunity cost.**

$$\textbf{Opportunity cost} = \textbf{Cost of the best opportunity forgone}$$

$$= \textbf{Rate of return on the best rejected project}$$

If one could predict the opportunity cost for some future period (like the next 12 months), this rate of return could be one way to judge whether to accept or reject any proposed capital expenditure.

EXAMPLE 15-2

Consider the situation represented by Figures 15-1 and 15-2. For a capital expenditure budget of $1.2 million ($1.2 × 10^6$), what is the opportunity cost?

SOLUTION

From Figure 15-2 we see that the 8 projects with a rate of return of 20% or more require a cumulative investment of $1.2 ($\times 10^6$). We would take on these projects and reject the other 4 (7, 11, 10, and 12) with rates of return of 18% or less. The best rejected project is 7, and it has an 18% rate of return. Thus the opportunity cost is 18%.

SELECTING A MINIMUM ATTRACTIVE RATE OF RETURN

Focusing on the three concepts on the cost of money (the cost of borrowed money, the cost of capital, and opportunity cost), which, if any, of these values should be used as the minimum attractive rate of return (MARR) in economic analyses?

Fundamentally, we know that unless the benefits of a project exceed its cost, we cannot add to the profitability of the firm. A lower boundary for the minimum attractive rate of return must be the cost of the money invested in the project. It would be unwise, for example, to borrow money at 8% and invest it in a project yielding a 6% rate of return.

Further, we know that no firm has an unlimited ability to borrow money. Bankers—and others who evaluate the limits of a firm's ability to borrow money—look at both the profitability of the firm and the relationship between the components in the firm's capital structure. This means that continued borrowing of money will require that additional shares must be sold to maintain an acceptable ratio between **ownership** and **debt.** In other words, borrowing for a particular investment project is only a block of money from the overall capital structure of the firm. This suggests that the MARR should not be less than the cost of capital. Finally, we know that the MARR should not be less than the rate of return on the best opportunity forgone. Stated simply,

Minimum attractive rate of return should be equal to the largest one of the following: cost of borrowed money, cost of capital, or opportunity cost.

ADJUSTING MARR TO ACCOUNT FOR RISK AND UNCERTAINTY

We know from our study of estimating the future that what actually happens is often different from the estimate. When we are fortunate enough to be able to assign probabilities to a set of possible future outcomes, we call this a **risk** situation. We saw in Chapter 10 that techniques like expected value and simulation may be used when the probabilities are known.

Uncertainty is the term used to describe the condition when the probabilities are *not* known. Thus, if the probabilities of future outcomes are known, we have *risk,* and if they are unknown we have *uncertainty*.

One way to reduce the likelihood of undertaking projects that do not produce satisfactory results is to pass up marginal projects. In other words, no matter what projects are undertaken, some will turn out better than anticipated and some worse. Some undesirable results can be prevented by choosing only the best projects and avoiding those whose expected results are closer to a minimum standard: then (in theory, at least) the projects selected will provide results *above* the minimum standard even if they do considerably worse than expected.

In projects accompanied by normal business risk and uncertainty, the MARR is used without adjustment. For projects with greater than average risk or uncertainty, some firms increase the MARR. A preferable way deals explicitly with the probabilities by using the techniques from Chapter 10. This may be more acceptable as an adjustment for uncertainty. When the interest rate (MARR) used in economic analysis calculations is raised to adjust for risk or uncertainty, greater emphasis is placed on immediate or short-term results and less emphasis on longer-term results.

EXAMPLE 15-3

Consider the two following alternatives: the MARR of Alt. B has been raised from 10% to 15% to take into account the greater risk and uncertainty that Alt. B's results may not be as favourable as indicated. What is the effect of this change of MARR on the decision?

Year	Alt. A	Alt. B
0	−$80	−$80
1–10	10	13.86
11–20	20	10

SOLUTION

		NPW		
Year	Alt. A	At 14.05%	At 10%	At 15%
0	−$80	−$80.00	−$80.00	−$80.00
1–10	10	52.05	61.45	50.19
11–20	20	27.95	47.38	24.81
		0	+28.83	−5.00

		NPW		
Year	Alt. B	At 15.48%	At 10%	At 15%
0	−$80	−$80.00	−$80.00	−$80.00
1–10	13.86	68.31	85.14	69.56
11–20	10	11.99	23.69	12.41
		0	+28.83	+1.97

Computations at MARR of 10% Ignoring Risk and Uncertainty

Both alternatives have the same positive NPW (+$28.83) at a MARR of 10%. Also, the differences in the benefits schedules (A − B) produce a 10% incremental rate of return. (The calculations are not shown here.) This must be true if NPW for the two alternatives is to remain constant at a MARR of 10%.

Considering Risk and Uncertainty with MARR of 10%

At 10%, both alternatives are equally desirable. Since Alt. B is believed to have greater risk and uncertainty, a logical conclusion is to select Alt. A rather than B.

Increase MARR to 15%

At a MARR of 15%, Alt. A has a negative NPW and Alt. B has a positive NPW. Alternative B is preferred under these circumstances.

Conclusion

On the basis of a business-risk MARR of 10%, the two alternatives are equivalent. Recognizing some greater risk of failure for Alt. B makes A the preferred alternative. If the MARR is increased to 15%, to add a margin of safety against risk and uncertainty, the computed decision is to select B. Since Alt. B has been shown to be less desirable than A, the decision, based on a MARR of 15%, may be an unfortunate one. The difficulty is that the same risk adjustment (increase the MARR by 5%) is applied to both alternatives even though they have different amounts of risk.

The conclusion to be drawn from Example 15-3 is that increasing the MARR to compensate for risk and uncertainty is only an approximate technique and may not always achieve the desired result. Nevertheless, it is common practice in industry to adjust the MARR upward to compensate for increased risk and uncertainty.

Inflation and the Cost of Borrowed Money

As inflation varies, what is its effect on the cost of borrowed money? A widely held view has been that interest rates on long-term borrowing, like 20-year Treasury bonds, will be about 3% more than the inflation rate. For borrowers this is the real—that is, after-inflation—cost of money, and for lenders the real return on loans. If inflation rates were to rise, it would follow that borrowing rates would also rise. All this suggests a rational and orderly situation, about as we might expect.

Unfortunately, things have not worked out this way. Figure 15-3 shows that the real interest rate has not always been about 3% and, in fact, there have been long periods

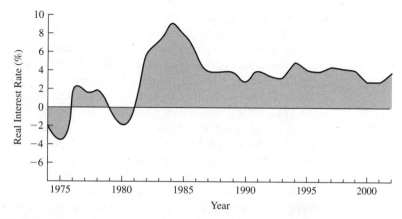

FIGURE 15-3 *The real interest rate.* The interest rate on 20-year US Treasury bonds *minus* the inflation rate, f, as measured by changes in the US Consumer Price Index.

during which the real interest rate was negative. Can this be possible? Would anyone invest money at an interest rate several percentage points below the inflation rate? Well, consider this: when the U.S. inflation rate was 12%, savings banks were paying $5\frac{1}{2}$% on regular passbook deposits. And there was a lot of money in those accounts. While there must be a relationship between interest rates and inflation, Figure 15-3 suggests that it is complex.

REPRESENTATIVE VALUES OF MARR USED IN INDUSTRY

We argued that the minimum attractive rate of return should be established at the highest one of the following: cost of borrowed money, cost of capital, or the opportunity cost.

The cost of borrowed money will vary from enterprise to enterprise, with the lowest rate being the prime interest rate. The prime rate may change several times in a year; it is widely reported in newspapers and business publications. As we pointed out, the interest rate for firms that do not qualify for the prime interest rate may be $\frac{1}{2}$% to several percentage points higher.

The cost of the capital of a firm is an elusive value. There is no widely accepted way to compute it; we know that as a *composite value* for the capital structure of the firm, it conventionally is higher than the cost of borrowed money. The cost of capital must consider the market valuation of the shares (common shares, etc.) of the firm, which may fluctuate widely, depending on the future earnings prospects of the firm. We cannot generalize on representative costs of capital.

Somewhat related to cost of capital is the computation of the return on total capital (long-term debt, capital stock, and retained earnings) actually achieved by firms. *Fortune* magazine, among others, does an annual analysis of the rate of return on total capital. The after-tax rate of return on total capital for individual firms ranges from 0% to about 40% and averages 8%. *Business Week* does a periodic survey of corporate performance. This magazine reports an after-tax rate of return on common shares and retained earnings. We would expect the values to be higher than the rate of return on total capital, and this is the case. The after-tax return on common shares and retained earnings ranges from 0% to about 65% with an average of 14%.

When discussing MARR, firms can usually be divided into two general groups. First, there are firms that are struggling along with an inadequate supply of investment capital or are in an unstable situation or unstable industry. These firms cannot or do not invest money in anything but the most critical projects with very high rates of return and a rapid return of the capital invested. Often these firms use a payback period of one year or less, before income taxes. For an investment project with a 5-year life, this corresponds to about a 60% after-tax rate of return. When these firms do rate-of-return analysis, they reduce the MARR to possibly 25% to 30% after income taxes. There can be a substantial difference between a one-year before-tax payback period and a 30% after-tax MARR, but this apparently does not disturb firms that specify this type of dual criterion.

The second group of firms represents the bulk of all enterprises. They are in a more stable situation and take a longer-range view of capital investments. Their greater money supply enables them to invest in capital investment projects that firms in the first group will reject. Like the first group, this group of firms also uses payback and rate-of-return analysis. When small capital investments (of about $500 or less) are considered, payback period is often the only analysis technique used. The criterion for accepting a proposal may be a before-tax payback period not exceeding 1 or 2 years. Larger investment projects are analyzed by rate of return. Where there is a normal level of business risk, an after-tax MARR of 12% to 15% appears to be widely used. The MARR is increased when there is greater risk involved.

In Chapter 9 we saw that payback period is not a proper method for the economic analysis of proposals. Thus, industrial use of payback criteria is *not* recommended. Fortunately, the trend in industry is toward greater use of accurate methods and less use of payback period.

Note that the values of MARR given earlier are approximations. But the values quoted appear to be opportunity costs, rather than the cost of borrowed money or capital. This indicates that firms cannot or do not obtain money to fund projects whose expected rates of return are nearer to the cost of borrowed money or cost of capital. While one could make a case that good projects are being rejected needlessly, one reason that firms operate as they do is that they can focus limited resources of people, management, and time on a smaller number of good projects.

One cannot leave this section without noting that the MARR used by enterprises is so much higher than can be obtained by individuals. (Where can you get a 30% after-tax rate of return without excessive risk?) The reason appears to be that businesses are not obliged to compete with the thousands of individuals in any region seeking a place to invest $2000 with safety, whereas the number of people who could or would want to invest $500,000 in a business is far smaller. This diminished competition, combined with a higher risk, appears to explain at least some of the difference.

Spreadsheets, Cumulative Investments, and the Opportunity Cost of Capital

As shown in earlier chapters, spreadsheets make computing rates of return dramatically easier. In addition, spreadsheets can be used to sort the projects by rate of return and then to calculate the cumulative first cost. This is accomplished through the following steps.

1. Enter or calculate each project's rate of return.
2. Select the data to be sorted. Do *not* include headings, but do include all information on the row that goes with each project.
3. Select the SORT tool (found in the menu under DATA), identify the rate-of-return column as the first key, and a sort order of descending. Also ensure that row sorting is selected. Sort.
4. Add a column for the cumulative first cost. This column is compared with the capital limit to identify the opportunity cost of capital and which projects should be funded.

Example 15-4 illustrates these steps.

EXAMPLE 15-4

A firm has a budget of $800,000 for projects this year. Which of the following projects should be accepted? What is the opportunity cost of capital?

Project	First Cost	Annual Benefit	Salvage Value	Life (years)
A	$200,000	$25,000	$50,000	15
B	250,000	47,000	−25,000	10
C	150,000	17,500	20,000	15
D	100,000	20,000	15,000	10
E	200,000	24,000	25,000	20
F	300,000	35,000	15,000	15
G	100,000	18,000	0	10
H	200,000	22,500	15,000	20
I	350,000	50,000	0	25

SOLUTION

The first step is to use the RATE function to find the rate of return for each project. The results of this step are shown in the top portion of Figure 15-4. Next the projects are sorted in descending

	A	B	C	D	E	F	G	H
1	Project	First Cost	Annual Benefit	Salvage Value	Life	IRR		
2	A	200,000	25,000	50,000	15	10.2%	=RATE(E2,C2,−B2,D2)	
3	B	250,000	47,000	−25,000	10	12.8%		
4	C	150,000	17,000	20,000	15	8.6%		
5	D	100,000	20,000	15,000	10	16.0%		
6	E	200,000	24,000	25,000	20	10.6%		
7	F	300,000	35,000	15,000	15	8.2%		
8	G	100,000	18,000	0	10	12.4%		
9	H	200,000	22,500	15,000	20	9.6%		
10	I	350,000	50,000	0	25	13.7%		
11	Projects Sorted by IRR						Cumulative First Cost	
12	D	100,000	20,000	15,000	10	16.0%	100,000	
13	I	350,000	50,000	0	25	13.7%	450,000	
14	B	250,000	47,000	−25,000	10	12.8%	700,000	
15	G	100,000	18,000	0	10	12.4%	800,000	
16	E	200,000	24,000	25,000	20	10.6%	1,000,000	
17	A	200,000	25,000	50,000	15	10.2%	1,200,000	
18	H	200,000	22,500	15,000	20	9.6%	1,400,000	
19	C	150,000	17,500	20,000	15	8.6%	1,550,000	
20	F	300,000	35,000	15,000	15	8.2%	1,850,000	

FIGURE 15-4 Spreadsheet for finding opportunity cost of capital.

order by their rates of return. Finally, the cumulative first cost is computed. Projects *D*, *I*, *B*, and *G* should be funded. The opportunity cost of capital is 12.4% if defined as the last project funded and 10.6% if defined as the first project rejected.

SUMMARY

There are four general sources of capital available to an enterprise. The most important one is money generated from the operation of the firm. This has two components: there is the portion of profit that is retained in the business; in addition, a profitable firm generates funds equal to its depreciation charges that are available for reinvestment.

The three other sources of capital are from outside the operation of the enterprise:

1. Borrowed money from banks, insurance companies, and so forth.
2. Longer-term borrowing from a lending institution or from the public in the form of mortgage bonds.
3. Sale of equity securities like common or preferred shares.

Retained profits and cash equal to depreciation charges are the primary sources of investment capital for most firms, and the only sources for many enterprises.

In selecting a value of MARR, three values are frequently considered:

1. Cost of borrowed money.
2. Cost of capital. This is a composite cost of the components of the overall capitalization of the enterprise.
3. Opportunity cost. This refers to the cost of the opportunity forgone; stated more simply, opportunity cost is the rate of return on the best investment project that is rejected.

The MARR should be equal to the highest one of these three values.

When there is a risk aspect to the problem (probabilities are known or reasonably estimated), this can be handled by techniques like expected value and simulation. Where there is uncertainty (probabilities of the various outcomes are not known), there are analytical techniques, but they are less satisfactory. A method commonly used to adjust for risk and uncertainty is to increase the MARR. This method has the effect of distorting the time-value-of-money relationship. The effect is to discount longer-term consequences more heavily than short-term consequences, something that may or may not be desirable. Other possibilities might be to adjust the discounted cash flows or the lives of the alternatives.

PROBLEMS

15-1 Examine the financial pages of your newspaper (or *The Globe and Mail* or *The Financial Post*) and determine the current interest rate on debenture bonds of two different industrial firms, and explain why the interest rates are different for these different bonds.

15-2   Consider four mutually exclusive alternatives:

	A	B	C	D
Initial cost	$0	$100	$50	$25
Uniform annual benefit	0	16.27	9.96	5.96
Computed rate of return	0%	10%	15%	20%

Each alternative has a 10-year useful life and no salvage value. Over what range of interest rates is C the preferred alternative?   (*Answer:* $4.5\% < i \leq 9.6\%$)

15-3   Frequently we read in the newspaper that one should lease a car rather than buying it. For a typical 24-month lease on a car costing $9400, the monthly lease charge is about $267. At the end of the 24 months, the car is returned to the leasing company (which owns the car). As an alternative, the same car could be bought with no down payment and 24 equal monthly payments, with interest at a 12% nominal annual percentage rate. At the end of 24 months the car is fully paid for. The car would then be worth about half its original cost.
   (*a*)   Over what range of nominal before-tax interest rates is leasing the preferred alternative?
   (*b*)   What are some of the reasons that would make leasing more desirable than is indicated in (*a*)?

15-4   Assume you have $2000 available for investment for a 5-year period. You wish to *invest* the money—not just spend it on things that are fun. There are obviously many alternatives available. You should be willing to assume a modest amount of risk of loss of some or all of the money if this is necessary, but not a great amount of risk (no investments in poker games or at horse races). How would you invest the money? What is your minimum attractive rate of return? Explain.

15-5   There are many venture capital syndicates that consist of a few (say, eight or ten) wealthy people who combine to make investments in small and (hopefully) growing businesses. Usually, the investors hire a young investment manager (often an engineer with an MBA) who seeks and analyzes investment opportunities for the group. Would you estimate that the MARR sought by this group is more or less than 12%? Explain.

15-6   A factory has a $100,000 capital budget. Determine which project(s) should be funded and the opportunity cost of capital.

Project	First Cost	Annual Benefits	Life (years)	Salvage Value
A	$50,000	$13,500	5	$5000
B	50,000	9,000	10	0
C	50,000	13,250	5	1000
D	50,000	9,575	8	6000

15-7   Chips USA is considering the following projects to improve its production process. Chips has a short life, and so a 3-year horizon is used in evaluation. Which projects should be done if the budget is $70,000? What is the opportunity cost of capital?

Project	First Cost	Benefit
1	$20,000	$11,000
2	30,000	14,000
3	10,000	6,000
4	5,000	2,400
5	25,000	13,000
6	15,000	7,000
7	40,000	21,000

15-8   National Motors's Rock Creek plant is considering the following projects to improve the company's production process. Which projects should be done if the budget is $500,000? What is the opportunity cost of capital?

Project	First Cost	Annual Benefit	Life (years)
1	$200,000	$50,000	15
2	300,000	70,000	10
3	100,000	40,000	5
4	50,000	12,500	10
5	250,000	75,000	5
6	150,000	32,000	20
7	400,000	125,000	5

15-9 The WhatZit Company has decided to fund 6 of 9 project proposals for the coming budget year. Determine the next capital budget for WhatZit. What is the MARR?

Project	First Cost	Annual Benefits	Life (years)
A	$15,000	$ 4,429	4
B	20,000	6,173	4
C	30,000	9,878	4
D	25,000	6,261	5
E	40,000	11,933	5
F	50,000	11,550	5
G	35,000	6,794	8
H	60,000	12,692	8
I	75,000	14,058	8

15-10 Which projects should be done if the budget is $100,000? What is the opportunity cost of capital?

Project	Life (years)	First Cost	Annual Benefit	Salvage Value
1	20	$20,000	$4000	
2	20	20,000	3200	$20,000
3	30	20,000	3300	10,000
4	15	20,000	4500	
5	25	20,000	4500	−20,000
6	10	20,000	5800	
7	15	20,000	4000	10,000

CHAPTER 16

Economic Analysis in the Public Sector

Hogs and Heavy Industry

Industrial facilities in North Carolina's Tar-Pamlico river basin knew they had a costly problem on their hands a few years ago. State environmental officials had discovered high levels of water pollutants flowing into a nearby estuary. Industrial plants would have to cut their pollution discharges significantly, the officials said. New regulations would be forthcoming.

There was just one snag in this plan: most of the pollutants weren't coming from industrial plants. They were coming from local farms. The Tar-Pamlico Basin's economy had a large agricultural component, including many hog and chicken farms. These farms were contributing substantial quantities of water pollutants in the form of runoff from manure, fertilizer, and pesticides.

So why didn't the environmental regulators go after the farmers? Simple: they weren't covered by the Clean Water Act, and the factories were.

The region's manufacturers had been heavily regulated for years and had already reduced their pollutants well below required levels. Trying to extract the last trace of pollutants from factory effluents would be extremely costly and would not do much to help the environment in any case. When regulators did the required cost analysis of their proposed regulations, they found that manufacturers would have to pay as much as $500 to eliminate one kilogram of targeted pollutants.

By contrast, farmers could cut pollutants significantly, at about one-tenth the cost to manufacturers, simply by instituting practices such as constructing retention ponds, which would help keep pollutants from running off their property into waterways.

After Completing This Chapter...

The student should be able to:

- Distinguish the unique objective and viewpoint of public decisions.
- Explain methods for determining the interest rates for evaluating public projects.
- Use the benefit-cost ratio to analysis projects.
- Distinguish between the conventional and modified versions of the benefit-cost ratio.
- Use an incremental benefit-cost ratio to evaluate a set of mutually exclusive projects.
- Discuss the impact of financing, duration, and politics in public investment analysis.

QUESTIONS TO CONSIDER

1. The situation described here is not unique to the Tar-Pamlico Basin. In fact, it exists to some degree right across the US. So why has Congress been slow to impose environmental requirements on the agricultural sector?
2. Manufacturers in the Tar-Pamlico Basin were eventually able to work out an agreement under which they paid local farmers to reduce pollutant runoff. What was the highest amount they were willing to pay per kilogram of pollutants avoided?
3. What was the lowest amount the farmers were willing to accept per kilogram of pollutant to sign on to the agreement?

So far we have considered economic analysis for companies in the private sector, where the main objectives are to generate profits for growth and to reward current shareholders. Investment decisions for private-sector companies involve evaluating the costs and benefits associated with prospective projects in terms of life-cycle cash flow streams. In earlier chapters we developed several methods for calculating measures of merit and making decisions. Public organizations, such as federal, provincial, and municipal governments, port authorities, and school districts also make investment decisions. For these decision-making bodies, economic analysis is complicated by several factors that do not affect companies in the private sector. These factors include the overall purpose of investment, project financing sources, expected project duration, effects of politics, beneficiaries of investment, and the multi-purpose nature of investments. The overall mission in the public sector is the same as that in the private sector—to make prudent investment decisions that promote the overall objectives of the organization.

The primary economic decision-making measure used in the public sector is the benefit-cost (B/C) ratio. This measure is calculated as a ratio of the equivalent worth of the benefits of investment in a project to the equivalent worth of costs. If the B/C ratio is *greater than 1.0,* the project under evaluation is accepted; if not, it is rejected. The B/C ratio is used to evaluate both single investments and sets of mutually exclusive projects (where the incremental B/C ratio is used). The uncertainties of quantifying cash flows, long project lives, and low interest rates all tend to lessen the reliability of the B/C ratio. There are two versions of the B/C ratio: *conventional* and *modified.* Both provide consistent recommendations to decision makers for single investment decisions and for decisions involving sets of mutually exclusive alternatives. The B/C ratio is a widely used and accepted measure in government economic analysis and decision making.

INVESTMENT OBJECTIVE

Organizations exist to promote the overall goals of those they serve. In private-sector companies, investment decisions are based on increasing the wealth and economic stability of the organization. Beneficiaries of investments generally are clearly identified as the owners or shareholders of the company.

In the public sector, the purpose of investment decisions is sometimes ambiguous. Theoretically the purpose of benefit-cost analysis is to measure (or quantify) a project so that one can determine whether or not it causes an increase (or decrease) in economic (and sometimes social) welfare. The idea is to measure the aggregate gain or loss from a particular decision. It concentrates on the efficient allocation of resources and not the distribution of income, so the concern is not the identification of winners and losers, but rather the 'general welfare' of society. In government economic analysis, it is not always easy to distinguish which investments promote the 'general welfare' and which do not.

Consider the case of a dam construction project to provide water, electricity, flood control, and recreational facilities. Such a project might seem to be advantageous for the entire population of the region. But on closer inspection, decision makers must consider that the dam will require the loss of land upstream due to backed-up water. Farmers will lose pasture or crop land, and nature lovers will lose woodlands. Or perhaps the land to be lost is a breeding ground for protected species, and environmentalists will oppose the project. The project may also have a negative impact on towns, cities, and regions downstream. How will it affect their water supply? Thus, a project initially deemed to have many benefits, on closer inspection, reveals many conflicting aspects. Such conflicting aspects are characteristic of investment and decision making in the public sector.

Public investment decisions are more difficult than those in the private sector owing to the many people, organizations, and political units that may be affected. Opposition to a proposal is more likely in public investment decisions than in those made by private-sector companies because for every group that benefits from a particular project, there is usually an opposing group. Many conflicts in opinion arise when the project involves the use of public lands, including industrial parks, housing developments, business districts, roadways, sewage plants, power plants, and landfills. Opposition may be based on the belief that development of *any* kind is bad or that the proposed development should not be near 'our' homes, schools, or businesses.

Consider the decision that a small town might face when considering whether to establish a municipal rose garden, seemingly a beneficial public investment with no adverse consequences. However, an economic analysis of the project must consider *all* effects of the project, including potential unforeseen outcomes. Where will visitors park their cars? Will increased travel around the park necessitate new traffic lights and signs? Will traffic and visitors to the park increase noise levels to adjacent homes? Will the garden's special varieties of roses create a disease hazard for local gardens? Will the garden require high levels of fertilizers and insecticides, and where will these substances wind up after they have been applied? Clearly, many issues must be addressed. What appeared to be a simple proposal for a city rose garden, in fact, brings up many aspects to be considered.

Our simple rose garden illustrates how effects on *all parties involved* must be identified, even for projects that seem very useful. Public decision makers must reach a compromise between the positive effects enjoyed by some groups and the negative effects on other groups. The overall objective is to make prudent decisions that *promote the general welfare,* but in the public sector the decision process is not so straightforward as in the private sector.

A major document in the development of benefit-cost analysis was the United States **Flood Act of 1936,** which specified that waterway improvements for flood control could be made as long as 'the benefits *to whomsoever they accrue* [italics added] are in excess of the estimated costs'. Perhaps the overall general objective of investment decision analysis in government should be a dual one: to promote the general welfare and to ensure that the value to those who can potentially benefit exceeds the overall costs to those who do not benefit.

VIEWPOINT FOR ANALYSIS

When governmental bodies do economic analysis, an important concern is the proper viewpoint of the analysis. A look at industry will help to explain how the viewpoint, or perspective, from which an analysis is conducted influences the final recommendation. Economic analysis, both governmental and industrial, must be based on a viewpoint. In the case of industry the viewpoint is obvious—a company in the private sector pays the costs and counts *its* benefits. Thus, both the costs and benefits are measured from the perspective of the firm.

Costs and benefits that occur outside the firm are referred to as external consequences (Figure 16-1). In years past, private-sector companies generally ignored the external consequences of their actions. Ask anyone who has lived near a cement plant, a slaughterhouse, or a steel mill about external consequences! More recently, governments have forced industry to reduce pollution and other undesirable external consequences, with the result that today many companies are evaluating the consequences of their action from a broader, or community-oriented, viewpoint.

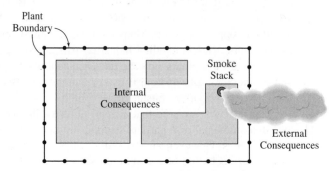

FIGURE 16-1 Internal and external consequences of an industrial plant.

The council members of a small town that levies taxes can be expected to take the 'viewpoint of the town' in making decisions: unless it can be shown that the money from taxes can be used *effectively*, the town council is unlikely to spend it. But what happens when the money is contributed to the town by the provincial government, as in 'revenue sharing' or by means of some other provincial grant? A provincial government may pay a share of project costs varying from 10 to 90%. Example 16-1 illustrates the viewpoint problem that is created.

EXAMPLE 16-1

A municipal project will cost $1 million. The provincial government will pay 50% of the cost if the project is undertaken. Although the original economic analysis showed that the PW of benefits was $1.5 million, a subsequent detailed analysis by the city engineer finds that a more realistic estimate of the PW of benefits is $750,000. The city council must decide whether to proceed with the project. What would you advise?

SOLUTION

From the viewpoint of the city, the project is still a good one. If the city puts up half the cost ($500,000) it will receive all the benefits ($750,000). On the other hand, from an *overall* viewpoint, the revised estimate of $750,000 of benefits does not justify the $1 million expenditure. This illustrates the dilemma caused by varying viewpoints. For economic efficiency, one does not want to encourage the expenditure of money, regardless of the source, unless the benefits at least equal the costs.

Possible viewpoints that may be taken include those of an individual, a business firm or corporation, a regional municipality, a city, a province, a nation, or a group of nations. To avoid sub-optimizing, the proper approach is to *take a viewpoint at least as broad as those who pay the costs and those who receive the benefits.* When the costs and benefits are totally confined to a town, for example, the town's viewpoint seems to be an appropriate basis for the analysis. But when the costs or the benefits are spread beyond the proposed viewpoint, then the viewpoint should be enlarged to this broader population.

Other than investments in defence and social programs, most of the benefits provided by government projects are realized at a regional or local level. Projects such as dams for electricity, flood control, and recreation; and transportation facilities such as roads, bridges, and harbours all benefit most those in the region in which they are constructed. Even smaller-scale projects, such as the municipal rose garden, although funded by public monies at a local or provincial level, provide most benefit to those nearby. As in the case of private decision making, it is important to adopt an appropriate and *consistent* viewpoint and to designate all the costs and benefits that arise from the prospective investment from that perspective. To shift perspective when quantifying costs and benefits could greatly skew the results of the analysis and subsequent decision, and thus such changes in perspective are misleading.

SELECTING AN INTEREST RATE

Several factors, not present for non-government firms, influence the selection of an interest rate for use in economic analysis in the government sector. Recall that for private-sector companies the overall objective is wealth maximization. An interest rate for use in evaluating projects is selected that is consistent with this goal. Many private-sector firms use *cost of capital* or *opportunity cost* concepts when setting an interest rate. The objective of public investment, on the other hand, involves the use of public resources to *promote the general welfare* and to secure the benefits of a given project *to whomsoever they may accrue*, as long as those benefits outweigh the costs. The setting of an interest rate for use in investment analysis is less clear-cut in this case. Several alternative concepts have been suggested for how government should set this rate; these are discussed next.

No Time-Value-of-Money Concept

In government, monies are obtained through taxation and spent about as quickly as they are obtained. Often, there is little delay between collecting money from taxpayers and spending it. (Remember that the federal government collects taxes every paycheque in the form of withholding tax.) The collection of taxes, like their disbursement, although based on an annual budget, is actually a continuous process. Using this line of reasoning, some would argue that there is little or no time lag between collecting and spending tax dollars. Thus, they would advocate the use of a 0% interest rate for economic analysis of public projects.

Cost of Capital Concept

Another consideration in the determination of interest rates in public investments is that most levels of government (federal, provincial, and local) borrow money for capital expenditures in addition to collecting taxes. Where money is borrowed for a specific project, one line of reasoning is to use an interest rate equal to the *cost of borrowed money.*

Opportunity Cost Concept

Opportunity cost, which is related to the interest rate on the best opportunity forgone, may take two forms in governmental economic analysis: government opportunity cost and taxpayer opportunity cost. In public decision making, if the interest rate is based on the opportunity cost to a government agency or other governing body, this interest rate is known as **government opportunity cost.** In this case the interest rate is set at that of the best prospective project for which funding is not available. One disadvantage of the government opportunity cost concept is that different agencies and subdivisions of government will have different opportunities. Therefore, political units could set different interest rates for use in economic analysis, and a project that may be rejected in one branch owing to an inadequate rate of return may be accepted in another. Differing interest rates lead to inconsistent evaluation and decision making across government.

Dollars used for public investments are generally gathered through taxation of the citizenry. The concept of **taxpayer opportunity cost** suggests that a correct interest rate to use in evaluating public investments is that which the *taxpayer* could have received if the government had not collected those dollars through taxation. This philosophy holds that through taxation the government is taking away the taxpayers' opportunity to use the same dollars for investment. The interest rate that the government requires should not be less than what the taxpayer would have received. This compelling argument is supported in general by the US government's Office of Management and Budget A94 directive that stipulates a 7% interest rate be used in economic analysis for a wide range of federal projects. It is not economically desirable to take money from a taxpayer with a 12% opportunity cost, for example, and invest it in a government project yielding 4%.

Recommended Concept

The general rule of thumb in setting an interest rate for government investments has been to select the *largest* of the cost of capital, the government opportunity cost, or the taxpayer opportunity cost interest rates. However, as is the case in the private sector, there is no hard and fast rule universally applied in all decision circumstances. Setting an interest rate for use in economic analysis is at the discretion of the government entity performing the analysis. Consider the seven government entities given in Table 16-1. The interest rates used by these decision-making bodies for evaluating investments could all potentially be set at

TABLE 16-1 Examples of Interest Rates Used in Government Economic Analysis

Government Entity	Interest Rate Used (%)
U.S. armed services	4
State agency	6
Federal agency for highway transportation safety	3
City port authority	5
City school board	6.5
State waterway commission	5
City of Anytown	8

different levels. Setting these interest rates would involve a management decision based on both objective (cost of capital, etc.) and subjective (risk attitudes, etc.) factors considered by each unit.

THE BENEFIT-COST RATIO

The benefit-cost ratio was described briefly in Chapter 9 as one of the economic analysis methods for evaluating prospective projects. This method is used almost exclusively in public investment analysis, and because of the magnitude of public dollars committed each year through such analysis, the benefit-cost ratio deserves our attention and understanding.

One of the primary reasons for the use of the benefit-cost ratio (B/C ratio) in public decision making is its simplicity. The ratio is formed by calculating the equivalent worth of the benefits accrued through investment in a project divided by the equivalent worth of the costs of the project. The benefit-cost ratio can be shown as follows:

$$\text{B/C ratio} = \frac{\text{Equivalent worth of net benefits}}{\text{Equivalent worth of costs}}$$

$$= \frac{\text{PW benefits}}{\text{PW costs}} = \frac{\text{FW benefits}}{\text{FW costs}} = \frac{\text{AW benefits}}{\text{AW costs}}$$

Notice that *any* of the equivalent worth methods (present, future, and annual) can be used to calculate this ratio. Each formulation of the ratio will produce an identical result, as illustrated in Example 16-2.

EXAMPLE 16-2

Demonstrate that for a highway expansion project with data given as follows, the same B/C ratio is obtained using the present, future, and annual worth formulations.

Initial costs of expansion	$1,500,000
Annual costs for operating/maintenance	65,000
Annual savings and benefits to travelers	225,000
Scrap value after useful life	300,000
Useful life of investment	30 years
Interest rate	8%

Using Present Worth

PW benefits $= 225,000(P/A, 8\%, 30) + 300,000(P/F, 8\%, 30) = \$2,563,000$

PW costs $= 1,500,000 + 65,000(P/A, 8\%, 30) = \$2,232,000$

Using Future Worth

FW benefits $= 225,000(F/A, 8\%, 30) + 300,000 = \$25,790,000$

FW costs $= 1,500,000(F/P, 8\%, 30) + 65,000(F/A, 8\%, 30) = \$22,460,000$

Using Annual Worth

AW benefits $= 225,000 + 300,000(A/F, 8\%, 30) = \$227,600$

AW costs $= 1,500,000(A/P, 8\%, 30) + 65,000 = \$198,200$

$$\text{B/C ratio} = \frac{2,563,000}{2,232,000} = \frac{25,790,000}{22,460,000} = \frac{227,600}{198,200} = 1.15$$

One can see that the ratio provided by each of these methods produces the same result: 1.15.

An economic analysis is performed to assist in the objective of making a decision. When one is using the B/C ratio, the decision rule is in two parts:

If the B/C ratio is > 1.0, then the decision should be to invest.

If the B/C ratio is < 1.0, then the decision should be not to invest.

Cases of a B/C ratio *just equal to* 1.0 are analogous to the case of a calculated net present worth of $0 or an IRR analysis that yields $i = $ MARR%. In other words, the decision measure is *just at* the break-even criteria. In such cases a detailed analysis of the input variables and their estimates is necessary, and one should consider the merits of other available opportunities for the targeted funds. But, if the B/C ratio is greater than or less than 1.0, the recommendation is clear.

The B/C ratio is a numerator/denominator relationship between the equivalent worths (EW) of *benefits* and *costs:*

$$\text{B/C ratio} = \frac{\text{EW of net benefits to whomsoever they may accrue}}{\text{EW of costs to the sponsors of the project}}$$

The numerator and denominator aspects of the ratio are sometimes interpreted and used in different fashions. For instance, the **conventional B/C ratio** defines the numerator and denominator as follows:

$$\text{Conventional B/C ratio} = \frac{\text{EW of net benefits}}{\text{EW of initial costs} + \text{EW of operating and maintenance costs}}$$

This is the formulation of the B/C ratio used in Example 16-2. However, there is another version of the ratio, called the **modified B/C ratio.** Using a *modified* version, the numerator and denominator are defined in a different manner: the *annual operating and maintenance* *costs* to the users are subtracted in the numerator, whereas they are added as a cost in the denominator. In this case the ratio becomes:

$$\text{Modified B/C ratio} = \frac{\text{EW of net benefits} - \text{EW of operating and maintenance costs}}{\text{EW of initial costs}}$$

For decision making, the two versions of the benefit-cost ratio will produce the same recommendation as to whether to *invest* or *not invest* in the project being considered. The *numeric B/C ratio* for the two versions will not always be the same, but the recommendation will be. This fact is illustrated in Example 16-3.

EXAMPLE 16-3

Consider the highway expansion project from Example 16-2. Let us use the present worth formulation of *conventional* and *modified* versions to calculate the B/C ratio.

SOLUTION

Using the *Conventional B/C Ratio*

$$\text{B/C ratio} = \frac{225,000(P/A, 8\%, 30) + 300,000(P/F, 8\%, 30)}{1,500,000 + 65,000(P/A, 8\%, 30)} = 1.15$$

Using the *Modified B/C Ratio*

$$\text{B/C ratio} = \frac{225,000(P/A, 8\%, 30) + 300,000(P/F, 8\%, 30) - 65,000(P/A, 8\%, 30)}{1,500,000} = 1.22$$

Whether the conventional or the modified ratio is used, the recommendation is to invest in the highway expansion project. The ratios are not identical in magnitude (1.15 vs 1.22), but the decision is the same.

It is important when one is using the conventional and modified B/C ratios not to directly compare the magnitudes of the two versions. Evaluating a project with one version may produce a higher ratio than is produced with the other version, but this does not imply that the project is somehow better.

The *net benefits to the users* of government projects are the difference between the expected *benefits* from investment minus the expected *disbenefits*. Disbenefits are the negative effects of government projects felt by some individuals or groups. For example, consider Canada's National Parks. Development projects by the skiing or lumber industries might provide enormous benefits to the recreation or construction sectors while creating simultaneous disbenefits for environmental groups. Table 16-2 illustrates some of the primary benefits and disbenefits of several example public investments.

TABLE 16-2 Example Benefits and Disbenefits for Public Investments

Public Project	Primary Benefits	Primary Disbenefits
New city airport just outside city	More flights, new businesses	Increased travel time to airport, more traffic in outer suburbs
Highway bypass around town	Shorter commuting times, reduced congestion on roads	Lost sales to businesses on roads, loss of agricultural lands
New subway	Faster commuting times, less pollution	Loss of jobs due to bus line closing, less access to service (fewer stops)
Creation of a city waste disposal facility versus sending waste to another region or province	Less costly, faster and more responsive to customers	Objectionable sights and smells, loss of market value for home-owners, loss of pristine forest land
Construction of a nuclear power plant	Lower energy costs, new industry in area	Environmental risks

INCREMENTAL BENEFIT-COST ANALYSIS

In Chapter 9 we discussed the use of the incremental benefit-cost ratio in economic decision analysis. As is the case with the internal rate of return (IRR) decision method, the incremental B/C ratio should be used in comparing *sets of mutually exclusive alternatives*. This method produces a result that is consistent with the result produced by optimizing the present worth of the decision alternatives over their respective life cycles. As with the incremental IRR method, it is *not* proper to simply calculate the B/C ratio for each alternative and choose the one with the highest value. Rather, an *incremental* approach is called for.

Elements of the Incremental Benefit-Cost Ratio Method

1. *Identify all relevant alternatives.* Decision makers identify the set of alternatives from which a choice is ultimately made. In this context it is important to identify *all* relevant and competitive alternatives. Decision rules or models can recommend a *best* course of action *only* from the set of identified alternatives. If a better alternative than those in the considered set exists, it will never be selected, and the solution will be sub-optimal. For benefit-cost ratio problems, the do-nothing option is always the 'base case' from which the incremental methodology proceeds.

2. *(Optional) Calculate the B/C ratio of each competing alternative in the set.* In this optional step one calculates the B/C ratio of each of the competing alternatives based on the *total cash flows for each alternative* by itself. Once the individual B/C ratios have been calculated, the alternatives with a ratio *less than 1.0* are eliminated from further consideration. The alternatives with a B/C ratio *greater than 1.0* remain in the set of feasible alternatives. This step gets the 'poor performers' out of the way before the incremental procedure is initiated. This step may be omitted, however, because the incremental analysis method will eliminate the subpar alternatives in due time.

3. *Rank order the projects.* The alternatives must be ordered according to increasing size of the *denominator of the B/C ratio.* (The rank order will be the same regardless of whether one uses the present worth, annual worth, or future worth of costs to form the *denominator.*) To form the list the *denominator* of the B/C ratio (cost portion of the ratio) is first calculated for each of the feasible alternatives and then placed into an ascending rank order from low to high cost. The do-nothing alternative always becomes the first on the ordered list.

4. *Identify the increment under consideration.* In this step the increment under consideration is identified. The first increment taken under consideration is always that of going from the do-nothing option to the first of the feasible alternatives. As the analysis proceeds, any identified increment is always in reference to some previously justified alternative.

5. *Calculate the B/C ratio on the considered incremental cash flows.* Once the increment under consideration has been identified, it is necessary to calculate the *incremental benefits* as well as the *incremental costs.* This is done by finding the cash flows that represent the difference (Δ) between the two alternatives under consideration. For two alternatives X and Y, the incremental benefits (ΔB) and incremental costs (ΔC) of going from Alternative X to Alternative Y must be determined. The increment can be written as (X $\rightarrow$ Y) to signify *going from* X *to* Y or as (Y $-$ X) to signify the *cash flows of* Y *minus* cash flows of X. Both modes identify the incremental costs and benefits of investing in alternative Y, where X is a previously justified (or base) alternative. The ΔB and ΔC values are used to calculate the overall *incremental B/C ratio* (ΔB/ΔC) of the increment.

6. *Use the incremental B/C ratio to make a decision.* The incremental B/C ratio (ΔB/ΔC) calculated in Step 5 is evaluated as follows: if the ratio is greater than 1.0, then the increment is desirable or justified; if the ratio is less than 1.0, it is not desirable, or is not justified. If an increment is accepted, the alternative associated with that additional increment of investment becomes the base from which the next increment is formed. In the case of an increment that is not justified, the alternative associated with the additional increment is rejected and the previously justified alternative is maintained as the base for formation of the next increment.

7. *Iterate to Step 4 until all increments (projects) have been considered.* The incremental method requires that the entire list of ranked feasible alternatives be evaluated. All pair-wise comparisons are made such that the additional increment being considered is examined with respect to a previously justified alternative. The incremental method continues until all alternatives have been evaluated.

8. *Select the best alternative from the set of mutually exclusive competing projects.* After all alternatives (and associated increments) have been considered, the incremental B/C ratio method calls for selection of the alternative that is *associated with the last justified increment.* In this way, it is assured that a maximum investment is made such that each ratio of equivalent worth of incremental benefits to equivalent worth of incremental costs is greater than 1.0. (A common error in applying the incremental B/C method is to choose the alternative with the *largest* incremental B/C ratio, which is inconsistent with the objective of maximizing investment size with incremental B/C ratios above 1.0.)

Both the conventional and modified versions of the B/C ratio can be used with the incremental B/C ratio method just described, but the two versions should not be mixed in the same problem. Such an approach could affect the rank order and cause confusion and errors. Instead, *one* of the two versions should be *consistently* used throughout the analysis.

Examples 16-4 and 16-5 illustrate the use of the incremental B/C ratio and show how the conventional and modified versions can be used with this procedure to evaluate sets of mutually exclusive alternatives.

EXAMPLE 16-4

A city is considering the construction and operation of facilities to provide electricity to several city-owned properties. Electricity will be provided by two coal-burning power plants and a distribution network wired to the properties targeted for conversion. A group studying the proposal has identified general cost and benefit categories for the project as follows.

Primary costs: Construction of the power plant facilities; cost of installing the power distribution network; life-cycle maintenance and operating costs.

Primary benefits: Elimination of payments to the current electricity provider; creation of jobs for construction, operation, and maintenance of the facilities and distribution network; revenue from selling excess power to utility companies; increased employment for city residents.

Assume that there have been four competing designs identified for the power plants. Each design affects costs and benefits in a unique way. Given the following data for the four mutually exclusive design alternatives, use the *conventional* B/C ratio method to recommend a course of action.

	Values ($\times$ \$10^4) for Competing Design Alternatives			
	I	II	III	IV
Project costs				
Plant construction cost	\$12,500	\$11,000	\$12,500	\$16,800
Annual operating and maintenance cost	120	480	450	145
Project benefits				
Annual savings from utility payments	580	700	950	1,300
Revenue from overcapacity	700	550	200	250
Annual effect of jobs created	400	750	150	500
Other data				
Project life, in years	45	45	45	45
Discounting rate (MARR)	8%	8%	8%	8%

SOLUTION

Alternatives I to IV constitute a set of *mutually exclusive* choices because we will select one and only one of the design options for the power plants. Therefore, an incremental B/C ratio method is used to obtain the solution. Let us use the incremental method as described above:

Step 1 The alternatives are do nothing, and designs I, II, III, and IV.

Step 2 In this optional step we calculate the *conventional* B/C ratio for each alternative based on individual cash flows. We will use the ratio of the PW of benefits to costs.

B/C ratio (I) $= (580 + 700 + 400)(P/A, 8\%, 45)/[12,500 + 120(P/A, 8\%, 45)] = 1.46$

B/C ratio (II) $= (700 + 550 + 750)(P/A, 8\%, 45)/[11,000 + 480(P/A, 8\%, 45)] = 1.44$

B/C ratio (III) $= (200 + 950 + 150)(P/A, 8\%, 45)/[12,500 + 325(P/A, 8\%, 45)] = 0.96$

B/C ratio (IV) $= (1300 + 250 + 500)(P/A, 8\%, 45)/[16,800 + 145(P/A, 8\%, 45)] = 1.34$

Alternative I, II, and IV all have B/C ratios greater than 1.0 and thus are included in the feasible set. Alternative III does not meet the acceptability criterion and should be eliminated from further consideration. However, to illustrate that Step 2 is optional, we will continue with all four design alternatives in the original feasible set.

Step 3 Here we calculate the PW of costs for each alternative in the feasible set (remember we are keeping alternative III along for the ride). The denominator of the *conventional* B/C ratio includes first cost and annual O&M costs. We calculate the PW of costs for each alternative as follows.

PW costs (I) $= 12,500 + 120(P/A, 8\%, 45) = \$13,953$

PW costs (II) $= 11,000 + 480(P/A, 8\%, 45) = \$16,812$

PW costs (III) $= 12,500 + 325(P/A, 8\%, 45) = \$16,435$

PW costs (IV) $= 16,800 + 145(P/A, 8\%, 45) = \$18,556$

The rank order from low to high value of the B/C ratio *denominator* is as follows: do nothing, I, III, II, IV.

Step 4 From the ranking list, the first increment considered is that of going from do nothing to Alternative I (do nothing $\rightarrow$ Alternative I). The analysis proceeds from this point.

Steps 5 and 6 We proceed through the analysis, designating the incremental cash flows and calculating Δ(B/C) until all feasible alternatives have been considered. Each additional increment taken under consideration must be based on the last justified increment.

Incremental Effect	Increment			
	(Do Nothing → I)	(I → III)	(I → II)	(II → IV)
ΔPlant construction cost	$12,500	$ 0	$−1500	$5800
ΔAnnual O&M cost	120	205	360	−335
PW of ΔCosts	13,953	2482	2859	1744
ΔAnnual utility payment savings	580	370	120	600
ΔAnnual overcapacity revenue	700	500	−150	−300
ΔAnnual benefits of new jobs	400	−250	350	−250
PW of ΔBenefits	20,342	−4601	3875	605
ΔB/C ratio (PW ΔB)/(PW ΔC)	1.46	−1.15	1.36	0.35
Is increment justified?	Yes	No	Yes	No

As an example of these calculations, consider the third increment (I → II).

ΔPlant construction cost	$= 11,000 - 12,500 = -\$1500$
ΔAnnual O&M cost	$= 480 - 120 = \$360$
PW of ΔCosts	$= -1500 + 360(P/A, 8\%, 45) = \$2859$
or	$= 16,812 - 13,953 = \$2859$
ΔAnnual utility payment savings	$= 700 - 580 = \$120$
ΔAnnual overcapacity revenue	$= 550 - 700 = -\$150$
ΔAnnual benefits of new jobs	$= 750 - 400 = \$350$
PW of ΔBenefits	$= (120 - 150 + 350)(P/A, 8\%, 45)$
	$= \$3875$
ΔB/C ratio (PW ΔB)/(PW ΔC)	$= 3875/2850 = 1.36$

Step 8 The analysis in the table proceeded as follows: do nothing to Alternative I was justified ($\Delta B/C$ ratio $= 1.46$), Alternative I became the new base; Alternative I to Alternative III was not justified (ΔB/C ratio $= -1.15$), Alternative I remained base; Alternative I to Alternative II was justified (ΔB/C ratio $= 1.36$), Alternative II became the base; Alternative II to Alternative IV was not justified (ΔB/C ratio $= 0.35$), and Alternative II became the recommended power plant design alternative because it is the one associated with the last justified increment. All alternatives in the feasible set were considered before the recommendation was produced. Notice that Alternative III, even though included in the feasible set, did not affect the recommendation and was eliminated through the incremental method. Notice also that the first increment considered (do nothing → I) was not selected even though it had the *largest* ΔB/C ratio (1.45). Selection is based on the alternative associated with the *last justified increment* (in this case, Alt. II).

EXAMPLE 16-5

Let us reconsider Example 16-4, this time using the modified B/C ratio to analyze the set of competing design alternatives. Again we will use the present worth method.

SOLUTION

Here we use the modified B/C ratio.

Step 1 The alternatives are still do nothing and designs I, II, III, and IV.

Step 2 We calculate the modified B/C ratio using the PW of benefits and costs.

B/C ratio (I) $= (580 + 700 + 400 - 120)(P/A, 8\%, 45)/(12,500) = 1.51$

B/C ratio (II) $= (700 + 550 + 750 - 480)(P/A, 8\%, 45)/(11,000) = 1.67$

B/C ratio (III) $= (200 + 950 + 150 - 325)(P/A, 8\%, 45)/(12,500) = 0.95$

B/C ratio (IV) $= (1300 + 250 + 500 - 145)(P/A, 8\%, 45)/(16,800) = 1.37$

Again Alternative III would be eliminated from further consideration because its B/C ratio is *less than 1.0*. In this case we will eliminate it from the feasible set, which now becomes do nothing and alternative designs I, II, and IV.

Step 3   The PW of costs for each alternative in the feasible set:

$$PW\ Costs\ (I)\ \ = \$12,500$$

$$PW\ Costs\ (II)\ = \$11,000$$

$$PW\ Costs\ (IV) = \$16,800$$

The right rank order is now do nothing, II, I, IV. Notice that the *modified* B/C ratio produces a rank order different from that yielded by the *conventional* version in Example 16-4.

Step 4   The first increment is now (do nothing → Alternative II). The method proceeds from this point.

Steps 5 and 6   These steps are done as follows.

	Increment		
Incremental Effects	**(Do Nothing → II)**	**(II → I)**	**(II → IV)**
ΔPlant construction cost	$11,000	$ 1500	$ 5800
PW of ΔCosts	11,000	1500	5800
ΔAnnual utility payment savings	700	−120	600
ΔAnnual overcapacity revenue	550	150	−300
ΔAnnual benefits of new jobs	750	−350	−250
ΔAnnual O&M disbenefit	480	−360	−335
PW of ΔBenefits	18,405	484	4662
ΔB/C ratio (PW ΔB)/(PW ΔC)	1.67	0.32	0.80
Is increment justified?	Yes	No	No

As an example of the calculations in the foregoing table, consider the third increment (II → IV).

ΔPlant construction cost	$= 16,800 - 11,000 = \$5800$
PW of ΔCosts	$= \$5800$
ΔAnnual utility payment savings	$= 1300 - 700 = \$600$
ΔAnnual overcapacity revenue	$= 250 - 550 = -\$300$
ΔAnnual benefits of new jobs	$= 500 - 750 = -\$250$
ΔAnnual O&M disbenefit	$= 145 - 480 = -\$335$
PW of ΔBenefits	$= (600 - 300 - 250 + 335)(P/A, 8\%, 45) = \$4662$
ΔB/C ratio, (PW ΔB)/(PW ΔC)	$= 4662/5800 = 0.80$

When the modified version of the B/C ratio is used, Alt. II emerges as the recommended power plant design—just as it did when we used the conventional B/C ratio.

OTHER EFFECTS OF PUBLIC PROJECTS

Three areas remain that merit discussion in describing the differences between government and nongovernment economic analysis: (1) financing government versus non-government projects, (2) the typical length of government versus non-government project lives, and (3) the general effects of politics on economic analysis.

Project Financing

Governmental and market-driven firms differ in the way investments in equipment, facilities, and other projects are financed. In general, firms rely on monies from individual investors (through stock and bond issuance), private lenders, and retained earnings from operations. These sources serve as the pool from which investment dollars for projects come. Management's job in the market-driven firm is to match financial resources with projects in a way that keeps the firm growing, produces an efficient and productive environment, and continues to attract investors and future lenders of capital.

On the other hand, the government sector often uses taxation and bonds as the source of investment capital. In government, taxation and revenue from operations is adequate to finance only modest projects. However, public projects tend to be large in scale (roadways, bridges, etc.), which means that for many public projects 100% of the investment costs must be borrowed—unlike those in the private sector.

In Canada, because of the division of powers between different levels of government, large public projects often have shared funding from federal, provincial, and municipal sources. Generally the projects are financed out of current revenues, but for large projects the governments have the right to borrow money through bond issues.

Another recent innovation is called P3 funding, which stands for public private partnership. In these arrangements, the need is identified by the government but the capital, engineering, construction, and operation of the resulting facility are undertaken by a private corporation. The corporation is repaid either by user fees or by payments from the government. The Confederation Bridge that links Prince Edward Island to the mainland is a P3 arrangement; the operating company collects and keeps the tolls according to a specified arrangement. The bypass freeway around Edmonton is another P3, but in this instance, there are to be no tolls and the provincial government will pay the P3 partner an annual fee.

Limitations on the use and sources of borrowed monies make funding projects in the public sector much different from this process in the private sector. Private-sector firms are seldom able to borrow 100% of required funds for projects, as can be done in the public sector, but at the same time, private entities do not face restrictions on debt retirement or the uncertainty of voter approval.

Project Duration

Another aspect distinguishing government projects from those in the private sector is the typical duration or project life of the investment. In the private sector, projects most often have a projected or intended life ranging between 5 and 15 years. On some occasions the project life is shorter and in others longer, but a majority of projects fall in this interval.

Complex advanced manufacturing technologies, like computer-aided manufacturing or flexible automated manufacturing cells, tend to have projects lives at the longer end of this range. In the 1980s, US manufacturing managers were criticized for short-sighted views of capital investments in such technologies. At that time short-sightedness and lack of investments were blamed for the overall loss of competitiveness in such key U.S. industries as textiles, steel, electronics, automotive, and machine tools.

Government projects usually have lives in the range of 20 to 50 years (or longer). Typical projects are highways, city water and sewer infrastructure, county dumps, and provincial museums. These projects, by nature, have a longer useful life than a typical project in the private sector. There are exceptions to this rule since private firms invest in facilities and other long-range projects, and government entities also invest in projects with shorter-term lives. But, in general, investment duration in the government sector is longer than in the private sector.

Government projects, because they tend to be long range and large scale, usually require substantial funding in the early stages. The highway, water and sewer, and library projects just mentioned could each cost millions of dollars for design, surveying, and construction. Therefore, it is in the best interest of decision makers who are advocates of such projects to spread that first cost over as many years as possible to reduce the annual cost of capital recovery. This tendency to use longer project lives to downplay the effects of a large first cost can affect the desirability of the project, as measured by the B/C ratio. Another aspect closely associated with managing the size of the capital recovery cost in a B/C ratio analysis is the interest rate used for discounting. Lower interest rates reduce the size of the capital recovery cost by reducing the penalty of having money tied up in a project. Example 16-6 illustrates the effects that project life and interest rate can have on the analysis and acceptability of a project.

EXAMPLE 16-6

Consider a project to build a new high school, needed because of increased (and projected) population growth. Information for the project is as follows:

Building first costs (design, planning, and construction)	$10,000,000
Initial cost for roadway and parking facilities	5,500,000
First cost to equip and furnish facility	500,000
Annual operating and maintenance costs	350,000
Annual savings from rented space	500,000
Annual benefits to community	1,500,000

With this project we examine the effect that varying project lives and interest rates have on the conventional B/C ratio. Project lives at 15, 30, and 60 years and interest rates at: 3, 10, and 15% are used to calculate the ratio for the investment. The ratio for each combination of project life and interest rate is tabulated as follows:

Conventional Benefit-Cost Ratio for Various
Combinations of Project Life and Interest Rate

	Interest		
Project Life (years)	**3%**	**10%**	**15%**
15	1.24	0.86	0.69
30	1.79	1.03	0.76
60	2.24	1.08	0.77

As an example of how the ratios are calculated, suppose that life $=$ 30 years and interest rate $=$ 10%.

$$\text{Conventional B/C ratio} = \frac{1,500,000 + 500,000}{(10,000 + 5,500,000 + 500,000)(A/P, 10\%, 30) + 350,000} = 1.03$$

From these numbers one can see the effect of project life and interest on the analysis and recommendation. At the lower interest rate, the project has B/C ratios above 1.0 in all cases of project life, while at the higher rate the ratios are all less than 1.0. At an interest rate of 10% the recommendation to invest changes from *no* at a life of 15 years to *yes* at 30 and 60 years. By manipulating these two parameters (project life and interest rate), it is possible to reach entirely different conclusions regarding the desirability of the project. The key point is that those advocating 'investment' would be well advised to use lower interest rates and longer project life in their example calculations.

Project Politics

To some degree political influences are felt in nearly every decision made in any organization. Predictably, some individual or group will support its own particular interests over competing views. This actuality exists in both government and market organizations. In government the effects of politics are continuously felt at all levels because of the large-scale and multi-purpose nature of projects and because government decision making involves the use of the citizens' common pool of money.

To illustrate on a small scale situations faced by government, compare the decision-making process a family may face when planning an evening out. As most families can attest, this decision is not always easy—even when Dad is footing the bill! Imagine the increased conflict that would arise if every member of the family were to contribute to the tab. Perhaps the choices would then be dine out, go to a hockey game, have a shopping spree at the mall, lend the money to Mom, or put it into the family bank account. This scenario characterizes government investment decisions—individuals and groups with different values and views spending a common *pool* of money. As with the family decision, the parties involved often have squabbles, form alliances, and manoeuvre politically.

The guideline for public decision making is to produce a net gain in economic or social welfare. However, it is impossible to please everyone all the time. Therefore benefit-cost analysis measures the variables in order to determine whether a particular policy or program has resulted in a net benefit.

As mentioned previously, government projects tend to be large in scale. Therefore, the time required to plan, design, fund, and construct such projects is usually several years. However, the political process tends to produce government leaders who support short-term decision making (because many government terms of office, either elected or appointed, are relatively short). Therein lies another difference between government and market decision making—short-term decision making, long-term projects.

Because government decision makers are in the public eye more than those in the private sector, governmental decisions are generally more affected by politics. Thus, the decisions that public officials make may not always be the best from an *overall* perspective. If a particular situation exposes a public official to ridicule, he may choose an expedient action to eliminate negative exposure (whereas a more careful analysis might have been better). Or, such a decision maker may placate a small, but vocal, political group over the interest of the majority of citizens by committing funds to a project favoured by the small group (at the expense of other better projects). Or, a public decision maker may avoid controversy by declining to make a decision on an important, but politically charged, issue (whereas it would be in the overall interest of the citizens if action were taken). Indeed, the role of politics in government decision making is more complex and far ranging than in the private sector.

EXAMPLE 16-7

Consider again Example 16-4, where we evaluated power plants designs. Remember that government projects are often opposed and supported by different groups in the populace. For that reason, decision makers become very aware of potential political aspects when they are considering such projects. For the electric power plant decision, several political considerations may affect any evaluation of funding this project.

- The mayor has been a strong advocate of workers' rights and has received abundant campaign support from organized labour (which is especially important in an industrial city). By championing this project, the mayor should be seen as pro-labour, thereby benefiting his bid for re-election, even if the project is not funded.
- The regulated electric utilities in the province are strongly against this project, claiming that it would compete directly with them and take away some of their biggest customers. The providers have a strong lobby and contacts in the provincial cabinet. The mayor of a neighbouring city has already protested that this project is the first step toward 'rampant socialism'.
- Business leaders in the municipalities where the two facilities would be constructed are in favour of the project because it would create more jobs and increase the tax base. These leaders promote the project as a win-win opportunity for government and industry, where the city can benefit by reducing costs, and the electric utilities can improve their service by focusing more effectively on residential customers and their needs.
- The Chamber of Commerce is promoting this project, proclaiming that it is an excellent example of 'initiating proactive and creative solutions to the problems that this city faces.'
- Federal and provincial regulatory agencies are watching this project closely with respect to environmental legislation. Speculation is that the plans are to use a high-sulphur grade of

local coal exclusively. Thus 'stack scrubbers' would be required, or the high-sulphur coal would have to be mixed with imported lower-sulphur coal to bring the overall air emissions in line with federal standards. The mayor is using this opportunity to make the point that 'the people of this city don't need regulators to tell us if we can use our own coal!'

- The coal operators and mining unions are very much in favour of this project. They see the increased demands for coal and the mayor's pro-labour advocacy as very positive. They plan to lobby strongly in favour of the project.
- Land preservation and environmental groups are strongly opposing the proposed project. They have studied the potential negative impacts of this project on the land and on water and air quality, as well as on the ecosystem and wildlife, in the areas where the two facilities would be constructed. Environmentalists have started a public awareness campaign urging the premier to act as the 'chief steward' of the natural beauty and resources of the province.

Will the project be funded? We can only guess. Clearly, however, we can see the competing influences that can be, and often are, part of decision making in the public sector.

SUMMARY

Economic analysis and decision making in government is notably different from these processes in the private sector because the basic objectives of the public and private sectors are fundamentally different. Government investments in projects seek to maximize benefits to the *greatest number of citizens,* while minimizing the *disbenefits to citizens* and *costs to the government.* Private firms, on the other hand, are focused primarily on maximizing shareholder wealth.

Several factors, not affecting private firms, enter into the decision-making process in government. The source of capital for public projects is limited primarily to taxes and bonds. Governments issuing bonds for project construction are subject to legislative restrictions on debt that do not apply to private firms. Also, raising tax and bond monies involves sometimes long and politically charged processes not found in the private sector. In addition, government projects tend to be larger than those of competitive firms and to affect many more people and groups in the population. All these factors slow down the process and make investment decision analysis more difficult for government decision makers than for those in the private sector. Another difference between the public and private sectors lies in how the interest rate (MARR) is set for economic studies. In the private sector, considerations for setting the rate include the cost of capital and opportunity costs. In government, establishing the interest rate is complicated by uncertainty in specifying the cost of capital and the issue of assigning opportunity costs to taxpayers or to the government.

The benefit-cost ratio is widely used to evaluate and justify government-funded projects. This measure of merit is the ratio of the equivalent worth of benefits to the equivalent worth of costs. This ratio can be calculated by PW, AW, or FW methods. A B/C ratio *greater than 1.0* indicates that a project should be invested in if funding sources are available. For considering *mutually exclusive alternatives,* an incremental method should be used to evaluate the merits of additional cost. This method results in the recommendation of the project with the highest investment cost that can be incrementally justified. Two versions of the B/C ratio, the *conventional* and *modified* B/C ratios, produce identical recommendations when single projects or sets of competing alternatives are being considered. The difference between the two ratios is in the way that annual operating and maintenance costs are handled—as an added cost in the denominator in the former, or as a subtracted benefit in the numerator in the latter.

PROBLEMS

16-1 Economic analysis and decision making in the public sector is often called 'a multi-actor or multi-stakeholder decision problem'. Explain what this phrase means.

16-2 Compare the general underlying objective of public decision making and private decision making.

16-3 This chapter describes a general recommendation regarding the scope of viewpoint that is appropriate in public decision making. What does it suggest? What example is given to highlight the dilemma of viewpoint in public decision making?

16-4 In government projects, what is meant by the phrase 'most of the benefits are local'. What conflict does this create for the federal government in the funding of projects from public monies?

16-5 Discuss the alternative concepts that can be employed when setting the discounting rate for economic analysis in the public sector. What is the final recommendation of this chapter for setting this rate?

16-6 What is the essential difference between the *conventional* and *modified* versions of the benefit-cost ratio? Is it possible for these two measures to provide conflicting recommendations regarding invest/do-not-invest decisions?

16-7 List the potential costs, benefits, and disbenefits that should be considered when one is evaluating a nuclear power plant construction.

16-8 Describe how a decision maker can use each of the following to skew the results of a B/C ratio analysis in favor of his or her own position on funding projects:
(*a*) Conventional versus modified ratios.
(*b*) Interest rates.
(*c*) Project duration.
(*d*) Benefits, costs, and disbenefits.

16-9 Think about a major government construction project under way in your province, city, or region. Are the de-

cision makers who originally analyzed and initiated the project currently in office? How can politicians use 'political posturing' with respect to government projects?

16-10 Consider the following investment opportunity:

Initial cost	$100,000
Additional cost at end of Year 1	150,000
Benefit at end of Year 1	0
Annual benefit per year at end of Years 2–10	20,000

With interest at 7%, what is the benefit-cost ratio for this project?   (*Answer:* 0.51)

16-11 A government agency has estimated that a flood control project has costs and benefits that are parabolic, according to the equation

$$(\text{Present worth of benefits})^2 - 22(\text{Present worth of cost}) + 44 = 0$$

where both benefits and costs are stated in millions of dollars. What is the present worth of cost for the optimal size project?

16-12 The Highridge region needs an additional supply of water from Steep Creek. The engineer has selected two plans for comparison:

Gravity plan: Divert water at a point 10 kilometres up Steep Creek and carry it through a pipeline by gravity to the district.

Pumping plan: Divert water at a point near the district and pump it through 2 kilometres of pipeline to the district. The pumping plant can be built in two stages, with half-capacity installed initially and the other half 10 years later.

Use a 40-year analysis period and 8% interest. Salvage values can be ignored. During the first 10 years, the average use of water will be less than during the remaining 30 years. Use the conventional benefit-cost ratio method to select the more economical plan.

	Gravity	Pumping
Initial investment	$2,800,000	$1,400,000
Additional investment in tenth year	0	200,000
Operation, maintenance, replacements, per year	10,000	25,000
Average power cost per year (first 10 years)	0	50,000
Average power cost per year (next 30 years)	0	100,000

(*Answer:* Pumping plan)

16-13 Calculate the conventional and modified benefit-cost ratio for the investment represented by the following data.

Required first costs	$1,200,000
Annual benefits to users	$500,000
Annual disbenefits to users	$25,000
Annual cost to government	$125,000
Project life	35 years
Interest rate	10%

16-14 For the data given in Problem 16-13, for handling benefits and costs, demonstrate that the calculated B/C ratio is the same using the each of the following methods: present worth, annual worth, and future worth.

16-15 Big City Carl, a local politician, is advancing a project for the construction of a new dock and pier system on the river to attract new commerce to the city. A committee appointed by the mayor (an opponent of Carl's) has developed the following estimates for the effects of the project.

Cost to wreck and remove current facilities	$ 750,000
Material, labour, and overhead for new construction	2,750,000
Annual operating and maintenance expenses	185,000
Annual benefits from new commerce	550,000
Annual disbenefits to sportsmen in area	35,000
Project life	20 years
Interest rate	8%

(*a*) Using the *conventional* B/C ratio, determine whether the project should be funded.

(*b*) After studying the numbers given by the committee, Big City Carl argued that the project life

should be *at least* 25 years and more likely closer to 30 years. How did he arrive at this estimate, and why is he making this statement?

16-16 Two different routes, which entail driving across a mountainous section, are being considered for a highway construction project. The first route (the **high road**) will require the building of several bridges, and it navigates around the highest mountain points, thus requiring more roadway. The second alternative (the **low road**) will require the construction of several tunnels, but takes a more direct approach through the mountainous area. Projected travel volume for this new section of road is 2500 cars per day. Given the following data, use the *modified* B/C ratio to determine which alternative should be recommended. Assume that project life is 45 years and $i = 6\%$.

	The High Road	The Low Road
Average construction cost per kilometre	$200,000	$450,000
Number of kilometres required	35	10
Annual benefit per car per kilometre	$0.015	$0.045
Annual O&M costs per kilometre	$2000	$10,000

16-17 The provincial government proposes to construct a multi-purpose water project to provide water for irrigation and municipal use. In addition, flood control and recreation benefits will be realized. The estimated benefits of the project computed for 10-year periods for the next 50 years are given in Table P16-17. The annual benefits may be assumed to be one-tenth of the decade benefits. The operation and maintenance cost of the project is estimated to be $15,000 per year. Assume a 50-year analysis period with no net project salvage value.

(*a*) If an interest rate of 5% is used, and a benefit-cost ratio of unity, what capital expenditure can be justified to build the water project now?

(*b*) If the interest rate is changed to 8%, how does this change the justified capital expenditure?

16-18 The city engineer has prepared two plans for the construction and maintenance of roads in the city park. Both plans are designed to meet the anticipated road and road maintenance requirements for the next 40 years. The minimum attractive rate of return used by the city is 7%.

TABLE P16-17 Data

	Decades				
Purpose	First	Second	Third	Fourth	Fifth
Municipal	$ 40,000	$ 50,000	$ 60,000	$ 70,000	$110,000
Irrigation	350,000	370,000	370,000	360,000	350,000
Flood control	150,000	150,000	150,000	150,000	150,000
Recreation	60,000	70,000	80,000	80,000	90,000
Totals	$600,000	$640,000	$660,000	$660,000	$700,000

Plan *A* is a three-stage development program: $300,000 is to be spent immediately, followed by $250,000 at the end of 15 years and $300,000 at the end of 30 years. Maintenance will be $75,000 per year for the first 15 years, $125,000 per year for the next 15 years, and $250,000 per year for the final 10 years.

Plan *B* is a two-stage program: $450,000 is required immediately (including money for special equipment), followed by $50,000 at the end of 15 years. Maintenance will be $100,000 per year for the first 15 years and $125,000 for each of the subsequent years. At the end of 40 years, it is believed that the equipment can be sold for $150,000.

(*a*) Use a conventional benefit-cost ratio analysis to determine which plan should be chosen.

(*b*) If you favoured Plan *B*, what value of MARR would you use in the computations? Explain.

16-19 The province is considering eliminating a railway level crossing by building an overpass. The new structure, together with the land needed, would cost $1.8 million. The analysis period is assumed to be 30 years based on the projection that either the railway or the highway above it will be relocated by then. Salvage value of the bridge (actually, the net value of the land on either side of the railway tracks) 30 years hence is estimated to be $100,000. A 6% interest rate is to be used.

At present, about 1000 vehicles per day are delayed by trains at the level crossing. Trucks represent 40%, and 60% are other vehicles. Time for truck drivers is valued at $18 per hour and for other drivers at $5 per hour. Average time saving per vehicle will be 2 minutes if the overpass is built. No time saving occurs for the railway.

The installation will save the railway an annual expense of $48,000 now spent for crossing guards. During the preceding 10-year period, the railway

has paid out $600,000 in settling lawsuits and accident cases related to the level crossing. The proposed project will entirely eliminate both these expenses. The province estimates that the new overpass will save it about $6000 per year in expenses due directly to the accidents. The overpass, if built, will belong to the province.

Should the overpass be built? If the overpass is built, how much should the railway be asked to contribute as its share of the $1,800,000 construction cost?

16-20 An existing two-lane highway between two cities, 10 miles apart, is to be converted to a four-lane divided highway. The average daily traffic (ADT) on the new highway is forecast to average 20,000 vehicles over the next 20 years. Trucks represent 5% of the total traffic. Annual maintenance on the existing highway is $1500 per lane-mile. The existing accident rate is 4.58 per million vehicle miles (MVM). Three alternative plans of improvement are now under consideration.

Plan A: Improve along the existing development by adding two lanes adjacent to the existing lanes at a cost of $450,000 per mile. It is estimated that this plan will reduce car travel time by 2 minutes a mile and truck travel time by 1 minute. The Plan A estimated accident rate is 2.50 per MVM. Annual maintenance is estimated to be $1250 per lanemile.

Plan B: Improve along the existing alignment with grade improvements at a cost of $650,000 per mile. Plan B would add two additional lanes, and it is estimated that this plan would reduce car and truck travel time by 3 minutes each. The accident rate on this improved road is estimated to be 2.40 per MVM. Annual maintenance is estimated to be $1000 per lanemile.

Plan C: Construct a new freeway on new alignment at a cost of $800,000 per mile. It is estimated that this plan would reduce car travel time by 5 minutes and truck travel time by 4 minutes. Plan C is 0.3 mile longer than A or B. The estimated accident rate for C is 2.30 per MVM. Annual maintenance is estimated to be $1000 per lane-mile. Plan C includes abandonment of the existing highway with no salvage value.

Incremental operating cost	
Cars	6¢ per mile
Trucks	18¢ per mile
Time saving	
Cars	3¢ per minute
Trucks	15¢ per minute
Average accident cost	$1200

If a 5% interest rate is used, which of the three proposed plans should be adopted? (*Answer:* Plan C)

16-21 The provincial highway department is preparing an economic analysis to see whether reconstruction of the pavement on a mountain road is justified. The number of vehicles travelling on the road increases each year; hence the benefits to the motoring public of the pavement reconstruction also increase. Based on a traffic count, the benefits are projected as follows:

Year	End-of-Year Benefit
2001	$10,000
2002	12,000
2003	14,000
2004	16,000
2005	18,000
2006	20,000
	and so on, increasing $2000 per year

The reconstructed pavement will cost $275,000 when it is installed and will have a 15-year useful life. The construction period is short, hence a beginning-of-year reconstruction will result in the end-of-year benefits listed in the table. Assume a 6% interest rate. The reconstruction, if done at all, must be done not later than 2006. Should it be done, and if so, in what year?

16-22 A section of a highway needs repair at a cost of $150,000, but the volume of traffic is so low that few motorists would benefit from the work. However, traffic is expected to increase. The repair work will produce benefits for 10 years after it is completed. The highway department is examining five mutually exclusive alternatives.

Should the road be repaired and, if so, when? Use a 15% MARR.

TABLE P16-22 Data

Year	Do Not Repair	Repair Now	Repair 2 Years Hence	Repair 4 Years Hence	Repair 5 Years Hence
0	0	−$150,000			
1	0	5,000			
2	0	10,000	−$150,000		
3	0	20,000	20,000		
4	0	30,000	30,000	−$150,000	
5	0	40,000	40,000	40,000	−$150,000
6	0	50,000	50,000	50,000	50,000
7	0	50,000	50,000	50,000	50,000
8	0	50,000	50,000	50,000	50,000
9	0	50,000	50,000	50,000	50,000
10	0	50,000	50,000	50,000	50,000
11	0	0	50,000	50,000	50,000
12	0	0	50,000	50,000	50,000
13	0	0	0	50,000	50,000
14	0	0	0	50,000	50,000
15	0	0	0	0	50,000

TABLE P16-24 Data

	Irish Fishery Design Alternatives			
	A	B	C	D
First cost	$9,500,000	$12,500,000	$14,000,000	$15,750,000
Annual benefits	2,200,000	1,500,000	1,000,000	2,500,000
Annual O&M costs	550,000	175,000	325,000	145,000
Annual disbenefits	350,000	150,000	75,000	700,000
Salvage value	1,000,000	6,000,000	3,500,000	7,500,000

16-23 A 50-metre tunnel must be constructed as part of a new sewer system for a city. Two alternatives are being considered. One is to build a full-capacity tunnel now for $500,000. The other alternative is to build a half-capacity tunnel now for $300,000 and then to build a second parallel half-capacity tunnel 20 years hence for $400,000. The cost to repair the tunnel lining every 10 years is estimated to be $20,000 for the full-capacity tunnel and $16,000 for each half-capacity tunnel.

Determine whether the full-capacity tunnel or the half-capacity tunnel should be constructed now. Solve the problem by the conventional benefit-cost ratio analysis, using a 5% interest rate and a 50-year analysis period. There will be no tunnel lining repair at the end of the 50 years.

16-24 The Fishery and Wildlife Agency of Ireland is considering four mutually exclusive design alternatives (Table P16-24) for a major salmon hatchery. This agency of the Irish government uses the following B/C ratio for decision making:

$$\text{B/C ratio} = \frac{\text{EW(Net benefits)}}{\text{EW(Capital recovery cost)} + \text{EW(O\&M cost)}}$$

Using an interest rate of 8% and a project life of 30 years, recommend which of the designs is best.

16-25 Six mutually exclusive investments have been identified for evaluation by means of the benefit-cost ratio method. Assume a MARR of 10%, an equal project life of 25 years for all alternatives, and the data in Table P16-25.

(a) Use annual worth and the B/C ratio to identify the better alternative.

(b) If this were a set of *independent* alternatives, how would you conduct a comparison?

TABLE P16-25 Data

	Mutually Exclusive Alternatives					
	1	2	3	4	5	6
Annualized net costs to sponsor	15.5	13.7	16.8	10.2	17.0	23.3
Annualized net benefits to users	20.0	16.0	15.0	13.7	22.0	25.0

16-26 Mr. D. O'Gratias, a top manager in his company, has been asked to consider the following mutually exclusive investment alternatives.

	A	B	C
Initial investment	$9500	$18,500	$22,000
Annual savings	3200	5,000	9,800
Annual costs	1000	2,750	6,400
Salvage value	6000	4,200	14,000
Project life, in years	15	15	15
MARR	12%	12%	12%

Answer the following questions.

(a) Use the *conventional* B/C ratio to evaluate the alternatives and make a recommendation.

(b) Use the *modified* B/C ratio to evaluate the alternatives and make a recommendation.

(c) Use a present worth analysis to evaluate the alternatives and make a recommendation.

(d) Use an internal rate of return analysis to evaluate the alternatives and make a recommendation.

(e) Use the simple payback period to evaluate the alternatives and make a recommendation.

CHAPTER 17

Rationing Capital among Competing Projects

The 2005 Federal Budget

Each year, the federal government releases a budget, which sets out where and how it plans to collect and invest the taxpayers' money. Budgets contain forecasts of the economy and estimates of how much the government will collect in taxes and duties. They also contain forecasts of spending needs in various areas. Some highlights of the 2005 proposed budget were the following:

- A projection of 2.9% and 3.1% growth of the Canadian economy in 2005 and 2006.
- Savings of nearly $9 billion through greater government efficiency.
- $5 billion in federal gas tax revenues for communities across Canada over the next five years for investments in sustainable infrastructure.
- $12.8 billion over the next five years for the armed forces.
- Elimination of the corporate surtax and lowering of the general corporate income tax rate from 21% to 19%.
- $700 million in 2005 and 2006 for a national child care program, with a total of $5 billion over the next five years.
- $170 million over five years to improve drug safety.
- $1 billion to combat climate change.

But since the ruling Liberal party did not have a majority in the House of Commons, to obtain enough votes to pass the budget, Prime Minister Martin made a deal with the New Democratic Party, who agreed to support the budget if Martin made certain changes, including a $4.6 billion increase in social program spending over two years. Martin said the deal—which enabled his government to survive—would be paid for with projected budget surpluses of $9 billion.

The deal between the Liberals and the NDP included:

- $1.6 billion for affordable housing.
- A $1.5-billion increase in transfers to provinces for tuition reduction and better job training.
- $900 million for the environment, with one more cent per litre of the federal gas tax going to public transit.
- $500 million for foreign aid to meet Canada's promised level of 0.7% of GDP.
- $100 million for a pension protection fund for workers.

Tax cuts for small and medium-sized businesses contained in the original budget would remain but cuts for large corporations would be deferred. Martin later announced that the business tax cuts would be withdrawn from the budget and would be introduced later as separate legislation.

The budget subsequently passed.

After Completing This Chapter...

The student should be able to:

- Identify capital expenditure project proposals of several types, including mutually exclusive alternatives and single project proposals.
- Identify and reject unattractive alternatives, and select the best alternative from each project proposal.
- Use the rate of return and present worth methods to ration capital among projects.
- Rank project proposals by MARR.

QUESTIONS TO CONSIDER

1. The government is proposing to share the gasoline tax (paid by road users) with the municipalities so that they can improve their roads and build new ones. What are the advantages and disadvantages of tying a revenue source and a service together like this?

2. In the budget, the government is seeking to combat climate change and encourage the protection of the natural environment. Specifically, it contains the following measures:

   - $1 billion over five years to encourage cost-effective initiatives that will reduce greenhouse gas emissions.
   - $225 million over five years to quadruple the number of homes retrofitted under the EnerGuide for Houses Retrofit Incentive program.
   - $300 million over five years in production incentives for wind power and other renewable sources of energy.
   - Almost $900 million over five years to protect our natural environment, including the Great Lakes, Canada's oceans, and the national parks.

   Is this the right way to deal with the 'external consequence' issue discussed in Chapter 16?

3. The government deferred the corporate tax cut to 2010. It is possible that by that time a different group of people will be in the government. If you are doing a problem that stretches past 2010, what tax rate should you use? What should you do if a risk analysis (see Chapter 10) shows that the tax rate is an important and possibly deciding variable?

We have until now dealt with situations where, at some interest rate, we choose each project's best mutually exclusive alternative. Thus, we were assuming that there is an ample amount of money to make all desired capital investments in these projects. But the concept of **scarcity of resources** is fundamental to a free market economy (not to mention government decision making). It is through this mechanism that more economically attractive activities are encouraged at the expense of less desirable activities. For industrial firms, there are often more ways of spending money than there is money available. The result is that we must select, from the available alternatives, the more attractive projects and reject—or, at least, delay—the less attractive projects.

The problem of rationing capital among competing projects is one part of our two-part problem called **capital budgeting**. In planning its capital expenditures, an industrial firm is faced with two questions: *Where will money for capital expenditures come from?* and *How shall we allocate the available money among the various competing projects?* In Chapter 15 we discussed the sources of money for capital expenditures as one part of choosing a suitable interest rate for economic analysis calculations. Thus, the first problem has been treated.

Throughout this book, we have examined for any given project two or more feasible alternatives, attempting to identify the most attractive one. For simplicity's sake, we have looked at these projects in an isolated setting, as if a firm had just one project it was considering. In the business world, this is rarely the case. A firm will find that there are many projects that are economically attractive. This situation raises two problems not previously considered:

1. How do you rank projects to show their order of economic attractiveness?
2. What do you do if there is not enough money to pay the costs of all economically attractive projects?

In this chapter we will look at the typical situation faced by a firm: multiple attractive projects, with an inadequate money supply to fund them all. To do this, we review our concepts of capital expenditure situations and available alternatives. Then we summarize the various techniques for determining whether a project is economically attractive. First we screen all alternatives to find those that merit further consideration; then we select the best alternative, assuming there is no shortage of money. The next step will be the addition of a budget constraint.

If we find that there is not enough money to fund the best alternative from each project, we will have to do what we can with the limited amount of money available. It will become important that we have a technique for accurately ranking the various competing projects in order of economic attractiveness. All this is designed to answer the question, *How shall we allocate available money among the various competing projects?*

CAPITAL EXPENDITURE PROJECT PROPOSALS

At the beginning of the book, we described decision making as the process of selecting the best alternative to achieve the desired objective in a given situation or problem. By carefully defining our objective and the model, we reduce a situation to one of selecting the best from the feasible alternatives. In this chapter we call the engineering decision-making process for a given situation or problem a **project proposal.** Associated with various project proposals are their particular available alternatives. For a firm with many project proposals, the following situation may result:

Capital Expenditure Proposals

Project	Alternatives
1. Acquire additional manufacturing facility	A Lease an existing building B Construct a new building C Contract for the manufacturing to be done overseas
2. Replace old grinding machine	A Purchase semiautomatic machine B Purchase automatic machine
3. Produce parts for the assembly line	A Make the parts in the plant B Buy the parts from a subcontractor

Our task is to apply economic analysis techniques to this more complex problem.

Mutually Exclusive Alternatives and Single Project Proposals

Until now we have dealt with mutually exclusive alternatives; that is, by choosing one alternative, we reject the other alternatives being considered. Even in the simplest problems encountered, the question was one of selection *between* alternatives. Should, for example, Machine A or Machine B be purchased to perform the necessary task? Clearly, the purchase of one of the machines meant that the other one would not be purchased. Since either machine would perform the task, the selection of one precludes the possibility of selecting the other one as well.

Even in the case of multiple alternatives, we have been considering mutually exclusive alternatives. A typical example was: What size of pipeline should be installed to supply water to a remote construction site? Only one alternative is to be selected. This is different from the situation for single project proposals, where only one course of action is outlined. Consider Example 17-1.

EXAMPLE 17-1

The general manager of a manufacturing plant has received the following project proposals from the various operating departments:

1. The foundry wishes to purchase a new ladle to speed up the casting operation.
2. The machine shop has asked for some new inspection equipment.
3. The painting department reports that improvements must be made to the spray booth to conform to new air pollution standards.
4. The office manager wants to buy a larger, more modern safe.

For each project there is a single course of action proposed. Note that the single project proposals are also independent, for there is no interrelationship or interdependence among them. The general manager can decide to allocate money for none, some, or all of the various project proposals.

SOLUTION

Each of the four project proposals has a single course of action. The general manager could, for example, buy the office manager a new safe *and* buy the inspection equipment for the machine shop. But he could also decide to *not* buy the office manager a safe *or* the equipment for the machine shop. There is, then, an alternative to buying the safe for the office manager: not to buy him the safe—to do nothing. Similarly, he could decide to do nothing about the request for the machine shop inspection equipment. Naturally, there are do-nothing alternatives for each of the four single project proposals:

1A. Purchase the foundry a new ladle.
1B. Do nothing. (Do not purchase a new ladle.)
2A. Obtain the inspection equipment for the machine shop.
2B. Do nothing. (Do not obtain the inspection equipment.)
3A. Make improvements to the spray booth in the painting department.
3B. Do nothing. (Make no improvements.)
4A. Buy a new safe for the office manager.
4B. Do nothing. (Let him use the old safe!)

One can adopt Alt. 1A (buy the ladle) or 1B (do not buy the ladle), but not both. We find that what we considered to be a single course of action is really a pair of mutually exclusive alternatives. Even Alt. 3 is in this category. The originally stated single proposal was:

> The painting department reports that improvements must be made to the spray booth to conform to new air pollution standards.

Since the painting department *must* make the improvements, is there actually another alternative? Although at first glance we might not think so, it may be possible to change the paint, or the spray equipment, and thereby solve the air pollution problem without any improvements to the spray booth. In this situation, there does not seem to be a practical do-nothing alternative, for failure to comply with the air pollution standards might result in large fines or even shutting down the plant. But if there is not a practical do-nothing alternative, there might be a number of do-something-else alternatives.

We conclude that all project proposals may be considered to have mutually exclusive alternatives.

Identifying and Rejecting Unattractive Alternatives

It is clear that no matter what the circumstances may be, we want to eliminate from further consideration any alternative that fails to meet the minimum level of economic attractiveness, provided one of the other alternatives does meet the criterion. Table 17-1 summarizes five techniques that may be used.

At first glance it appears that many calculations are required, but this appearance is misleading. *Any* of the five techniques listed in Table 17-1 may be used to determine whether to reject an alternative. Each will produce the same decision regarding *Reject/Don't reject*.

TABLE 17-1 Criteria for Rejecting Unattractive Alternatives

For Each Alternative Compute	Reject Alternative When	Do Not Reject Alternative When
Rate of return, i	$i <$ MARR	$i \geq$ MARR
Present worth, PW	PW of benefits $<$ PW of costs	PW of benefits $\geq$ PW of costs
Annual cost, EUAC	EUAC $>$ EUAB	EUAC $\leq$ EUAB
Annual benefit, EUAB		
Benefit-cost ratio, B/C	B/C $<$ 1	B/C $\geq$ 1
Net present worth, NPW	NPW $<$ 0	NPW $\geq$ 0

Selecting the Best Alternative from Each Project Proposal

The task of selecting the best alternative from among two or more mutually exclusive alternatives has been a primary subject of this book. Since a project proposal is the same form of problem, we may use any of the several methods discussed in Chapters 5 to 9. The criteria are summarized in Table 17-2.

TABLE 17-2 Criteria for Choosing the Best Alternative from among Mutually Exclusive Alternatives

Analysis Method	Situations		
	Fixed Input (The cost of each alternative is the same.)	**Fixed Output** (The benefits from each alternative are the same.)	**Neither Input Nor Output Fixed** (Neither the costs nor the benefits for the alternatives given are the same.)
Present worth	Maximize present worth of benefits	Minimize present worth of cost	Maximize net present worth
Annual cash flow	Maximize equivalent uniform annual benefits	Minimize equivalent uniform annual cost	Maximize EAUB − EAUC
Benefit-cost ratio	Maximize benefit-cost ratio	Maximize benefit-cost ratio	Incremental benefit-cost ratio analysis is required
Rate of return	Incremental rate of return analysis is required		

RATIONING CAPITAL BY RATE OF RETURN

One way of looking at the capital rationing problem is through the use of rate of return. The technique for selecting from among independent projects is illustrated by Example 17-2. These projects are independent, so their only interdependence is the budget constraint on the total accepted set of projects.

EXAMPLE 17-2

Nine independent projects are being considered. Figure 17-1 may be prepared from the following data.

Project	Cost (thousands)	Uniform Annual Benefit (thousands)	Useful Life (years)	Salvage Value (thousands)	Computed Rate of Return
1	$100	$23.85	10	$ 0	20%
2	200	39.85	10	0	15
3	50	34.72	2	0	25
4	100	20.00	6	100	20
5	100	20.00	10	100	20
6	100	18.00	10	100	18
7	300	94.64	4	0	10
8	300	47.40	10	100	12
9	50	7.00	10	50	14

If a capital budget of $650,000 is available, which projects should be selected?

SOLUTION

Looking at the nine projects, we see that some are expected to produce a larger rate of return than others. It is natural that if we are to select from among them, we will pick those with a higher rate of return. When the projects are arrayed by rate of return, as in Figure 17-1, the correct choice (of Projects 3, 1, 4, 5, 6, and 2) is readily apparent.

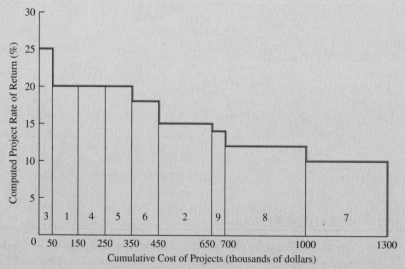

FIGURE 17-1 Cumulative cost of projects versus rate of return.

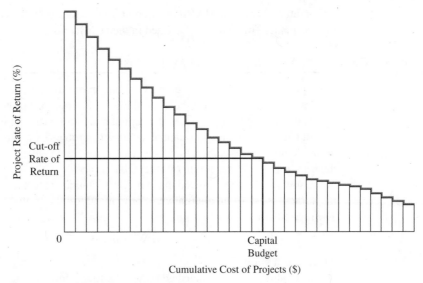

FIGURE 17-2 Location of the cut-off rate of return.

In Example 17-2, the rate of return was computed for each project and then the projects were arranged in order of decreasing rate of return. For a fixed amount of money in the capital budget, the projects are selected by going down the list until the money is exhausted. Thus when we use this procedure, we cut off approving projects at the point where the money runs out. This point is called the **cut-off rate of return.** Figure 17-2 illustrates the general situation.

For any set of ranked projects and any capital budget, the rate of return at which the budget is exhausted is the cut-off rate of return. In Figure 17-2 the cost of each individual project is small in comparison to the capital budget. The cumulative cost curve is a relatively smooth curve producing a specific cut-off rate of return. Looking back at Figure 17-1, we see the curve is actually a step function. For Example 17-2, the cut-off rate of return is between 14 and 15% for a capital budget of $650,000.

Significance of the Cut-off Rate of Return

Cut-off rate of return is determined by comparison of an established capital budget and the available projects. One must examine all the projects and all the money for some period of time (like one year) to compute the cut-off rate of return. It is a computation relating known projects to a known money supply. For this period of time, the cut-off rate of return is the opportunity cost (rate of return on the opportunity or project forgone) and also the minimum attractive rate of return. In other words, the minimum attractive rate of return to get a project accomplished *is* the cut-off rate of return.

MARR = Cut-off rate of return = Opportunity cost

We generally use the minimum attractive rate of return to decide whether to approve an individual project even though we do not know exactly what other projects will be proposed during the year. In this situation, we cannot know whether the MARR is equal to the cut-off rate of return. When the MARR is different from the cut-off rate of return, incorrect decisions may be made. This will be illustrated in the next section.

RATIONING CAPITAL BY PRESENT WORTH METHODS

Throughout this book we have chosen from among project alternatives to maximize net present worth. If we can do the same thing for a group of projects, and we do not exceed the available money supply, then the capital budgeting problem is solved.

But more frequently in capital budgeting problem we will be unable to accept all desirable projects. We, therefore, have a task not previously encountered. We must choose the best from the larger group of acceptable projects.

Lorie and Savage[1] showed that a proper technique is to use a multiplier, p, to decrease the attractiveness of an alternative in proportion to its use of the scarce supply of money. The revised criterion is

$$NPW - p(PW \text{ of cost}) \tag{17-1}$$

where p is a multiplier.

If a value of p were selected (say, 0.1), then some alternatives with a positive NPW will have a negative [NPW $- p$(PW of cost)]. This new criterion will reduce the number of favourable alternatives and thereby reduce the combined cost of the projects meeting this more severe criterion. By trial and error, the multiplier p is adjusted until the total cost of the projects meeting the [NPW $- p$(PW of cost)] criterion equals the available money supply—the capital budget.

EXAMPLE 17-3

Use the present worth method to determine which of the nine independent projects of Example 17-2 should be included in a capital budget of $650,000. The minimum attractive rate of return has been set at 8%.

Project	Cost (thousands)	Uniform Annual Benefit (thousands)	Useful Life (years)	Salvage Value (thousands)	Computed NPW (thousands)
1	$100	$23.85	10	$ 0	$60.04
2	200	39.85	10	0	67.40
3	50	34.72	2	0	11.91
4	100	20.00	6	100	55.48
5	100	20.00	10	100	80.52
6	100	18.00	10	100	67.10
7	300	94.64	4	0	13.46
8	300	47.40	10	100	64.38
9	50	7.00	10	50	20.13

---

[1] Lorie, J. and L. Savage, 'Three Problems in Rationing Capital', *Journal of Business,* October 1955, pp. 229–239.

SOLUTION

In locating a value of p in [NPW − p(PW of cost)] by trial and error, we use the following table, where all amounts are in thousands of dollars.

Project	Cost	Computed NPW	Trial $p = 0.20$ [NPW − p(PW of cost)]	Cost	Trial $p = 0.25$ [NPW − p(PW of cost)]	Cost
1	$ 100	$60.04	$40.04	$ 100	$35.04	$100
2	200	67.40	27.40	200	17.40	200
3	50	11.91	1.91	50	−0.59	
4	100	55.48	35.48	100	30.48	100
5	100	80.52	60.52	100	55.52	100
6	100	67.10	47.10	100	42.10	100
7	300	13.46	−46.54		−61.54	
8	300	64.38	4.38	300	−10.62	
9	50	20.13	10.13	50	7.63	50
	$1300			$1000		$650

For a value of p equal to 0.25, the best selection order is computed to be Projects 1, 2, 4, 5, 6, and 9. (*Note:* the smallest value for p is 0.239.)

Alternative Formation of Example 17-3

The preceding answer does not agree with the solution obtained in Example 17-2. The difficulty is that the interest rate used in the present worth calculations is not equal to the computed cut-off rate of return. In Example 17-2 the cut-off rate of return was between 14% and 15%, say 14.5%. We will recompute the present worth solution using MARR = 14.5%, again with amounts in thousands of dollars.

Project	Cost	Computed NPW at 14.5%	Cost of Projects with Positive NPW
1	$100	$22.01	$100
2	200	3.87	200
3	50	6.81	50
4	100	21.10	100
5	100	28.14	100
6	100	17.91	100
7	300	−27.05	
8	300	−31.69	
9	50	−1.28	
			$650

SOLUTION

At a MARR of 14.5% the best set of projects is the same as computed in Example 17-2, namely, Projects 1, 2, 3, 4, 5, and 6, and their cost equals the capital budget. It can be seen that only projects with a rate of return greater than MARR can have a positive NPW at this interest rate. With MARR equal to the cut-off rate of return, we *must* obtain the same solution by either the rate of return or present worth method.

Example 17-4 outlines the present worth method for the more elaborate case of independent projects with mutually exclusive alternatives.

EXAMPLE 17-4

A company is preparing its capital budget for next year according to the steps in the flow chart shown in Figure 17-3. The amount has been set at $250,000 by the Board of Directors. The MARR of 8% is believed to be close to the cut-off rate of return. The following project proposals are being considered.

Project Proposals	Cost (thousands)	Uniform Annual Benefit (thousands)	Salvage Value (thousands)	Useful Life (years)	Computed NPW (thousands)
Proposal 1					
Alt. A	$100	$23.85	$ 0	10	$60.04
Alt. B	150	32.20	0	10	66.06
Alt. C	200	39.85	0	10	67.40
Alt. D	0	0			0
Proposal 2					
Alt. A	50	14.92	0	5	9.57
Alt. B	0	0			0
Proposal 3					
Alt. A	100	18.69	25	10	36.99
Alt. B	150	19.42	125	10	38.21
Alt. C	0	0			0

Which project alternatives should be selected, based on present worth methods?

SOLUTION

The following tabulation shows that to maximize NPW, we would choose Alternatives 1C, 2A, and 3B. The total cost of these three projects is $400,000. Since the capital budget is only $250,000, we cannot fund these projects. To penalize all projects in proportion to their cost, we will use Equation 17-1 with its multiplier, p. As a first trial, a value of $p = 0.10$ is selected and the alternatives with the largest [NPW, $-p$(PW of cost)] selected.

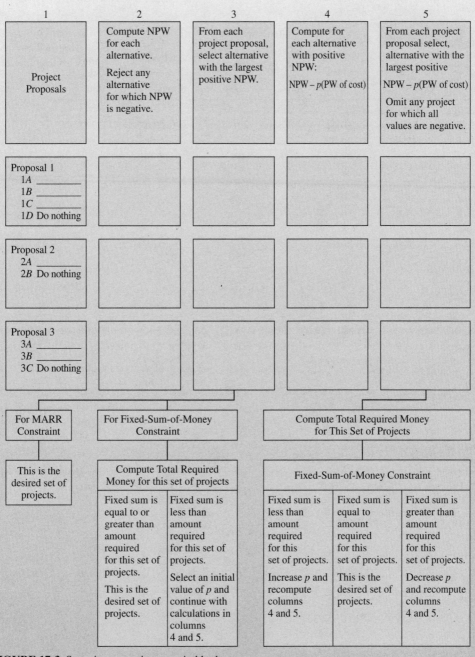

1	2	3	4	5
Project Proposals	Compute NPW for each alternative. Reject any alternative for which NPW is negative.	From each project proposal, select alternative with the largest positive NPW.	Compute for each alternative with positive NPW: NPW – p(PW of cost)	From each project proposal select, alternative with the largest positive NPW – p(PW of cost) Omit any project for which all values are negative.
Proposal 1 1A _____ 1B _____ 1C _____ 1D Do nothing				
Proposal 2 2A _____ 2B Do nothing				
Proposal 3 3A _____ 3B _____ 3C Do nothing				

For MARR Constraint	For Fixed-Sum-of-Money Constraint	Compute Total Required Money for This Set of Projects

This is the desired set of projects.	Compute Total Required Money for this set of projects	Fixed-Sum-of-Money Constraint

	Fixed sum is equal to or greater than amount required for this set of projects. This is the desired set of projects.	Fixed sum is less than amount required for this set of projects. Select an initial value of p and continue with calculations in columns 4 and 5.	Fixed sum is less than amount required for this set of projects. Increase p and recompute columns 4 and 5.	Fixed sum is equal to amount required for this set of projects. This is the desired set of projects.	Fixed sum is greater than amount required for this set of projects. Decrease p and recompute columns 4 and 5.

FIGURE 17-3 Steps in computing a capital budget.

			Alternative with Largest Positive NPW (thousands)		$p=0.10$ [NPW − p(PW of cost)]	Alternative with Largest Positive [NPW − p(PW of cost)] (thousands)	
Project Proposals	Cost (thousands)	NPW (thousands)	Alt.	Cost	(thousands)	Alt.	Cost
Proposal 1							
Alt. A	$100	$60.04			$50.04		
Alt. B	150	66.06			51.06	1B	$150
Alt. C	200	67.40	1C	$200	47.40		
Alt. D	0	0			0		
Proposal 2							
Alt. A	50	9.57	2A	50	4.57	2A	50
Alt. B	0	0			0		
Proposal 3							
Alt. A	100	36.99			26.99	3A	100
Alt. B	150	38.21	3B	150	23.21		
Alt. C	0	0			0		
				$400			$300

The first trial with $p = 0.10$ selects Alternatives 1B, 2A, and 3A with a total cost of $300,000. This still is greater than the $250,000 capital budget. Another trial is needed with a larger value of p. Select $p = 0.15$ and recompute.

				$p=0.15$ Alternative with Largest Positive [NPW − p(PW of cost)] (thousands)	
Project Proposals	Cost (thousands)	NPW (thousands)	[NPW − p(PW of cost)] (thousands)	Alt.	Cost
Proposal 1					
Alt. A	$100	$60.04	$45.04	1A	$100
Alt. B	150	66.06	43.56		
Alt. C	200	67.40	37.40		
Alt. D	0	0	0		
Proposal 2					
Alt. A	50	9.57	2.07	2A	50
Alt. B	0	0	0		
Proposal 3					
Alt. A	100	36.99	21.99	3A	100
Alt. B	150	38.21	15.71		
Alt. C	0	0	0		
					$250

The second trial, with $p = 0.15$, points to Alternatives 1A, 2A, and 3A for a total cost of $250,000. This equals the capital budget, hence is the desired set of projects.

EXAMPLE 17-5

Solve Example 17-4 by the rate of return method. For project proposals with two or more alternatives, incremental rate of return analysis is required. The data from Example 17-4 and the computed rate of return for each alternative and each increment of investment are as follows:

SOLUTION

					Incremental Analysis			
Combination of Alternatives	Cost (thousands)	Uniform Annual Benefit (thousands)	Salvage Value (thousands)	Computed Rate of Return	Cost (thousands)	Uniform Annual Benefit (thousands)	Salvage Value (thousands)	Computed Rate of Return
Proposal 1								
A	$100	$23.85	$ 0	20.0%				
B − A					$ 50	$ 8.35	$ 0	10.6%
B	150	32.20	0	17.0				
C − B					50	7.65	0	8.6
C − A					100	16.00	0	9.6
C	200	39.85	0	15.0				
D	0	0	0	0				
Proposal 2								
A	50	14.92	0	15.0				
B	0	0	0	0				
Proposal 3								
A	100	18.69	25	15.0				
B − A					50	0.73	100	8.3
B	150	19.42	125	12.0				
C	0	0	0	0				

The various separable increments of investment may be ranked by rate of return. They are plotted in a graph of cumulative cost versus rate of return in Figure 17-4. The ranking of projects by rate of return gives the following:

Project
1A
2A
3A
1B in place of 1A
1C in place of 1B
3B in place of 3A

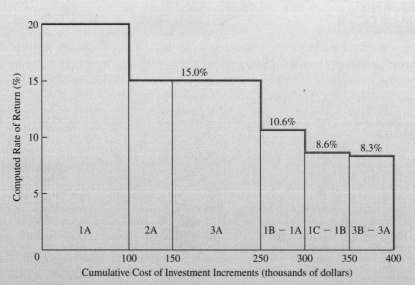

FIGURE 17-4 Cumulative cost versus incremental rate of return.

For a budget of $250,000, the selected projects are 1A, 2A, and 3A. Note that if a budget of $300,000 were available, 1B would replace 1A, making the proper set of projects 1B, 2A, and 3A. At a budget of $400,000, 1C would replace 1B; and 3B would replace 3A, making the selected projects 1C, 2A, and 3B. These answers agree with the computations in Example 17-4.

RANKING PROJECT PROPOSALS

Closely related to the problem of capital budgeting is the matter of ranking project proposals. We will first examine a method of ranking by present worth methods and then show that project rate of return is not a suitable method of ranking projects.

Anyone who has ever bought firecrackers probably used the practical ranking criterion of 'biggest bang for the buck' in making a selection. This same criterion—stated more elegantly—may be used to correctly rank independent projects.

Rank independent projects according to their value of net present worth divided by the present worth of cost. The appropriate interest rate is MARR (as a reasonable estimate of the cut-off rate of return).

Example 17-6 illustrates the method of computation.

EXAMPLE 17-6

Rank the following nine independent projects in their order of desirability, based on a 14.5% minimum attractive rate of return. (To facilitate matters, the necessary computations are included in the tabulation.)

Project	Cost (thousands)	Uniform Annual Benefit (thousands)	Useful Life (years)	Salvage Value (thousands)	Computed Rate of Return	Computed NPW at 14.5% (thousands)	Computed NPW/Cost (thousands)
1	$100	$23.85	10	$ 0	20%	$22.01	0.2201
2	200	39.85	10	0	15	3.87	0.0194
3	50	34.72	2	0	25	6.81	0.1362
4	100	20.00	6	100	20	21.10	0.2110
5	100	20.00	10	100	20	28.14	0.2814
6	100	18.00	10	100	18	17.91	0.1791
7	300	94.64	4	0	10	−27.05	−0.0902
8	300	47.40	10	100	12	−31.69	−0.1056
9	50	7.00	10	50	14	−1.28	−0.0256

SOLUTION

Ranked by NPW/PW of cost, the projects are listed as follows:

Project	NPW/PW of Cost	Rate of Return
5	0.2814	20%
1	0.2201	20
4	0.2110	20
6	0.1791	18
3	0.1362	25
2	0.0194	15
9	−0.0256	14
7	−0.0902	10
8	−0.1056	12

The rate of return tabulation illustrates that it is not a satisfactory ranking criterion and would have given a different ranking from the present worth criterion.

In Example 17-6, the projects are ranked according to the ratio NPW/PW of cost. In Figure 17-3, the criterion used is [NPW − p(PW of cost)]. If one were to compute the value of p at which [NPW − p(PW of cost)] = 0, we would obtain p = (NPW/PW of cost). Thus the multiplier p is the ranking criterion at the point at which [NPW − p(PW of cost)] = 0.

If independent projects can be ranked in their order of desirability, then the selection of projects to be included in a capital budget is a simple task. One may proceed down the list of ranked projects until the capital budget has been exhausted. The only difficulty with this scheme occurs, occasionally, when the capital budget is more than enough for n projects but too little for $n + 1$ projects.

In Example 17-6, a capital budget of $300,000 is just right to fund the top three projects. But a capital budget of $550,000 is more than enough for the top five projects (sum = $450,000) but not enough for the top six projects (sum = $650,000). When we have this lumpiness problem, it may not be possible to say with certainty that the best use of a capital budget of $550,000 is to fund the top five projects. There may be some other set of projects that makes better use of the available $550,000. While some trial-and-error computations may indicate the proper set of projects, more elaborate techniques are needed to prove optimality.

As a practical matter, a capital budget probably has some flexibility. If in Example 17-6 the tentative capital budget is $550,000, then a careful examination of Project 2 will dictate whether to expand the capital budget to $650,000 (to be able to include Project 2) or to drop back to $450,000 (and leave Project 2 out of the capital budget).

SUMMARY

Before this chapter we had assumed that all worthwhile projects are approved and implemented. But industrial firms, like individuals and governments, are often faced with more good projects than can be funded with the money available. The task is to select the best projects and reject, or at least delay, the rest.

Alternatives are mutually exclusive when the acceptance of one effectively prevents the adoption of the other. This could be because the alternatives perform the same function (like Pump A as opposed to Pump B) or would occupy the same physical location (like a gas station rather than a hamburger stand). If a project has a single alternative of doing something, we know there is likely to be a mutually exclusive alternative of doing nothing— or possibly doing something else. A project proposal may be thought of as having two or more mutually exclusive alternatives. Projects are assumed in this chapter to be independent.

Capital may be rationed among competing investment opportunities by either rate of return or present worth methods. The results may not always be the same for these two methods in many practical situations.

If projects are ranked by rate of return, a proper procedure is to go down the list until the capital budget has been exhausted. The rate of return at this point is the cut-off rate of return. This procedure gives the best group of projects but does not necessarily have them in the proper priority order.

Maximizing NPW is a suitable present worth selection rule when the available projects do not exhaust the money supply. But if the amount of money needed for the best alternative from each project exceeds the available money, a more severe criterion is imposed: adopt only alternatives and projects that have a positive NPW − p(PW of cost). The value of the multiplier p is chosen by trial and error until the alternatives and projects meeting the criterion just equal the available capital budget money.

It has been shown in earlier chapters that the usual business objective is to maximize NPW, and this is not necessarily the same as maximizing rate of return. One suitable procedure is to use the ratio (NPW/PW of cost) to rank the projects. This present worth ranking method will order the projects so that, for a limited capital budget, NPW will be maximized. We know that MARR must be adjusted from time to time to reasonably balance the cost of the projects that meet the MARR criterion and the available supply of money. This adjustment of the MARR to equal the cut-off rate of return is essential for the rate of return and present worth methods to yield compatible results.

Another way of ranking is by incremental rate of return analysis. Once a ranking has been made, we can go down the list and accept the projects until the money runs out. There is a theoretical difficulty if the capital budget contains more money than is needed for n projects, but not enough for one more ($n+1$ projects). As a practical matter, capital budgets are seldom inflexible, with the result that some additional money may be allocated if the $(n+1)$st project looks like it should be included.

PROBLEMS

17-1 Each of the following 10 independent projects has a 10-year life and no salvage value.

Project	Cost (thousands)	Uniform Annual Benefits (thousands)	Computed Rate of Return
1	$ 5	$1.03	16%
2	15	3.22	17
3	10	1.77	12
4	30	4.88	10
5	5	1.19	20
6	20	3.83	14
7	5	1.00	15
8	20	3.69	13
9	5	1.15	19
10	10	2.23	18

The projects have been proposed by the staff of the Ace Card Company. The MARR of Ace has been 12% for several years.

(a) If there is ample money available, what projects should Ace approve?

(b) Rank order all the acceptable projects in their order of desirability.

(c) If only $55,000 is available, which projects should be approved?

17-2 At Red Deer Products, four project proposals (three with mutually exclusive alternatives) are being considered. All the alternatives have a 10-year useful life and no salvage value.

Project Proposal	Cost (thousands)	Uniform Annual Benefits (thousands)	Computed Rate of Return
Project 1			
Alt. A	$25	$4.61	13%
Alt. B	50	9.96	15
Alt. C	10	2.39	20
Project 2			
Alt. A	20	4.14	16
Alt. B	35	6.71	14
Project 3			
Alt. A	25	5.56	18
Alt. B	10	2.15	17
Project 4	10	1.70	11

(a) Use rate of return methods to determine which set of projects should be undertaken if the MARR is 10%.

(b) Use rate of return methods to determine which set of projects should be undertaken if the capital budget is limited to $100,000.

(c) For a budget of $100,000, what interest rate should be used in rationing capital by present worth methods? (Limit your answer to a value for which there is a compound interest table available in the appendix).

(d) Using the interest rate determined in part (c), rank order the eight different investment opportunities by means of the present worth method.

(e) For a budget of $100,000 and the ranking in part (d), which of the investment opportunities should be selected?

TABLE P17-3 Data

Prospective Gift	'Oh' Rating of Gift If Given to Various Family Members						
	Father	Mother	Sister	Brother	Aunt	Uncle	Cousin
1. $20 box of candy	4	4	2	1	5	2	3
2. $12 box of cigars	3	0	0	1	0	1	2
3. $16 necktie	2	0	0	3	0	3	2
4. $20 shirt or blouse	5	3	4	4	4	1	4
5. $24 sweater	3	4	5	4	3	4	2
6. $30 camera	1	5	2	5	1	2	0
7. $ 6 calendar	0	0	1	0	1	0	1
8. $16 magazine subscription	4	3	4	4	3	1	3
9. $18 book	3	4	2	3	4	0	3
10. $16 game	2	2	3	2	2	1	2

17-3 Al Dale is planning his Christmas shopping for seven people. To quantify how much his various relatives would enjoy receiving items from a list of prospective gifts, Al has assigned appropriateness units (called 'ohs') for each gift if given to each of the seven people. A rating of 5 ohs represents a gift that the recipient would really like. A rating of 4 ohs indicates the recipient would like it four-fifths as much; 3 ohs, three-fifths as much, and so forth. A zero rating indicates an inappropriate gift that cannot be given to that person.

The objective is to select the most appropriate set of gifts for the seven people (that is, maximize total ohs) that can be obtained with the selected budget.

(a) How much will it cost to buy the seven gifts the people would like best, if there is ample money for Christmas shopping?

(b) If the Christmas shopping budget is set at $112, which gifts should be purchased, and what is their total appropriateness rating in ohs?

(c) If the Christmas shopping budget must be cut to $90, which gifts should be purchased, and what is their total appropriateness rating in ohs?

(*Answer:* (*a*), $168)

The following facts are to be used in solving Problems 17-4 through 17-7 In assembling data for the Peabody Company annual capital budget, five independent projects are being considered. Detailed examination by the staff has resulted in the identification of three to six mutually exclusive do-something alternatives for each project. In addition,

each project has a do-nothing alternative. The projects and their alternatives are listed at the top of next page.

Each project concerns operations at Peabody's major brewery. The plant was leased from another firm many years ago, and the lease expires 16 years from now. For this reason, the analysis period for all projects is 16 years. Peabody considers 12% to be the minimum attractive rate of return.

In solving the Peabody problems, an important assumption concerns the situation at the end of the useful life of an alternative when the alternative has a useful life shorter than the 16-year analysis period. Two replacement possibilities are listed.

Assumption 1: When an alternative has a useful life of less than 16 years, it will be replaced by a new alternative with the same useful life as the original. This may need to occur more than once. The new alternative will have a 12% computed rate of return and, hence, a NPW = 0 at 12%.

Assumption 2: When an alternative has a useful life of less than 16 years, it will be replaced at the end of its useful life by an identical alternative (one with the same cost, uniform annual benefit, useful life, and salvage value as the original alternative).

17-4 For an unlimited supply of money, and replacement Assumption 1, which project alternatives should Peabody select? Solve the problem by present worth methods. (*Answer:* Alternatives 1B, 2A, 3F, 4A, and 5A)

Project Proposal	Cost (thousands)	Uniform Annual Benefit (thousands)	Useful Life (years)	End-of-Useful-Life Salvage Value (thousands)	Computed Rate of Return
Project 1					
Alt. A	$40	$13.52	2	$20	10%
Alt. B	10	1.87	16	5	18
Alt. C	55	18.11	4	0	12
Alt. D	30	6.69	8	0	15
Alt. E	15	3.75	2	15	25
Project 2					
Alt. A	10	1.91	16	2	18
Alt. B	5	1.30	8	0	20
Alt. C	5	0.97	8	2	15
Alt. D	15	5.58	4	0	18
Project 3					
Alt. A	20	2.63	16	10	12
Alt. B	5	0.84	16	0	15
Alt. C	10	1.28	16	0	10
Alt. D	15	2.52	16	0	15
Alt. E	10	3.50	4	0	15
Alt. F	15	2.25	16	15	15
Project 4					
Alt. A	10	2.61	8	0	20
Alt. B	5	0.97	16	0	18
Alt. C	5	0.90	16	5	18
Alt. D	15	3.34	8	0	15
Project 5					
Alt. A	5	0.75	8	5	15
Alt. B	10	3.50	4	0	15
Alt. C	15	2.61	8	5	12

17-5 For an unlimited supply of money, and replacement Assumption 2, which project alternatives should Peabody select? Solve the problem by present worth methods.

17-6 For an unlimited supply of money, and replacement Assumption 2, which project alternatives should Peabody select? Solve the problem by rate of return methods. (*Hint:* By careful inspection of the alternatives, you should be able to reject about half of them. Even then the problem requires lengthy calculations.)

17-7 For a capital budget of $55,000, and replacement Assumption 2, which project alternatives should Peabody select? (*Answer:* Alternatives 1E, 2A, 3F, 4A, and 5A)

17-8 A financier has a staff of three people whose job it is to examine possible business ventures for him.

Periodically they present their findings concerning business opportunities. On a particular occasion, they presented the following investment opportunities:

Project A: This is a project for the use of the commercial land the financier already owns. There are three mutually exclusive alternatives.

- A1. Sell the land for $500,000.
- A2. Lease the property for a car-washing business. An annual income, after all costs (property taxes, etc.) of $98,700 would be received at the end of each year for 20 years. It is believed that at the end of the 20 years, the property could be sold for $750,000.
- A3. Construct an office building on the land. The building will cost $4.5 million to construct and will not produce any net income for the first 2 years. The probabilities of various levels of rental income,

after all expenses, for the subsequent 18 years are as follows:

Annual Rental Income	Probability
$1,000,000	0.1
1,100,000	0.3
1,200,000	0.4
1,900,000	0.2

The property (building and land) probably can be sold for $3 million at the end of 20 years.

Project B: An insurance company is seeking to borrow money for 90 days at 13 3/4% per annum, compounded continuously.

Project C: A financier owns a manufacturing company. The firm desires additional working capital to allow it to increase its inventories of raw materials and finished products. An investment of $2 million will allow the company to obtain sales that in the past the company had to forgo. The additional capital will increase company profits by $500,000 a year. The financier can recover this additional investment by ordering the company to reduce its inventories and to return the $2 million. For planning purposes, assume the additional investment will be returned at the end of 10 years.

Project D: The owners of *Sunrise* magazine are seeking a loan of $500,000 for 10 years at a 16% interest rate.

Project E: The Galveston Bank has indicated a willingness to accept a deposit of any sum of money over $100,000, for any desired duration, at a 14.06% interest rate, compounded monthly. It seems likely that this interest rate will be available from Galveston, or some other bank, for the next several years.

Project F: A car rental company is seeking a loan of $2 million to expand its fleet of cars. The Company offers to repay the loan by paying $1 million at the end of Year 1 and $1,604,800 at the end of Year 2.

• If there is $4 million available for investment now (or $4.5 million if the Project A land is sold), which projects should be selected? What is the MARR in this situation?
• If there is $9 million available for investment now (or $9.5 million if the Project A land is sold), which projects should be selected?

17-9 The Raleigh Soap Company has been offered a 5-year contract to manufacture and package a leading brand of soap for Taker Bros. It is understood that the contract will not be extended past the 5 years because Taker Bros. plans to build its own plant nearby. The contract calls for 10,000 metric tons (one metric ton equals 1000 kg) of soap a year. Raleigh normally produces 12,000 metric tons of soap a year, and so production for the 5-year period would be increased to 22,000 metric tons. Raleigh must decide what changes, if any, to make to accommodate this increased production. Five projects are under consideration.

Project 1: Increase liquid storage capacity. Raleigh has been forced to buy caustic soda in tank truck quantities owing to inadequate storage capacity. If another liquid caustic soda tank is installed to hold 1000 cubic metres, the caustic soda may be purchased in railroad tank car quantities at a more favourable price. The result would be a saving of 0.1 cent per kilogram of soap. The tank, which would cost $83,400, has no net salvage value.

Project 2: Acquire another sulphonation unit. The present capacity of the plant is limited by the sulphonation unit. The additional 12,000 metric tons of soap cannot be produced without an additional sulphonation unit. Another unit can be installed for $320,000.

Project 3: Expand the packaging department. With the new contract, the packaging department must either work two 8-hour shifts or have another packaging line installed. If the two-shift operation is used, a 20% wage premium must be paid for the second shift. This premium would amount to $35,000 a year. The second packaging line could be installed for $150,000. It would have a $42,000 salvage value at the end of 5 years.

Project 4: Build a new warehouse. The existing warehouse will be inadequate for the greater production. It is estimated that 400 square metres of additional warehouse is needed. A new warehouse can be built on a lot beside the existing warehouse for $225,000, including the land. The annual taxes, insurance, and other ownership costs would be $5000 a year. It is believed the warehouse could be sold at the end of 5 years for $200,000.

Project 5: Lease a warehouse. An alternative to building an additional warehouse would be to lease warehouse space. A suitable warehouse one mile away could be leased for $15,000 per year. The $15,000 includes taxes, insurance, and so forth. The annual cost of moving materials to this more remote warehouse would be $34,000 a year.

The contract offered by Taker Bros. is a favourable one, which Raleigh Soap plans to accept. Raleigh management has set a 15% before-tax minimum attractive rate of return as the criterion for any of the projects. Which projects should be undertaken?

17-10 Ten capital spending proposals have been made to the budget committee as the members prepare the annual budget for their firm. Each independent project has a 5-year life and no salvage value.

Project	Initial Cost (thousands)	Uniform Annual Benefit (thousands)	Computed Rate of Return
A	$10	$2.98	15%
B	15	5.58	25
C	5	1.53	16
D	20	5.55	12
E	15	4.37	14
F	30	9.81	19
G	25	7.81	17
H	10	3.49	22
I	5	1.67	20
J	10	3.20	18

(a) On the basis of a MARR of 14%, which projects should be approved?
(b) Rank-order all the projects in order of desirability.
(c) If only $85,000 is available, which projects should be approved?

17-11 Mike Moore's microbrewery is considering production of a new ale called Mike's Honey Harvest Brew. To produce this new offering he is considering two independent projects. Each of these projects has two mutually exclusive alternatives, and each alternative has a useful life of 10 years and no salvage value. Mike's MARR is 8%. Information regarding the projects and alternatives are given in the following table:

Project or Alternative	Cost	Annual Benefit
Project 1. Purchase new fermenting tanks		
Alt. A: 5000-gallon tank	$ 5000	$1192
Alt. B: 15,000-gallon tank	10,000	1992
Project 2. Purchase bottle filler and capper		
Alt. A: 2500-bottle/hour machine	15,000	3337
Alt. B: 5000-bottle/hour machine	25,000	4425

Use incremental rate of return analysis to complete the following worksheet.

Proj./Alt.	Cost, P	Annual Benefit, A	$A/P, i, 10$	IRR
1A	$ 5,000	$1192	0.2385	20%
1B–1A	5,000	800	0.1601	
2A	15,000	3337		
2B–2A	10,000			

Use this information to determine:
(a) Which projects should be funded if only $15,000 is available.
(b) The cut-off rate of return if only $15,000 is available.
(c) Which projects should be funded if $25,000 is available.

CHAPTER 18

Accounting and Engineering Economy

The Two Faces of ABB

In the late 1990s, ABB Ltd was flying high. The European conglomerate was an engineering giant with a global network of operations and forward-looking management.

Unlike many of its competitors, ABB was also committed to modernizing its accounting system. ABB had previously followed the traditional practice of assigning overhead costs to its divisions on a roughly equal basis. But this practice tended to obscure the fact that some activities incurred far more costs than others.

So the company spent substantial time and resources switching over to an activity-based costing (ABC) system, which assigns costs to the activities that actually produce them. The results were quite positive, allowing ABB to zero in on areas where it could cut costs most effectively.

But in other respects, ABB's accounting practices were less than proactive.

One of its most serious problems arose at Combustion Engineering, an American subsidiary that was exposed to numerous asbestos liability claims. For years, ABB had downplayed the extent of this potential liability, despite warnings from outside analysts. Finally, in late 2002, ABB admitted that its asbestos liability exceeded the subsidiary's total asset value.

But there was still more bad news. The company also issued a very poor third-quarter earnings report, after earlier assuring investors that ABB was on target to improve its earnings and decrease its debt. When incredulous investors asked how this could have happened, ABB management blamed 'poor internal reporting'.

By this point, ABB's stock price had nose-dived, and credit rating agencies viewed the company's bonds as little better than junk.

Today, ABB is struggling to survive. It has sent Combustion Engineering into bankruptcy in an attempt to limit asbestos claims and has sold off other subsidiaries. It has also slashed thousands of jobs and embarked on a wide array of cost-cutting projects.

ABB now claims to have put the worst behind it. Investors must certainly be hoping that is true.

After Completing This Chapter...

The students should be able to:

- Describe the links between engineering economy and accounting.
- Describe the objectives of general accounting, explain what financial transactions are, and show how they are important.
- Use a firm's balance sheet and associated financial ratios to evaluate the firm's health.
- Use a firm's income statement and associated financial ratios to evaluate the firm's performance.
- Use traditional absorption costing to calculate product costs.
- Understand the greater accuracy in product costs available with activity-based costing (ABC).

QUESTIONS TO CONSIDER

1. ABB's adoption of activity-based costing received widespread publicity and boosted the company's reputation for innovative management. How may this have affected investors' assumptions about the company's other accounting practices?
2. Outside analysts estimated that ABB lost over $690 million in 2001. Yet many investors were still stunned by its poor showing in late 2002. What does this say about the relationship between financial accounting and investor confidence?
3. ABB was not alone in its financial misery, of course. How did ABB's accounting problems compare with those of well-known American companies such as Enron and WorldCom?

Engineering economy focuses on the financial aspects of projects, while accounting focuses on the financial aspects of firms. Thus the application of engineering economy is much easier if one has some understanding of accounting principles. In fact, one important accounting topic, depreciation, was the subject of an earlier chapter.

THE ROLE OF ACCOUNTING

Accounting data are used to value capital equipment, to decide whether to make or buy a part, to determine costs and set prices, to set indirect cost rates, and to make product mix decisions. Accounting is used in private-sector firms and public-sector agencies, but for simplicity this chapter uses 'the firm' to designate both. Accountants track the costs of projects and products, which are the basis for estimating future costs and revenues.

The engineering economy, accounting, and managerial functions are interdependent. As shown in Figure 18-1, data and communications flow between them. Whether carried out by a single person in a small firm or by distinct divisions in a large firm, all are needed.

- Engineering economy analyzes the economic impact of design alternatives and projects over their life cycles.
- Accounting determines the dollar impact of past decisions, reports on the economic viability of a unit or firm, and evaluates potential funding sources.
- Management allocates available investment funds to projects, evaluates unit and firm performance, allocates resources, and selects and directs personnel.

Accounting	Management	Engineering Economy
About past	About past and future	About future
Analyzing	Capital budgeting	Feasibility of alternatives
Summarizing	Decision making	Collecting and analyzing data
Reporting	Setting goals	Estimating
Financial indicators	Assessing impacts	Evaluating projects
Economic trends	Analyzing risk	Recommending
Cost acquisitions	Planning	Auditing
	Controlling	Identifying needs
	Record keeping	Trade-offs and constraints
← Data and Communication →	← Data and Communication →	
← Budgeting	Data and Communication	Estimating →

FIGURE 18-1 The accounting, managerial, and engineering economy functions.

Accounting for Business Transactions

A business transaction involves two parties and the exchange of dollars (or the promise of dollars) for a product or service. Each day, millions of transactions occur between firms and their customers, suppliers, vendors, and employees. Transactions are the lifeblood of the business world and are most often stated in monetary terms. The accounting function records, analyzes, and reports these exchanges.

Transactions can be as simple as payment for a water bill, or as complex as the international transfer of millions of dollars worth of buildings, land, equipment, inventory, and other assets. Also, with transactions, one business event may lead to another—all of which need to be accounted for. Consider, for example, the process of selling a robot or bulldozer. This simple act involves several related transactions: (1) equipment released from inventory, (2) equipment shipped to the purchaser, (3) invoicing the purchaser, and finally (4) collecting from the purchaser.

Transaction accounting involves more than just reporting: it includes finding, synthesizing, summarizing, and analyzing data. For the engineering economist, historical data housed in the accounting function are the foundation for estimates of future costs and revenues.

Most accounting is done in nominal or *stable* dollars. Higher market values and costs due to inflation are less objective than cost data, and with a going concern, accountants have decided that objectivity should be maintained. Similarly, most assets are valued at their acquisition cost adjusted for depreciation and improvements. To be conservative, when market value is lower than this adjusted cost, the lower value is used. This restrains the interests of management in maximizing a firm's apparent value. If a firm must be liquidated, then current market value must be estimated.

The accounting function provides data for *general accounting* and *cost accounting*. This chapter's presentation begins with the balance sheet and income statement, which are the two key summaries of financial transactions for general accounting. This discussion includes some of the basic financial ratios used for short- and long-term evaluations. The chapter concludes with a key topic in cost accounting—allocating indirect expenses.

THE BALANCE SHEET

The primary accounting statements are the **balance sheet** and **income statement.** The **balance sheet** describes the firm's financial condition at a specific time, while the income statement describes the firm's performance over a period of time—usually a year.

The balance sheet lists the firm's assets, liabilities, and equity on a specified date. This is a picture of the organization's financial health or a snap shot in time. Usually, balance sheets are taken at the end of the quarter and fiscal year. The balance sheet is based on the **fundamental accounting equation:**

$$\text{Assets} = \text{Liabilities} + \text{Equity} \qquad (18\text{-}1)$$

Figure 18-2 illustrates the basic format of the balance sheet. Notice in the balance sheet, as in Equation 18-1, that **assets** are listed on the left-hand side and **liabilities** and **equity** are on the right-hand side. The fact that the firm's resources are *balanced* by the sources of funds is the basis for the name of the balance sheet.

Assets

In Equation 18-1 and Figure 18-2, **assets** are owned by the firm and have monetary value. **Liabilities** are the dollar claims against the firm. **Equity** represents funding from the firm and its owners (the shareholders). In Equation 18-1, assets are always balanced by the sum of the liabilities and the equity. Retained earnings are set so that equity equals assets minus liabilities.

On a balance sheet, assets are listed in order of decreasing liquidity, that is, according to how quickly each one can be converted to cash. Thus, *current assets* are listed first, and

Balance Sheet for Engineered Industries, December 31, 2004 (all amounts in $1000s)

Assets		Liabilities	
Current assets		Current liabilities	
Cash	1940	Accounts payable	1150
Accounts receivable	950	Notes payable	80
Securities	4100		
Inventories	1860	Accrued expense	950
(*minus*) Bad debt provision	−80	Total current liabilities	2180
Total current assets	8770		
		Long-term liabilities	1200
Fixed assets			
Land	335		
Plant and equipment	6500		
(*minus*) Accumulated depr.	−2350		
Total fixed assets	4485	**Equity**	
		Preferred shares	110
Other assets		Common shares	650
Prepays/deferred charges	140	Capital surplus	930
Intangibles	420	Retained earnings	8745
Total other assets	560	Total equity	10,435
Total assets	**13,815**	**Total liabilities and equity**	**13,815**

FIGURE 18-2 Sample balance sheet.

within that category in order of decreasing liquidity are listed cash, receivables, securities, and inventories. *Fixed assets,* or *property, plant, and equipment,* are used to produce and deliver goods and/or services, and they are not intended for sale. Lastly, items such as prepayments and intangibles such as patents are listed.

The term 'receivables' comes from the manner of handling billing and payment for most business sales. Rather than requesting immediate payment for every transaction by check or credit card, most businesses record each transaction and then once a month bill for all transactions. The total that has been billed less payments already received is called accounts receivable, or receivables.

Liabilities

On the balance sheet, liabilities are divided into two major classifications—short term and long term. The *short-term* or *current liabilities* are expenses, notes, and other payable accounts that are due within one year from the balance sheet date. *Long-term liabilities* include mortgages, bonds, and loans with later due dates. For Engineered Industries in Figure 18-2, total current and long-term liabilities are $2,180,000 and $1,200,000 respectively. Often in performing engineering economic analyses, the **working capital** for a project must be estimated. The total amount of working capital available may be calculated with Equation 18-2 as the difference between current assets and current liabilities.

$$\text{Working capital} = \text{Current assets} - \text{Current liabilities} \qquad (18\text{-}2)$$

For Engineered Industries, there would be $4,670,000 - $2,180,000 = $2,490,000$ available in working capital.

Equity

Equity is also called *owner's equity* or *net worth.* It includes the par value of the owners' shareholdings and the capital surplus, which are the excess dollars brought in over par value when the shares were issued. Retained earnings are dollars a firm chooses to retain rather than pay out as dividends to shareholders. Equity is the dollar quantity that always brings the balance sheet, and thus the fundamental accounting equation, into balance. For Engineered Industries, *total equity* value is listed at $10,435,000. From Equation 18-1 and the assets, liabilities, and equity values in Figure 18-2, we can write the balance as follows:

$$\text{Assets} = \text{Liabilities} + \text{Equity}$$

$$\text{Assets (current, fixed, other)} = \text{Liabilities (current and long-term)} + \text{Equity}$$

$$(4,670,000 + 4,485,000 + 560,000) = (2,180,000 + 1,200,000) + 10,435,000$$

$$\$13,815,000 = \$13,815,000$$

An example of owner's equity is ownership of a home. Most homes are purchased by means of a mortgage loan that is paid off at a certain interest rate over 15 to 30 years. At any point in time, the difference between what is owed to the bank (the remaining balance on the mortgage) and what the house is worth (its appraised market value) is the *owner's equity.* In this case, the loan balance is the *liability,* and the home's value is the *asset*—with *equity* being the difference. Over time, as the house loan is paid off, the owner's equity increases.

The balance sheet is a very useful tool that shows one view of the firm's financial condition at a particular point in time.

Financial Ratios Derived from Balance Sheet Data

One common way to evaluate the firm's health is through ratios of quantities on the balance sheet. Firms in a particular industry will usually have similar values, and exceptions will often indicate firms with better or worse performance. Two common ratios used to analyze the firm's current position are the current ratio and the acid-test ratio.

A firm's **current ratio** is the ratio of current assets to current liabilities, as in Equation 18-3.

$$\text{Current ratio} = \text{Current assets/Current liabilities} \tag{18-3}$$

This ratio provides insight into the firm's solvency over the short term by indicating its ability to cover current liabilities. Historically, firms aim to be at or above a ratio of 2.0; however, this depends heavily on the industry as well as the individual firm's management practices and philosophies. The current ratio for Engineered Industries in Figure 18-2 is above 2 ($4,670,000/2,180,000 = 2.14$).

Both working capital and the current ratio indicate the firm's ability to meet currently maturing obligations. However, neither describes the type of assets owned. The **acid-test**

ratio or **quick ratio** becomes important when one wishes to consider the firm's ability to pay debt 'instantly'. The acid-test ratio is computed by dividing a firm's **quick assets** (cash, receivables, and market securities) by total current liabilities, as in Equation 18-4.

$$\text{Acid-test ratio} = \text{Quick assets}/\text{Total current liabilities} \qquad (18\text{-}4)$$

Current inventories are excluded from quick assets because of the time required to sell these inventories, collect the receivables, and subsequently have the cash on hand to reduce debt. For Engineered Industries in Figure 18-2, the calculated acid-test ratio is well below the current ratio [(1,940,000 + 950,000)/2,180,000 = 1.33].

Working capital, current ratio, and acid-test ratio are all indications of the firm's financial health (status). A thorough financial evaluation would consider all three, including comparisons with values from previous periods and with broad-based industry standards.

THE INCOME STATEMENT

The **income statement** or **profit and loss statement** summarizes the firm's revenues and expenses over a month, quarter, or year. Rather than being a snapshot like the balance sheet, the income statement encompasses a *period* of business activity. The income statement is used to evaluate revenue and expenses that occur in the interval *between* consecutive balance sheet statements. The income statement reports the firm's *net income (profit)* or *loss* by subtracting expenses from revenues. If revenues minus expenses is positive in Equation 18-5, there has been a profit, if negative a loss has occurred.

$$\text{Revenues} - \text{Expenses} = \text{Net profit (Loss)} \qquad (18\text{-}5)$$

Revenue, as in Equation 18-5, serves to increase ownership in a firm, while expenses serve to decrease ownership. Figure 18-3 is an example of an income statement.

To aid in analyzing performance, the income statement in Figure 18-3 separates operating and non-operating activities and shows revenues and expenses for each. Operating revenues are made up of sales revenues (minus returns and allowances), while non-operating revenues come from rents and interest receipts.

Operating expenses produce the products and services that generate the firm's revenue stream of cash flows. Typical operating expenses include cost of goods sold, selling and promotion costs, depreciation, general and administrative costs, and lease payments. *Cost of goods sold (COGS)* includes the labour, materials, and indirect costs of production.

Engineers design production systems, and they are involved in labour loading, specifying materials, and make or buy decisions. All these items affect a firm's cost of goods sold. Good engineering design focuses not only on technical functionality but also on cost-effectiveness as the design *integrates* the entire production system. Also of interest to the engineering economist is *depreciation* (see Chapter 11)—which is the systematic 'writing off' of a capital expense over a period of years. This non-cash expense is important because it represents a decrease in value in the firm's capital assets.

The operating revenues and expenses are shown first, so that the firm's operating income from its products and services can be calculated. Also shown on the income statement are non-operating expenses such as interest payments on debt in the form of loans or bonds.

Income Statement for Engineered Industries for End of Year 2005
(all amounts in $1000)

Operating revenues and expenses	
Operating Revenues	
Sales	28,900
(*minus*) Returns and allowances	−870
Total operating revenues	**28,030**
Operating expenses	
Cost of goods and services sold	
Labour	6140
Materials	4640
Indirect cost	2280
Selling and promotion	930
Depreciation	1850
General and administrative	900
Lease payments	510
Total operating expense	**17,250**
Total operating income	**10,780**
Non-operating revenues and expenses	
Rents	400
Interest receipts	180
Interest payments	−120
Total Non-operating income	**460**
Net income before taxes	**11,240**
Income taxes	3930
Net profit (loss) for Year 2005	**7310**

FIGURE 18-3 Sample income statement.

From the data in Figure 18-3, Engineered Industries has total expenses (operating $=$ $17,250,000$ and non-operating $= \$120,000$) of $17,370,000$. Total revenues are $28,610,000$ ($= \$28,030,000 + \$400,000 + \$180,000$). The net after-tax profit for year 2005 shown in Figure 18-3 as $7,310,000$, but it can also be calculated using Equation 18-5 as:

$$\text{Net profits (Loss)} = \text{Revenues} - \text{Expenses [before taxes]}$$

$$\$11,240,000 = 28,610,000 - 17,370,000 \text{ [before taxes] and with}$$

$$\$3,930,000 \text{ taxes paid}$$

thus

$$\$7,310,000 = 11,240,000 - 3,930,000 \text{ [after taxes]}$$

Financial Ratios Derived from Income Statement Data

Interest coverage, as given in Equation 18-6, is calculated as the ratio of total income to interest payments—where *total income* is total revenues minus all expenses except interest payments.

$$\text{Interest coverage} = \text{Total income}/\text{Interest payments} \qquad (18\text{-}6)$$

The interest coverage ratio (which for industrial firms should be at least 3.0) indicates how much revenue must drop to affect the firm's ability to finance its debt. With an interest coverage ratio of 3.0, a firm's revenue would have to decrease by two-thirds (unlikely) before it became impossible to pay the interest on the debt. The larger the interest coverage ratio the better. Engineered Industries in Figure 18-3 has an interest coverage ratio of

$$(28,610,000 - 17,250,000)/120,000 = 94.7$$

Another important financial ratio based on the income statement is the **net profit ratio.** This ratio (Equation 18-7) equals net profits divided by net sales revenue. Net sales revenue equals sales minus returns and allowances.

$$\text{Net profit ratio} = \text{Net profit}/\text{Net sales revenue} \qquad (18\text{-}7)$$

This ratio provides insight into the cost efficiency of operations as well as a firm's ability to convert sales into profits. For Engineered Industries in Figure 18-3, the net profit ratio is $7,310,000/28,030,000 = 0.261 = 26.1\%$. Like other financial measures, the net profit ratio is best evaluated by comparisons with other time periods and industry benchmarks.

Linking the Balance Sheet, Income Statement, and Capital Transactions

The balance sheet and the income statement are separate but linked documents. Understanding how the two are linked together helps clarify each. Accounting describes these links as the *articulation* between these reports.

The balance sheet shows a firm's assets, liabilities, and equity at a particular point in time, whereas the income statement summarizes revenues and expenses over a time interval. These tabulations can be visualized as a snapshot at the period's beginning (a balance sheet), a video summary over the period (the income statement), and a snapshot at the period's end (another balance sheet). The income statement and changes in the balance sheets summarize the business transactions that have occurred during that period.

There are many links between these statements and the cash flows that make up business transactions, but for engineering economic analysis the following are the most important.

1. Overall profit or loss (income statement) and the starting and ending equity (balance sheets).
2. Acquisition of capital assets.
3. Depreciation of capital assets.

The overall profit or loss during the year (shown on the income statement) is reflected in the change in retained earnings between the balance sheets at the beginning and end of the year. To find the change in retained earnings (RE), one must also subtract any dividends distributed to the owners and add the value of any new capital stock sold:

$$RE_{beg} + \text{Net income/Loss} + \text{New stock} - \text{Dividends} = RE_{end}$$

When capital equipment is purchased, the balance sheet changes, but the income statement does not. If cash is paid, then the cash asset account decrease equals the increase in the capital equipment account—there is no change in total assets. If a loan is used, then the capital equipment account increases, and so does the liability item for loans. In both cases the equity accounts and the income statement are unchanged.

The depreciation of capital equipment is shown as a line on the income statement. The depreciation for that year equals the change in accumulated depreciation between the beginning and the end of the year—after subtraction of the accumulated depreciation for any asset that is sold or disposed of during that year.

Example 18-1 applies these relationships to with the data in Figures 18-1 and 18-2.

EXAMPLE 18-1

For simplicity, assume that Engineered Industries will not pay dividends in 2005 and did not sell any capital equipment. It did purchase $4 million in capital equipment. What can be said about the values on the balance sheet at the end of 2005, using the linkages just described?

SOLUTION

First, the net profit of $7,310,000 will be added to the retained earnings from the end of 2004 to find the new retained earnings at the end of 2005:

$$RE_{12/31/2005} = \$7,310,000 + \$8,745,000 = \$16,055,000$$

Second, the fixed assets shown at the end of 2005 would increase from $6,500,000 to $10,500,000. (*Note:* This is a major investment of the retained earnings in the firm's physical assets.)

Third, the accumulated depreciation would increase by the $1,850,000 in depreciation shown in the 2005 income statement from the $2,350,000 shown in the 2004 balance sheet. The new accumulated depreciation on the 2005 balance sheet would be $4,200,000. Combined with the change in the amount of capital equipment, the new fixed asset total for 2005 would equal:

$$\$335,000 + \$10,500,000 - \$4,200,000 = \$6,635,000$$

TRADITIONAL COST ACCOUNTING

A firm's *cost-accounting system* collects, analyzes, and reports operational performance data (costs, utilization rates, etc.). Cost-accounting data are used to develop product costs, to determine the mix of labour, materials, and other costs in a production setting, and to evaluate instances of outsourcing or subcontracting.

Direct and Indirect Costs

Costs incurred to produce a product or service are traditionally classified as either *direct* or *indirect (overhead)*. Direct costs come from activities directly associated with the final product or service produced. Examples include material costs and labour costs for engineering design, component assembly, painting, and drilling.

Some organizational activities are difficult to link to specific projects, products, or services. For example, the receiving and shipping areas of a manufacturing plant are used by all incoming materials and all outgoing products. Materials and products differ in their weight, size, fragility, value, number of units, packaging, and so on, and the receiving and shipping costs depend on all these factors. Also, different materials arrive and different products are shipped together, so these costs are intermingled and often cannot be tied directly to each product or material.

Other costs, such as the organization's management, sales, and administrative expenses, are difficult to link directly to individual products or services. These indirect or overhead expenses also include machine depreciation, engineering and technical support, and customer warranties.

Indirect Cost Allocation

To allocate indirect costs to different departments, products, and services, accountants use quantities such as direct-labour hours, direct-labour costs, material costs, and total direct cost. One of these is chosen to be the burden vehicle. The total of all indirect or overhead costs is divided by the total for the burden vehicle. For example, if direct-labour hours is the burden vehicle, then overhead will be allocated on the basis of overhead dollar per direct-labour hour. Then each product, project, or department will *absorb* (or be allocated) overhead costs, based on the number of direct-labour hours each has.

This is the basis for calling traditional costing systems **absorption costing.** For decision making, the problem is that the absorbed costs represent average, not incremental, performance.

Four common ways of allocating overhead are direct-labour hours, direct-labour cost, direct-materials cost, and total direct cost. The first two differ significantly only if the cost per hour of labour differs for different products. Example 18-2 uses direct-labour and direct-materials cost to illustrate the difference choices of burden vehicle.

EXAMPLE 18-2

Industrial Robots does not manufacture its own motors or computer chips. Its premium product differs from its standard product in having heavier-duty motors and more computer chips for greater flexibility.

As a result, Industrial Robots manufactures a higher fraction of the standard product's value itself, and it purchases a higher fraction of the premium product's value. Use the following data to allocate $850,000 in overhead on the basis of labour cost and materials cost.

	Standard	Premium
Number of units per year	750	400
Labour cost (each)	$400	$500
Materials cost (each)	$550	$900

SOLUTION

First, the labour and material costs for the standard product, the premium product, and in total are calculated.

	Standard	Premium	Total
Number of units per year	750	400	
Labour cost (each)	$ 400	$ 500	
Materials cost (each)	550	900	
Labour cost	300,000	200,000	$500,000
Materials cost	412,500	360,000	772,500

Then the allocated cost per labour dollar, $1.70, is found by dividing the $850,000 in overhead by the $500,000 in total labour cost. The allocated cost per material dollar, $1.100324, is found by dividing the $850,000 in overhead by the $772,500 in materials cost. Now, the $850,000 in allocated overhead is split between the two products on the basis of labour costs and material costs.

	Standard	Premium	Total
Labour cost	$300,000	$200,000	$500,000
Overhead/labour	1.70	1.70	
Allocation by labour	510,000	340,000	850,000
Material cost	412,500	360,000	0
Overhead/material	1.100324	1.100324	
Allocation by material	453,884	396,117	850,000

If labour cost is the burden vehicle, then 60% of the $850,000 in overhead is allocated to the standard product. If material cost is the burden vehicle, then 53.4% is allocated to the standard product. In both cases, the $850,000 has been split between the two products. Using total direct costs would produce another overhead allocation between these two values. However, for decision making about product mix and product prices, incremental overhead costs must be analyzed. All the allocation or burden vehicles are based on an average cost of overhead per unit of burden vehicle.

Problems with Traditional Cost Accounting

Allocation of indirect costs can distort product costs and the decisions based on those costs. To be accurate, the analyst must determine which indirect or overhead expenses will be changed because of an engineering project. In other words, what are the incremental cash

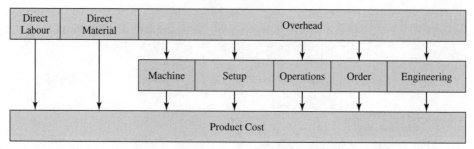

FIGURE 18-4 Activity-based costing versus traditional overhead allocation. (Based on an example by Kim LaScola Needy.)

flows? For example, vacation and sick leave accrual may be part of overhead, but will they change if the labour content is changed? The changes in costs incurred must be estimated. Loadings, or allocations, of overhead expenses cannot be used.

This issue has become very important because in some firms, automation has reduced direct-labour content to less than 5% of the product's cost. Yet in some of these firms, the basis for allocating overhead is still direct-labour hours or cost.

Other firms are shifting to activity-based costing (ABC),[1] where each activity is linked to specific cost drivers, and the number of dollars allocated as overhead is minimized. Figure 18-4 illustrates the difference between activity-based costing and traditional overhead allocations.[2]

Other Problems to Watch For

Centralized accounting systems have often been accused by project managers of being too slow or being 'untimely'. Because engineering economy is not concerned with the problem of daily project control, this is a less critical issue. However, if an organization establishes multiple files and systems so that project managers (and others) have the timely data they need, then the level of accuracy in one or all systems may be low. As a result, analysts making cost estimates will have to consider other internal data sources.

There are several cases in which data on equipment or inventory values may be questionable. When inventory is valued on a 'last in, first out' basis, the remaining inventory may be valued too low. Similarly, land valued at its acquisition cost is likely to be significantly undervalued. Finally, capital equipment may be valued at either a low or a high value, depending on allowable depreciation techniques and company policy.

---

[1]Liggett, Hampton R., Jaime Trevino, and Jerome P. Lavelle, 'Activity-Based Cost Management Systems, in an Advanced Manufacturing Environment'. In *Economic and Financial Justification of Advanced Manufacturing Technologies,* 1992, H. R. Parsaei (editor), Elsevier Science.

[2]Tippet, Donald D., and Peter Hoekstra, 'Activity-Based Costing: a Manufacturing Management Decision-Making Aid', *Engineering Management Journal,* Vol. 5, No. 2, June 1993, American Society for Engineering Management, pp. 37–42.

PROBLEMS

18-1 Why is it important for engineers and managers to understand accounting principles? Name a few ways that they can do so.

18-2 Explain the accounting function within a firm. What does this function do, and why is it important? What types of data does it provide?

18-3 Develop short definitions for the following terms: balance sheet, income statement, and fundamental accounting equation.

18-4 Explain the difference between short-term and long-term liabilities.

18-5 List the two primary general accounting statements. What is each used for and how do they differ? Which is most important?

18-6 What is the advantage of comparing financial statements across periods or against industry benchmarks rather than looking at statements associated with a single date or period?

18-7 If a firm has a current ratio less than 2.0 and an acid-test ratio less than 1.0, will the company eventually go bankrupt and out of business? Explain your answer.

18-8 Calculate the equity of the Gravel Construction Company if it has $1 million worth of assets. Gravel has $127,000 in current liabilities and $210,000 in long-term liabilities.

18-9 Scarmack's Paint Company has annual sales of $500,000 per year. If there is a profit of $1000 per day, 6 days per week operation, what is the total yearly business expense? All calculations are on a before-tax basis. (*Answer:* $188,000)

18-10 Mama L's Baby Monitor Company has current assets of $5 million and current liabilities of $2 million. Give the company's working capital and current ratio. (*Answer:* $3 million, 2.50)

18-11 Laila's Surveying Inc. had revenues of $100,000 in 2004. Expenses totalled $60,000. What was her net profit (or loss)?

18-12 From the following data, taken from the balance sheet of Petey's Widget Factory, determine the working capital, current ratio, and acid-test ratio.

Cash	$ 90,000
Net accounts and notes receivable	175,000
Retailers' inventories	210,000
Prepaid expenses	6,000
Accounts and notes payable (short term)	322,000
Accumulated liabilities	87,000

18-13 The general ledger of the Fly-Buy-Nite (FBN) Engineering Company contained the following account balances (partial listing) at the end of June, 2004. Construct an income statement. What is the net income before taxes and the net profit (or loss) after taxes? FBN has a tax rate of 27%.

	Amount (thousands)
Administrative expenses	$ 2,750
Subcontracted services	18,000
Development expenses	900
Interest expense	200
Sales revenue	30,000
Selling expenses	4,500

18-14 For Gee-Whiz Devices calculate the following: working capital, current ratio, and acid-test ratio.

Gee-Whiz Devices Balance Sheet Data

Cash	$100,000
Market securities	45,000
Net accounts and notes receivable	150,000
Retailers' inventories	200,000
Prepaid expenses	8,000
Accounts and notes payable (short term)	315,000
Accumulated liabilities to date	90,000

(*Answer:* $90,000; 1.22; 0.73)

18-15 For Evergreen Environmental Engineering (EEE), determine the working capital, current ratio, and acid-test ratio. Evaluate the company's economic situation with respect to its ability to pay off debt.

EEE Balance Sheet Data (thousands)

Cash	$110,000
Securities	40,000
Accounts receivable	160,000
Retailers' inventories	250,000
Prepaid expenses	3,000
Accounts payable	351,000
Accumulated liabilities	89,000

18-16 For Magdalen Industries, compute the net income before taxes and net profit (or loss). Taxes for the year were $1 million. Calculate the interest coverage and net profit ratio. Is the interest coverage acceptable? Explain why or why not.

Magdalen Industries Balance Sheet Data (millions)

Revenues	
Total operating revenue	$51
(including sales of $48 million)	
Total non-operating revenue	35
Expenses	
Total operating expenses	70
Total non-operating expenses	7
(interest payments)	

18-17 Find the net income of Turbo Start given the following data.

Turbo Start Balance Sheet Data (thousands)

Accounts payable	$ 1,000
Selling expense	5,000
Sales revenue	50,000
Owner's equity	4,500
Income taxes	2,000
Cost of goods sold	30,000
Accounts receivable	15,000

(*Answer:* $13,000,000)

18-18 Turbo Start (Problem 18-17) has current assets totaling $1.5 million (this includes $500,000 in current inventory) and current liabilities totaling $50,000. Find the current ratio and acid-test ratio. Are the ratios at desirable levels? Explain.

18-19 For J&W Graphics Supply, compute the current ratio. Is this a financially healthy company? Explain.

J&W Graphics Supply Balance Sheet Data (thousands)

Assets	
Cash	$1740
Inventories	900
Accounts receivable	2500
Bad debt provision	−75
Liabilities	
Notes payable	500
Accrued expenses	125
Accounts payable	1050

18-20 For Sutton Manufacturing, determine the current ratio and the acid-test ratio. Are these values acceptable? Why or why not?

Sutton Manufacturing Balance Sheet Data (thousands)

Assets		**Liabilities**	
Current assets		Current liabilities	
Cash	$ 870	Notes payable	$ 500
Accounts receivable	450	Accounts payable	600
Inventory	1200	Accruals	200
Prepaid expenses	50	Taxes payable	30
Other expenses	10	Current portion long-term debt	100
Total current assets	2670	Total current liabilities	1430
		Long-term debt	2000
		Officer debt (subordinated)	200
Net fixed assets		Total liabilities	3630
Land	1200		
Plant and equipment	2800		
		Common stock	1670
Notes receivable	200	Capital surplus	400
Intangibles	20	Retained earnings	1200
Other assets	100	Total net worth	3270
Total assets	6900	Total liabilities and net worth	6900

18-21 For Andrew's Electronic Instruments, calculate the interest coverage and net profit ratio. Is Andrew's business healthy?

Income Statement for Andrew's Electronics for End of Year 2004 (thousands)

Revenues

Operating revenues	
Sales	$395
(*minus*) Returns	−15
Total operating revenues	380
Non-operating revenues	
Interest receipts	50
Stock revenues	25
Total non-operating revenues	75
Total revenues, R	455

Expenses

Operating expenses	
Cost of goods and services sold	
Labour	200
Materials	34
Indirect cost	68
Selling and promotion	20
Depreciation	30
General and administrative	10
Lease payments	10
Total operating expenses	372
Non-operating expenses	
Interest payments	22
Total non-operating expenses	22
Total expenses, E	394
Net income before Taxes, R − E	61
Incomes taxes	30
Net profit (Loss) for the year 2004	31

18-22 LeGaroutte Industries makes industrial pipe manufacturing equipment. Use direct-labour hours as the burden vehicle, and compute the total cost per unit for each model given in the table. Total manufacturing indirect costs are $15,892,000, and there are 100,000 units manufactured per year for Model S, 50,000 for Model M, and 82,250 for Model G.

Item	Model S	Model M	Model G
Direct-material costs	$3,800,000	$1,530,000	$2,105,000
Direct-labour costs	600,000	380,000	420,000
Direct-labour hours	64,000	20,000	32,000

(*Answer:* $132; $93; $84)

18-23 Par Golf Equipment Company produces two types of golf bag: the standard and deluxe models. The total indirect cost to be allocated to the two bags is $35,000. Determine the net revenue that Par Golf can expect from the sale of each bag.

(*a*) Use direct-labour cost to allocate indirect costs.

(*b*) Use direct-materials cost to allocate indirect costs.

Data Item	Standard	Deluxe
Direct-labour cost	$50,000	$65,000
Direct-material cost	35,000	47,500
Selling price	60	95
Units produced	1800	1400

18-24 RLW-II Enterprises estimated that indirect manufacturing costs for the year would be $60 million and that 12,000 machine-hours would be used.

(*a*) Compute the predetermined indirect cost application rate using machine hours as the burden vehicle.

(*b*) Determine the total cost of production for a product with direct material costs of $1 million, direct-labour costs of $600,000, and 200 machine-hours.

(*Answers:* $5000, $2.6 million)

18-25 Categorize each of the following costs as direct or indirect. Assume that a traditional costing system is in place.

Machine run costs	Cost to market the product
Machine depreciation	Cost of storage
Material handling costs	Insurance costs
Cost of materials	Cost of product sales force
Overtime expenses	Engineering drawings
Machine operator wages	Machine labour
Utility costs	Cost of tooling and fixtures
Support (administrative) staff salaries	

Introduction to Spreadsheets

Computerized spreadsheets, which are available nearly everywhere, can be easily applied to economic analysis. In fact, spreadsheets were originally developed to analyze financial data, and they are often credited with initiating the explosive growth in demand for desktop computing.

A spreadsheet is a two-dimensional table whose cells can contain numerical values, labels, or formulas. The software automatically updates the table when an entry is changed, and it has powerful tools for copying formulas, creating graphs, and formatting results.

THE ELEMENTS OF A SPREADSHEET

A spreadsheet is a two-dimensional table that labels the columns in alphabetical order A to Z, AA to AZ, BA to BZ, up to IV (256 columns). The rows are numbered from 1 to 65,536. Thus a *cell* of the spreadsheet is specified by its column letter and row number. For example, A3 is the third row in column A and AA6 is the sixth row in the twenty-seventh column. Each cell can contain a label, a numerical value, or a formula.

A *label* is any cell where the contents should be treated as text. Arithmetic cannot be performed on labels. Labels are used for names of variables, row and column headings, and explanatory notes. In Excel any cell that contains more than a simple number, such as 3.14159, is treated as a label, unless it begins with an =, which is the signal for a formula. Thus 2*3 and B1+B2 are labels. Meaningful labels can be wider than a normal column. One solution is to allow those cells to 'wrap' text, which is one of the 'alignment' options. In the table heading row (row 8) in Example A-1, the user has turned this on by selecting row 8, right clicking on the row, and selecting wrap text under the alignment tab.

A *numerical value* is any number. Acceptable formats for entry or display include percentages, currency, accounting, scientific, fractions, date, and time. In addition the number of decimal digits, the display of $ symbols, and commas for 'thousands' separators can be adjusted. The format for cells can be changed by selecting a cell, a block of cells, a row, a column, or the entire spreadsheet. Then right-click on the selected area, and a menu that includes 'format cells' will appear. Then number formats, alignment, borders, fonts, and patterns can be selected.

Formulas must begin with an =, such as =3*4^2 or =B1+B2. They can include many functions—financial, statistical, trigonometric, etc. (and others can be defined by the user). The formula for the 'current' cell is displayed in the formula bar at the top of the spreadsheet. The value resulting from the formula is displayed in the cell in the spreadsheet.

Often the printed-out spreadsheet will be part of a report or a homework assignment and the formulas must be explained. Here is an easy way to place a copy of the formula in an adjacent or nearby cell. (1) Convert the cell with the formula to a label by inserting a space before the = sign. (2) Copy that label to an adjacent cell by using cut and paste. Do not drag the cell to copy it, because any formula ending with a number (even an address like B4) will have the number automatically incremented. (3) Convert the original formula back into a formula by deleting the space.

DEFINING VARIABLES IN A DATA BLOCK

The cell A1, top left corner, is the HOME cell for a spreadsheet. Thus, the top left area is where the data block should be placed. This data block should have every variable in the spreadsheet with an adjacent label for each. This data block supports a basic principle of good spreadsheet modelling, which is to use variables in your models.

The data block in Example A-1 contains *entered data*—the loan amount (A2), the number of payments (A3), and the interest rate (A4), and *computed data*—the payment (A6). Then instead of using the loan amount of $5000 in a formula, you use the cell reference A2. Even if a value is referred to only once, it is better to include it in the data block. By using one location to define each variable, you can change any value at one place in the spreadsheet and have the entire spreadsheet instantly recomputed.

For the following reasons, you should use a data block, even for simple homework problems.

1. You may be able to use it for another problem.
2. Solutions to simple problems may grow into solutions for complex problems.
3. Good habits, like using data blocks, are easy to maintain once they are established.
4. It makes the assumptions clear if you've estimated a value or for grading.

In the real world, data blocks are even more important. Most problems are solved more than once as more and more accurate values are estimated. Often the spreadsheet is revised to add other variables, time periods, locations, etc. Without data blocks, it is hard to change a spreadsheet and the likelihood of missing a required change skyrockets.

If you want your formulas to be easier to read, you can name your variables. Note: In current versions of Excel, the location or name of the cell is displayed at the left of the formula bar. Variable names can be entered here. They will then be applied automatically if cell addresses are entered by the point and click method. If cell addresses are entered as A2, then A2 is what is displayed. To change a displayed A2 to the name of the cell (LoanAmount), the process is to click on insert, click on name, click on apply, and then select the names to be applied.

COPY COMMAND

The copy command and relative/absolute addressing make spreadsheet models easy to build. If the range of cells to be copied contains only labels, numbers, and functions, then the copy command is easy to use and understand. For example, the formula =EXP(1.9) would be copied unchanged to a new location. However, cell addresses are usually part of the range being copied, and their absolute and relative addresses are treated differently.

An *absolute address* is denoted by adding $ signs before the column and/or row. For example in Figure A-1a, $A$4 is the absolute address for the interest rate. When an absolute address is copied, the column and/or row that is fixed is copied unchanged. Thus $A$4 is completely fixed, $A4 fixes the column, and A$4 fixes the row. One common use for absolute addresses is any data block entry, such as the interest rate. When entering or editing a formula, changing between A4, $A$4, A$4, $A4, and A4 is most easily done with the F4 key, which scrolls an address through the choices.

In contrast a *relative address* is best interpreted as directions from one cell to another. For example in Figure A-1a, the balance due in year t equals the balance due in year $t - 1$ minus the principal payment in year t. Specifically for the balance due in year 1, D10 contains =D9−C10. From cell D10, cell D9 is one row up and C10 is one column to the left, so the formula is really (contents of 1 up) minus (contents of 1 to the left). When a cell containing a relative address is copied to a new location, it is these directions that are copied to determine any new relative addresses. So if cell D10 is copied to cell F14, the formula is =F13−E14.

Thus to calculate a loan repayment schedule, as in Figure A-1, the row of formulas is created and then copied for the remaining years.

EXAMPLE A-1

Table 3-1 shows four repayment schedules for a loan of $5000 to be repaid over 5 years at an interest rate of 8%. Use a spreadsheet to calculate the amortization schedule for the constant principal payment option.

SOLUTION

The first step is to enter the loan amount, number of periods, and interest rate into a data block in the top left part of the spreadsheet. The next step is to calculate the constant principal payment amount, which was given as $1252.28 in Table 3-1. The factor approach to finding this value is given in Chapter 3, and the spreadsheet function is explained in Chapter 4.

The next step is to identify the columns for the amortization schedule. These are the year, interest owed, principal payment, and balance due. Because some of these labels are wider than a normal column, the cells are formatted so that the text wraps (row height increases automatically). The initial balance is shown in the year 0 row.

Next, the formulas for the first year are written, as shown in Figure A-1a. The interest owed (cell B10) equals the interest rate ($A$4) times the balance due for year 0 (D9). The principal payment (cell C10) equals the annual payment ($A$6) minus the interest owed and paid (B10). Finally, the balance due (cell D10) equals the balance due for the previous year (D9) minus the principal payment (C10). The results are shown in Figure A-1a.

Now cells A10 to D10 are selected for year 1. By dragging down on the right corner of D10, you can copy the entire row for years 2 to 5. Note that if cut and paste is used, it is necessary to complete the year column separately (dragging increments the year, but cutting and pasting does not). The results are shown in Figure A-1b.

	A	B	C	D	E
1	Entered Data				
2	5000	Loan Amount			
3	5	Number of Payments			
4	8%	Interest Rate			
5	Computer Data				
6	$1,252.28	Loan Payment			
7					
8	Year	Interest Owed	Principal Payment	Balance Due	
9	0			5000.00	
10	1	400.00	852.28	4147.72	=D9−C10
11					
12		=$A$4*D9		=$A$6−B10	

(a)

	A	B	C	D	E
8	Year	Interest Owed	Principal Payment	Balance Due	
9	0			5000.00	
10	1	400.00	852.28	4147.72	
11	2	331.82	920.46	3227.25	
12	3	258.18	994.10	2233.15	
13	4	178.65	1073.63	1159.52	
14	5	92.76	1159.52	0.00	=D13−C14
15					
16		=$A$4*D13		=$A$6−B14	

(b)

FIGURE A-1 (a) Year 1 amortization schedule. (b) Completed amortization schedule.

This appendix has introduced the basics of spreadsheets. Chapter 2 uses spreadsheets and simple bar charts to draw cash flow diagrams. Chapters 4–15 each have spreadsheet sections. These are designed to develop spreadsheet modelling skills and to reinforce your understanding of engineering economy. As current spreadsheet packages are built around the use of mice to click on cells and items in charts, there is usually a logical connection between what you would like to do and how to do it. It seems the best way to learn how to use the spreadsheet package is simply to play around with it. In addition, as you look at the menu choices, you will find new commands that you hadn't thought of but that you will find useful.

Compound Interest Tables

Values of Interest Factors When *N* Equals Infinity

Single Payment:

$(F/P, i, \infty) = \infty$

$(P/F, i, \infty) = 0$

Arithmetic Gradient Series:

$(A/G, i, \infty) = 1/i$

$(P/G, i, \infty) = 1/i^2$

Uniform Payment Series:

$(A/F, i, \infty) = 0$

$(A/P, i, \infty) = i$

$(F/A, i, \infty) = \infty$

$(P/A, i, \infty) = 1/i$

$1/4\%$ — Compound Interest Factors — $1/4\%$

	Single Payment		Uniform Payment Series				Arithmetic Gradient		
	Compound Amount Factor	Present Worth Factor	Sinking Fund Factor	Capital Recovery Factor	Compound Amount Factor	Present Worth Factor	Gradient Uniform Series	Gradient Present Worth	
	Find F Given P	Find P Given F	Find A Given F	Find A Given P	Find F Given A	Find P Given A	Find A Given G	Find P Given G	
n	F/P	P/F	A/F	A/P	F/A	P/A	A/G	P/G	n
1	1.003	.9975	1.0000	1.0025	1.000	0.998	0	0	1
2	1.005	.9950	.4994	.5019	2.003	1.993	0.504	1.005	2
3	1.008	.9925	.3325	.3350	3.008	2.985	1.005	2.999	3
4	1.010	.9901	.2491	.2516	4.015	3.975	1.501	5.966	4
5	1.013	.9876	.1990	.2015	5.025	4.963	1.998	9.916	5
6	1.015	.9851	.1656	.1681	6.038	5.948	2.498	14.861	6
7	1.018	.9827	.1418	.1443	7.053	6.931	2.995	20.755	7
8	1.020	.9802	.1239	.1264	8.070	7.911	3.490	27.611	8
9	1.023	.9778	.1100	.1125	9.091	8.889	3.987	35.440	9
10	1.025	.9753	.0989	.1014	10.113	9.864	4.483	44.216	10
11	1.028	.9729	.0898	.0923	11.139	10.837	4.978	53.950	11
12	1.030	.9705	.0822	.0847	12.167	11.807	5.474	64.634	12
13	1.033	.9681	.0758	.0783	13.197	12.775	5.968	76.244	13
14	1.036	.9656	.0703	.0728	14.230	13.741	6.464	88.826	14
15	1.038	.9632	.0655	.0680	15.266	14.704	6.957	102.301	15
16	1.041	.9608	.0613	.0638	16.304	15.665	7.451	116.716	16
17	1.043	.9584	.0577	.0602	17.344	16.624	7.944	132.063	17
18	1.046	.9561	.0544	.0569	18.388	17.580	8.437	148.319	18
19	1.049	.9537	.0515	.0540	19.434	18.533	8.929	165.492	19
20	1.051	.9513	.0488	.0513	20.482	19.485	9.421	183.559	20
21	1.054	.9489	.0464	.0489	21.534	20.434	9.912	202.531	21
22	1.056	.9465	.0443	.0468	22.587	21.380	10.404	222.435	22
23	1.059	.9442	.0423	.0448	23.644	22.324	10.894	243.212	23
24	1.062	.9418	.0405	.0430	24.703	23.266	11.384	264.854	24
25	1.064	.9395	.0388	.0413	25.765	24.206	11.874	287.407	25
26	1.067	.9371	.0373	.0398	26.829	25.143	12.363	310.848	26
27	1.070	.9348	.0358	.0383	27.896	26.078	12.852	335.150	27
28	1.072	.9325	.0345	.0370	28.966	27.010	13.341	360.343	28
29	1.075	.9301	.0333	.0358	30.038	27.940	13.828	386.366	29
30	1.078	.9278	.0321	.0346	31.114	28.868	14.317	413.302	30
36	1.094	.9140	.0266	.0291	37.621	34.387	17.234	592.632	36
40	1.105	.9049	.0238	.0263	42.014	38.020	19.171	728.882	40
48	1.127	.8871	.0196	.0221	50.932	45.179	23.025	1 040.22	48
50	1.133	.8826	.0188	.0213	53.189	46.947	23.984	1 125.96	50
52	1.139	.8782	.0180	.0205	55.458	48.705	24.941	1 214.76	52
60	1.162	.8609	.0155	.0180	64.647	55.653	28.755	1 600.31	60
70	1.191	.8396	.0131	.0156	76.395	64.144	33.485	2 147.87	70
72	1.197	.8355	.0127	.0152	78.780	65.817	34.426	2 265.81	72
80	1.221	.8189	.0113	.0138	88.440	72.427	38.173	2 764.74	80
84	1.233	.8108	.0107	.0132	93.343	75.682	40.037	3 030.06	84
90	1.252	.7987	.00992	.0124	100.789	80.504	42.820	3 447.19	90
96	1.271	.7869	.00923	.0117	108.349	85.255	45.588	3 886.62	96
100	1.284	.7790	.00881	.0113	113.451	88.383	47.425	4 191.60	100
104	1.297	.7713	.00843	.0109	118.605	91.480	49.256	4 505.93	104
120	1.349	.7411	.00716	.00966	139.743	103.563	56.512	5 852.52	120
240	1.821	.5492	.00305	.00555	328.306	180.312	107.590	19 399.75	240
360	2.457	.4070	.00172	.00422	582.745	237.191	152.894	36 264.96	360
480	3.315	.3016	.00108	.00358	926.074	279.343	192.673	53 821.93	480

$1/_2$%   Compound Interest Factors $1/_2$%

	Single Payment		Uniform Payment Series				Arithmetic Gradient		
	Compound Amount Factor	Present Worth Factor	Sinking Fund Factor	Capital Recovery Factor	Compound Amount Factor	Present Worth Factor	Gradient Uniform Series	Gradient Present Worth	
n	Find F Given P F/P	Find P Given F P/F	Find A Given F A/F	Find A Given P A/P	Find F Given A F/A	Find P Given A P/A	Find A Given G A/G	Find P Given G P/G	n
1	1.005	.9950	1.0000	1.0050	1.000	0.995	0	0	1
2	1.010	.9901	.4988	.5038	2.005	1.985	0.499	0.991	2
3	1.015	.9851	.3317	.3367	3.015	2.970	0.996	2.959	3
4	1.020	.9802	.2481	.2531	4.030	3.951	1.494	5.903	4
5	1.025	.9754	.1980	.2030	5.050	4.926	1.990	9.803	5
6	1.030	.9705	.1646	.1696	6.076	5.896	2.486	14.660	6
7	1.036	.9657	.1407	.1457	7.106	6.862	2.980	20.448	7
8	1.041	.9609	.1228	.1278	8.141	7.823	3.474	27.178	8
9	1.046	.9561	.1089	.1139	9.182	8.779	3.967	34.825	9
10	1.051	.9513	.0978	.1028	10.228	9.730	4.459	43.389	10
11	1.056	.9466	.0887	.0937	11.279	10.677	4.950	52.855	11
12	1.062	.9419	.0811	.0861	12.336	11.619	5.441	63.218	12
13	1.067	.9372	.0746	.0796	13.397	12.556	5.931	74.465	13
14	1.072	.9326	.0691	.0741	14.464	13.489	6.419	86.590	14
15	1.078	.9279	.0644	.0694	15.537	14.417	6.907	99.574	15
16	1.083	.9233	.0602	.0652	16.614	15.340	7.394	113.427	16
17	1.088	.9187	.0565	.0615	17.697	16.259	7.880	128.125	17
18	1.094	.9141	.0532	.0582	18.786	17.173	8.366	143.668	18
19	1.099	.9096	.0503	.0553	19.880	18.082	8.850	160.037	19
20	1.105	.9051	.0477	.0527	20.979	18.987	9.334	177.237	20
21	1.110	.9006	.0453	.0503	22.084	19.888	9.817	195.245	21
22	1.116	.8961	.0431	.0481	23.194	20.784	10.300	214.070	22
23	1.122	.8916	.0411	.0461	24.310	21.676	10.781	233.680	23
24	1.127	.8872	.0393	.0443	25.432	22.563	11.261	254.088	24
25	1.133	.8828	.0377	.0427	26.559	23.446	11.741	275.273	25
26	1.138	.8784	.0361	.0411	27.692	24.324	12.220	297.233	26
27	1.144	.8740	.0347	.0397	28.830	25.198	12.698	319.955	27
28	1.150	.8697	.0334	.0384	29.975	26.068	13.175	343.439	28
29	1.156	.8653	.0321	.0371	31.124	26.933	13.651	367.672	29
30	1.161	.8610	.0310	.0360	32.280	27.794	14.127	392.640	30
36	1.197	.8356	.0254	.0304	39.336	32.871	16.962	557.564	36
40	1.221	.8191	.0226	.0276	44.159	36.172	18.836	681.341	40
48	1.270	.7871	.0185	.0235	54.098	42.580	22.544	959.928	48
50	1.283	.7793	.0177	.0227	56.645	44.143	23.463	1 035.70	50
52	1.296	.7716	.0169	.0219	59.218	45.690	24.378	1 113.82	52
60	1.349	.7414	.0143	.0193	69.770	51.726	28.007	1 448.65	60
70	1.418	.7053	.0120	.0170	83.566	58.939	32.468	1 913.65	70
72	1.432	.6983	.0116	.0166	86.409	60.340	33.351	2 012.35	72
80	1.490	.6710	.0102	.0152	98.068	65.802	36.848	2 424.65	80
84	1.520	.6577	.00961	.0146	104.074	68.453	38.576	2 640.67	84
90	1.567	.6383	.00883	.0138	113.311	72.331	41.145	2 976.08	90
96	1.614	.6195	.00814	.0131	122.829	76.095	43.685	3 324.19	96
100	1.647	.6073	.00773	.0127	129.334	78.543	45.361	3 562.80	100
104	1.680	.5953	.00735	.0124	135.970	80.942	47.025	3 806.29	104
120	1.819	.5496	.00610	.0111	163.880	90.074	53.551	4 823.52	120
240	3.310	.3021	.00216	.00716	462.041	139.581	96.113	13 415.56	240
360	6.023	.1660	.00100	.00600	1 004.5	166.792	128.324	21 403.32	360
480	10.957	.0913	.00050	.00550	1 991.5	181.748	151.795	27 588.37	480

$3/4\%$ **Compound Interest Factors** $3/4\%$

	Single Payment		Uniform Payment Series				Arithmetic Gradient		
	Compound Amount Factor	Present Worth Factor	Sinking Fund Factor	Capital Recovery Factor	Compound Amount Factor	Present Worth Factor	Gradient Uniform Series	Gradient Present Worth	
	Find F Given P	Find P Given F	Find A Given F	Find A Given P	Find F Given A	Find P Given A	Find A Given G	Find P Given G	
n	F/P	P/F	A/F	A/P	F/A	P/A	A/G	P/G	n
1	1.008	.9926	1.0000	1.0075	1.000	0.993	0	0	1
2	1.015	.9852	.4981	.5056	2.008	1.978	0.499	0.987	2
3	1.023	.9778	.3308	.3383	3.023	2.956	0.996	2.943	3
4	1.030	.9706	.2472	.2547	4.045	3.926	1.492	5.857	4
5	1.038	.9633	.1970	.2045	5.076	4.889	1.986	9.712	5
6	1.046	.9562	.1636	.1711	6.114	5.846	2.479	14.494	6
7	1.054	.9490	.1397	.1472	7.160	6.795	2.971	20.187	7
8	1.062	.9420	.1218	.1293	8.213	7.737	3.462	26.785	8
9	1.070	.9350	.1078	.1153	9.275	8.672	3.951	34.265	9
10	1.078	.9280	.0967	.1042	10.344	9.600	4.440	42.619	10
11	1.086	.9211	.0876	.0951	11.422	10.521	4.927	51.831	11
12	1.094	.9142	.0800	.0875	12.508	11.435	5.412	61.889	12
13	1.102	.9074	.0735	.0810	13.602	12.342	5.897	72.779	13
14	1.110	.9007	.0680	.0755	14.704	13.243	6.380	84.491	14
15	1.119	.8940	.0632	.0707	15.814	14.137	6.862	97.005	15
16	1.127	.8873	.0591	.0666	16.932	15.024	7.343	110.318	16
17	1.135	.8807	.0554	.0629	18.059	15.905	7.822	124.410	17
18	1.144	.8742	.0521	.0596	19.195	16.779	8.300	139.273	18
19	1.153	.8676	.0492	.0567	20.339	17.647	8.777	154.891	19
20	1.161	.8612	.0465	.0540	21.491	18.508	9.253	171.254	20
21	1.170	.8548	.0441	.0516	22.653	19.363	9.727	188.352	21
22	1.179	.8484	.0420	.0495	23.823	20.211	10.201	206.170	22
23	1.188	.8421	.0400	.0475	25.001	21.053	10.673	224.695	23
24	1.196	.8358	.0382	.0457	26.189	21.889	11.143	243.924	24
25	1.205	.8296	.0365	.0440	27.385	22.719	11.613	263.834	25
26	1.214	.8234	.0350	.0425	28.591	23.542	12.081	284.421	26
27	1.224	.8173	.0336	.0411	29.805	24.360	12.548	305.672	27
28	1.233	.8112	.0322	.0397	31.029	25.171	13.014	327.576	28
29	1.242	.8052	.0310	.0385	32.261	25.976	13.479	350.122	29
30	1.251	.7992	.0298	.0373	33.503	26.775	13.942	373.302	30
36	1.309	.7641	.0243	.0318	41.153	31.447	16.696	525.038	36
40	1.348	.7416	.0215	.0290	46.447	34.447	18.507	637.519	40
48	1.431	.6986	.0174	.0249	57.521	40.185	22.070	886.899	48
50	1.453	.6882	.0166	.0241	60.395	41.567	22.949	953.911	50
52	1.475	.6780	.0158	.0233	63.312	42.928	23.822	1 022.64	52
60	1.566	.6387	.0133	.0208	75.425	48.174	27.268	1 313.59	60
70	1.687	.5927	.0109	.0184	91.621	54.305	31.465	1 708.68	70
72	1.713	.5839	.0105	.0180	95.008	55.477	32.289	1 791.33	72
80	1.818	.5500	.00917	.0167	109.074	59.995	35.540	2 132.23	80
84	1.873	.5338	.00859	.0161	116.428	62.154	37.137	2 308.22	84
90	1.959	.5104	.00782	.0153	127.881	65.275	39.496	2 578.09	90
96	2.049	.4881	.00715	.0147	139.858	68.259	41.812	2 854.04	96
100	2.111	.4737	.00675	.0143	148.147	70.175	43.332	3 040.85	100
104	2.175	.4597	.00638	.0139	156.687	72.035	44.834	3 229.60	104
120	2.451	.4079	.00517	.0127	193.517	78.942	50.653	3 998.68	120
240	6.009	.1664	.00150	.00900	667.901	111.145	85.422	9 494.26	240
360	14.731	.0679	.00055	.00805	1 830.8	124.282	107.115	13 312.50	360
480	36.111	.0277	.00021	.00771	4 681.5	129.641	119.662	15 513.16	480

1%

Compound Interest Factors

1%

	Single Payment		Uniform Payment Series				Arithmetic Gradient		
	Compound Amount Factor	Present Worth Factor	Sinking Fund Factor	Capital Recovery Factor	Compound Amount Factor	Present Worth Factor	Gradient Uniform Series	Gradient Present Worth	
	Find F Given P F/P	Find P Given F P/F	Find A Given F A/F	Find A Given P A/P	Find F Given A F/A	Find P Given A P/A	Find A Given G A/G	Find P Given G P/G	
n									n
1	1.010	.9901	1.0000	1.0100	1.000	0.990	0	0	1
2	1.020	.9803	.4975	.5075	2.010	1.970	0.498	0.980	2
3	1.030	.9706	.3300	.3400	3.030	2.941	0.993	2.921	3
4	1.041	.9610	.2463	.2563	4.060	3.902	1.488	5.804	4
5	1.051	.9515	.1960	.2060	5.101	4.853	1.980	9.610	5
6	1.062	.9420	.1625	.1725	6.152	5.795	2.471	14.320	6
7	1.072	.9327	.1386	.1486	7.214	6.728	2.960	19.917	7
8	1.083	.9235	.1207	.1307	8.286	7.652	3.448	26.381	8
9	1.094	.9143	.1067	.1167	9.369	8.566	3.934	33.695	9
10	1.105	.9053	.0956	.1056	10.462	9.471	4.418	41.843	10
11	1.116	.8963	.0865	.0965	11.567	10.368	4.900	50.806	11
12	1.127	.8874	.0788	.0888	12.682	11.255	5.381	60.568	12
13	1.138	.8787	.0724	.0824	13.809	12.134	5.861	71.112	13
14	1.149	.8700	.0669	.0769	14.947	13.004	6.338	82.422	14
15	1.161	.8613	.0621	.0721	16.097	13.865	6.814	94.481	15
16	1.173	.8528	.0579	.0679	17.258	14.718	7.289	107.273	16
17	1.184	.8444	.0543	.0643	18.430	15.562	7.761	120.783	17
18	1.196	.8360	.0510	.0610	19.615	16.398	8.232	134.995	18
19	1.208	.8277	.0481	.0581	20.811	17.226	8.702	149.895	19
20	1.220	.8195	.0454	.0554	22.019	18.046	9.169	165.465	20
21	1.232	.8114	.0430	.0530	23.239	18.857	9.635	181.694	21
22	1.245	.8034	.0409	.0509	24.472	19.660	10.100	198.565	22
23	1.257	.7954	.0389	.0489	25.716	20.456	10.563	216.065	23
24	1.270	.7876	.0371	.0471	26.973	21.243	11.024	234.179	24
25	1.282	.7798	.0354	.0454	28.243	22.023	11.483	252.892	25
26	1.295	.7720	.0339	.0439	29.526	22.795	11.941	272.195	26
27	1.308	.7644	.0324	.0424	30.821	23.560	12.397	292.069	27
28	1.321	.7568	.0311	.0411	32.129	24.316	12.852	312.504	28
29	1.335	.7493	.0299	.0399	33.450	25.066	13.304	333.486	29
30	1.348	.7419	.0287	.0387	34.785	25.808	13.756	355.001	30
36	1.431	.6989	.0232	.0332	43.077	30.107	16.428	494.620	36
40	1.489	.6717	.0205	.0305	48.886	32.835	18.178	596.854	40
48	1.612	.6203	.0163	.0263	61.223	37.974	21.598	820.144	48
50	1.645	.6080	.0155	.0255	64.463	39.196	22.436	879.417	50
52	1.678	.5961	.0148	.0248	67.769	40.394	23.269	939.916	52
60	1.817	.5504	.0122	.0222	81.670	44.955	26.533	1 192.80	60
70	2.007	.4983	.00993	.0199	100.676	50.168	30.470	1 528.64	70
72	2.047	.4885	.00955	.0196	104.710	51.150	31.239	1 597.86	72
80	2.217	.4511	.00822	.0182	121.671	54.888	34.249	1 879.87	80
84	2.307	.4335	.00765	.0177	130.672	56.648	35.717	2 023.31	84
90	2.449	.4084	.00690	.0169	144.863	59.161	37.872	2 240.56	90
96	2.599	.3847	.00625	.0163	159.927	61.528	39.973	2 459.42	96
100	2.705	.3697	.00587	.0159	170.481	63.029	41.343	2 605.77	100
104	2.815	.3553	.00551	.0155	181.464	64.471	42.688	2 752.17	104
120	3.300	.3030	.00435	.0143	230.039	69.701	47.835	3 334.11	120
240	10.893	.0918	.00101	.0110	989.254	90.819	75.739	6 878.59	240
360	35.950	.0278	.00029	.0103	3 495.0	97.218	89.699	8 720.43	360
480	118.648	.00843	.00008	.0101	11 764.8	99.157	95.920	9 511.15	480

$1\frac{1}{4}\%$ Compound Interest Factors $1\frac{1}{4}\%$

	Single Payment		Uniform Payment Series				Arithmetic Gradient		
	Compound Amount Factor	Present Worth Factor	Sinking Fund Factor	Capital Recovery Factor	Compound Amount Factor	Present Worth Factor	Gradient Uniform Series	Gradient Present Worth	
	Find F Given P F/P	Find P Given F P/F	Find A Given F A/F	Find A Given P A/P	Find F Given A F/A	Find P Given A P/A	Find A Given G A/G	Find P Given G P/G	
n									n
1	1.013	.9877	1.0000	1.0125	1.000	0.988	0	0	1
2	1.025	.9755	.4969	.5094	2.013	1.963	0.497	0.976	2
3	1.038	.9634	.3292	.3417	3.038	2.927	0.992	2.904	3
4	1.051	.9515	.2454	.2579	4.076	3.878	1.485	5.759	4
5	1.064	.9398	.1951	.2076	5.127	4.818	1.976	9.518	5
6	1.077	.9282	.1615	.1740	6.191	5.746	2.464	14.160	6
7	1.091	.9167	.1376	.1501	7.268	6.663	2.951	19.660	7
8	1.104	.9054	.1196	.1321	8.359	7.568	3.435	25.998	8
9	1.118	.8942	.1057	.1182	9.463	8.462	3.918	33.152	9
10	1.132	.8832	.0945	.1070	10.582	9.346	4.398	41.101	10
11	1.146	.8723	.0854	.0979	11.714	10.218	4.876	49.825	11
12	1.161	.8615	.0778	.0903	12.860	11.079	5.352	59.302	12
13	1.175	.8509	.0713	.0838	14.021	11.930	5.827	69.513	13
14	1.190	.8404	.0658	.0783	15.196	12.771	6.299	80.438	14
15	1.205	.8300	.0610	.0735	16.386	13.601	6.769	92.058	15
16	1.220	.8197	.0568	.0693	17.591	14.420	7.237	104.355	16
17	1.235	.8096	.0532	.0657	18.811	15.230	7.702	117.309	17
18	1.251	.7996	.0499	.0624	20.046	16.030	8.166	130.903	18
19	1.266	.7898	.0470	.0595	21.297	16.849	8.628	145.119	19
20	1.282	.7800	.0443	.0568	22.563	17.599	9.088	159.940	20
21	1.298	.7704	.0419	.0544	23.845	18.370	9.545	175.348	21
22	1.314	.7609	.0398	.0523	25.143	19.131	10.001	191.327	22
23	1.331	.7515	.0378	.0503	26.458	19.882	10.455	207.859	23
24	1.347	.7422	.0360	.0485	27.788	20.624	10.906	224.930	24
25	1.364	.7330	.0343	.0468	29.136	21.357	11.355	242.523	25
26	1.381	.7240	.0328	.0453	30.500	22.081	11.803	260.623	26
27	1.399	.7150	.0314	.0439	31.881	22.796	12.248	279.215	27
28	1.416	.7062	.0300	.0425	32.280	23.503	12.691	298.284	28
29	1.434	.6975	.0288	.0413	34.696	24.200	13.133	317.814	29
30	1.452	.6889	.0277	.0402	36.129	24.889	13.572	337.792	30
36	1.564	.6394	.0222	.0347	45.116	28.847	16.164	466.297	36
40	1.644	.6084	.0194	.0319	51.490	31.327	17.852	559.247	40
48	1.845	.5509	.0153	.0278	65.229	35.932	21.130	759.248	48
50	1.861	.5373	.0145	.0270	68.882	37.013	21.930	811.692	50
52	1.908	.5242	.0138	.0263	72.628	38.068	22.722	864.960	52
60	2.107	.4746	.0113	.0238	88.575	42.035	25.809	1 084.86	60
70	2.386	.4191	.00902	.0215	110.873	46.470	29.492	1 370.47	70
72	2.446	.4088	.00864	.0211	115.675	47.293	30.205	1 428.48	72
80	2.701	.3702	.00735	.0198	136.120	50.387	32.983	1 661.89	80
84	2.839	.3522	.00680	.0193	147.130	51.822	34.326	1 778.86	84
90	3.059	.3269	.00607	.0186	164.706	53.846	36.286	1 953.85	90
96	3.296	.3034	.00545	.0179	183.643	55.725	38.180	2 127.55	96
100	3.463	.2887	.00507	.0176	197.074	56.901	39.406	2 242.26	100
104	3.640	.2747	.00474	.0172	211.190	58.021	40.604	2 355.90	104
120	4.440	.2252	.00363	.0161	275.220	61.983	45.119	2 796.59	120
240	19.716	.0507	.00067	.0132	1 497.3	75.942	67.177	5 101.55	240
360	87.543	.0114	.00014	.0126	6 923.4	79.086	75.840	5 997.91	360
480	388.713	.00257	.00003	.0125	31 017.1	79.794	78.762	6 284.74	480

$1^1/_2\%$

Compound Interest Factors

$1^1/_2\%$

	Single Payment		Uniform Payment Series				Arithmetic Gradient		
	Compound Amount Factor	Present Worth Factor	Sinking Fund Factor	Capital Recovery Factor	Compound Amount Factor	Present Worth Factor	Gradient Uniform Series	Gradient Present Worth	
	Find F Given P F/P	Find P Given F P/F	Find A Given F A/F	Find A Given P A/P	Find F Given A F/A	Find P Given A P/A	Find A Given G A/G	Find P Given G P/G	
n									n
1	1.015	.9852	1.0000	1.0150	1.000	0.985	0	0	1
2	1.030	.9707	.4963	.5113	2.015	1.956	0.496	0.970	2
3	1.046	.9563	.3284	.3434	3.045	2.912	0.990	2.883	3
4	1.061	.9422	.2444	.2594	4.091	3.854	1.481	5.709	4
5	1.077	.9283	.1941	.2091	5.152	4.783	1.970	9.422	5
6	1.093	.9145	.1605	.1755	6.230	5.697	2.456	13.994	6
7	1.110	.9010	.1366	.1516	7.323	6.598	2.940	19.400	7
8	1.126	.8877	.1186	.1336	8.433	7.486	3.422	25.614	8
9	1.143	.8746	.1046	.1196	9.559	8.360	3.901	32.610	9
10	1.161	.8617	.0934	.1084	10.703	9.222	4.377	40.365	10
11	1.178	.8489	.0843	.0993	11.863	10.071	4.851	48.855	11
12	1.196	.8364	.0767	.0917	13.041	10.907	5.322	58.054	12
13	1.214	.8240	.0702	.0852	14.237	11.731	5.791	67.943	13
14	1.232	.8118	.0647	.0797	15.450	12.543	6.258	78.496	14
15	1.250	.7999	.0599	.0749	16.682	13.343	6.722	89.694	15
16	1.269	.7880	.0558	.0708	17.932	14.131	7.184	101.514	16
17	1.288	.7764	.0521	.0671	19.201	14.908	7.643	113.937	17
18	1.307	.7649	.0488	.0638	20.489	15.673	8.100	126.940	18
19	1.327	.7536	.0459	.0609	21.797	16.426	8.554	140.505	19
20	1.347	.7425	.0432	.0582	23.124	17.169	9.005	154.611	20
21	1.367	.7315	.0409	.0559	24.470	17.900	9.455	169.241	21
22	1.388	.7207	.0387	.0537	25.837	18.621	9.902	184.375	22
23	1.408	.7100	.0367	.0517	27.225	19.331	10.346	199.996	23
24	1.430	.6995	.0349	.0499	28.633	20.030	10.788	216.085	24
25	1.451	.6892	.0333	.0483	30.063	20.720	11.227	232.626	25
26	1.473	.6790	.0317	.0467	31.514	21.399	11.664	249.601	26
27	1.495	.6690	.0303	.0453	32.987	22.068	12.099	266.995	27
28	1.517	.6591	.0290	.0440	34.481	22.727	12.531	284.790	28
29	1.540	.6494	.0278	.0428	35.999	23.376	12.961	302.972	29
30	1.563	.6398	.0266	.0416	37.539	24.016	13.388	321.525	30
36	1.709	.5851	.0212	.0362	47.276	27.661	15.901	439.823	36
40	1.814	.5513	.0184	.0334	54.268	29.916	17.528	524.349	40
48	2.043	.4894	.0144	.0294	69.565	34.042	20.666	703.537	48
50	2.105	.4750	.0136	.0286	73.682	35.000	21.428	749.955	50
52	2.169	.4611	.0128	.0278	77.925	35.929	22.179	796.868	52
60	2.443	.4093	.0104	.0254	96.214	39.380	25.093	988.157	60
70	2.835	.3527	.00817	.0232	122.363	43.155	28.529	1 231.15	70
72	2.921	.3423	.00781	.0228	128.076	43.845	29.189	1 279.78	72
80	3.291	.3039	.00655	.0215	152.710	46.407	31.742	1 473.06	80
84	3.493	.2863	.00602	.0210	166.172	47.579	32.967	1 568.50	84
90	3.819	.2619	.00532	.0203	187.929	49.210	34.740	1 709.53	90
96	4.176	.2395	.00472	.0197	211.719	50.702	36.438	1 847.46	96
100	4.432	.2256	.00437	.0194	228.802	51.625	37.529	1 937.43	100
104	4.704	.2126	.00405	.0190	246.932	52.494	38.589	2 025.69	104
120	5.969	.1675	.00302	.0180	331.286	55.498	42.518	2 359.69	120
240	35.632	.0281	.00043	.0154	2 308.8	64.796	59.737	3 870.68	240
360	212.700	.00470	.00007	.0151	14 113.3	66.353	64.966	4 310.71	360
480	1 269.7	.00079	.00001	.0150	84 577.8	66.614	66.288	4 415.74	480

$1^3/_4\%$ Compound Interest Factors $1^3/_4\%$

	Single Payment		Uniform Payment Series				Arithmetic Gradient		
	Compound Amount Factor	Present Worth Factor	Sinking Fund Factor	Capital Recovery Factor	Compound Amount Factor	Present Worth Factor	Gradient Uniform Series	Gradient Present Worth	
n	Find F Given P F/P	Find P Given F P/F	Find A Given F A/F	Find A Given P A/P	Find F Given A F/A	Find P Given A P/A	Find A Given G A/G	Find P Given G P/G	n
1	1.018	.9828	1.0000	1.0175	1.000	0.983	0	0	1
2	1.035	.9659	.4957	.5132	2.018	1.949	0.496	0.966	2
3	1.053	.9493	.3276	.3451	3.053	2.898	0.989	2.865	3
4	1.072	.9330	.2435	.2610	4.106	3.831	1.478	5.664	4
5	1.091	.9169	.1931	.2106	5.178	4.748	1.965	9.332	5
6	1.110	.9011	.1595	.1770	6.269	5.649	2.450	13.837	6
7	1.129	.8856	.1355	.1530	7.378	6.535	2.931	19.152	7
8	1.149	.8704	.1175	.1350	8.508	7.405	3.409	25.245	8
9	1.169	.8554	.1036	.1211	9.656	8.261	3.885	32.088	9
10	1.189	.8407	.0924	.1099	10.825	9.101	4.357	39.655	10
11	1.210	.8263	.0832	.1007	12.015`	9.928	4.827	47.918	11
12	1.231	.8121	.0756	.0931	13.225	10.740	5.294	56.851	12
13	1.253	.7981	.0692	.0867	14.457	11.538	5.758	66.428	13
14	1.275	.7844	.0637	.0812	15.710	12.322	6.219	76.625	14
15	1.297	.7709	.0589	.0764	16.985	13.093	6.677	87.417	15
16	1.320	.7576	.0547	.0722	18.282	13.851	7.132	98.782	16
17	1.343	.7446	.0510	.0685	19.602	14.595	7.584	110.695	17
18	1.367	.7318	.0477	.0652	20.945	15.327	8.034	123.136	18
19	1.390	.7192	.0448	.0623	22.311	16.046	8.481	136.081	19
20	1.415	.7068	.0422	.0597	23.702	16.753	8.924	149.511	20
21	1.440	.6947	.0398	.0573	25.116	17.448	9.365	163.405	21
22	1.465	.6827	.0377	.0552	26.556	18.130	9.804	177.742	22
23	1.490	.6710	.0357	.0532	28.021	18.801	10.239	192.503	23
24	1.516	.6594	.0339	.0514	29.511	19.461	10.671	207.671	24
25	1.543	.6481	.0322	.0497	31.028	20.109	11.101	223.225	25
26	1.570	.6369	.0307	.0482	32.571	20.746	11.528	239.149	26
27	1.597	.6260	.0293	.0468	34.141	21.372	11.952	255.425	27
28	1.625	.6152	.0280	.0455	35.738	21.987	12.373	272.036	28
29	1.654	.6046	.0268	.0443	37.363	22.592	12.791	288.967	29
30	1.683	.5942	.0256	.0431	39.017	23.186	13.206	306.200	30
36	1.867	.5355	.0202	.0377	49.566	26.543	15.640	415.130	36
40	2.002	.4996	.0175	.0350	57.234	28.594	17.207	492.017	40
48	2.300	.4349	.0135	.0310	74.263	32.294	20.209	652.612	48
50	2.381	.4200	.0127	.0302	78.903	33.141	20.932	693.708	50
52	2.465	.4057	.0119	.0294	83.706	33.960	21.644	735.039	52
60	2.832	.3531	.00955	.0271	104.676	36.964	24.389	901.503	60
70	3.368	.2969	.00739	.0249	135.331	40.178	27.586	1 108.34	70
72	3.487	.2868	.00704	.0245	142.127	40.757	28.195	1 149.12	72
80	4.006	.2496	.00582	.0233	171.795	42.880	30.533	1 309.25	80
84	4.294	.2329	.00531	.0228	188.246	43.836	31.644	1 387.16	84
90	4.765	.2098	.00465	.0221	215.166	45.152	33.241	1 500.88	90
96	5.288	.1891	.00408	.0216	245.039	46.337	34.756	1 610.48	96
100	5.668	.1764	.00375	.0212	266.753	47.062	35.721	1 681.09	100
104	6.075	.1646	.00345	.0209	290.028	47.737	36.652	1 749.68	104
120	8.019	.1247	.00249	.0200	401.099	50.017	40.047	2 003.03	120
240	64.308	.0156	.00028	.0178	3 617.6	56.254	53.352	3 001.27	240
360	515.702	.00194	.00003	.0175	29 411.5	57.032	56.443	3 219.08	360
480	4 135.5	.00024		.0175	236 259.0	57.129	57.027	3 257.88	480

2%

Compound Interest Factors

2%

	Single Payment		Uniform Payment Series				Arithmetic Gradient		
	Compound Amount Factor	Present Worth Factor	Sinking Fund Factor	Capital Recovery Factor	Compound Amount Factor	Present Worth Factor	Gradient Uniform Series	Gradient Present Worth	
	Find F Given P F/P	Find P Given F P/F	Find A Given F A/F	Find A Given P A/P	Find F Given A F/A	Find P Given A P/A	Find A Given G A/G	Find P Given G P/G	
n									n
1	1.020	.9804	1.0000	1.0200	1.000	0.980	0	0	1
2	1.040	.9612	.4951	.5151	2.020	1.942	0.495	0.961	2
3	1.061	.9423	.3268	.3468	3.060	2.884	0.987	2.846	3
4	1.082	.9238	.2426	.2626	4.122	3.808	1.475	5.617	4
5	1.104	.9057	.1922	.2122	5.204	4.713	1.960	9.240	5
6	1.126	.8880	.1585	.1785	6.308	5.601	2.442	13.679	6
7	1.149	.8706	.1345	.1545	7.434	6.472	2.921	18.903	7
8	1.172	.8535	.1165	.1365	8.583	7.325	3.396	24.877	8
9	1.195	.8368	.1025	.1225	9.755	8.162	3.868	31.571	9
10	1.219	.8203	.0913	.1113	10.950	8.983	4.337	38.954	10
11	1.243	.8043	.0822	.1022	12.169	9.787	4.802	46.996	11
12	1.268	.7885	.0746	.0946	13.412	10.575	5.264	55.669	12
13	1.294	.7730	.0681	.0881	14.680	11.348	5.723	64.946	13
14	1.319	.7579	.0626	.0826	15.974	12.106	6.178	74.798	14
15	1.346	.7430	.0578	.0778	17.293	12.849	6.631	85.200	15
16	1.373	.7284	.0537	.0737	18.639	13.578	7.080	96.127	16
17	1.400	.7142	.0500	.0700	20.012	14.292	7.526	107.553	17
18	1.428	.7002	.0467	.0667	21.412	14.992	7.968	119.456	18
19	1.457	.6864	.0438	.0638	22.840	15.678	8.407	131.812	19
20	1.486	.6730	.0412	.0612	24.297	16.351	8.843	144.598	20
21	1.516	.6598	.0388	.0588	25.783	17.011	9.276	157.793	21
22	1.546	.6468	.0366	.0566	27.299	17.658	9.705	171.377	22
23	1.577	.6342	.0347	.0547	28.845	18.292	10.132	185.328	23
24	1.608	.6217	.0329	.0529	30.422	18.914	10.555	199.628	24
25	1.641	.6095	.0312	.0512	32.030	19.523	10.974	214.256	25
26	1.673	.5976	.0297	.0497	33.671	20.121	11.391	229.196	26
27	1.707	.5859	.0283	.0483	35.344	20.707	11.804	244.428	27
28	1.741	.5744	.0270	.0470	37.051	21.281	12.214	259.936	28
29	1.776	.5631	.0258	.0458	38.792	21.844	12.621	275.703	29
30	1.811	.5521	.0247	.0447	40.568	22.396	13.025	291.713	30
36	2.040	.4902	.0192	.0392	51.994	25.489	15.381	392.036	36
40	2.208	.4529	.0166	.0366	60.402	27.355	16.888	461.989	40
48	2.587	.3865	.0126	.0326	79.353	30.673	19.755	605.961	48
50	2.692	.3715	.0118	.0318	84.579	31.424	20.442	642.355	50
52	2.800	.3571	.0111	.0311	90.016	32.145	21.116	678.779	52
60	3.281	.3048	.00877	.0288	114.051	34.761	23.696	823.692	60
70	4.000	.2500	.00667	.0267	149.977	37.499	26.663	999.829	70
72	4.161	.2403	.00633	.0263	158.056	37.984	27.223	1 034.050	72
80	4.875	.2051	.00516	.0252	193.771	39.744	29.357	1 166.781	80
84	5.277	.1895	.00468	.0247	213.865	40.525	30.361	1 230.413	84
90	5.943	.1683	.00405	.0240	247.155	41.587	31.793	1 322.164	90
96	6.693	.1494	.00351	.0235	284.645	42.529	33.137	1 409.291	96
100	7.245	.1380	.00320	.0232	312.230	43.098	33.986	1 464.747	100
104	7.842	.1275	.00292	.0229	342.090	43.624	34.799	1 518.082	104
120	10.765	.0929	.00205	.0220	488.255	45.355	37.711	1 710.411	120
240	115.887	.00863	.00017	.0202	5 744.4	49.569	47.911	2 374.878	240
360	1 247.5	.00080	.00002	.0200	62 326.8	49.960	49.711	2 483.567	360
480	13 429.8	.00007		.0200	671 442.0	49.996	49.964	2 498.027	480

$2^1/_2\%$

Compound Interest Factors

$2^1/_2\%$

	Single Payment		Uniform Payment Series				Arithmetic Gradient		
	Compound Amount Factor	Present Worth Factor	Sinking Fund Factor	Capital Recovery Factor	Compound Amount Factor	Present Worth Factor	Gradient Uniform Series	Gradient Present Worth	
	Find F Given P F/P	Find P Given F P/F	Find A Given F A/F	Find A Given P A/P	Find F Given A F/A	Find P Given A P/A	Find A Given G A/G	Find P Given G P/G	
n									n
1	1.025	.9756	1.0000	1.0250	1.000	0.976	0	0	1
2	1.051	.9518	.4938	.5188	2.025	1.927	0.494	0.952	2
3	1.077	.9286	.3251	.3501	3.076	2.856	0.984	2.809	3
4	1.104	.9060	.2408	.2658	4.153	3.762	1.469	5.527	4
5	1.131	.8839	.1902	.2152	5.256	4.646	1.951	9.062	5
6	1.160	.8623	.1566	.1816	6.388	5.508	2.428	13.374	6
7	1.189	.8413	.1325	.1575	7.547	6.349	2.901	18.421	7
8	1.218	.8207	.1145	.1395	8.736	7.170	3.370	24.166	8
9	1.249	.8007	.1005	.1255	9.955	7.971	3.835	30.572	9
10	1.280	.7812	.0893	.1143	11.203	8.752	4.296	37.603	10
11	1.312	.7621	.0801	.1051	12.483	9.514	4.753	45.224	11
12	1.345	.7436	.0725	.0975	13.796	10.258	5.206	53.403	12
13	1.379	.7254	.0660	.0910	15.140	10.983	5.655	62.108	13
14	1.413	.7077	.0605	.0855	16.519	11.691	6.100	71.309	14
15	1.448	.6905	.0558	.0808	17.932	12.381	6.540	80.975	15
16	1.485	.6736	.0516	.0766	19.380	13.055	6.977	91.080	16
17	1.522	.6572	.0479	.0729	20.865	13.712	7.409	101.595	17
18	1.560	.6412	.0447	.0697	22.386	14.353	7.838	112.495	18
19	1.599	.6255	.0418	.0668	23.946	14.979	8.262	123.754	19
20	1.639	.6103	.0391	.0641	25.545	15.589	8.682	135.349	20
21	1.680	.5954	.0368	.0618	27.183	16.185	9.099	147.257	21
22	1.722	.5809	.0346	.0596	28.863	16.765	9.511	159.455	22
23	1.765	.5667	.0327	.0577	30.584	17.332	9.919	171.922	23
24	1.809	.5529	.0309	.0559	32.349	17.885	10.324	184.638	24
25	1.854	.5394	.0293	.0543	34.158	18.424	10.724	197.584	25
26	1.900	.5262	.0278	.0528	36.012	18.951	11.120	210.740	26
27	1.948	.5134	.0264	.0514	37.912	19.464	11.513	224.088	27
28	1.996	.5009	.0251	.0501	39.860	19.965	11.901	237.612	28
29	2.046	.4887	.0239	.0489	41.856	20.454	12.286	251.294	29
30	2.098	.4767	.0228	.0478	43.903	20.930	12.667	265.120	30
31	2.150	.4651	.0217	.0467	46.000	21.395	13.044	279.073	31
32	2.204	.4538	.0208	.0458	48.150	24.849	13.417	293.140	32
33	2.259	.4427	.0199	.0449	50.354	22.292	13.786	307.306	33
34	2.315	.4319	.0190	.0440	52.613	22.724	14.151	321.559	34
35	2.373	.4214	.0182	.0432	54.928	23.145	14.512	335.886	35
40	2.685	.3724	.0148	.0398	67.402	25.103	16.262	408.221	40
45	3.038	.3292	.0123	.0373	81.516	26.833	17.918	480.806	45
50	3.437	.2909	.0103	.0353	97.484	28.362	19.484	552.607	50
55	3.889	.2572	.00865	.0337	115.551	29.714	20.961	622.827	55
60	4.400	.2273	.00735	.0324	135.991	30.909	22.352	690.865	60
65	4.978	.2009	.00628	.0313	159.118	31.965	23.660	756.280	65
70	5.632	.1776	.00540	.0304	185.284	32.898	24.888	818.763	70
75	6.372	.1569	.00465	.0297	214.888	33.723	26.039	878.114	75
80	7.210	.1387	.00403	.0290	248.382	34.452	27.117	934.217	80
85	8.157	.1226	.00349	.0285	286.278	35.096	28.123	987.026	85
90	9.229	.1084	.00304	.0280	329.154	35.666	29.063	1 036.54	90
95	10.442	.0958	.00265	.0276	377.663	36.169	29.938	1 082.83	95
100	11.814	.0846	.00231	.0273	432.548	36.614	30.752	1 125.97	100

3%

Compound Interest Factors

3%

	Single Payment		Uniform Payment Series				Arithmetic Gradient		
	Compound Amount Factor	Present Worth Factor	Sinking Fund Factor	Capital Recovery Factor	Compound Amount Factor	Present Worth Factor	Gradient Uniform Series	Gradient Present Worth	
	Find F Given P F/P	Find P Given F P/F	Find A Given F A/F	Find A Given P A/P	Find F Given A F/A	Find P Given A P/A	Find A Given G A/G	Find P Given G P/G	
n									n
1	1.030	.9709	1.0000	1.0300	1.000	0.971	0	0	1
2	1.061	.9426	.4926	.5226	2.030	1.913	0.493	0.943	2
3	1.093	.9151	.3235	.3535	3.091	2.829	0.980	2.773	3
4	1.126	.8885	.2390	.2690	4.184	3.717	1.463	5.438	4
5	1.159	.8626	.1884	.2184	5.309	4.580	1.941	8.889	5
6	1.194	.8375	.1546	.1846	6.468	5.417	2.414	13.076	6
7	1.230	.8131	.1305	.1605	7.662	6.230	2.882	17.955	7
8	1.267	.7894	.1125	.1425	8.892	7.020	3.345	23.481	8
9	1.305	.7664	.0984	.1284	10.159	7.786	3.803	29.612	9
10	1.344	.7441	.0872	.1172	11.464	8.530	4.256	36.309	10
11	1.384	.7224	.0781	.1081	12.808	9.253	4.705	43.533	11
12	1.426	.7014	.0705	.1005	14.192	9.954	5.148	51.248	12
13	1.469	.6810	.0640	.0940	15.618	10.635	5.587	59.419	13
14	1.513	.6611	.0585	.0885	17.086	11.296	6.021	68.014	14
15	1.558	.6419	.0538	.0838	18.599	11.938	6.450	77.000	15
16	1.605	.6232	.0496	.0796	20.157	12.561	6.874	86.348	16
17	1.653	.6050	.0460	.0760	21.762	13.166	7.294	96.028	17
18	1.702	.5874	.0427	.0727	23.414	13.754	7.708	106.014	18
19	1.754	.5703	.0398	.0698	25.117	14.324	8.118	116.279	19
20	1.806	.5537	.0372	.0672	26.870	14.877	8.523	126.799	20
21	1.860	.5375	.0349	.0649	28.676	15.415	8.923	137.549	21
22	1.916	.5219	.0327	.0627	30.537	15.937	9.319	148.509	22
23	1.974	.5067	.0308	.0608	32.453	16.444	9.709	159.656	23
24	2.033	.4919	.0290	.0590	34.426	16.936	10.095	170.971	24
25	2.094	.4776	.0274	.0574	36.459	17.413	10.477	182.433	25
26	2.157	.4637	.0259	.0559	38.553	17.877	10.853	194.026	26
27	2.221	.4502	.0246	.0546	40.710	18.327	11.226	205.731	27
28	2.288	.4371	.0233	.0533	42.931	18.764	11.593	217.532	28
29	2.357	.4243	.0221	.0521	45.219	19.188	11.956	229.413	29
30	2.427	.4120	.0210	.0510	47.575	19.600	12.314	241.361	30
31	2.500	.4000	.0200	.0500	50.003	20.000	12.668	253.361	31
32	2.575	.3883	.0190	.0490	52.503	20.389	13.017	265.399	32
33	2.652	.3770	.0182	.0482	55.078	20.766	13.362	277.464	33
34	2.732	.3660	.0173	.0473	57.730	21.132	13.702	289.544	34
35	2.814	.3554	.0165	.0465	60.462	21.487	14.037	301.627	35
40	3.262	.3066	.0133	.0433	75.401	23.115	15.650	361.750	40
45	3.782	.2644	.0108	.0408	92.720	24.519	17.156	420.632	45
50	4.384	.2281	.00887	.0389	112.797	25.730	18.558	477.480	50
55	5.082	.1968	.00735	.0373	136.072	26.774	19.860	531.741	55
60	5.892	.1697	.00613	.0361	163.053	27.676	21.067	583.052	60
65	6.830	.1464	.00515	.0351	194.333	28.453	22.184	631.201	65
70	7.918	.1263	.00434	.0343	230.594	29.123	23.215	676.087	70
75	9.179	.1089	.00367	.0337	272.631	29.702	24.163	717.698	75
80	10.641	.0940	.00311	.0331	321.363	30.201	25.035	756.086	80
85	12.336	.0811	.00265	.0326	377.857	30.631	25.835	791.353	85
90	14.300	.0699	.00226	.0323	443.349	31.002	26.567	823.630	90
95	16.578	.0603	.00193	.0319	519.272	31.323	27.235	853.074	95
100	19.219	.0520	.00165	.0316	607.287	31.599	27.844	879.854	100

$3^1/_2\%$

Compound Interest Factors

$3^1/_2\%$

	Single Payment		Uniform Payment Series				Arithmetic Gradient		
	Compound Amount Factor	Present Worth Factor	Sinking Fund Factor	Capital Recovery Factor	Compound Amount Factor	Present Worth Factor	Gradient Uniform Series	Gradient Present Worth	
	Find F Given P	Find P Given F	Find A Given F	Find A Given P	Find F Given A	Find P Given A	Find A Given G	Find P Given G	
n	F/P	P/F	A/F	A/P	F/A	P/A	A/G	P/G	n
1	1.035	.9662	1.0000	1.0350	1.000	0.966	0	0	1
2	1.071	.9335	.4914	.5264	2.035	1.900	0.491	0.933	2
3	1.109	.9019	.3219	.3569	3.106	2.802	0.977	2.737	3
4	1.148	.8714	.2373	.2723	4.215	3.673	1.457	5.352	4
5	1.188	.8420	.1865	.2215	5.362	4.515	1.931	8.719	5
6	1.229	.8135	.1527	.1877	6.550	5.329	2.400	12.787	6
7	1.272	.7860	.1285	.1635	7.779	6.115	2.862	17.503	7
8	1.317	.7594	.1105	.1455	9.052	6.874	3.320	22.819	8
9	1.363	.7337	.0964	.1314	10.368	7.608	3.771	28.688	9
10	1.411	.7089	.0852	.1202	11.731	8.317	4.217	35.069	10
11	1.460	.6849	.0761	.1111	13.142	9.002	4.657	41.918	11
12	1.511	.6618	.0685	.1035	14.602	9.663	5.091	49.198	12
13	1.564	.6394	.0621	.0971	16.113	10.303	5.520	56.871	13
14	1.619	.6178	.0566	.0916	17.677	10.921	5.943	64.902	14
15	1.675	.5969	.0518	.0868	19.296	11.517	6.361	73.258	15
16	1.734	.5767	.0477	.0827	20.971	12.094	6.773	81.909	16
17	1.795	.5572	.0440	.0790	22.705	12.651	7.179	90.824	17
18	1.857	.5384	.0408	.0758	24.500	13.190	7.580	99.976	18
19	1.922	.5202	.0379	.0729	26.357	13.710	7.975	109.339	19
20	1.990	.5026	.0354	.0704	28.280	14.212	8.365	118.888	20
21	2.059	.4856	.0330	.0680	30.269	14.698	8.749	128.599	21
22	2.132	.4692	.0309	.0659	32.329	15.167	9.128	138.451	22
23	2.206	.4533	.0290	.0640	34.460	15.620	9.502	148.423	23
24	2.283	.4380	.0273	.0623	36.666	16.058	9.870	158.496	24
25	2.363	.4231	.0257	.0607	38.950	16.482	10.233	168.652	25
26	2.446	.4088	.0242	.0592	41.313	16.890	10.590	178.873	26
27	2.532	.3950	.0229	.0579	43.759	17.285	10.942	189.143	27
28	2.620	.3817	.0216	.0566	46.291	17.667	11.289	199.448	28
29	2.712	.3687	.0204	.0554	48.911	18.036	11.631	209.773	29
30	2.807	.3563	.0194	.0544	51.623	18.392	11.967	220.105	30
31	2.905	.3442	.0184	.0534	54.429	18.736	12.299	230.432	31
32	3.007	.3326	.0174	.0524	57.334	19.069	12.625	240.742	32
33	3.112	.3213	.0166	.0516	60.341	19.390	12.946	251.025	33
34	3.221	.3105	.0158	.0508	63.453	19.701	13.262	261.271	34
35	3.334	.3000	.0150	.0500	66.674	20.001	13.573	271.470	35
40	3.959	.2526	.0118	.0468	84.550	21.355	15.055	321.490	40
45	4.702	.2127	.00945	.0445	105.781	22.495	16.417	369.307	45
50	5.585	.1791	.00763	.0426	130.998	23.456	17.666	414.369	50
55	6.633	.1508	.00621	.0412	160.946	24.264	18.808	456.352	55
60	7.878	.1269	.00509	.0401	196.516	24.945	19.848	495.104	60
65	9.357	.1069	.00419	.0392	238.762	25.518	20.793	530.598	65
70	11.113	.0900	.00346	.0385	288.937	26.000	21.650	562.895	70
75	13.199	.0758	.00287	.0379	348.529	26.407	22.423	592.121	75
80	15.676	.0638	.00238	.0374	419.305	26.749	23.120	618.438	80
85	18.618	.0537	.00199	.0370	503.365	27.037	23.747	642.036	85
90	22.112	.0452	.00166	.0367	603.202	27.279	24.308	663.118	90
95	26.262	.0381	.00139	.0364	721.778	27.483	24.811	681.890	95
100	31.191	.0321	.00116	.0362	862.608	27.655	25.259	698.554	100

4% Compound Interest Factors **4%**

	Single Payment		Uniform Payment Series				Arithmetic Gradient		
	Compound Amount Factor	Present Worth Factor	Sinking Fund Factor	Capital Recovery Factor	Compound Amount Factor	Present Worth Factor	Gradient Uniform Series	Gradient Present Worth	
	Find F Given P F/P	Find P Given F P/F	Find A Given F A/F	Find A Given P A/P	Find F Given A F/A	Find P Given A P/A	Find A Given G A/G	Find P Given G P/G	
n									n
1	1.040	.9615	1.0000	1.0400	1.000	0.962	0	0	1
2	1.082	.9246	.4902	.5302	2.040	1.886	0.490	0.925	2
3	1.125	.8890	.3203	.3603	3.122	2.775	0.974	2.702	3
4	1.170	.8548	.2355	.2755	4.246	3.630	1.451	5.267	4
5	1.217	.8219	.1846	.2246	5.416	4.452	1.922	8.555	5
6	1.265	.7903	.1508	.1908	6.633	5.242	2.386	12.506	6
7	1.316	.7599	.1266	.1666	7.898	6.002	2.843	17.066	7
8	1.369	.7307	.1085	.1485	9.214	6.733	3.294	22.180	8
9	1.423	.7026	.0945	.1345	10.583	7.435	3.739	27.801	9
10	1.480	.6756	.0833	.1233	12.006	8.111	4.177	33.881	10
11	1.539	.6496	.0741	.1141	13.486	8.760	4.609	40.377	11
12	1.601	.6246	.0666	.1066	15.026	9.385	5.034	47.248	12
13	1.665	.6006	.0601	.1001	16.627	9.986	5.453	54.454	13
14	1.732	.5775	.0547	.0947	18.292	10.563	5.866	61.962	14
15	1.801	.5553	.0499	.0899	20.024	11.118	6.272	69.735	15
16	1.873	.5339	.0458	.0858	21.825	11.652	6.672	77.744	16
17	1.948	.5134	.0422	.0822	23.697	12.166	7.066	85.958	17
18	2.026	.4936	.0390	.0790	25.645	12.659	7.453	94.350	18
19	2.107	.4746	.0361	.0761	27.671	13.134	7.834	102.893	19
20	2.191	.4564	.0336	.0736	29.778	13.590	8.209	111.564	20
21	2.279	.4388	.0313	.0713	31.969	14.029	8.578	120.341	21
22	2.370	.4220	.0292	.0692	34.248	14.451	8.941	129.202	22
23	2.465	.4057	.0273	.0673	36.618	14.857	9.297	138.128	23
24	2.563	.3901	.0256	.0656	39.083	15.247	9.648	147.101	24
25	2.666	.3751	.0240	.0640	41.646	15.622	9.993	156.104	25
26	2.772	.3607	.0226	.0626	44.312	15.983	10.331	165.121	26
27	2.883	.3468	.0212	.0612	47.084	16.330	10.664	174.138	27
28	2.999	.3335	.0200	.0600	49.968	16.663	10.991	183.142	28
29	3.119	.3207	.0189	.0589	52.966	16.984	11.312	192.120	29
30	3.243	.3083	.0178	.0578	56.085	17.292	11.627	201.062	30
31	3.373	.2965	.0169	.0569	59.328	17.588	11.937	209.955	31
32	3.508	.2851	.0159	.0559	62.701	17.874	12.241	218.792	32
33	3.648	.2741	.0151	.0551	66.209	18.148	12.540	227.563	33
34	3.794	.2636	.0143	.0543	69.858	18.411	12.832	236.260	34
35	3.946	.2534	.0136	.0536	73.652	18.665	13.120	244.876	35
40	4.801	.2083	.0105	.0505	95.025	19.793	14.476	286.530	40
45	5.841	.1712	.00826	.0483	121.029	20.720	15.705	325.402	45
50	7.107	.1407	.00655	.0466	152.667	21.482	16.812	361.163	50
55	8.646	.1157	.00523	.0452	191.159	22.109	17.807	393.689	55
60	10.520	.0951	.00420	.0442	237.990	22.623	18.697	422.996	60
65	12.799	.0781	.00339	.0434	294.968	23.047	19.491	449.201	65
70	15.572	.0642	.00275	.0427	364.290	23.395	20.196	472.479	70
75	18.945	.0528	.00223	.0422	448.630	23.680	20.821	493.041	75
80	23.050	.0434	.00181	.0418	551.244	23.915	21.372	511.116	80
85	28.044	.0357	.00148	.0415	676.089	24.109	21.857	526.938	85
90	34.119	.0293	.00121	.0412	827.981	24.267	22.283	540.737	90
95	41.511	.0241	.00099	.0410	1 012.8	24.398	22.655	552.730	95
100	50.505	.0198	.00081	.0408	1 237.6	24.505	22.980	563.125	100

$4^1/_2\%$ Compound Interest Factors $4^1/_2\%$

	Single Payment		Uniform Payment Series				Arithmetic Gradient		
	Compound Amount Factor	Present Worth Factor	Sinking Fund Factor	Capital Recovery Factor	Compound Amount Factor	Present Worth Factor	Gradient Uniform Series	Gradient Present Worth	
	Find F Given P	Find P Given F	Find A Given F	Find A Given P	Find F Given A	Find P Given A	Find A Given G	Find P Given G	
n	F/P	P/F	A/F	A/P	F/A	P/A	A/G	P/G	n
1	1.045	.9569	1.0000	1.0450	1.000	0.957	0	0	1
2	1.092	.9157	.4890	.5340	2.045	1.873	0.489	0.916	2
3	1.141	.8763	.3188	.3638	3.137	2.749	0.971	2.668	3
4	1.193	.8386	.2337	.2787	4.278	3.588	1.445	5.184	4
5	1.246	.8025	.1828	.2278	5.471	4.390	1.912	8.394	5
6	1.302	.7679	.1489	.1939	6.717	5.158	2.372	12.233	6
7	1.361	.7348	.1247	.1697	8.019	5.893	2.824	16.642	7
8	1.422	.7032	.1066	.1516	9.380	6.596	3.269	21.564	8
9	1.486	.6729	.0926	.1376	10.802	7.269	3.707	26.948	9
10	1.553	.6439	.0814	.1264	12.288	7.913	4.138	32.743	10
11	1.623	.6162	.0722	.1172	13.841	8.529	4.562	38.905	11
12	1.696	.5897	.0647	.1097	15.464	9.119	4.978	45.391	12
13	1.772	.5643	.0583	.1033	17.160	9.683	5.387	52.163	13
14	1.852	.5400	.0528	.0978	18.932	10.223	5.789	59.182	14
15	1.935	.5167	.0481	.0931	20.784	10.740	6.184	66.416	15
16	2.022	.4945	.0440	.0890	22.719	11.234	6.572	73.833	16
17	2.113	.4732	.0404	.0854	24.742	11.707	6.953	81.404	17
18	2.208	.4528	.0372	.0822	26.855	12.160	7.327	89.102	18
19	2.308	.4333	.0344	.0794	29.064	12.593	7.695	96.901	19
20	2.412	.4146	.0319	.0769	31.371	13.008	8.055	104.779	20
21	2.520	.3968	.0296	.0746	33.783	13.405	8.409	112.715	21
22	2.634	.3797	.0275	.0725	36.303	13.784	8.755	120.689	22
23	2.752	.3634	.0257	.0707	38.937	14.148	9.096	128.682	23
24	2.876	.3477	.0240	.0690	41.689	14.495	9.429	136.680	24
25	3.005	.3327	.0224	.0674	44.565	14.828	9.756	144.665	25
26	3.141	.3184	.0210	.0660	47.571	15.147	10.077	152.625	26
27	3.282	.3047	.0197	.0647	50.711	15.451	10.391	160.547	27
28	3.430	.2916	.0185	.0635	53.993	15.743	10.698	168.420	28
29	3.584	.2790	.0174	.0624	57.423	16.022	10.999	176.232	29
30	3.745	.2670	.0164	.0614	61.007	16.289	11.295	183.975	30
31	3.914	.2555	.0154	.0604	64.752	16.544	11.583	191.640	31
32	4.090	.2445	.0146	.0596	68.666	16.789	11.866	199.220	32
33	4.274	.2340	.0137	.0587	72.756	17.023	12.143	206.707	33
34	4.466	.2239	.0130	.0580	77.030	17.247	12.414	214.095	34
35	4.667	.2143	.0123	.0573	81.497	17.461	12.679	221.380	35
40	5.816	.1719	.00934	.0543	107.030	18.402	13.917	256.098	40
45	7.248	.1380	.00720	.0522	138.850	19.156	15.020	287.732	45
50	9.033	.1107	.00560	.0506	178.503	19.762	15.998	316.145	50
55	11.256	.0888	.00439	.0494	227.918	20.248	16.860	341.375	55
60	14.027	.0713	.00345	.0485	289.497	20.638	17.617	363.571	60
65	17.481	.0572	.00273	.0477	366.237	20.951	18.278	382.946	65
70	21.784	.0459	.00217	.0472	461.869	21.202	18.854	399.750	70
75	27.147	.0368	.00172	.0467	581.043	21.404	19.354	414.242	75
80	33.830	.0296	.00137	.0464	729.556	21.565	19.785	426.680	80
85	42.158	.0237	.00109	.0461	914.630	21.695	20.157	437.309	85
90	52.537	.0190	.00087	.0459	1 145.3	21.799	20.476	446.359	90
95	65.471	.0153	.00070	.0457	1 432.7	21.883	20.749	454.039	95
100	81.588	.0123	.00056	.0456	1 790.9	21.950	20.981	460.537	100

5%   Compound Interest Factors   5%

	Single Payment		Uniform Payment Series				Arithmetic Gradient		
	Compound Amount Factor	Present Worth Factor	Sinking Fund Factor	Capital Recovery Factor	Compound Amount Factor	Present Worth Factor	Gradient Uniform Series	Gradient Present Worth	
	Find F Given P	Find P Given F	Find A Given F	Find A Given P	Find F Given A	Find P Given A	Find A Given G	Find P Given G	
n	F/P	P/F	A/F	A/P	F/A	P/A	A/G	P/G	n
1	1.050	.9524	1.0000	1.0500	1.000	0.952	0	0	1
2	1.102	.9070	.4878	.5378	2.050	1.859	0.488	0.907	2
3	1.158	.8638	.3172	.3672	3.152	2.723	0.967	2.635	3
4	1.216	.8227	.2320	.2820	4.310	3.546	1.439	5.103	4
5	1.276	.7835	.1810	.2310	5.526	4.329	1.902	8.237	5
6	1.340	.7462	.1470	.1970	6.802	5.076	2.358	11.968	6
7	1.407	.7107	.1228	.1728	8.142	5.786	2.805	16.232	7
8	1.477	.6768	.1047	.1547	9.549	6.463	3.244	20.970	8
9	1.551	.6446	.0907	.1407	11.027	7.108	3.676	26.127	9
10	1.629	.6139	.0795	.1295	12.578	7.722	4.099	31.652	10
11	1.710	.5847	.0704	.1204	14.207	8.306	4.514	37.499	11
12	1.796	.5568	.0628	.1128	15.917	8.863	4.922	43.624	12
13	1.886	.5303	.0565	.1065	17.713	9.394	5.321	49.988	13
14	1.980	.5051	.0510	.1010	19.599	9.899	5.713	56.553	14
15	2.079	.4810	.0463	.0963	21.579	10.380	6.097	63.288	15
16	2.183	.4581	.0423	.0923	23.657	10.838	6.474	70.159	16
17	2.292	.4363	.0387	.0887	25.840	11.274	6.842	77.140	17
18	2.407	.4155	.0355	.0855	28.132	11.690	7.203	84.204	18
19	2.527	.3957	.0327	.0827	30.539	12.085	7.557	91.327	19
20	2.653	.3769	.0302	.0802	33.066	12.462	7.903	98.488	20
21	2.786	.3589	.0280	.0780	35.719	12.821	8.242	105.667	21
22	2.925	.3419	.0260	.0760	38.505	13.163	8.573	112.846	22
23	3.072	.3256	.0241	.0741	41.430	13.489	8.897	120.008	23
24	3.225	.3101	.0225	.0725	44.502	13.799	9.214	127.140	24
25	3.386	.2953	.0210	.0710	47.727	14.094	9.524	134.227	25
26	3.556	.2812	.0196	.0696	51.113	14.375	9.827	141.258	26
27	3.733	.2678	.0183	.0683	54.669	14.643	10.122	148.222	27
28	3.920	.2551	.0171	.0671	58.402	14.898	10.411	155.110	28
29	4.116	.2429	.0160	.0660	62.323	15.141	10.694	161.912	29
30	4.322	.2314	.0151	.0651	66.439	15.372	10.969	168.622	30
31	4.538	.2204	.0141	.0641	70.761	15.593	11.238	175.233	31
32	4.765	.2099	.0133	.0633	75.299	15.803	11.501	181.739	32
33	5.003	.1999	.0125	.0625	80.063	16.003	11.757	188.135	33
34	5.253	.1904	.0118	.0618	85.067	16.193	12.006	194.416	34
35	5.516	.1813	.0111	.0611	90.320	16.374	12.250	200.580	35
40	7.040	.1420	.00828	.0583	120.799	17.159	13.377	229.545	40
45	8.985	.1113	.00626	.0563	159.699	17.774	14.364	255.314	45
50	11.467	.0872	.00478	.0548	209.347	18.256	15.223	277.914	50
55	14.636	.0683	.00367	.0537	272.711	18.633	15.966	297.510	55
60	18.679	.0535	.00283	.0528	353.582	18.929	16.606	314.343	60
65	23.840	.0419	.00219	.0522	456.795	19.161	17.154	328.691	65
70	30.426	.0329	.00170	.0517	588.525	19.343	17.621	340.841	70
75	38.832	.0258	.00132	.0513	756.649	19.485	18.018	351.072	75
80	49.561	.0202	.00103	.0510	971.222	19.596	18.353	359.646	80
85	63.254	.0158	.00080	.0508	1 245.1	19.684	18.635	366.800	85
90	80.730	.0124	.00063	.0506	1 594.6	19.752	18.871	372.749	90
95	103.034	.00971	.00049	.0505	2 040.7	19.806	19.069	377.677	95
100	131.500	.00760	.00038	.0504	2 610.0	19.848	19.234	381.749	100

6%

Compound Interest Factors

6%

	Single Payment		Uniform Payment Series				Arithmetic Gradient		
	Compound Amount Factor	Present Worth Factor	Sinking Fund Factor	Capital Recovery Factor	Compound Amount Factor	Present Worth Factor	Gradient Uniform Series	Gradient Present Worth	
n	Find F Given P F/P	Find P Given F P/F	Find A Given F A/F	Find A Given P A/P	Find F Given A F/A	Find P Given A P/A	Find A Given G A/G	Find P Given G P/G	n
1	1.060	.9434	1.0000	1.0600	1.000	0.943	0	0	1
2	1.124	.8900	.4854	.5454	2.060	1.833	0.485	0.890	2
3	1.191	.8396	.3141	.3741	3.184	2.673	0.961	2.569	3
4	1.262	.7921	.2286	.2886	4.375	3.465	1.427	4.945	4
5	1.338	.7473	.1774	.2374	5.637	4.212	1.884	7.934	5
6	1.419	.7050	.1434	.2034	6.975	4.917	2.330	11.459	6
7	1.504	.6651	.1191	.1791	8.394	5.582	2.768	15.450	7
8	1.594	.6274	.1010	.1610	9.897	6.210	3.195	19.841	8
9	1.689	.5919	.0870	.1470	11.491	6.802	3.613	24.577	9
10	1.791	.5584	.0759	.1359	13.181	7.360	4.022	29.602	10
11	1.898	.5268	.0668	.1268	14.972	7.887	4.421	34.870	11
12	2.012	.4970	.0593	.1193	16.870	8.384	4.811	40.337	12
13	2.133	.4688	.0530	.1130	18.882	8.853	5.192	45.963	13
14	2.261	.4423	.0476	.1076	21.015	9.295	5.564	51.713	14
15	2.397	.4173	.0430	.1030	23.276	9.712	5.926	57.554	15
16	2.540	.3936	.0390	.0990	25.672	10.106	6.279	63.459	16
17	2.693	.3714	.0354	.0954	28.213	10.477	6.624	69.401	17
18	2.854	.3503	.0324	.0924	30.906	10.828	6.960	75.357	18
19	3.026	.3305	.0296	.0896	33.760	11.158	7.287	81.306	19
20	3.207	.3118	.0272	.0872	36.786	11.470	7.605	87.230	20
21	3.400	.2942	.0250	.0850	39.993	11.764	7.915	93.113	21
22	3.604	.2775	.0230	.0830	43.392	12.042	8.217	98.941	22
23	3.820	.2618	.0213	.0813	46.996	12.303	8.510	104.700	23
24	4.049	.2470	.0197	.0797	50.815	12.550	8.795	110.381	24
25	4.292	.2330	.0182	.0782	54.864	12.783	9.072	115.973	25
26	4.549	.2198	.0169	.0769	59.156	13.003	9.341	121.468	26
27	4.822	.2074	.0157	.0757	63.706	13.211	9.603	126.860	27
28	5.112	.1956	.0146	.0746	68.528	13.406	9.857	132.142	28
29	5.418	.1846	.0136	.0736	73.640	13.591	10.103	137.309	29
30	5.743	.1741	.0126	.0726	79.058	13.765	10.342	142.359	30
31	6.088	.1643	.0118	.0718	84.801	13.929	10.574	147.286	31
32	6.453	.1550	.0110	.0710	90.890	14.084	10.799	152.090	32
33	6.841	.1462	.0103	.0703	97.343	14.230	11.017	156.768	33
34	7.251	.1379	.00960	.0696	104.184	14.368	11.228	161.319	34
35	7.686	.1301	.00897	.0690	111.435	14.498	11.432	165.743	35
40	10.286	.0972	.00646	.0665	154.762	15.046	12.359	185.957	40
45	13.765	.0727	.00470	.0647	212.743	15.456	13.141	203.109	45
50	18.420	.0543	.00344	.0634	290.335	15.762	13.796	217.457	50
55	24.650	.0406	.00254	.0625	394.171	15.991	14.341	229.322	55
60	32.988	.0303	.00188	.0619	533.126	16.161	14.791	239.043	60
65	44.145	.0227	.00139	.0614	719.080	16.289	15.160	246.945	65
70	59.076	.0169	.00103	.0610	967.928	16.385	15.461	253.327	70
75	79.057	.0126	.00077	.0608	1 300.9	16.456	15.706	258.453	75
80	105.796	.00945	.00057	.0606	1 746.6	16.509	15.903	262.549	80
85	141.578	.00706	.00043	.0604	2 343.0	16.549	16.062	265.810	85
90	189.464	.00528	.00032	.0603	3 141.1	16.579	16.189	268.395	90
95	253.545	.00394	.00024	.0602	4 209.1	16.601	16.290	270.437	95
100	339.300	.00295	.00018	.0602	5 638.3	16.618	16.371	272.047	100

7%   Compound Interest Factors   7%

	Single Payment		Uniform Payment Series				Arithmetic Gradient		
	Compound Amount Factor	Present Worth Factor	Sinking Fund Factor	Capital Recovery Factor	Compound Amount Factor	Present Worth Factor	Gradient Uniform Series	Gradient Present Worth	
	Find F Given P F/P	Find P Given F P/F	Find A Given F A/F	Find A Given P A/P	Find F Given A F/A	Find P Given A P/A	Find A Given G A/G	Find P Given G P/G	
n									n
1	1.070	.9346	1.0000	1.0700	1.000	0.935	0	0	1
2	1.145	.8734	.4831	.5531	2.070	1.808	0.483	0.873	2
3	1.225	.8163	.3111	.3811	3.215	2.624	0.955	2.506	3
4	1.311	.7629	.2252	.2952	4.440	3.387	1.416	4.795	4
5	1.403	.7130	.1739	.2439	5.751	4.100	1.865	7.647	5
6	1.501	.6663	.1398	.2098	7.153	4.767	2.303	10.978	6
7	1.606	.6227	.1156	.1856	8.654	5.389	2.730	14.715	7
8	1.718	.5820	.0975	.1675	10.260	5.971	3.147	18.789	8
9	1.838	.5439	.0835	.1535	11.978	6.515	3.552	23.140	9
10	1.967	.5083	.0724	.1424	13.816	7.024	3.946	27.716	10
11	2.105	.4751	.0634	.1334	15.784	7.499	4.330	32.467	11
12	2.252	.4440	.0559	.1259	17.888	7.943	4.703	37.351	12
13	2.410	.4150	.0497	.1197	20.141	8.358	5.065	42.330	13
14	2.579	.3878	.0443	.1143	22.551	8.745	5.417	47.372	14
15	2.759	.3624	.0398	.1098	25.129	9.108	5.758	52.446	15
16	2.952	.3387	.0359	.1059	27.888	9.447	6.090	57.527	16
17	3.159	.3166	.0324	.1024	30.840	9.763	6.411	62.592	17
18	3.380	.2959	.0294	.0994	33.999	10.059	6.722	67.622	18
19	3.617	.2765	.0268	.0968	37.379	10.336	7.024	72.599	19
20	3.870	.2584	.0244	.0944	40.996	10.594	7.316	77.509	20
21	4.141	.2415	.0223	.0923	44.865	10.836	7.599	82.339	21
22	4.430	.2257	.0204	.0904	49.006	11.061	7.872	87.079	22
23	4.741	.2109	.0187	.0887	53.436	11.272	8.137	91.720	23
24	5.072	.1971	.0172	.0872	58.177	11.469	8.392	96.255	24
25	5.427	.1842	.0158	.0858	63.249	11.654	8.639	100.677	25
26	5.807	.1722	.0146	.0846	68.677	11.826	8.877	104.981	26
27	6.214	.1609	.0134	.0834	74.484	11.987	9.107	109.166	27
28	6.649	.1504	.0124	.0824	80.698	12.137	9.329	113.227	28
29	7.114	.1406	.0114	.0814	87.347	12.278	9.543	117.162	29
30	7.612	.1314	.0106	.0806	94.461	12.409	9.749	120.972	30
31	8.145	.1228	.00980	.0798	102.073	12.532	9.947	124.655	31
32	8.715	.1147	.00907	.0791	110.218	12.647	10.138	128.212	32
33	9.325	.1072	.00841	.0784	118.934	12.754	10.322	131.644	33
34	9.978	.1002	.00780	.0778	128.259	12.854	10.499	134.951	34
35	10.677	.0937	.00723	.0772	138.237	12.948	10.669	138.135	35
40	14.974	.0668	.00501	.0750	199.636	13.332	11.423	152.293	40
45	21.002	.0476	.00350	.0735	285.750	13.606	12.036	163.756	45
50	29.457	.0339	.00246	.0725	406.530	13.801	12.529	172.905	50
55	41.315	.0242	.00174	.0717	575.930	13.940	12.921	180.124	55
60	57.947	.0173	.00123	.0712	813.523	14.039	13.232	185.768	60
65	81.273	.0123	.00087	.0709	1 146.8	14.110	13.476	190.145	65
70	113.990	.00877	.00062	.0706	1 614.1	14.160	13.666	193.519	70
75	159.877	.00625	.00044	.0704	2 269.7	14.196	13.814	196.104	75
80	224.235	.00446	.00031	.0703	3 189.1	14.222	13.927	198.075	80
85	314.502	.00318	.00022	.0702	4 478.6	14.240	14.015	199.572	85
90	441.105	.00227	.00016	.0702	6 287.2	14.253	14.081	200.704	90
95	618.673	.00162	.00011	.0701	8 823.9	14.263	14.132	201.558	95
100	867.720	.00115	.00008	.0701	12 381.7	14.269	14.170	202.200	100

8%   **Compound Interest Factors**   **8%**

	Single Payment		Uniform Payment Series				Arithmetic Gradient		
	Compound Amount Factor	Present Worth Factor	Sinking Fund Factor	Capital Recovery Factor	Compound Amount Factor	Present Worth Factor	Gradient Uniform Series	Gradient Present Worth	
	Find F Given P F/P	Find P Given F P/F	Find A Given F A/F	Find A Given P A/P	Find F Given A F/A	Find P Given A P/A	Find A Given G A/G	Find P Given G P/G	
n									n
1	1.080	.9259	1.0000	1.0800	1.000	0.926	0	0	1
2	1.166	.8573	.4808	.5608	2.080	1.783	0.481	0.857	2
3	1.260	.7938	.3080	.3880	3.246	2.577	0.949	2.445	3
4	1.360	.7350	.2219	.3019	4.506	3.312	1.404	4.650	4
5	1.469	.6806	.1705	.2505	5.867	3.993	1.846	7.372	5
6	1.587	.6302	.1363	.2163	7.336	4.623	2.276	10.523	6
7	1.714	.5835	.1121	.1921	8.923	5.206	2.694	14.024	7
8	1.851	.5403	.0940	.1740	10.637	5.747	3.099	17.806	8
9	1.999	.5002	.0801	.1601	12.488	6.247	3.491	21.808	9
10	2.159	.4632	.0690	.1490	14.487	6.710	3.871	25.977	10
11	2.332	.4289	.0601	.1401	16.645	7.139	4.240	30.266	11
12	2.518	.3971	.0527	.1327	18.977	7.536	4.596	34.634	12
13	2.720	.3677	.0465	.1265	21.495	7.904	4.940	39.046	13
14	2.937	.3405	.0413	.1213	24.215	8.244	5.273	43.472	14
15	3.172	.3152	.0368	.1168	27.152	8.559	5.594	47.886	15
16	3.426	.2919	.0330	.1130	30.324	8.851	5.905	52.264	16
17	3.700	.2703	.0296	.1096	33.750	9.122	6.204	56.588	17
18	3.996	.2502	.0267	.1067	37.450	9.372	6.492	60.843	18
19	4.316	.2317	.0241	.1041	41.446	9.604	6.770	65.013	19
20	4.661	.2145	.0219	.1019	45.762	9.818	7.037	69.090	20
21	5.034	.1987	.0198	.0998	50.423	10.017	7.294	73.063	21
22	5.437	.1839	.0180	.0980	55.457	10.201	7.541	76.926	22
23	5.871	.1703	.0164	.0964	60.893	10.371	7.779	80.673	23
24	6.341	.1577	.0150	.0950	66.765	10.529	8.007	84.300	24
25	6.848	.1460	.0137	.0937	73.106	10.675	8.225	87.804	25
26	7.396	.1352	.0125	.0925	79.954	10.810	8.435	91.184	26
27	7.988	.1252	.0114	.0914	87.351	10.935	8.636	94.439	27
28	8.627	.1159	.0105	.0905	95.339	11.051	8.829	97.569	28
29	9.317	.1073	.00962	.0896	103.966	11.158	9.013	100.574	29
30	10.063	.0994	.00883	.0888	113.283	11.258	9.190	103.456	30
31	10.868	.0920	.00811	.0881	123.346	11.350	9.358	106.216	31
32	11.737	.0852	.00745	.0875	134.214	11.435	9.520	108.858	32
33	12.676	.0789	.00685	.0869	145.951	11.514	9.674	111.382	33
34	13.690	.0730	.00630	.0863	158.627	11.587	9.821	113.792	34
35	14.785	.0676	.00580	.0858	172.317	11.655	9.961	116.092	35
40	21.725	.0460	.00386	.0839	259.057	11.925	10.570	126.042	40
45	31.920	.0313	.00259	.0826	386.506	12.108	11.045	133.733	45
50	46.902	.0213	.00174	.0817	573.771	12.233	11.411	139.593	50
55	68.914	.0145	.00118	.0812	848.925	12.319	11.690	144.006	55
60	101.257	.00988	.00080	.0808	1 253.2	12.377	11.902	147.300	60
65	148.780	.00672	.00054	.0805	1 847.3	12.416	12.060	149.739	65
70	218.607	.00457	.00037	.0804	2 720.1	12.443	12.178	151.533	70
75	321.205	.00311	.00025	.0802	4 002.6	12.461	12.266	152.845	75
80	471.956	.00212	.00017	.0802	5 887.0	12.474	12.330	153.800	80
85	693.458	.00144	.00012	.0801	8 655.7	12.482	12.377	154.492	85
90	1 018.9	.00098	.00008	.0801	12 724.0	12.488	12.412	154.993	90
95	1 497.1	.00067	.00005	.0801	18 701.6	12.492	12.437	155.352	95
100	2 199.8	.00045	.00004	.0800	27 484.6	12.494	12.455	155.611	100

9% <div align="center">Compound Interest Factors</div> **9%**

	Single Payment		Uniform Payment Series				Arithmetic Gradient		
	Compound Amount Factor	Present Worth Factor	Sinking Fund Factor	Capital Recovery Factor	Compound Amount Factor	Present Worth Factor	Gradient Uniform Series	Gradient Present Worth	
	Find F Given P F/P	Find P Given F P/F	Find A Given F A/F	Find A Given P A/P	Find F Given A F/A	Find P Given A P/A	Find A Given G A/G	Find P Given G P/G	
n									n
1	1.090	.9174	1.0000	1.0900	1.000	0.917	0	0	1
2	1.188	.8417	.4785	.5685	2.090	1.759	0.478	0.842	2
3	1.295	.7722	.3051	.3951	3.278	2.531	0.943	2.386	3
4	1.412	.7084	.2187	.3087	4.573	3.240	1.393	4.511	4
5	1.539	.6499	.1671	.2571	5.985	3.890	1.828	7.111	5
6	1.677	.5963	.1329	.2229	7.523	4.486	2.250	10.092	6
7	1.828	.5470	.1087	.1987	9.200	5.033	2.657	13.375	7
8	1.993	.5019	.0907	.1807	11.028	5.535	3.051	16.888	8
9	2.172	.4604	.0768	.1668	13.021	5.995	3.431	20.571	9
10	2.367	.4224	.0658	.1558	15.193	6.418	3.798	24.373	10
11	2.580	.3875	.0569	.1469	17.560	6.805	4.151	28.248	11
12	2.813	.3555	.0497	.1397	20.141	7.161	4.491	32.159	12
13	3.066	.3262	.0436	.1336	22.953	7.487	4.818	36.073	13
14	3.342	.2992	.0384	.1284	26.019	7.786	5.133	39.963	14
15	3.642	.2745	.0341	.1241	29.361	8.061	5.435	43.807	15
16	3.970	.2519	.0303	.1203	33.003	8.313	5.724	47.585	16
17	4.328	.2311	.0270	.1170	36.974	8.544	6.002	51.282	17
18	4.717	.2120	.0242	.1142	41.301	8.756	6.269	54.886	18
19	5.142	.1945	.0217	.1117	46.019	8.950	6.524	58.387	19
20	5.604	.1784	.0195	.1095	51.160	9.129	6.767	61.777	20
21	6.109	.1637	.0176	.1076	56.765	9.292	7.001	65.051	21
22	6.659	.1502	.0159	.1059	62.873	9.442	7.223	68.205	22
23	7.258	.1378	.0144	.1044	69.532	9.580	7.436	71.236	23
24	7.911	.1264	.0130	.1030	76.790	9.707	7.638	74.143	24
25	8.623	.1160	.0118	.1018	84.701	9.823	7.832	76.927	25
26	9.399	.1064	.0107	.1007	93.324	9.929	8.016	79.586	26
27	10.245	.0976	.00973	.0997	102.723	10.027	8.191	82.124	27
28	11.167	.0895	.00885	.0989	112.968	10.116	8.357	84.542	28
29	12.172	.0822	.00806	.0981	124.136	10.198	8.515	86.842	29
30	13.268	.0754	.00734	.0973	136.308	10.274	8.666	89.028	30
31	14.462	.0691	.00669	.0967	149.575	10.343	8.808	91.102	31
32	15.763	.0634	.00610	.0961	164.037	10.406	8.944	93.069	32
33	17.182	.0582	.00556	.0956	179.801	10.464	9.072	94.931	33
34	18.728	.0534	.00508	.0951	196.983	10.518	9.193	96.693	34
35	20.414	.0490	.00464	.0946	215.711	10.567	9.308	98.359	35
40	31.409	.0318	.00296	.0930	337.883	10.757	9.796	105.376	40
45	48.327	.0207	.00190	.0919	525.860	10.881	10.160	110.556	45
50	74.358	.0134	.00123	.0912	815.085	10.962	10.430	114.325	50
55	114.409	.00874	.00079	.0908	1 260.1	11.014	10.626	117.036	55
60	176.032	.00568	.00051	.0905	1 944.8	11.048	10.768	118.968	60
65	270.847	.00369	.00033	.0903	2 998.3	11.070	10.870	120.334	65
70	416.731	.00240	.00022	.0902	4 619.2	11.084	10.943	121.294	70
75	641.193	.00156	.00014	.0901	7 113.3	11.094	10.994	121.965	75
80	986.555	.00101	.00009	.0901	10 950.6	11.100	11.030	122.431	80
85	1 517.9	.00066	.00006	.0901	16 854.9	11.104	11.055	122.753	85
90	2 335.5	.00043	.00004	.0900	25 939.3	11.106	11.073	122.976	90
95	3 593.5	.00028	.00003	.0900	39 916.8	11.108	11.085	123.129	95
100	5 529.1	.00018	.00002	.0900	61 422.9	11.109	11.093	123.233	100

10%   Compound Interest Factors   10%

	Single Payment		Uniform Payment Series				Arithmetic Gradient		
	Compound Amount Factor	Present Worth Factor	Sinking Fund Factor	Capital Recovery Factor	Compound Amount Factor	Present Worth Factor	Gradient Uniform Series	Gradient Present Worth	
	Find F Given P	Find P Given F	Find A Given F	Find A Given P	Find F Given A	Find P Given A	Find A Given G	Find P Given G	
n	F/P	P/F	A/F	A/P	F/A	P/A	A/G	P/G	n
1	1.100	.9091	1.0000	1.1000	1.000	0.909	0	0	1
2	1.210	.8264	.4762	.5762	2.100	1.736	0.476	0.826	2
3	1.331	.7513	.3021	.4021	3.310	2.487	0.937	2.329	3
4	1.464	.6830	.2155	.3155	4.641	3.170	1.381	4.378	4
5	1.611	.6209	.1638	.2638	6.105	3.791	1.810	6.862	5
6	1.772	.5645	.1296	.2296	7.716	4.355	2.224	9.684	6
7	1.949	.5132	.1054	.2054	9.487	4.868	2.622	12.763	7
8	2.144	.4665	.0874	.1874	11.436	5.335	3.004	16.029	8
9	2.358	.4241	.0736	.1736	13.579	5.759	3.372	19.421	9
10	2.594	.3855	.0627	.1627	15.937	6.145	3.725	22.891	10
11	2.853	.3505	.0540	.1540	18.531	6.495	4.064	26.396	11
12	3.138	.3186	.0468	.1468	21.384	6.814	4.388	29.901	12
13	3.452	.2897	.0408	.1408	24.523	7.103	4.699	33.377	13
14	3.797	.2633	.0357	.1357	27.975	7.367	4.996	36.801	14
15	4.177	.2394	.0315	.1315	31.772	7.606	5.279	40.152	15
16	4.595	.2176	.0278	.1278	35.950	7.824	5.549	43.416	16
17	5.054	.1978	.0247	.1247	40.545	8.022	5.807	46.582	17
18	5.560	.1799	.0219	.1219	45.599	8.201	6.053	49.640	18
19	6.116	.1635	.0195	.1195	51.159	8.365	6.286	52.583	19
20	6.728	.1486	.0175	.1175	57.275	8.514	6.508	55.407	20
21	7.400	.1351	.0156	.1156	64.003	8.649	6.719	58.110	21
22	8.140	.1228	.0140	.1140	71.403	8.772	6.919	60.689	22
23	8.954	.1117	.0126	.1126	79.543	8.883	7.108	63.146	23
24	9.850	.1015	.0113	.1113	88.497	8.985	7.288	65.481	24
25	10.835	.0923	.0102	.1102	98.347	9.077	7.458	67.696	25
26	11.918	.0839	.00916	.1092	109.182	9.161	7.619	69.794	26
27	13.110	.0763	.00826	.1083	121.100	9.237	7.770	71.777	27
28	14.421	.0693	.00745	.1075	134.210	9.307	7.914	73.650	28
29	15.863	.0630	.00673	.1067	148.631	9.370	8.049	75.415	29
30	17.449	.0573	.00608	.1061	164.494	9.427	8.176	77.077	30
31	19.194	.0521	.00550	.1055	181.944	9.479	8.296	78.640	31
32	21.114	.0474	.00497	.1050	201.138	9.526	8.409	80.108	32
33	23.225	.0431	.00450	.1045	222.252	9.569	8.515	81.486	33
34	25.548	.0391	.00407	.1041	245.477	9.609	8.615	82.777	34
35	28.102	.0356	.00369	.1037	271.025	9.644	8.709	83.987	35
40	45.259	.0221	.00226	.1023	442.593	9.779	9.096	88.953	40
45	72.891	.0137	.00139	.1014	718.905	9.863	9.374	92.454	45
50	117.391	.00852	.00086	.1009	1 163.9	9.915	9.570	94.889	50
55	189.059	.00529	.00053	.1005	1 880.6	9.947	9.708	96.562	55
60	304.482	.00328	.00033	.1003	3 034.8	9.967	9.802	97.701	60
65	490.371	.00204	.00020	.1002	4 893.7	9.980	9.867	98.471	65
70	789.748	.00127	.00013	.1001	7 887.5	9.987	9.911	98.987	70
75	1 271.9	.00079	.00008	.1001	12 709.0	9.992	9.941	99.332	75
80	2 048.4	.00049	.00005	.1000	20 474.0	9.995	9.961	99.561	80
85	3 299.0	.00030	.00003	.1000	32 979.7	9.997	9.974	99.712	85
90	5 313.0	.00019	.00002	.1000	53 120.3	9.998	9.983	99.812	90
95	8 556.7	.00012	.00001	.1000	85 556.9	9.999	9.989	99.877	95
100	13 780.6	.00007	.00001	.1000	137 796.3	9.999	9.993	99.920	100

12%

Compound Interest Factors

12%

	Single Payment		Uniform Payment Series				Arithmetic Gradient		
	Compound Amount Factor	Present Worth Factor	Sinking Fund Factor	Capital Recovery Factor	Compound Amount Factor	Present Worth Factor	Gradient Uniform Series	Gradient Present Worth	
	Find F Given P	Find P Given F	Find A Given F	Find A Given P	Find F Given A	Find P Given A	Find A Given G	Find P Given G	
n	F/P	P/F	A/F	A/P	F/A	P/A	A/G	P/G	n
1	1.120	.8929	1.0000	1.1200	1.000	0.893	0	0	1
2	1.254	.7972	.4717	.5917	2.120	1.690	0.472	0.797	2
3	1.405	.7118	.2963	.4163	3.374	2.402	0.925	2.221	3
4	1.574	.6355	.2092	.3292	4.779	3.037	1.359	4.127	4
5	1.762	.5674	.1574	.2774	6.353	3.605	1.775	6.397	5
6	1.974	.5066	.1232	.2432	8.115	4.111	2.172	8.930	6
7	2.211	.4523	.0991	.2191	10.089	4.564	2.551	11.644	7
8	2.476	.4039	.0813	.2013	12.300	4.968	2.913	14.471	8
9	2.773	.3606	.0677	.1877	14.776	5.328	3.257	17.356	9
10	3.106	.3220	.0570	.1770	17.549	5.650	3.585	20.254	10
11	3.479	.2875	.0484	.1684	20.655	5.938	3.895	23.129	11
12	3.896	.2567	.0414	.1614	24.133	6.194	4.190	25.952	12
13	4.363	.2292	.0357	.1557	28.029	6.424	4.468	28.702	13
14	4.887	.2046	.0309	.1509	32.393	6.628	4.732	31.362	14
15	5.474	.1827	.0268	.1468	37.280	6.811	4.980	33.920	15
16	6.130	.1631	.0234	.1434	42.753	6.974	5.215	36.367	16
17	6.866	.1456	.0205	.1405	48.884	7.120	5.435	38.697	17
18	7.690	.1300	.0179	.1379	55.750	7.250	5.643	40.908	18
19	8.613	.1161	.0158	.1358	63.440	7.366	5.838	42.998	19
20	9.646	.1037	.0139	.1339	72.052	7.469	6.020	44.968	20
21	10.804	.0926	.0122	.1322	81.699	7.562	6.191	46.819	21
22	12.100	.0826	.0108	.1308	92.503	7.645	6.351	48.554	22
23	13.552	.0738	.00956	.1296	104.603	7.718	6.501	50.178	23
24	15.179	.0659	.00846	.1285	118.155	7.784	6.641	51.693	24
25	17.000	.0588	.00750	.1275	133.334	7.843	6.771	53.105	25
26	19.040	.0525	.00665	.1267	150.334	7.896	6.892	54.418	26
27	21.325	.0469	.00590	.1259	169.374	7.943	7.005	55.637	27
28	23.884	.0419	.00524	.1252	190.699	7.984	7.110	56.767	28
29	26.750	.0374	.00466	.1247	214.583	8.022	7.207	57.814	29
30	29.960	.0334	.00414	.1241	241.333	8.055	7.297	58.782	30
31	33.555	.0298	.00369	.1237	271.293	8.085	7.381	59.676	31
32	37.582	.0266	.00328	.1233	304.848	8.112	7.459	60.501	32
33	42.092	.0238	.00292	.1229	342.429	8.135	7.530	61.261	33
34	47.143	.0212	.00260	.1226	384.521	8.157	7.596	61.961	34
35	52.800	.0189	.00232	.1223	431.663	8.176	7.658	62.605	35
40	93.051	.0107	.00130	.1213	767.091	8.244	7.899	65.116	40
45	163.988	.00610	.00074	.1207	1 358.2	8.283	8.057	66.734	45
50	289.002	.00346	.00042	.1204	2 400.0	8.304	8.160	67.762	50
55	509.321	.00196	.00024	.1202	4 236.0	8.317	8.225	68.408	55
60	897.597	.00111	.00013	.1201	7 471.6	8.324	8.266	68.810	60
65	1 581.9	.00063	.00008	.1201	13 173.9	8.328	8.292	69.058	65
70	2 787.8	.00036	.00004	.1200	23 223.3	8.330	8.308	69.210	70
75	4 913.1	.00020	.00002	.1200	40 933.8	8.332	8.318	69.303	75
80	8 658.5	.00012	.00001	.1200	72 145.7	8.332	8.324	69.359	80
85	15 259.2	.00007	.00001	.1200	127 151.7	8.333	8.328	69.393	85
90	26 891.9	.00004		.1200	224 091.1	8.333	8.330	69.414	90
95	47 392.8	.00002		.1200	394 931.4	8.333	8.331	69.426	95
100	83 522.3	.00001		.1200	696 010.5	8.333	8.332	69.434	100

15% Compound Interest Factors 15%

	Single Payment		Uniform Payment Series				Arithmetic Gradient		
	Compound Amount Factor	Present Worth Factor	Sinking Fund Factor	Capital Recovery Factor	Compound Amount Factor	Present Worth Factor	Gradient Uniform Series	Gradient Present Worth	
	Find F Given P	Find P Given F	Find A Given F	Find A Given P	Find F Given A	Find P Given A	Find A Given G	Find P Given G	
n	F/P	P/F	A/F	A/P	F/A	P/A	A/G	P/G	n
1	1.150	.8696	1.0000	1.1500	1.000	0.870	0	0	1
2	1.322	.7561	.4651	.6151	2.150	1.626	0.465	0.756	2
3	1.521	.6575	.2880	.4380	3.472	2.283	0.907	2.071	3
4	1.749	.5718	.2003	.3503	4.993	2.855	1.326	3.786	4
5	2.011	.4972	.1483	.2983	6.742	3.352	1.723	5.775	5
6	2.313	.4323	.1142	.2642	8.754	3.784	2.097	7.937	6
7	2.660	.3759	.0904	.2404	11.067	4.160	2.450	10.192	7
8	3.059	.3269	.0729	.2229	13.727	4.487	2.781	12.481	8
9	3.518	.2843	.0596	.2096	16.786	4.772	3.092	14.755	9
10	4.046	.2472	.0493	.1993	20.304	5.019	3.383	16.979	10
11	4.652	.2149	.0411	.1911	24.349	5.234	3.655	19.129	11
12	5.350	.1869	.0345	.1845	29.002	5.421	3.908	21.185	12
13	6.153	.1625	.0291	.1791	34.352	5.583	4.144	23.135	13
14	7.076	.1413	.0247	.1747	40.505	5.724	4.362	24.972	14
15	8.137	.1229	.0210	.1710	47.580	5.847	4.565	26.693	15
16	9.358	.1069	.0179	.1679	55.717	5.954	4.752	28.296	16
17	10.761	.0929	.0154	.1654	65.075	6.047	4.925	29.783	17
18	12.375	.0808	.0132	.1632	75.836	6.128	5.084	31.156	18
19	14.232	.0703	.0113	.1613	88.212	6.198	5.231	32.421	19
20	16.367	.0611	.00976	.1598	102.444	6.259	5.365	33.582	20
21	18.822	.0531	.00842	.1584	118.810	6.312	5.488	34.645	21
22	21.645	.0462	.00727	.1573	137.632	6.359	5.601	35.615	22
23	24.891	.0402	.00628	.1563	159.276	6.399	5.704	36.499	23
24	28.625	.0349	.00543	.1554	184.168	6.434	5.798	37.302	24
25	32.919	.0304	.00470	.1547	212.793	6.464	5.883	38.031	25
26	37.857	.0264	.00407	.1541	245.712	6.491	5.961	38.692	26
27	43.535	.0230	.00353	.1535	283.569	6.514	6.032	39.289	27
28	50.066	.0200	.00306	.1531	327.104	6.534	6.096	39.828	28
29	57.575	.0174	.00265	.1527	377.170	6.551	6.154	40.315	29
30	66.212	.0151	.00230	.1523	434.745	6.566	6.207	40.753	30
31	76.144	.0131	.00200	.1520	500.957	6.579	6.254	41.147	31
32	87.565	.0114	.00173	.1517	577.100	6.591	6.297	41.501	32
33	100.700	.00993	.00150	.1515	664.666	6.600	6.336	41.818	33
34	115.805	.00864	.00131	.1513	765.365	6.609	6.371	42.103	34
35	133.176	.00751	.00113	.1511	881.170	6.617	6.402	42.359	35
40	267.864	.00373	.00056	.1506	1 779.1	6.642	6.517	43.283	40
45	538.769	.00186	.00028	.1503	3 585.1	6.654	6.583	43.805	45
50	1 083.7	.00092	.00014	.1501	7 217.7	6.661	6.620	44.096	50
55	2 179.6	.00046	.00007	.1501	14 524.1	6.664	6.641	44.256	55
60	4 384.0	.00023	.00003	.1500	29 220.0	6.665	6.653	44.343	60
65	8 817.8	.00011	.00002	.1500	58 778.6	6.666	6.659	44.390	65
70	17 735.7	.00006	.00001	.1500	118 231.5	6.666	6.663	44.416	70
75	35 672.9	.00003		.1500	237 812.5	6.666	6.665	44.429	75
80	71 750.9	.00001		.1500	478 332.6	6.667	6.666	44.436	80
85	144 316.7	.00001		.1500	962 104.4	6.667	6.666	44.440	85

18%   **Compound Interest Factors**   **18%**

	Single Payment		Uniform Payment Series				Arithmetic Gradient		
	Compound Amount Factor	Present Worth Factor	Sinking Fund Factor	Capital Recovery Factor	Compound Amount Factor	Present Worth Factor	Gradient Uniform Series	Gradient Present Worth	
	Find F Given P	Find P Given F	Find A Given F	Find A Given P	Find F Given A	Find P Given A	Find A Given G	Find P Given G	
n	F/P	P/F	A/F	A/P	F/A	P/A	A/G	P/G	n
1	1.180	.8475	1.0000	1.1800	1.000	0.847	0	0	1
2	1.392	.7182	.4587	.6387	2.180	1.566	0.459	0.718	2
3	1.643	.6086	.2799	.4599	3.572	2.174	0.890	1.935	3
4	1.939	.5158	.1917	.3717	5.215	2.690	1.295	3.483	4
5	2.288	.4371	.1398	.3198	7.154	3.127	1.673	5.231	5
6	2.700	.3704	.1059	.2859	9.442	3.498	2.025	7.083	6
7	3.185	.3139	.0824	.2624	12.142	3.812	2.353	8.967	7
8	3.759	.2660	.0652	.2452	15.327	4.078	2.656	10.829	8
9	4.435	.2255	.0524	.2324	19.086	4.303	2.936	12.633	9
10	5.234	.1911	.0425	.2225	23.521	4.494	3.194	14.352	10
11	6.176	.1619	.0348	.2148	28.755	4.656	3.430	15.972	11
12	7.288	.1372	.0286	.2086	34.931	4.793	3.647	17.481	12
13	8.599	.1163	.0237	.2037	42.219	4.910	3.845	18.877	13
14	10.147	.0985	.0197	.1997	50.818	5.008	4.025	20.158	14
15	11.974	.0835	.0164	.1964	60.965	5.092	4.189	21.327	15
16	14.129	.0708	.0137	.1937	72.939	5.162	4.337	22.389	16
17	16.672	.0600	.0115	.1915	87.068	5.222	4.471	23.348	17
18	19.673	.0508	.00964	.1896	103.740	5.273	4.592	24.212	18
19	23.214	.0431	.00810	.1881	123.413	5.316	4.700	24.988	19
20	27.393	.0365	.00682	.1868	146.628	5.353	4.798	25.681	20
21	32.324	.0309	.00575	.1857	174.021	5.384	4.885	26.300	21
22	38.142	.0262	.00485	.1848	206.345	5.410	4.963	26.851	22
23	45.008	.0222	.00409	.1841	244.487	5.432	5.033	27.339	23
24	53.109	.0188	.00345	.1835	289.494	5.451	5.095	27.772	24
25	62.669	.0160	.00292	.1829	342.603	5.467	5.150	28.155	25
26	73.949	.0135	.00247	.1825	405.272	5.480	5.199	28.494	26
27	87.260	.0115	.00209	.1821	479.221	5.492	5.243	28.791	27
28	102.966	.00971	.00177	.1818	566.480	5.502	5.281	29.054	28
29	121.500	.00823	.00149	.1815	669.447	5.510	5.315	29.284	29
30	143.370	.00697	.00126	.1813	790.947	5.517	5.345	29.486	30
31	169.177	.00591	.00107	.1811	934.317	5.523	5.371	29.664	31
32	199.629	.00501	.00091	.1809	1 103.5	5.528	5.394	29.819	32
33	235.562	.00425	.00077	.1808	1 303.1	5.532	5.415	29.955	33
34	277.963	.00360	.00065	.1806	1 538.7	5.536	5.433	30.074	34
35	327.997	.00305	.00055	.1806	1 816.6	5.539	5.449	30.177	35
40	750.377	.00133	.00024	.1802	4 163.2	5.548	5.502	30.527	40
45	1 716.7	.00058	.00010	.1801	9 531.6	5.552	5.529	30.701	45
50	3 927.3	.00025	.00005	.1800	21 813.0	5.554	5.543	30.786	50
55	8 984.8	.00011	.00002	.1800	49 910.1	5.555	5.549	30.827	55
60	20 555.1	.00005	.00001	.1800	114 189.4	5.555	5.553	30.846	60
65	47 025.1	.00002		.1800	261 244.7	5.555	5.554	30.856	65
70	107 581.9	.00001		.1800	597 671.7	5.556	5.555	30.860	70

20%   Compound Interest Factors   20%

	Single Payment		Uniform Payment Series				Arithmetic Gradient		
	Compound Amount Factor	Present Worth Factor	Sinking Fund Factor	Capital Recovery Factor	Compound Amount Factor	Present Worth Factor	Gradient Uniform Series	Gradient Present Worth	
	Find F Given P	Find P Given F	Find A Given F	Find A Given P	Find F Given A	Find P Given A	Find A Given G	Find P Given G	
n	F/P	P/F	A/F	A/P	F/A	P/A	A/G	P/G	n
1	1.200	.8333	1.0000	1.2000	1.000	0.833	0	0	1
2	1.440	.6944	.4545	.6545	2.200	1.528	0.455	0.694	2
3	1.728	.5787	.2747	.4747	3.640	2.106	0.879	1.852	3
4	2.074	.4823	.1863	.3863	5.368	2.589	1.274	3.299	4
5	2.488	.4019	.1344	.3344	7.442	2.991	1.641	4.906	5
6	2.986	.3349	.1007	.3007	9.930	3.326	1.979	6.581	6
7	3.583	.2791	.0774	.2774	12.916	3.605	2.290	8.255	7
8	4.300	.2326	.0606	.2606	16.499	3.837	2.576	9.883	8
9	5.160	.1938	.0481	.2481	20.799	4.031	2.836	11.434	9
10	6.192	.1615	.0385	.2385	25.959	4.192	3.074	12.887	10
11	7.430	.1346	.0311	.2311	32.150	4.327	3.289	14.233	11
12	8.916	.1122	.0253	.2253	39.581	4.439	3.484	15.467	12
13	10.699	.0935	.0206	.2206	48.497	4.533	3.660	16.588	13
14	12.839	.0779	.0169	.2169	59.196	4.611	3.817	17.601	14
15	15.407	.0649	.0139	.2139	72.035	4.675	3.959	18.509	15
16	18.488	.0541	.0114	.2114	87.442	4.730	4.085	19.321	16
17	22.186	.0451	.00944	.2094	105.931	4.775	4.198	20.042	17
18	26.623	.0376	.00781	.2078	128.117	4.812	4.298	20.680	18
19	31.948	.0313	.00646	.2065	154.740	4.843	4.386	21.244	19
20	38.338	.0261	.00536	.2054	186.688	4.870	4.464	21.739	20
21	46.005	.0217	.00444	.2044	225.026	4.891	4.533	22.174	21
22	55.206	.0181	.00369	.2037	271.031	4.909	4.594	22.555	22
23	66.247	.0151	.00307	.2031	326.237	4.925	4.647	22.887	23
24	79.497	.0126	.00255	.2025	392.484	4.937	4.694	23.176	24
25	95.396	.0105	.00212	.2021	471.981	4.948	4.735	23.428	25
26	114.475	.00874	.00176	.2018	567.377	4.956	4.771	23.646	26
27	137.371	.00728	.00147	.2015	681.853	4.964	4.802	23.835	27
28	164.845	.00607	.00122	.2012	819.223	4.970	4.829	23.999	28
29	197.814	.00506	.00102	.2010	984.068	4.975	4.853	24.141	29
30	237.376	.00421	.00085	.2008	1 181.9	4.979	4.873	24.263	30
31	284.852	.00351	.00070	.2007	1 419.3	4.982	4.891	24.368	31
32	341.822	.00293	.00059	.2006	1 704.1	4.985	4.906	24.459	32
33	410.186	.00244	.00049	.2005	2 045.9	4.988	4.919	24.537	33
34	492.224	.00203	.00041	.2004	2 456.1	4.990	4.931	24.604	34
35	590.668	.00169	.00034	.2003	2 948.3	4.992	4.941	24.661	35
40	1 469.8	.00068	.00014	.2001	7 343.9	4.997	4.973	24.847	40
45	3 657.3	.00027	.00005	.2001	18 281.3	4.999	4.988	24.932	45
50	9 100.4	.00011	.00002	.2000	45 497.2	4.999	4.995	24.970	50
55	22 644.8	.00004	.00001	.2000	113 219.0	5.000	4.998	24.987	55
60	56 347.5	.00002		.2000	281 732.6	5.000	4.999	24.994	60

25%   Compound Interest Factors   25%

	Single Payment		Uniform Payment Series				Arithmetic Gradient		
	Compound Amount Factor	Present Worth Factor	Sinking Fund Factor	Capital Recovery Factor	Compound Amount Factor	Present Worth Factor	Gradient Uniform Series	Gradient Present Worth	
	Find F Given P	Find P Given F	Find A Given F	Find A Given P	Find F Given A	Find P Given A	Find A Given G	Find P Given G	
n	F/P	P/F	A/F	A/P	F/A	P/A	A/G	P/G	n
1	1.250	.8000	1.0000	1.2500	1.000	0.800	0	0	1
2	1.563	.6400	.4444	.6944	2.250	1.440	0.444	0.640	2
3	1.953	.5120	.2623	.5123	3.813	1.952	0.852	1.664	3
4	2.441	.4096	.1734	.4234	5.766	2.362	1.225	2.893	4
5	3.052	.3277	.1218	.3718	8.207	2.689	1.563	4.204	5
6	3.815	.2621	.0888	.3388	11.259	2.951	1.868	5.514	6
7	4.768	.2097	.0663	.3163	15.073	3.161	2.142	6.773	7
8	5.960	.1678	.0504	.3004	19.842	3.329	2.387	7.947	8
9	7.451	.1342	.0388	.2888	25.802	3.463	2.605	9.021	9
10	9.313	.1074	.0301	.2801	33.253	3.571	2.797	9.987	10
11	11.642	.0859	.0235	.2735	42.566	3.656	2.966	10.846	11
12	14.552	.0687	.0184	.2684	54.208	3.725	3.115	11.602	12
13	18.190	.0550	.0145	.2645	68.760	3.780	3.244	12.262	13
14	22.737	.0440	.0115	.2615	86.949	3.824	3.356	12.833	14
15	28.422	.0352	.00912	.2591	109.687	3.859	3.453	13.326	15
16	35.527	.0281	.00724	.2572	138.109	3.887	3.537	13.748	16
17	44.409	.0225	.00576	.2558	173.636	3.910	3.608	14.108	17
18	55.511	.0180	.00459	.2546	218.045	3.928	3.670	14.415	18
19	69.389	.0144	.00366	.2537	273.556	3.942	3.722	14.674	19
20	86.736	.0115	.00292	.2529	342.945	3.954	3.767	14.893	20
21	108.420	.00922	.00233	.2523	429.681	3.963	3.805	15.078	21
22	135.525	.00738	.00186	.2519	538.101	3.970	3.836	15.233	22
23	169.407	.00590	.00148	.2515	673.626	3.976	3.863	15.362	23
24	211.758	.00472	.00119	.2512	843.033	3.981	3.886	15.471	24
25	264.698	.00378	.00095	.2509	1 054.8	3.985	3.905	15.562	25
26	330.872	.00302	.00076	.2508	1 319.5	3.988	3.921	15.637	26
27	413.590	.00242	.00061	.2506	1 650.4	3.990	3.935	15.700	27
28	516.988	.00193	.00048	.2505	2 064.0	3.992	3.946	15.752	28
29	646.235	.00155	.00039	.2504	2 580.9	3.994	3.955	15.796	29
30	807.794	.00124	.00031	.2503	3 227.2	3.995	3.963	15.832	30
31	1 009.7	.00099	.00025	.2502	4 035.0	3.996	3.969	15.861	31
32	1 262.2	.00079	.00020	.2502	5 044.7	3.997	3.975	15.886	32
33	1 577.7	.00063	.00016	.2502	6 306.9	3.997	3.979	15.906	33
34	1 972.2	.00051	.00013	.2501	7 884.6	3.998	3.983	15.923	34
35	2 465.2	.00041	.00010	.2501	9 856.8	3.998	3.986	15.937	35
40	7 523.2	.00013	.00003	.2500	30 088.7	3.999	3.995	15.977	40
45	22 958.9	.00004	.00001	.2500	91 831.5	4.000	3.998	15.991	45
50	70 064.9	.00001		.2500	280 255.7	4.000	3.999	15.997	50
55	213 821.2			.2500	855 280.7	4.000	4.000	15.999	55

30% Compound Interest Factors 30%

	Single Payment		Uniform Payment Series				Arithmetic Gradient		
	Compound Amount Factor	Present Worth Factor	Sinking Fund Factor	Capital Recovery Factor	Compound Amount Factor	Present Worth Factor	Gradient Uniform Series	Gradient Present Worth	
	Find F Given P	Find P Given F	Find A Given F	Find A Given P	Find F Given A	Find P Given A	Find A Given G	Find P Given G	
n	F/P	P/F	A/F	A/P	F/A	P/A	A/G	P/G	n
1	1.300	.7692	1.0000	1.3000	1.000	0.769	0	0	1
2	1.690	.5917	.4348	.7348	2.300	1.361	0.435	0.592	2
3	2.197	.4552	.2506	.5506	3.990	1.816	0.827	1.502	3
4	2.856	.3501	.1616	.4616	6.187	2.166	1.178	2.552	4
5	3.713	.2693	.1106	.4106	9.043	2.436	1.490	3.630	5
6	4.827	.2072	.0784	.3784	12.756	2.643	1.765	4.666	6
7	6.275	.1594	.0569	.3569	17.583	2.802	2.006	5.622	7
8	8.157	.1226	.0419	.3419	23.858	2.925	2.216	6.480	8
9	10.604	.0943	.0312	.3312	32.015	3.019	2.396	7.234	9
10	13.786	.0725	.0235	.3235	42.619	3.092	2.551	7.887	10
11	17.922	.0558	.0177	.3177	56.405	3.147	2.683	8.445	11
12	23.298	.0429	.0135	.3135	74.327	3.190	2.795	8.917	12
13	30.287	.0330	.0102	.3102	97.625	3.223	2.889	9.314	13
14	39.374	.0254	.00782	.3078	127.912	3.249	2.969	9.644	14
15	51.186	.0195	.00598	.3060	167.286	3.268	3.034	9.917	15
16	66.542	.0150	.00458	.3046	218.472	3.283	3.089	10.143	16
17	86.504	.0116	.00351	.3035	285.014	3.295	3.135	10.328	17
18	112.455	.00889	.00269	.3027	371.518	3.304	3.172	10.479	18
19	146.192	.00684	.00207	.3021	483.973	3.311	3.202	10.602	19
20	190.049	.00526	.00159	.3016	630.165	3.316	3.228	10.702	20
21	247.064	.00405	.00122	.3012	820.214	3.320	3.248	10.783	21
22	321.184	.00311	.00094	.3009	1 067.3	3.323	3.265	10.848	22
23	417.539	.00239	.00072	.3007	1 388.5	3.325	3.278	10.901	23
24	542.800	.00184	.00055	.3006	1 806.0	3.327	3.289	10.943	24
25	705.640	.00142	.00043	.3004	2 348.8	3.329	3.298	10.977	25
26	917.332	.00109	.00033	.3003	3 054.4	3.330	3.305	11.005	26
27	1 192.5	.00084	.00025	.3003	3 971.8	3.331	3.311	11.026	27
28	1 550.3	.00065	.00019	.3002	5 164.3	3.331	3.315	11.044	28
29	2 015.4	.00050	.00015	.3001	6 714.6	3.332	3.319	11.058	29
30	2 620.0	.00038	.00011	.3001	8 730.0	3.332	3.322	11.069	30
31	3 406.0	.00029	.00009	.3001	11 350.0	3.332	3.324	11.078	31
32	4 427.8	.00023	.00007	.3001	14 756.0	3.333	3.326	11.085	32
33	5 756.1	.00017	.00005	.3001	19 183.7	3.333	3.328	11.090	33
34	7 483.0	.00013	.00004	.3000	24 939.9	3.333	3.329	11.094	34
35	9 727.8	.00010	.00003	.3000	32 422.8	3.333	3.330	11.098	35
40	36 118.8	.00003	.00001	.3000	120 392.6	3.333	3.332	11.107	40
45	134 106.5	.00001		.3000	447 018.3	3.333	3.333	11.110	45

35%

Compound Interest Factors

35%

	Single Payment		Uniform Payment Series				Arithmetic Gradient		
	Compound Amount Factor	Present Worth Factor	Sinking Fund Factor	Capital Recovery Factor	Compound Amount Factor	Present Worth Factor	Gradient Uniform Series	Gradient Present Worth	
	Find F Given P	Find P Given F	Find A Given F	Find A Given P	Find F Given A	Find P Given A	Find A Given G	Find P Given G	
n	F/P	P/F	A/F	A/P	F/A	P/A	A/G	P/G	n
1	1.350	.7407	1.0000	1.3500	1.000	0.741	0	0	1
2	1.822	.5487	.4255	.7755	2.350	1.289	0.426	0.549	2
3	2.460	.4064	.2397	.5897	4.173	1.696	0.803	1.362	3
4	3.322	.3011	.1508	.5008	6.633	1.997	1.134	2.265	4
5	4.484	.2230	.1005	.4505	9.954	2.220	1.422	3.157	5
6	6.053	.1652	.0693	.4193	14.438	2.385	1.670	3.983	6
7	8.172	.1224	.0488	.3988	20.492	2.508	1.881	4.717	7
8	11.032	.0906	.0349	.3849	28.664	2.598	2.060	5.352	8
9	14.894	.0671	.0252	.3752	39.696	2.665	2.209	5.889	9
10	20.107	.0497	.0183	.3683	54.590	2.715	2.334	6.336	10
11	27.144	.0368	.0134	.3634	74.697	2.752	2.436	6.705	11
12	36.644	.0273	.00982	.3598	101.841	2.779	2.520	7.005	12
13	49.470	.0202	.00722	.3572	138.485	2.799	2.589	7.247	13
14	66.784	.0150	.00532	.3553	187.954	2.814	2.644	7.442	14
15	90.158	.0111	.00393	.3539	254.739	2.825	2.689	7.597	15
16	121.714	.00822	.00290	.3529	344.897	2.834	2.725	7.721	16
17	164.314	.00609	.00214	.3521	466.611	2.840	2.753	7.818	17
18	221.824	.00451	.00158	.3516	630.925	2.844	2.776	7.895	18
19	299.462	.00334	.00117	.3512	852.748	2.848	2.793	7.955	19
20	404.274	.00247	.00087	.3509	1 152.2	2.850	2.808	8.002	20
21	545.769	.00183	.00064	.3506	1 556.5	2.852	2.819	8.038	21
22	736.789	.00136	.00048	.3505	2 102.3	2.853	2.827	8.067	22
23	994.665	.00101	.00035	.3504	2 839.0	2.854	2.834	8.089	23
24	1 342.8	.00074	.00026	.3503	3 833.7	2.855	2.839	8.106	24
25	1 812.8	.00055	.00019	.3502	5 176.5	2.856	2.843	8.119	25
26	2 447.2	.00041	.00014	.3501	6 989.3	2.856	2.847	8.130	26
27	3 303.8	.00030	.00011	.3501	9 436.5	2.856	2.849	8.137	27
28	4 460.1	.00022	.00008	.3501	12 740.3	2.857	2.851	8.143	28
29	6 021.1	.00017	.00006	.3501	17 200.4	2.857	2.852	8.148	29
30	8 128.5	.00012	.00004	.3500	23 221.6	2.857	2.853	8.152	30
31	10 973.5	.00009	.00003	.3500	31 350.1	2.857	2.854	8.154	31
32	14 814.3	.00007	.00002	.3500	42 323.7	2.857	2.855	8.157	32
33	19 999.3	.00005	.00002	.3500	57 137.9	2.857	2.855	8.158	33
34	26 999.0	.00004	.00001	.3500	77 137.2	2.857	2.856	8.159	34
35	36 448.7	.00003	.00001	.3500	104 136.3	2.857	2.856	8.160	35

40%   Compound Interest Factors   **40%**

	Single Payment		Uniform Payment Series				Arithmetic Gradient		
	Compound Amount Factor	Present Worth Factor	Sinking Fund Factor	Capital Recovery Factor	Compound Amount Factor	Present Worth Factor	Gradient Uniform Series	Gradient Present Worth	
	Find F Given P	Find P Given F	Find A Given F	Find A Given P	Find F Given A	Find P Given A	Find A Given G	Find P Given G	
n	F/P	P/F	A/F	A/P	F/A	P/A	A/G	P/G	n
1	1.400	.7143	1.0000	1.4000	1.000	0.714	0	0	1
2	1.960	.5102	.4167	.8167	2.400	1.224	0.417	0.510	2
3	2.744	.3644	.2294	.6294	4.360	1.589	0.780	1.239	3
4	3.842	.2603	.1408	.5408	7.104	1.849	1.092	2.020	4
5	5.378	.1859	.0914	.4914	10.946	2.035	1.358	2.764	5
6	7.530	.1328	.0613	.4613	16.324	2.168	1.581	3.428	6
7	10.541	.0949	.0419	.4419	23.853	2.263	1.766	3.997	7
8	14.758	.0678	.0291	.4291	34.395	2.331	1.919	4.471	8
9	20.661	.0484	.0203	.4203	49.153	2.379	2.042	4.858	9
10	28.925	.0346	.0143	.4143	69.814	2.414	2.142	5.170	10
11	40.496	.0247	.0101	.4101	98.739	2.438	2.221	5.417	11
12	56.694	.0176	.00718	.4072	139.235	2.456	2.285	5.611	12
13	79.371	.0126	.00510	.4051	195.929	2.469	2.334	5.762	13
14	111.120	.00900	.00363	.4036	275.300	2.478	2.373	5.879	14
15	155.568	.00643	.00259	.4026	386.420	2.484	2.403	5.969	15
16	217.795	.00459	.00185	.4018	541.988	2.489	2.426	6.038	16
17	304.913	.00328	.00132	.4013	759.783	2.492	2.444	6.090	17
18	426.879	.00234	.00094	.4009	1 064.7	2.494	2.458	6.130	18
19	597.630	.00167	.00067	.4007	1 419.6	2.496	2.468	6.160	19
20	836.682	.00120	.00048	.4005	2 089.2	2.497	2.476	6.183	20
21	1 171.4	.00085	.00034	.4003	2 925.9	2.498	2.482	6.200	21
22	1 639.9	.00061	.00024	.4002	4 097.2	2.498	2.487	6.213	22
23	2 295.9	.00044	.00017	.4002	5 737.1	2.499	2.490	6.222	23
24	3 214.2	.00031	.00012	.4001	8 033.0	2.499	2.493	6.229	24
25	4 499.9	.00022	.00009	.4001	11 247.2	2.499	2.494	6.235	25
26	6 299.8	.00016	.00006	.4001	15 747.1	2.500	2.496	6.239	26
27	8 819.8	.00011	.00005	.4000	22 046.9	2.500	2.497	6.242	27
28	12 347.7	.00008	.00003	.4000	30 866.7	2.500	2.498	6.244	28
29	17 286.7	.00006	.00002	.4000	43 214.3	2.500	2.498	6.245	29
30	24 201.4	.00004	.00002	.4000	60 501.0	2.500	2.499	6.247	30
31	33 882.0	.00003	.00001	.4000	84 702.5	2.500	2.499	6.248	31
32	47 434.8	.00002	.00001	.4000	118 584.4	2.500	2.499	6.248	32
33	66 408.7	.00002	.00001	.4000	166 019.2	2.500	2.500	6.249	33
34	92 972.1	.00001		.4000	232 427.9	2.500	2.500	6.249	34
35	130 161.0	.00001		.4000	325 400.0	2.500	2.500	6.249	35

45%   Compound Interest Factors   **45%**

	Single Payment		Uniform Payment Series				Arithmetic Gradient		
	Compound Amount Factor	Present Worth Factor	Sinking Fund Factor	Capital Recovery Factor	Compound Amount Factor	Present Worth Factor	Gradient Uniform Series	Gradient Present Worth	
	Find F Given P F/P	Find P Given F P/F	Find A Given F A/F	Find A Given P A/P	Find F Given A F/A	Find P Given A P/A	Find A Given G A/G	Find P Given G P/G	
n									n
1	1.450	.6897	1.0000	1.4500	1.000	0.690	0	0	1
2	2.103	.4756	.4082	.8582	2.450	1.165	0.408	0.476	2
3	3.049	.3280	.2197	.6697	4.553	1.493	0.758	1.132	3
4	4.421	.2262	.1316	.5816	7.601	1.720	1.053	1.810	4
5	6.410	.1560	.0832	.5332	12.022	1.876	1.298	2.434	5
6	9.294	.1076	.0543	.5043	18.431	1.983	1.499	2.972	6
7	13.476	.0742	.0361	.4861	27.725	2.057	1.661	3.418	7
8	19.541	.0512	.0243	.4743	41.202	2.109	1.791	3.776	8
9	28.334	.0353	.0165	.4665	60.743	2.144	1.893	4.058	9
10	41.085	.0243	.0112	.4612	89.077	2.168	1.973	4.277	10
11	59.573	.0168	.00768	.4577	130.162	2.185	2.034	4.445	11
12	86.381	.0116	.00527	.4553	189.735	2.196	2.082	4.572	12
13	125.252	.00798	.00362	.4536	276.115	2.204	2.118	4.668	13
14	181.615	.00551	.00249	.4525	401.367	2.210	2.145	4.740	14
15	263.342	.00380	.00172	.4517	582.982	2.214	2.165	4.793	15
16	381.846	.00262	.00118	.4512	846.325	2.216	2.180	4.832	16
17	553.677	.00181	.00081	.4508	1 228.2	2.218	2.191	4.861	17
18	802.831	.00125	.00056	.4506	1 781.8	2.219	2.200	4.882	18
19	1 164.1	.00086	.00039	.4504	2 584.7	2.220	2.206	4.898	19
20	1 688.0	.00059	.00027	.4503	3 748.8	2.221	2.210	4.909	20
21	2 447.5	.00041	.00018	.4502	5 436.7	2.221	2.214	4.917	21
22	3 548.9	.00028	.00013	.4501	7 884.3	2.222	2.216	4.923	22
23	5 145.9	.00019	.00009	.4501	11 433.2	2.222	2.218	4.927	23
24	7 461.6	.00013	.00006	.4501	16 579.1	2.222	2.219	4.930	24
25	10 819.3	.00009	.00004	.4500	24 040.7	2.222	2.220	4.933	25
26	15 688.0	.00006	.00003	.4500	34 860.1	2.222	2.221	4.934	26
27	22 747.7	.00004	.00002	.4500	50 548.1	2.222	2.221	4.935	27
28	32 984.1	.00003	.00001	.4500	73 295.8	2.222	2.221	4.936	28
29	47 826.9	.00002	.00001	.4500	106 279.9	2.222	2.222	4.937	29
30	69 349.1	.00001	.00001	.4500	154 106.8	2.222	2.222	4.937	30
31	100 556.1	.00001		.4500	223 455.9	2.222	2.222	4.938	31
32	145 806.4	.00001		.4500	324 012.0	2.222	2.222	4.938	32
33	211 419.3			.4500	469 818.5	2.222	2.222	4.938	33
34	306 558.0			.4500	681 237.8	2.222	2.222	4.938	34
35	444 509.2			.4500	987 795.9	2.222	2.222	4.938	35

50%   Compound Interest Factors   **50%**

	Single Payment		Uniform Payment Series				Arithmetic Gradient		
	Compound Amount Factor	Present Worth Factor	Sinking Fund Factor	Capital Recovery Factor	Compound Amount Factor	Present Worth Factor	Gradient Uniform Series	Gradient Present Worth	
n	Find F Given P F/P	Find P Given F P/F	Find A Given F A/F	Find A Given P A/P	Find F Given A F/A	Find P Given A P/A	Find A Given G A/G	Find P Given G P/G	n
1	1.500	.6667	1.0000	1.5000	1.000	0.667	0	0	1
2	2.250	.4444	.4000	.9000	2.500	1.111	0.400	0.444	2
3	3.375	.2963	.2105	.7105	4.750	1.407	0.737	1.037	3
4	5.063	.1975	.1231	.6231	8.125	1.605	1.015	1.630	4
5	7.594	.1317	.0758	.5758	13.188	1.737	1.242	2.156	5
6	11.391	.0878	.0481	.5481	20.781	1.824	1.423	2.595	6
7	17.086	.0585	.0311	.5311	32.172	1.883	1.565	2.947	7
8	25.629	.0390	.0203	.5203	49.258	1.922	1.675	3.220	8
9	38.443	.0260	.0134	.5134	74.887	1.948	1.760	3.428	9
10	57.665	.0173	.00882	.5088	113.330	1.965	1.824	3.584	10
11	86.498	.0116	.00585	.5058	170.995	1.977	1.871	3.699	11
12	129.746	.00771	.00388	.5039	257.493	1.985	1.907	3.784	12
13	194.620	.00514	.00258	.5026	387.239	1.990	1.933	3.846	13
14	291.929	.00343	.00172	.5017	581.859	1.993	1.952	3.890	14
15	437.894	.00228	.00114	.5011	873.788	1.995	1.966	3.922	15
16	656.814	.00152	.00076	.5008	1 311.7	1.997	1.976	3.945	16
17	985.261	.00101	.00051	.5005	1 968.5	1.998	1.983	3.961	17
18	1 477.9	.00068	.00034	.5003	2 953.8	1.999	1.988	3.973	18
19	2 216.8	.00045	.00023	.5002	4 431.7	1.999	1.991	3.981	19
20	3 325.3	.00030	.00015	.5002	6 648.5	1.999	1.994	3.987	20
21	4 987.9	.00020	.00010	.5001	9 973.8	2.000	1.996	3.991	21
22	7 481.8	.00013	.00007	.5001	14 961.7	2.000	1.997	3.994	22
23	11 222.7	.00009	.00004	.5000	22 443.5	2.000	1.998	3.996	23
24	16 834.1	.00006	.00003	.5000	33 666.2	2.000	1.999	3.997	24
25	25 251.2	.00004	.00002	.5000	50 500.3	2.000	1.999	3.998	25
26	37 876.8	.00003	.00001	.5000	75 751.5	2.000	1.999	3.999	26
27	56 815.1	.00002	.00001	.5000	113 628.3	2.000	2.000	3.999	27
28	85 222.7	.00001	.00001	.5000	170 443.4	2.000	2.000	3.999	28
29	127 834.0	.00001		.5000	255 666.1	2.000	2.000	4.000	29
30	191 751.1	.00001		.5000	383 500.1	2.000	2.000	4.000	30
31	287 626.6			.5000	575 251.2	2.000	2.000	4.000	31
32	431 439.9			.5000	862 877.8	2.000	2.000	4.000	32

60% Compound Interest Factors **60%**

	Single Payment		Uniform Payment Series				Arithmetic Gradient		
	Compound Amount Factor	Present Worth Factor	Sinking Fund Factor	Capital Recovery Factor	Compound Amount Factor	Present Worth Factor	Gradient Uniform Series	Gradient Present Worth	
n	Find F Given P F/P	Find P Given F P/F	Find A Given F A/F	Find A Given P A/P	Find F Given A F/A	Find P Given A P/A	Find A Given G A/G	Find P Given G P/G	n
1	1.600	.6250	1.0000	1.6000	1.000	0.625	0	0	1
2	2.560	.3906	.3846	.9846	2.600	1.016	0.385	0.391	2
3	4.096	.2441	.1938	.7938	5.160	1.260	0.698	0.879	3
4	6.554	.1526	.1080	.7080	9.256	1.412	0.946	1.337	4
5	10.486	.0954	.0633	.6633	15.810	1.508	1.140	1.718	5
6	16.777	.0596	.0380	.6380	26.295	1.567	1.286	2.016	6
7	26.844	.0373	.0232	.6232	43.073	1.605	1.396	2.240	7
8	42.950	.0233	.0143	.6143	69.916	1.628	1.476	2.403	8
9	68.719	.0146	.00886	.6089	112.866	1.642	1.534	2.519	9
10	109.951	.00909	.00551	.6055	181.585	1.652	1.575	2.601	10
11	175.922	.00568	.00343	.6034	291.536	1.657	1.604	2.658	11
12	281.475	.00355	.00214	.6021	467.458	1.661	1.624	2.697	12
13	450.360	.00222	.00134	.6013	748.933	1.663	1.638	2.724	13
14	720.576	.00139	.00083	.6008	1 199.3	1.664	1.647	2.742	14
15	1 152.9	.00087	.00052	.6005	1 919.9	1.665	1.654	2.754	15
16	1 844.7	.00054	.00033	.6003	3 072.8	1.666	1.658	2.762	16
17	2 951.5	.00034	.00020	.6002	4 917.5	1.666	1.661	2.767	17
18	4 722.4	.00021	.00013	.6001	7 868.9	1.666	1.663	2.771	18
19	7 555.8	.00013	.00008	.6011	12 591.3	1.666	1.664	2.773	19
20	12 089.3	.00008	.00005	.6000	20 147.1	1.667	1.665	2.775	20
21	19 342.8	.00005	.00003	.6000	32 236.3	1.667	1.666	2.776	21
22	30 948.5	.00003	.00002	.6000	51 579.2	1.667	1.666	2.777	22
23	49 517.6	.00002	.00001	.6000	82 527.6	1.667	1.666	2.777	23
24	79 228.1	.00001	.00001	.6000	132 045.2	1.667	1.666	2.777	24
25	126 765.0	.00001		.6000	211 273.4	1.667	1.666	2.777	25
26	202 824.0			.6000	338 038.4	1.667	1.667	2.778	26
27	324 518.4			.6000	540 862.4	1.667	1.667	2.778	27
28	519 229.5			.6000	865 380.9	1.667	1.667	2.778	28

Continuous Compounding—Single Payment Factors

rn	Compound Amount Factor e^{rn} Find F Given P F/P	Present Worth Factor e^{-rn} Find P Given F P/F	rn	Compound Amount Factor e^{rn} Find F Given P F/P	Present Worth Factor e^{-rn} Find P Given F P/F
.01	1.0101	.9900	.51	1.6653	.6005
.02	1.0202	.9802	.52	1.6820	.5945
.03	1.0305	.9704	.53	1.6989	.5886
.04	1.0408	.9608	.54	1.7160	.5827
.05	1.0513	.9512	.55	1.7333	.5769
.06	1.0618	.9418	.56	1.7507	.5712
.07	1.0725	.9324	.57	1.7683	.5655
.08	1.0833	.9231	.58	1.7860	.5599
.09	1.0942	.9139	.59	1.8040	.5543
.10	1.1052	.9048	.60	1.8221	.5488
.11	1.1163	.8958	.61	1.8404	.5434
.12	1.1275	.8869	.62	1.8589	.5379
.13	1.1388	.8781	.63	1.8776	.5326
.14	1.1503	.8694	.64	1.8965	.5273
.15	1.1618	.8607	.65	1.9155	.5220
.16	1.1735	.8521	.66	1.9348	.5169
.17	1.1853	.8437	.67	1.9542	.5117
.18	1.1972	.8353	.68	1.9739	.5066
.19	1.2092	.8270	.69	1.9937	.5016
.20	1.2214	.8187	.70	2.0138	.4966
.21	1.2337	.8106	.71	2.0340	.4916
.22	1.2461	.8025	.72	2.0544	.4868
.23	1.2586	.7945	.73	2.0751	.4819
.24	1.2712	.7866	.74	2.0959	.4771
.25	1.2840	.7788	.75	2.1170	.4724
.26	1.2969	.7711	.76	2.1383	.4677
.27	1.3100	.7634	.77	2.1598	.4630
.28	1.3231	.7558	.78	2.1815	.4584
.29	1.3364	.7483	.79	2.2034	.4538
.30	1.3499	.7408	.80	2.2255	.4493
.31	1.3634	.7334	.81	2.2479	.4449
.32	1.3771	.7261	.82	2.2705	.4404
.33	1.3910	.7189	.83	2.2933	.4360
.34	1.4049	.7118	.84	2.3164	.4317
.35	1.4191	.7047	.85	2.3396	.4274
.36	1.4333	.6977	.86	2.3632	.4232
.37	1.4477	.6907	.87	2.3869	.4190
.38	1.4623	.6839	.88	2.4109	.4148
.39	1.4770	.6771	.89	2.4351	.4107
.40	1.4918	.6703	.90	2.4596	.4066
.41	1.5068	.6637	.91	2.4843	.4025
.42	1.5220	.6570	.92	2.5093	.3985
.43	1.5373	.6505	.93	2.5345	.3946
.44	1.5527	.6440	.94	2.5600	.3906
.45	1.5683	.6376	.95	2.5857	.3867
.46	1.5841	.6313	.96	2.6117	.3829
.47	1.6000	.6250	.97	2.6379	.3791
.48	1.6161	.6188	.98	2.6645	.3753
.49	1.6323	.6126	.99	2.6912	.3716
.50	1.6487	.6065	1.00	2.7183	.3679

References

AASHTO. *A Manual on User Benefit Analysis of Highway and Bus-Transit Improvements.* Washington, D.C.: American Association of State Highway and Transportation Officials, 1978.

American Telephone and Telegraph Co. *Engineering Economy*, 3rd edn. McGraw-Hill, 1977.

Baasel, W.D. *Preliminary Chemical Engineering Plant Design*, 2nd edn. Van Nostrand Reinhold, 1990.

Benford, H.A. *Naval Architect's Introduction to Engineering Economics.* Ann Arbor: Dept. of Naval Architecture and Marine Engineering Report 282, 1983.

Bernhard, R.H. 'A Comprehensive Comparison and Critique of Discounting Indices Proposed for Capital Investment Evaluation', *The Engineering Economist.* Vol. 16, No. 3, 1971, pp. 157–86.

Bussey, L.E., and T.G. Eschenbach. *The Economic Analysis of Industrial Projects* 2nd edn. Prentice-Hall, 1992.

The Engineering Economist. A quarterly journal of the Engineering Economy Divisions of ASEE and IIE.

Eschenbach, T. *Engineering Economy: Applying Theory to Practice,* 2nd edn. Oxford University Press, 2003.

Eschenbach, T.G., and J.P. Lavelle. 'Technical Note: MACRS Depreciation with a Spreadsheet Function: A Teaching and Practice Note', *The Engineering Economist.* Vol. 46, No. 2, 2001, pp. 153–61.

———. 'How Risk and Uncertainty Are/Could/Should Be Presented in Engineering Economy', *Proceedings of the 11th Industrial Engineering Research Conference.* IIE, Orlando, May 2002, CD.

Fish, J.C.L. *Engineering Economics.* McGraw-Hill, 1915.

Jones, B.W. *Inflation in Engineering Economic Analysis.* John Wiley & Sons, 1982.

Lavelle, J.P., H.R. Liggett, and H.R. Parsaei, eds. *Economic Evaluation of Advanced Technologies: Techniques and Case Studies.* Taylor & Francis, 2002.

Liggett, H.R., J. Trevino, and J.P. Lavelle. 'Activity-Based Cost Management Systems in an Advanced Manufacturing Environment', *Economic and Financial Justification of Advanced Manufacturing Technologies.* H.R. Parsaei, W.G. Sullivan, and T.R. Hanley, eds. Elsevier Science, 1992.

Lorie, J.H., and L.J. Savage. 'Three Problems in Rationing Capital', *The Journal of Business.* Vol. 28, No. 4, 1955, pp. 229–39.

Newnan, D.G. 'Determining Rate of Return by Means of Payback Period and Useful Life', *The Engineering Economist.* Vol. 15, No. 1, 1969, pp. 29–39.

Peters, M.S., K.D. Timmerhaus, M. Peters, and R.E. West. *Plant Design and Economics for Chemical Engineers,* 5th edn. McGraw-Hill, 2002.

Perry, H.P., D.W. Green, and J.O. Maloney. *Perry's Chemical Engineers' Handbook,* 7th edn. McGraw-Hill, 1997.

Sprague, J.C., and J.D. Whittaker. *Economic Analysis for Engineers and Managers—Canadian Edition*. Prentice Hall, 1985.

Steiner, H.M. *Public and Private Investments: Socioeconomic Analysis.* John Wiley & Sons, 1980.

Sundaram, M., ed. *Engineering Economy Exam File.* Oxford University Press, 2003.

Terborgh, G. *Business Investment Management.* Machinery and Allied Products Institute, 1967.

Tippet, D.D., and P. Hoekstra. 'Activity-Based Costing: A Manufacturing Management Decision-Making Aid', *Engineering Management Journal.* Vol. 5, No. 2, June 1993. American Society for Engineering Management, pp. 3742.

Wellington, A.M. *The Economic Theory of Railway Location.* John Wiley & Sons, 1887.

Index